Probability and Mathematical Statistics (Conti

WILLIAMS • Diffusions, Markov Proce
Foundations

ZACKS • Theory of Statistical Inference

Applied Probability and Statistics

ABRAHAM and LEDOLTER • Statistical Methods for Forecasting

AICKIN • Linear Statistical Analysis of Discrete Data

ANDERSON, AUQUIER, HAUCK, OAKES, VANDAELE, and WEISBERG • Statistical Methods for Comparative Studies

ARTHANARI and DODGE • Mathematical Programming in Statistics

BAILEY • The Elements of Stochastic Processes with Applications to the Natural Sciences

BAILEY • Mathematics, Statistics and Systems for Health

BARNETT • Interpreting Multivariate Data

BARNETT and LEWIS • Outliers in Statistical Data

BARTHOLOMEW • Stochastic Models for Social Processes, *Third Edition*

BARTHOLOMEW and FORBES • Statistical Techniques for Manpower Planning

BECK and ARNOLD • Parameter Estimation in Engineering and Science

BELSLEY, KUH, and WELSCH • Regression Diagnostics: Identifying Influential Data and Sources of Collinearity

BHAT • Elements of Applied Stochastic Processes

BLOOMFIELD • Fourier Analysis of Time Series: An Introduction

BOX • R. A. Fisher, The Life of a Scientist

BOX and DRAPER • Evolutionary Operation: A Statistical Method for Process Improvement

BOX, HUNTER, and HUNTER • Statistics for Experimenters: An Introduction to Design, Data Analysis, and Model Building

BROWN and HOLLANDER • Statistics: A Biomedical Introduction

BROWNLEE • Statistical Theory and Methodology in Science and Engineering, *Second Edition*

BURY • Statistical Models in Applied Science

CHAMBERS • Computational Methods for Data Analysis

CHATTERJEE and PRICE • Regression Analysis by Example

CHOW • Analysis and Control of Dynamic Economic Systems

CHOW • Econometric Analysis by Control Methods

COCHRAN • Sampling Techniques, *Third Edition*

COCHRAN and COX • Experimental Designs, *Second Edition*

CONOVER • Practical Nonparametric Statistics, *Second Edition*

CONOVER and IMAN • Introduction to Modern Business Statistics

CORNELL • Experiments with Mixtures: Designs, Models and The Analysis of Mixture Data

COX • Planning of Experiments

DANIEL • Biostatistics: A Foundation for Analysis in the Health Sciences, *Third Edition*

DANIEL • Applications of Statistics to Industrial Experimentation

DANIEL and WOOD • Fitting Equations to Data: Computer Analysis of Multifactor Data, *Second Edition*

DAVID • Order Statistics, *Second Edition*

DAVISON • Multidimensional Scaling

DEMING • Sample Design in Business Research

DODGE and ROMIG • Sampling Inspection Tables, *Second Edition*

DOWDY and WEARDEN • Statistics for Research

DRAPER and SMITH • Applied Regression Analysis, *Second Edition*

DUNN • Basic Statistics: A Primer for the Biomedical Sciences, *Second Edition*

DUNN and CLARK • Applied Statistics: Analysis of Variance and Regression

ELANDT-JOHNSON • Probability Models and Statistical Methods in Genetics

ELANDT-JOHNSON and JOHNSON • Survival Models and Data Analysis

continued on back

Statistical Methods for Forecasting

Statistical Methods for Forecasting

BOVAS ABRAHAM
University of Waterloo
JOHANNES LEDOLTER
University of Iowa

John Wiley & Sons
New York • Chichester • Brisbane • Toronto • Singapore

Library of Congress Cataloging in Publication Data:

Abraham, Bovas, 1942–
Statistical methods for forecasting.

(Wiley series in probability and mathematical statistics, ISSN 0271-6356. Applied probability and statistics)
Bibliography: p.
Includes index.
1. Prediction theory. 2. Regression analysis.
3. Time series analysis. I. Ledolter, Johannes.
II. Title. III. Series.

QA279.2.A27 1983 519.5′4 83-7006
ISBN 0-471-86764-0

Printed in the United States of America

20 19 18 17 16 15 14 13 12 11

To our families

Preface

Forecasting is an important part of decision making, and many of our decisions are based on predictions of future unknown events. Many books on forecasting and time series analysis have been published recently. Some of them are introductory and just describe the various methods heuristically. Certain others are very theoretical and focus on only a few selected topics.

This book is about the statistical methods and models that can be used to produce short-term forecasts. Our objective is to provide an intermediate-level discussion of a variety of statistical forecasting methods and models, to explain their interconnections, and to bridge the gap between theory and practice.

Forecast systems are introduced in Chapter 1. Various aspects of regression models are discussed in Chapter 2, and special problems that occur when fitting regression models to time series data are considered. Chapters 3 and 4 apply the regression and smoothing approach to predict a single time series. A brief introduction to seasonal adjustment methods is also given. Parametric models for nonseasonal and seasonal time series are explained in Chapters 5 and 6. Procedures for building such models and generating forecasts are discussed. Chapter 7 describes the relationships between the forecasts produced from exponential smoothing and those produced from parametric time series models. Several advanced topics, such as transfer function modeling, state space models, Kalman filtering, Bayesian forecasting, and methods for forecast evaluation, comparison, and control are given in Chapter 8. Exercises are provided in the back of the book for each chapter.

This book evolved from lecture notes for an MBA forecasting course and from notes for advanced undergraduate and beginning graduate statistics courses we have taught at the University of Waterloo and at the University of Iowa. It is oriented toward advanced undergraduate and beginning graduate students in statistics, business, engineering, and the social sciences.

A calculus background, some familiarity with matrix algebra, and an intermediate course in mathematical statistics are sufficient prerequisites.

Most business schools require their doctoral students to take courses in regression, forecasting, and time series analysis, and most offer courses in forecasting as an elective for MBA students. Courses in regression and in applied time series at the advanced undergraduate and beginning graduate level are also part of most statistics programs. This book can be used in several ways. It can serve as a text for a two-semester sequence in regression, forecasting, and time series analysis for Ph.D. business students, for MBA students with an area of concentration in quantitative methods, and for advanced undergraduate or beginning graduate students in applied statistics. It can also be used as a text for a one-semester course in forecasting (emphasis on Chapters 3 to 7), for a one-semester course in applied time series analysis (Chapters 5 to 8), or for a one-semester course in regression analysis (Chapter 2, and parts of Chapters 3 and 4). In addition, the book should be useful for the professional forecast practitioner.

We are grateful to a number of friends who helped in the preparation of this book. We are glad to record our thanks to Steve Brier, Bob Hogg, Paul Horn, and K. Vijayan, who commented on various parts of the manuscript. Any errors and omissions in this book are, of course, ours. We appreciate the patience and careful typing of the secretarial staff at the College of Business Administration, University of Iowa and of Marion Kaufman and Lynda Hohner at the Department of Statistics, University of Waterloo. We are thankful for the many suggestions we received from our students in forecasting, regression, and time series courses. We are also grateful to the *Biometrika* trustees for permission to reprint condensed and adapted versions of Tables 8, 12 and 18 from *Biometrika Tables for Statisticians*, edited by E. S. Pearson and H. O. Hartley.

We are greatly indebted to George Box who taught us time series analysis while we were graduate students at the University of Wisconsin. We wish to thank him for his guidance and for the wisdom which he shared so freely. It is also a pleasure to acknowledge George Tiao for his warm encouragement. His enthusiasm and enlightenment has been a constant source of inspiration.

We could not possibly discuss every issue in statistical forecasting. However, we hope that this volume provides the background that will allow the reader to adapt the methods included here to his or her particular needs.

B. ABRAHAM
J. LEDOLTER

Waterloo, Ontario
Iowa City, Iowa
June 1983

Contents

Statistical Methods for Forecasting

CHAPTER 1

Introduction and Summary

Webster's dictionary defines *forecasting* as an activity "to calculate or predict some future event or condition, usually as a result of rational study or analysis of pertinent data."

1.1. IMPORTANCE OF GOOD FORECASTS

The ability to form good forecasts has been highly valued throughout history. Even today various types of fortune-tellers claim to have the power to predict future events. Frequently their predictions turn out to be false. However, occasionally their predictions come true; apparently often enough to secure a living for these forecasters.

We all make forecasts, although we may not recognize them as forecasts. For example, a person waiting for a bus or parents expecting a telephone call from their children may not consider themselves forecasters. However, from past experience and from reading the bus schedule, the person waiting for the bus expects it to arrive at a certain time or within a certain time interval. Parents who have usually received calls from their children every weekend expect to receive one during the coming weekend also.

These people form expectations, and they make forecasts. So does a bank manager who predicts the cash flow for the next quarter, or a control engineer who adjusts certain input variables to maintain the future value of some output variable as close as possible to a specified target, or a company manager who predicts sales or estimates the number of man-hours required to meet a given production schedule. All make statements about future events, patterning the forecasts closely on previous occurrences and assuming that the future will be similar to the past.

Since future events involve uncertainty, the forecasts are usually not perfect. The objective of forecasting is to reduce the forecast error: to produce forecasts that are seldom incorrect and that have small forecast errors. In business, industry, and government, policymakers must anticipate the future behavior of many critical variables before they make decisions. Their decisions depend on forecasts, and they expect these forecasts to be accurate; a forecast system is needed to make such predictions. Each situation that requires a forecast comes with its own unique set of problems, and the solutions to one are by no means the solutions in another situation. However, certain general principles are common to most forecasting problems and should be incorporated into any forecast system.

1.2. CLASSIFICATION OF FORECAST METHODS

Forecast methods may be broadly classified into *qualitative* and *quantitative* techniques. *Qualitative* or *subjective* forecast methods are intuitive, largely educated guesses that may or may not depend on past data. Usually these forecasts cannot be reproduced by someone else, since the forecaster does not specify explicitly how the available information is incorporated into the forecast. Even though subjective forecasting is a nonrigorous approach, it may be quite appropriate and the only reasonable method in certain situations.

Forecasts that are based on mathematical or statistical models are called *quantitative*. Once the underlying model or technique has been chosen, the corresponding forecasts are determined automatically; they are fully reproducible by any forecaster. Quantitative methods or models can be further classified as deterministic or probabilistic (also known as stochastic or statistical).

In *deterministic* models the relationship between the variable of interest, Y, and the explanatory or predictor variables $X_1, \ldots, X_p$ is determined exactly:

$$Y = f(X_1, \ldots, X_p; \beta_1, \ldots, \beta_m) \tag{1.1}$$

The function f and the coefficients $\beta_1, \ldots, \beta_m$ are known with certainty. The traditional "laws" in the physical sciences are examples of such deterministic relationships.

In the social sciences, however, the relationships are usually *stochastic*. Measurement errors and variability from other uncontrolled variables introduce random (stochastic) components. This leads to *probabilistic* or *stochastic models* of the form

$$Y = f(X_1, \ldots, X_p; \beta_1, \ldots, \beta_m) + \text{noise} \tag{1.2}$$

where the noise or error component is a realization from a certain probability distribution.

Frequently the functional form f and the coefficients are not known and have to be determined from past data. Usually the data occur in time-ordered sequences referred to as *time series*. Statistical models in which the available observations are used to determine the model form are also called *empirical* and are the main subject of this book. In particular, we discuss regression and single-variable prediction methods. In *single-variable forecasting*, we use the past history of the series, let's say z_t, where t is the time index, and extrapolate it into the future. For example, we may study the features in a series of monthly Canadian consumer price indices and extrapolate the pattern over the coming months. Smoothing methods or parametric time series models may be used for this purpose. In *regression forecasting*, we make use of the relationships between the variable to be forecast and the other variables that explain its variation. For example, we may forecast monthly beer sales from the price of beer, consumers' disposable income, and seasonal temperature; or predict the sales of a cereal product by its price (relative to the industry), its advertising, and the availability of its coupons. The standard regression models measure instantaneous effects. However, there are often lag effects, where the variable of interest depends on present and past values of the independent (i.e., predictor) variables. Such relationships can be studied by combining regression and time series models.

1.3. CONCEPTUAL FRAMEWORK OF A FORECAST SYSTEM

In this book we focus our attention exclusively on quantitative forecast methods. In general, a quantitative forecast system consists of two major components, as illustrated in Figure 1.1. At the first stage, the *model-building stage*, a forecasting model is constructed from pertinent data and available theory. In some instances, theory (for example, economic theory) may suggest particular models; in other cases, such theory may not exist or may be incomplete, and historical data must be used to specify an appropriate model. The tentatively entertained model usually contains unknown parameters; an estimation approach, such as least squares, can be used to determine these constants. Finally, the forecaster must check the adequacy of the fitted model. It could be inadequate for a number of reasons; for example, it could include inappropriate variables or it could have misspecified the functional relationship. If the model is unsatisfactory, it has to be respecified, and the iterative cycle of model specification–estimation–diagnostic checking must be repeated until a satisfactory model is found.

At the second stage, the *forecasting stage*, the final model is used to obtain the forecasts. Since these forecasts depend on the specified model,

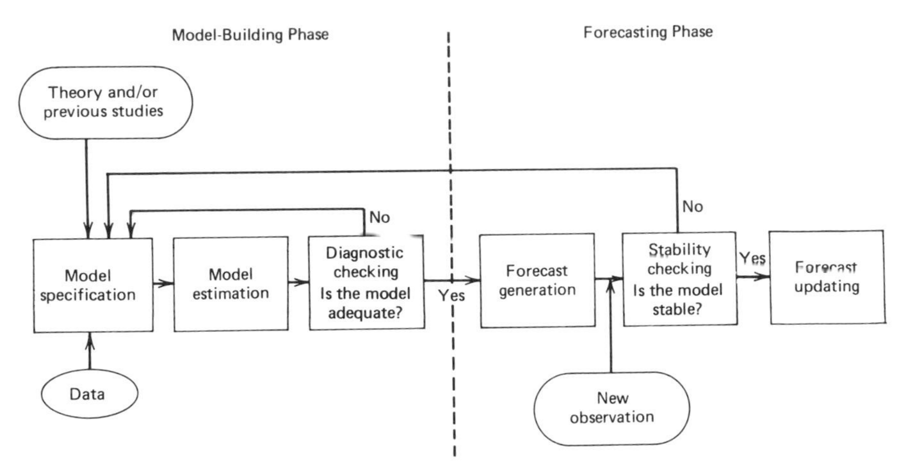

Figure 1.1. Conceptual framework of a forecasting system.

one has to make sure that the model and its parameters stay constant during the forecast period. The stability of the forecast model can be assessed by checking the forecasts against the new observations. Forecast errors can be calculated, and possible changes in the model can be detected. For example, particular functions of these forecast errors can indicate a bias in the forecasts (i.e., consistent over- or underpredictions). The most recent observation can also be used to update the forecasts. Since observations are recorded sequentially in time, updating procedures that can be applied routinely and that avoid the computation of each forecast from first principles are very desirable.

1.4. CHOICE OF A PARTICULAR FORECAST MODEL

Among many other forecast criteria, the choice of the forecast model or technique depends on (1) what degree of accuracy is required, (2) what the forecast horizon is, (3) how high a cost for producing the forecasts can be tolerated, (4) what degree of complexity is required, and (5) what data are available.

Sometimes only crude forecasts are needed; in other instances great accuracy is essential. In some applications, inaccuracy can be very costly; for example, inaccurate forecasts of an economic indicator could force the Federal Reserve Board to boost its lending rate, thus creating a chain of undesirable events. However, increasing the accuracy usually raises substantially the costs of data acquisition, computer time, and personnel. If a small loss in accuracy is not too critical, and if it lowers costs substantially, the simpler but less accurate model may be preferable to the more accurate but more complex one.

The forecast horizon is also essential, since the methods that produce short-term and long-term forecasts differ. For example, a manufacturer may wish to predict the sales of a product for the next 3 months, while an electric utility may wish to predict the demand for electricity over the next 25 years.

A forecaster should try building simple models, which are easy to understand, use, and explain. An elaborate model may lead to more accurate forecasts but may be more costly and difficult to implement. Ockham's razor, also known as the principle of parsimony, says that in a choice among competing models, other things being equal, the simplest is preferable.

Another important consideration in the choice of an appropriate forecast method is the availability of suitable data; one cannot expect to construct accurate empirical forecast models from a limited and incomplete data base.

1.5. FORECAST CRITERIA

The most important criterion for choosing a forecast method is its accuracy, or how closely the forecast predicts the actual event. Let us denote the actual observation at time t with z_t and its forecast, which uses the information up to and including time $t - 1$, with $z_{t-1}(1)$. Then the objective is to find a forecast such that the future forecast error $z_t - z_{t-1}(1)$ is as small as possible. However, note that this is a future forecast error and, since z_t has not yet been observed, its value is unknown; we can talk only about its expected value, conditional on the observed history up to and including time $t - 1$. If both negative (overprediction) and positive (underprediction) forecast errors are equally undesirable, it would make sense to choose the forecast such that the *mean absolute error* $E|z_t - z_{t-1}(1)|$, or the *mean square error* $E[z_t - z_{t-1}(1)]^2$ is minimized. The forecasts that minimize the mean square error are called *minimum mean square error (MMSE) forecasts*. The mean square error criterion is used here since it leads to simpler mathematical solutions.

1.6. OUTLINE OF THE BOOK

This book is about the statistical methods and models that can be used to produce short-term forecasts. It consists of four major parts: regression, smoothing methods, time series models, and selected special topics.

In Chapter 2 we discuss the *regression model*, which describes the relationship between a dependent or response variable and a set of independent or predictor variables. We discuss how regression models are built from historical data, how their parameters are estimated, and how they can be used for forecasting. We describe the matrix representation of regression models and cover such topics as transformations, multicollinearity, and the special problems that occur in fitting regression models to time series data. We discuss how to detect serial correlation in the errors, the consequences of such correlation, and generalizations of regression models that take account of this correlation explicitly.

In Chapters 3 through 7 we discuss how to forecast a single time series, without information from other, possibly related, series. In Chapter 3 we review regression and smoothing as methods of forecasting nonseasonal time series. Nonseasonal series are characterized by time trends and uncorrelated error or noise components. The trend component is usually a polynomial in time; constant, linear, and quadratic trends are special cases.

In models with stable, nonchanging trend components and uncorrelated errors, the parameters can be estimated by least squares. If the trend components change with time, *discounted least squares*, also known as *general exponential smoothing*, can be used to estimate the parameters and derive future forecasts. There the influence of the observations on the parameter estimates diminishes with the age of the observations. Special cases lead to simple, double, and triple exponential smoothing and are discussed in detail.

In Chapter 4 we apply the regression and smoothing methods to forecast seasonal series. Seasonal time series are decomposed into trend, seasonal, and error components. The seasonal component is expressed as a sum of either trigonometric functions or seasonal indicators. We describe the regression approach for series with stable trend and seasonal components, and discuss general exponential smoothing and Winters' additive and multiplicative methods for series with time-changing components. In addition, seasonal adjustment is introduced, with emphasis on the Census X-11 method.

A stochastic modeling or time series analysis approach to forecast a single time series is given in Chapters 5 and 6. In Chapter 5 we discuss the class of *autoregressive integrated moving average* (*ARIMA*) *models*, which can represent many stationary and nonstationary stochastic processes. A stationary stochastic process is characterized by its mean, its variance, and its autocorrelation function. Transformations, in particular successive differences, transform nonstationary series with changing means into stationary series. The patterns in the autocorrelation functions implied by specific ARIMA models are analyzed in detail; to simplify the model-specification we also introduce partial autocorrelations and describe their patterns. We explain Box and Jenkins' (1976) three-stage iterative model-building strategy, which consists of model specification, parameter estimation, and diagnostic checking. We show how forecasts from ARIMA models can be derived; how minimum mean square error forecasts and corresponding prediction intervals can be easily calculated; and how the implied forecast functions, which consist of exponential and polynomial functions of the forecast horizon, adapt as new observations are observed.

The class of ARIMA models is extended to include seasonal time series models. Multiplicative seasonal ARIMA models are described in Chapter 6, and their implied autocorrelation and partial autocorrelation functions are discussed. The minimum mean square error forecasts from such models are illustrated with several examples. The forecast functions include polynomial trends and trigonometric seasonal components that, unlike those in the seasonal regression model described in Chapter 4, adapt as new observations become available.

In Chapter 7 we discuss the relationships between the forecasts from exponential smoothing and the forecasts derived from ARIMA time series models. We show that the time series approach actually includes general exponential smoothing as a special case; exponential smoothing forecast procedures are implied by certain restricted ARIMA models. Implications of these relationships for the forecast practitioner are discussed.

In Chapter 8 we introduce several more advanced forecast techniques. We describe *transfer function models*, which relate an output series to present and past values of an input series; *intervention time series modeling*, which can be used to assess the effect of an exogenous intervention; *Bayesian forecasting* and *Kalman filtering*; *time series models with time-varying coefficients*; *adaptive filtering*; and *post-sample forecast evaluation* and *tracking signals*.

Throughout the book we emphasize a model-based approach to forecasting. We discuss how models are built to generate forecasts, and match commonly used forecast procedures to models within which they generate optimal forecasts. We stress the importance of checking the adequacy of a model before using it for forecasting. An appropriate forecast system has to produce uncorrelated one-step-ahead forecast errors, since correlations among these forecast errors would indicate that there is information in the data that has not yet been used.

Many actual examples are presented in the book. The data sets are real and have been obtained from published articles and consulting projects. Exercises are given in the back of the book for each chapter.

CHAPTER 2

The Regression Model and Its Application in Forecasting

Regression analysis is concerned with modeling the relationships among variables. It quantifies how a response (or dependent) variable is related to a set of explanatory (independent, predictor) variables. For example, a manager might be interested in knowing how the sales of a particular product are related to its price, the prices of competitive products, and the amount spent for advertising. Or an economist might be interested in knowing how a change in per capita income affects consumption, how a price change in gasoline affects gasoline demand, or how the gross national product is related to government spending. Engineers might be interested in knowing how the yield of a particular chemical process depends on reaction time, temperature, and the type of catalyst used.

If the true relationships among the variables were known exactly, the investigator (manager, economist, engineer) would be in a position to understand, predict, and control the response. For example, the economist could predict gasoline sales for any fixed price (forecasting) or could choose the price to keep the gasoline sales at a fixed level (control).

The true relationships among the studied variables, however, will rarely be known, and one must rely on empirical evidence to develop approximations. In addition, the responses will vary, even if the experiment is repeated under apparently identical conditions. This variation that occurs from one repetition to the next is called *noise*, *experimental variation*, *experimental error*, or merely *error*. The variation can come from many sources; it is usually due to measurement error and variation in other, uncontrollable variables. To take explicit account of it, we have to consider probabilistic (or statistical) models.

In this chapter we introduce one such probabilistic model, namely the regression model. We discuss how to construct such a model from empirical data and how to use it in forecasting. In our discussion, we emphasize general principles of statistical model building, such as the specification of models, the estimation of unknown coefficients (parameters), and diagnostic procedures to check the adequacy of the considered model.

2.1. THE REGRESSION MODEL

Let us study the relationship between a dependent variable Y and p independent variables $X_1, X_2, \ldots, X_p$. An index t is introduced to denote the dependent variable at time t (or for subject t) by y_t and the p independent variables by $x_{t1}, x_{t2}, \ldots, x_{tp}$. For observations that occur in natural time order, the index t stands for time. In situations where there is no such ordering, t is just an arbitrary index. For example, in cross-sectional data, where we get observations on different subjects (companies, counties, etc.), the particular ordering has no meaning.

In its most general form, the regression model can be written as

$$y_t = f(\mathbf{x}_t; \boldsymbol{\beta}) + \varepsilon_t \tag{2.1}$$

where $f(\mathbf{x}_t; \boldsymbol{\beta})$ is a mathematical function of the p independent variables $\mathbf{x}_t = (x_{t1}, \ldots, x_{tp})'$ and unknown parameters $\boldsymbol{\beta} = (\beta_1, \ldots, \beta_m)'$. In the following discussion we assume that, apart from the unknown parameters, the functional form of the model is known.

The model in (2.1) is probabilistic, since the error term ε_t is a random variable. It is assumed that:

1. Its mean, $E(\varepsilon_t) = 0$, and its variance, $V(\varepsilon_t) = \sigma^2$, are constant and do not depend on t.
2. The errors ε_t are uncorrelated; that is, $\text{Cov}(\varepsilon_t, \varepsilon_{t-k}) = E(\varepsilon_t \varepsilon_{t-k}) = 0$ for all t and $k \neq 0$.
3. The errors come from a normal distribution. Then assumption 2 implies independence among the errors.

Due to the random nature of the error terms ε_t, the dependent variable y_t itself is a random variable. The model in Equation (2.1) can therefore also be expressed in terms of the conditional distribution of y_t given $\mathbf{x}_t =$

$(x_{t1}, \ldots, x_{tp})'$. In this context the assumptions can be rewritten:

1. The conditional mean, $E(y_t|\mathbf{x}_t) = f(\mathbf{x}_t; \boldsymbol{\beta})$, depends on the independent variables $\mathbf{x}_t$ and the parameters $\boldsymbol{\beta}$, and the variance $V(y_t|\mathbf{x}_t) = \sigma^2$ is independent of $\mathbf{x}_t$ and time.
2. The dependent variables y_t and y_{t-k} for different time periods (or subjects) are uncorrelated:
$$\text{Cov}(y_t, y_{t-k}) = E[y_t - f(\mathbf{x}_t; \boldsymbol{\beta})][y_{t-k} - f(\mathbf{x}_{t-k}; \boldsymbol{\beta})] = 0.$$
3. Conditional on $\mathbf{x}_t$, y_t follows a normal distribution with mean $f(\mathbf{x}_t; \boldsymbol{\beta})$ and variance σ^2; this is denoted by $N(f(\mathbf{x}_t; \boldsymbol{\beta}), \sigma^2)$.

These assumptions imply that the mean of the conditional distribution of y_t is a function of the independent variables $\mathbf{x}_t$. This relationship, however, is not deterministic, as for each fixed $\mathbf{x}_t$ the corresponding y_t will scatter around its mean. The variation of this scatter does not depend on t or on the levels of the independent variables. Furthermore, the error ε_t cannot be predicted from other errors.

In our discussion we assume that the independent variables are fixed and nonstochastic. This simplifies the derivations in this chapter. However, most results also hold if the predictor variables are random.

Several special cases of the general regression model are given below:

1. $y_t = \beta_0 + \varepsilon_t$ (constant mean model)
2. $y_t = \beta_0 + \beta_1 x_t + \varepsilon_t$ (simple linear regression model)
3. $y_t = \beta_0 \exp(\beta_1 x_t) + \varepsilon_t$ (exponential growth model)
4. $y_t = \beta_0 + \beta_1 x_t + \beta_2 x_t^2 + \varepsilon_t$ (quadratic model)
5. $y_t = \beta_0 + \beta_1 x_{t1} + \beta_2 x_{t2} + \varepsilon_t$ (linear model with two independent variables)
6. $y_t = \beta_0 + \beta_1 x_{t1} + \beta_2 x_{t2} + \beta_{11} x_{t1}^2 + \beta_{22} x_{t2}^2 + \beta_{12} x_{t1} x_{t2} + \varepsilon_t$ (quadratic model with two independent variables)

In Figure 2.1 we have plotted $E(y_t|\mathbf{x}_t) = f(\mathbf{x}_t; \boldsymbol{\beta})$ for the models given above. An individual y_t will vary around its mean function according to a normal distribution with constant variance σ^2. This is illustrated in Figure 2.1 for model 2.

2.1.1. Linear and Nonlinear Models

All models except the third are *linear in the parameters*, which means that the derivatives of $f(\mathbf{x}_t; \boldsymbol{\beta})$ with respect to the parameters in $\boldsymbol{\beta}$ do not depend

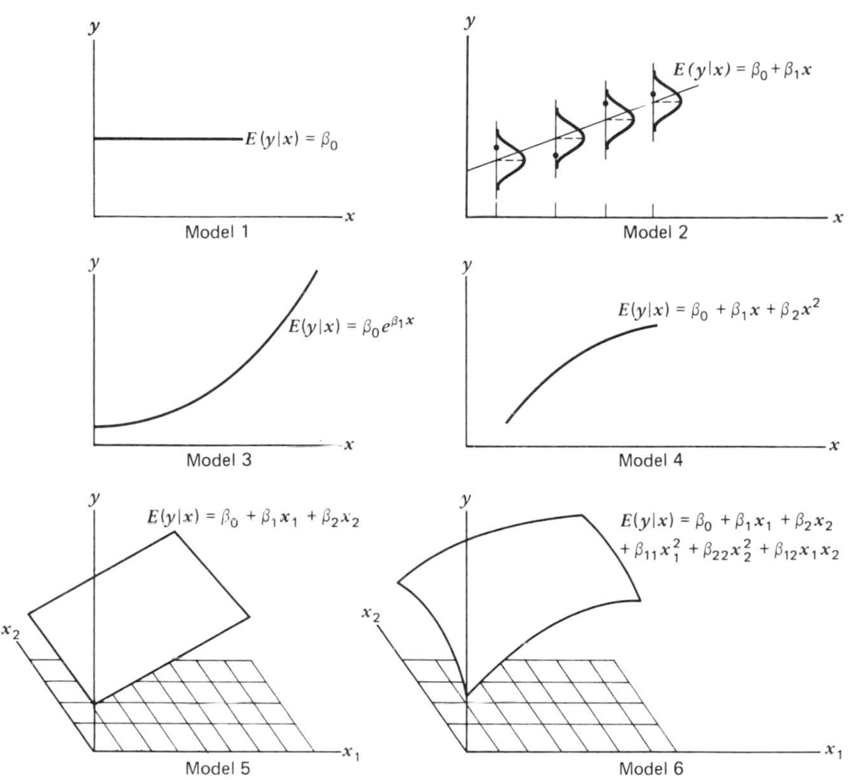

Figure 2.1. Graphical representation of the regression models 1–6. The dots in model 2 represent possible realizations.

on $\boldsymbol{\beta}$. Model 3 is nonlinear in the parameters, since the derivatives

$$\frac{\partial}{\partial \beta_0} \beta_0 e^{\beta_1 x_t} = e^{\beta_1 x_t} \qquad \text{and} \qquad \frac{\partial}{\partial \beta_1} \beta_0 e^{\beta_1 x_t} = \beta_0 e^{\beta_1 x_t} x_t$$

depend on β_0 and β_1.

Models that are linear in the parameters can always be written as

$$y_t = \beta_0 + \beta_1 x_{t1} + \cdots + \beta_p x_{tp} + \varepsilon_t \tag{2.2}$$

The set of the p independent variables $X_1, \ldots, X_p$ can be either original predictor variables or functions thereof. For example, in model 6 there are $p = 5$ independent variables; the last three variables are the squares and the cross-product of X_1 and X_2.

Models 3, 4, and 6 are nonlinear in the independent variables. Models 1, 2, and 5 are *linear in the independent variables*, since the derivatives of $f(\mathbf{x}_t; \boldsymbol{\beta})$ with respect to elements in $\mathbf{x}_t$ are constant and do not depend on $\mathbf{x}_t$. In model 5, for example, β_1 measures the effect of a unit change in variable X_1 on the response Y. Since the model is linear in the independent variables, this effect is additive and does not depend on the level of the other variable, X_2. If the cross-product (interaction) term $X_1 X_2$ is introduced, that is, if

$$y_t = \beta_0 + \beta_1 x_{t1} + \beta_2 x_{t2} + \beta_{12} x_{t1} x_{t2} + \varepsilon_t$$

the effect of a unit change in X_1 is given by $\beta_1 + \beta_{12} X_2$, which does depend on the level of the second variable. If, in addition, squared terms (model 6) are introduced, the relationship becomes quadratic.

2.2. PREDICTION FROM REGRESSION MODELS WITH KNOWN COEFFICIENTS

If the parameters $\boldsymbol{\beta}$ and σ are known, the conditional distribution $p(y_k|\mathbf{x}_k)$ of a future y_k, for given settings of the independent variables $\mathbf{x}_k = (x_{k1}, \ldots, x_{kp})'$, is normal with mean $E(y_k|\mathbf{x}_k) = f(\mathbf{x}_k; \boldsymbol{\beta})$ and variance $V(y_k|\mathbf{x}_k) = \sigma^2$. Thus the conditional mean $f(\mathbf{x}_k; \boldsymbol{\beta})$ can be taken as the forecast for the unknown y_k.

Let us denote the prediction of y_k by y_k^{pred} and the corresponding forecast error by $e_k = y_k - y_k^{\text{pred}}$. Then the expected value of the squared forecast error is

$$E(e_k^2) = E\left[f(\mathbf{x}_k; \boldsymbol{\beta}) + \varepsilon_k - y_k^{\text{pred}}\right]^2 = \sigma^2 + E\left[f(\mathbf{x}_k; \boldsymbol{\beta}) - y_k^{\text{pred}}\right]^2$$

This is minimized if

$$y_k^{\text{pred}} = f(\mathbf{x}_k; \boldsymbol{\beta}) \tag{2.3}$$

Thus the conditional mean in (2.3) is in fact a *minimum mean square error* (*MMSE*) *forecast*. Furthermore, this forecast is unbiased, in the sense that $E(e_k) = 0$, and the variance of the forecast error is given by $V(e_k) = \sigma^2$. Thus a $100(1-\alpha)$ percent prediction interval for the future value y_k is given by

$$\left[f(\mathbf{x}_k; \boldsymbol{\beta}) - u_{\alpha/2}\sigma;\ f(\mathbf{x}_k; \boldsymbol{\beta}) + u_{\alpha/2}\sigma\right] \tag{2.4}$$

where $u_{\alpha/2}$ is the $100(1-\alpha/2)$ percentage point of the $N(0,1)$ distribution.

The parameters $\boldsymbol{\beta}$ and σ in regression model (2.1) will rarely be known. They have to be estimated from past data on the dependent and independent variables $(y_1, \mathbf{x}_1), (y_2, \mathbf{x}_2), \ldots, (y_n, \mathbf{x}_n)$. If $\hat{\boldsymbol{\beta}}$ denotes the estimate of $\boldsymbol{\beta}$, then $\hat{y}_k^{\text{pred}} = f(\mathbf{x}_k; \hat{\boldsymbol{\beta}})$ can be taken as a forecast for y_k. The properties of this prediction and the corresponding prediction intervals depend on the properties of the estimator $\hat{\boldsymbol{\beta}}$. These will be discussed next.

2.3. LEAST SQUARES ESTIMATES OF UNKNOWN COEFFICIENTS

The problem of estimating parameters $\boldsymbol{\beta}$ from historical data is one of choosing parameter estimates $\hat{\boldsymbol{\beta}} = (\hat{\beta}_1, \ldots, \hat{\beta}_m)'$, such that the fitted function $f(\mathbf{x}_t; \hat{\boldsymbol{\beta}})$ is "close" to the observations. The squared distance $[y_t - f(\mathbf{x}_t; \boldsymbol{\beta})]^2$ is usually chosen to measure "closeness."

The parameter estimates that minimize the sum of the squared deviations

$$S(\boldsymbol{\beta}) = \sum_{t=1}^{n} [y_t - f(\mathbf{x}_t; \boldsymbol{\beta})]^2 \tag{2.5}$$

are called the *least squares estimates* and are denoted by $\hat{\boldsymbol{\beta}}$.

2.3.1. Some Examples

For example, in the *simple linear regression model through the origin*, $y_t = \beta x_t + \varepsilon_t$, the least squares estimate of β is found by minimizing

$$S(\beta) = \sum_{t=1}^{n} (y_t - \beta x_t)^2$$

The minimum can be found analytically. From the first-order condition for finding extreme values,

$$\frac{\partial S(\beta)}{\partial \beta} = -2 \sum_{t=1}^{n} (y_t - \beta x_t) x_t = 0$$

we find that the least squares estimate is given by

$$\hat{\beta} = \frac{\sum_{t=1}^{n} x_t y_t}{\sum_{t=1}^{n} x_t^2} \tag{2.6}$$

Furthermore, it can be shown that the second-order condition for a minimum is satisfied.

In the *simple linear regression model*, $y_t = \beta_0 + \beta_1 x_t + \varepsilon_t$, the estimates of β_0 and β_1 are found by minimizing

$$S(\beta_0, \beta_1) = \sum_{t=1}^{n} (y_t - \beta_0 - \beta_1 x_t)^2$$

Setting the first partial derivatives equal to zero,

$$\frac{\partial}{\partial \beta_0} S(\beta_0, \beta_1) = -2\Sigma(y_t - \beta_0 - \beta_1 x_t) = 0$$

$$\frac{\partial}{\partial \beta_1} S(\beta_0, \beta_1) = -2\Sigma(y_t - \beta_0 - \beta_1 x_t)x_t = 0$$

and solving this equation system for β_0 and β_1 leads to the least squares estimates

$$\hat{\beta}_1 = \frac{\Sigma(x_t - \bar{x})y_t}{\Sigma(x_t - \bar{x})^2} = \frac{n\Sigma x_t y_t - (\Sigma x_t)(\Sigma y_t)}{n\Sigma x_t^2 - (\Sigma x_t)^2}$$

$$\hat{\beta}_0 = \bar{y} - \hat{\beta}_1 \bar{x} \tag{2.7}$$

where $\bar{y} = (1/n)\Sigma y_t$ and $\bar{x} = (1/n)\Sigma x_t$ are the sample means of y and x, respectively. For simplicity we have suppressed the limits ($t = 1, \ldots, n$) in the summations.

Example 2.1: Running Performance

For illustration, consider the data listed in Table 2.1. In this example we are interested in predicting the racing performance of trained female distance runners. The measured variables include the running performance in a 10-km road race; body composition such as height, weight, skinfold sum, and relative body fat; and maximal aerobic power. The data were taken from a study by Conley, Krahenbuhl, Burkett, and Millar (1981). The subjects for their study were 14 trained and competition-experienced female runners who had placed among the top 20 women in a 10-km road race with more than 300 female entrants. The laboratory data were collected during the week following the competitive run.

For the moment we are interested only in the relationship between the racing performance (running time) and maximal aerobic power (volume of

Table 2.1. Physical and Performance Characteristics of 14 Female Runners[a]

X_1	X_2	X_3	X_4	X_5	Y
163	53.6	76.4	17.9	61.32	39.37
167	56.4	62.1	15.2	55.29	39.80
166	58.1	65.0	17.0	52.83	40.03
157	43.1	44.9	12.6	57.94	41.32
150	44.8	59.7	13.9	53.31	42.03
151	39.5	59.3	19.2	51.32	42.37
162	52.1	98.7	19.6	52.18	43.93
168	58.8	73.1	19.6	52.37	44.90
152	44.3	59.2	17.4	57.91	44.90
161	47.4	51.5	14.4	53.93	45.12
161	47.8	61.4	7.9	47.88	45.60
165	49.1	62.5	10.5	47.41	46.03
157	50.4	60.3	12.6	47.17	47.83
154	46.4	76.7	19.6	51.05	48.55

[a] X_1, height; X_2, weight; X_3, skinfold sum; X_4, relative body fat; X_5, V_{O_2}; Y, running time.

Source: Conley et al. (1981).

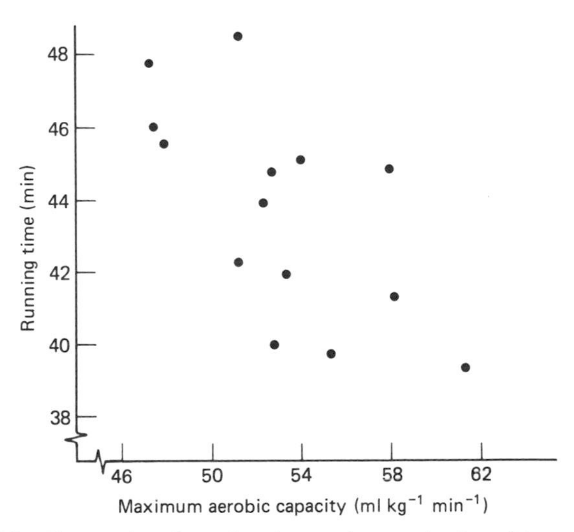

Figure 2.2. Scatter plot of running time against maximal aerobic capacity.

maximal oxygen uptake; V_{O_2}). The scatter plot of Y_t = running time versus $X_t = V_{O_2}$ in Figure 2.2 shows that the relationship is approximately linear, which suggests a model of the form

$$y_t = \beta_0 + \beta_1 x_t + \varepsilon_t$$

For this example, $n = 14$, $\Sigma x_t = 741.91$, $\Sigma x_t^2 = 39{,}539.57$, $\Sigma y_t = 611.78$, $\Sigma y_t^2 = 26{,}846.13$, and $\Sigma x_t y_t = 32{,}316.01$. It follows from the expressions in (2.7) that

$$\hat{\beta}_1 = -0.468 \qquad \text{and} \qquad \hat{\beta}_0 = 68.494$$

This means that over the studied range of aerobic capacity a one-unit increase in V_{O_2} (ml kg^{-1} min^{-1}) reduces the running time by 0.468 sec.

2.3.2. Estimation in the General Linear Regression Model

In Equations (2.7) we have given a closed form expression for the least squares estimates in the simple linear regression model. In fact, closed form expressions of the least squares estimates can be derived for all models that are linear in the parameters. Only in models that are nonlinear in the parameters can closed form solutions not be found. In this latter case the parameters have to be estimated using either a grid search or an iterative minimization procedure to minimize the error sum of squares. For a discussion of nonlinear least squares, we refer the interested reader to the book by Draper and Smith (1981), Chapter 10.

In the remainder of this chapter we discuss linear regression models for which closed form expressions of the least squares estimates can be derived. Since many nonlinear relationships can be approximated, at least locally, by linear models, the assumption of linearity is not too restrictive.

Linear regression models can be written as

$$y_t = \beta_0 + \beta_1 x_{t1} + \beta_2 x_{t2} + \cdots + \beta_p x_{tp} + \varepsilon_t \tag{2.8}$$

or

$$y_t = \mathbf{x}_t' \boldsymbol{\beta} + \varepsilon_t$$

where $\mathbf{x}_t = (1, x_{t1}, \ldots, x_{tp})'$ and $\boldsymbol{\beta} = (\beta_0, \beta_1, \ldots, \beta_p)'$. If β_0 is not in the model, $\mathbf{x}_t = (x_{t1}, \ldots, x_{tp})'$. To simplify the calculation of the least squares estimates for the general linear regression model, we introduce matrix notation and define the $n \times 1$ column vectors $\mathbf{y} = (y_1, \ldots, y_n)'$; $\boldsymbol{\varepsilon} =$

$(\varepsilon_1, \ldots, \varepsilon_n)'$ and the $n \times (p+1)$ matrix

$$\mathbf{X} = \begin{bmatrix} 1 & x_{11} & \cdots & x_{1p} \\ 1 & x_{21} & \cdots & x_{2p} \\ \vdots & \vdots & & \vdots \\ 1 & x_{n1} & \cdots & x_{np} \end{bmatrix}$$

Then we can write the regression model for $1 \leqslant t \leqslant n$ as

$$\mathbf{y} = \mathbf{X}\boldsymbol{\beta} + \boldsymbol{\varepsilon} \tag{2.9}$$

The assumptions about the error terms ε_t in $\boldsymbol{\varepsilon} = (\varepsilon_1, \ldots, \varepsilon_n)'$ have been discussed earlier. Restating these assumptions in vector notation implies that the vector $\boldsymbol{\varepsilon}$ has mean

$$E(\boldsymbol{\varepsilon}) = \begin{bmatrix} E(\varepsilon_1) \\ \vdots \\ E(\varepsilon_n) \end{bmatrix} = \begin{bmatrix} 0 \\ \vdots \\ 0 \end{bmatrix} = \mathbf{0}$$

and symmetric covariance matrix

$$\mathbf{V}(\boldsymbol{\varepsilon}) = E(\boldsymbol{\varepsilon}\boldsymbol{\varepsilon}') = \begin{bmatrix} V(\varepsilon_1) & \operatorname{Cov}(\varepsilon_1, \varepsilon_2) & \cdots & \operatorname{Cov}(\varepsilon_1, \varepsilon_n) \\ \operatorname{Cov}(\varepsilon_1, \varepsilon_2) & V(\varepsilon_2) & \cdots & \operatorname{Cov}(\varepsilon_2, \varepsilon_n) \\ \vdots & \vdots & & \vdots \\ \operatorname{Cov}(\varepsilon_1, \varepsilon_n) & \operatorname{Cov}(\varepsilon_2, \varepsilon_n) & \cdots & V(\varepsilon_n) \end{bmatrix} = \sigma^2 \mathbf{I}$$

since the covariances $\operatorname{Cov}(\varepsilon_i, \varepsilon_j)$ between ε_i and ε_j $(i \neq j)$ are assumed zero; $\mathbf{I}$ is the $n \times n$ identity matrix.

We also assume that the matrix $\mathbf{X}$, which is usually called the design matrix, is fixed (nonstochastic) and of full column rank. This means that the columns of the $\mathbf{X}$ matrix are not linearly related. A violation of this assumption would indicate that some of the independent variables are redundant, since at least one of the variables would contain the same information as a linear combination of the others.

Equivalently, the regression model can be thought of as a random vector $\mathbf{y}$, with mean $E(\mathbf{y}) = \mathbf{X}\boldsymbol{\beta}$, which depends on the settings of the independent variables, and the covariance matrix

$$\mathbf{V}(\mathbf{y}) = E[\mathbf{y} - E(\mathbf{y})][\mathbf{y} - E(\mathbf{y})]' = \sigma^2 \mathbf{I}$$

In matrix notation the least squares criterion can be expressed as minimizing

$$S(\boldsymbol{\beta}) = \sum_{t=1}^{n} (y_t - \mathbf{x}_t'\boldsymbol{\beta})^2 = (\mathbf{y} - \mathbf{X}\boldsymbol{\beta})'(\mathbf{y} - \mathbf{X}\boldsymbol{\beta})$$

As shown below, the minimization of $S(\boldsymbol{\beta})$ leads to the least squares estimator $\hat{\boldsymbol{\beta}}$, which satisfies the $p + 1$ equations

$$(\mathbf{X}'\mathbf{X})\hat{\boldsymbol{\beta}} = \mathbf{X}'\mathbf{y} \tag{2.10}$$

These are referred to as the *normal equations*. Since we have assumed that the design matrix $\mathbf{X}$ is of full column rank, the inverse of $\mathbf{X}'\mathbf{X}$ can be calculated. The solution of the normal equations is then given by

$$\hat{\boldsymbol{\beta}} = (\mathbf{X}'\mathbf{X})^{-1}\mathbf{X}'\mathbf{y} \tag{2.11}$$

To prove the result in (2.11), we show that any other estimate of $\boldsymbol{\beta}$ will lead to a larger value of $S(\boldsymbol{\beta})$. We can express $S(\boldsymbol{\beta})$ as

$$\begin{aligned} S(\boldsymbol{\beta}) &= (\mathbf{y} - \mathbf{X}\boldsymbol{\beta})'(\mathbf{y} - \mathbf{X}\boldsymbol{\beta}) \\ &= (\mathbf{y} - \mathbf{X}\hat{\boldsymbol{\beta}} + \mathbf{X}\hat{\boldsymbol{\beta}} - \mathbf{X}\boldsymbol{\beta})'(\mathbf{y} - \mathbf{X}\hat{\boldsymbol{\beta}} + \mathbf{X}\hat{\boldsymbol{\beta}} - \mathbf{X}\boldsymbol{\beta}) \\ &= (\mathbf{y} - \mathbf{X}\hat{\boldsymbol{\beta}})'(\mathbf{y} - \mathbf{X}\hat{\boldsymbol{\beta}}) + (\hat{\boldsymbol{\beta}} - \boldsymbol{\beta})'\mathbf{X}'\mathbf{X}(\hat{\boldsymbol{\beta}} - \boldsymbol{\beta}) \end{aligned}$$

since

$$\begin{aligned} (\hat{\boldsymbol{\beta}} - \boldsymbol{\beta})'\mathbf{X}'(\mathbf{y} - \mathbf{X}\hat{\boldsymbol{\beta}}) &= (\hat{\boldsymbol{\beta}} - \boldsymbol{\beta})'(\mathbf{X}'\mathbf{y} - \mathbf{X}'\mathbf{X}\hat{\boldsymbol{\beta}}) \\ &= (\hat{\boldsymbol{\beta}} - \boldsymbol{\beta})'\left[\mathbf{X}'\mathbf{y} - (\mathbf{X}'\mathbf{X})(\mathbf{X}'\mathbf{X})^{-1}\mathbf{X}'\mathbf{y}\right] \\ &= (\hat{\boldsymbol{\beta}} - \boldsymbol{\beta})'(\mathbf{X}'\mathbf{y} - \mathbf{X}'\mathbf{y}) = 0 \end{aligned}$$

Therefore we can write

$$\begin{aligned} S(\boldsymbol{\beta}) &= S(\hat{\boldsymbol{\beta}}) + (\hat{\boldsymbol{\beta}} - \boldsymbol{\beta})'\mathbf{X}'\mathbf{X}(\hat{\boldsymbol{\beta}} - \boldsymbol{\beta}) \\ &= S(\hat{\boldsymbol{\beta}}) + \mathbf{c}'\mathbf{c} \end{aligned}$$

where the $n \times 1$ vector $\mathbf{c} = \mathbf{X}(\hat{\boldsymbol{\beta}} - \boldsymbol{\beta})$. Now since

$$\mathbf{c}'\mathbf{c} = \sum_{i=1}^{n} c_i^2 \geqslant 0$$

it follows that $S(\boldsymbol{\beta}) \geqslant S(\hat{\boldsymbol{\beta}})$; the equality holds if and only if $\boldsymbol{\beta} = \hat{\boldsymbol{\beta}}$. This completes the proof.

We illustrate this general result on two special cases.

1. *Simple linear regression through the origin*:

$$y_t = \beta x_t + \varepsilon_t \qquad \text{for} \quad 1 \leqslant t \leqslant n$$

In this case $\mathbf{y}' = (y_1, \ldots, y_n)$ and $\mathbf{X}$ is the $n \times 1$ vector $(x_1, x_2, \ldots, x_n)'$. The normal equation is given by

$$\left(\sum_{t=1}^{n} x_t^2 \right) \hat{\beta} = \sum_{t=1}^{n} x_t y_t$$

leading to the least squares estimate

$$\hat{\beta} = \frac{\sum_{t=1}^{n} x_t y_t}{\sum_{t=1}^{n} x_t^2}$$

This is the same as in (2.6), which we had derived from first principles by minimizing $S(\beta) = \Sigma(y_t - \beta x_t)^2$ directly.

2. *Simple linear regression model*:

$$y_t = \beta_0 + \beta_1 x_t + \varepsilon_t \qquad \text{for} \quad 1 \leqslant t \leqslant n$$

In this case

$$\mathbf{y} = \begin{bmatrix} y_1 \\ y_2 \\ \vdots \\ y_n \end{bmatrix} \qquad \mathbf{X} = \begin{bmatrix} 1 & x_1 \\ 1 & x_2 \\ \vdots & \vdots \\ 1 & x_n \end{bmatrix} \qquad \boldsymbol{\beta} = \begin{bmatrix} \beta_0 \\ \beta_1 \end{bmatrix}$$

The normal equations in (2.10) (there are $p + 1 = 2$ equations) are given by

$$\begin{bmatrix} n & \Sigma x_t \\ \Sigma x_t & \Sigma x_t^2 \end{bmatrix} \begin{bmatrix} \hat{\beta}_0 \\ \hat{\beta}_1 \end{bmatrix} = \begin{bmatrix} \Sigma y_t \\ \Sigma x_t y_t \end{bmatrix}$$

and the least squares estimates $\hat{\beta}_0, \hat{\beta}_1$ are

$$\begin{bmatrix} \hat{\beta}_0 \\ \hat{\beta}_1 \end{bmatrix} = \begin{bmatrix} n & \Sigma x_t \\ \Sigma x_t & \Sigma x_t^2 \end{bmatrix}^{-1} \begin{bmatrix} \Sigma y_t \\ \Sigma x_t y_t \end{bmatrix}$$

Inverting the 2×2 matrix and making some algebraic simplifications leads to the least squares estimates given in (2.7).

2.4. PROPERTIES OF LEAST SQUARES ESTIMATORS

The least squares estimate $\hat{\boldsymbol{\beta}}$ in (2.11) is a function of past data $(y_1, \mathbf{x}_1), \ldots, (y_n, \mathbf{x}_n)$. In fact, it is a linear combination of the elements in $\mathbf{y}$ and can be written as $\hat{\boldsymbol{\beta}} = \mathbf{A}\mathbf{y}$, where $\mathbf{A}$ is the $(p+1) \times n$ matrix $\mathbf{A} = (\mathbf{X}'\mathbf{X})^{-1}\mathbf{X}'$. The weights in these linear combinations are functions of the elements in the design matrix $\mathbf{X}$, which are taken to be fixed. The vector $\mathbf{y}$, however, is random and changes from sample to sample. Thus the least squares estimator $\hat{\boldsymbol{\beta}}$ is also a random variable. To construct confidence intervals for the parameters and to formulate hypothesis tests, we have to know its sampling distribution.

Since the least squares estimator $\hat{\boldsymbol{\beta}}$ is a linear function of the random variables $\mathbf{y}$, it is straightforward to derive its properties. In the derivation we use standard results about the distribution of linear functions of random variables and the distribution of certain quadratic forms. These are summarized in Appendix 2. Properties of the least squares estimator and related issues are described below.

1. The least squares estimator in (2.11) is *unbiased*: $E(\hat{\boldsymbol{\beta}}) = \boldsymbol{\beta}$. This follows immediately from the definition of $\hat{\boldsymbol{\beta}}$ and the model in (2.9), since

$$E(\hat{\boldsymbol{\beta}}) = E\left[(\mathbf{X}'\mathbf{X})^{-1}\mathbf{X}'\mathbf{y}\right] = (\mathbf{X}'\mathbf{X})^{-1}\mathbf{X}'E(\mathbf{y}) = (\mathbf{X}'\mathbf{X})^{-1}\mathbf{X}'\mathbf{X}\boldsymbol{\beta} = \boldsymbol{\beta}$$

2. The $(p+1) \times (p+1)$ covariance matrix of $\hat{\boldsymbol{\beta}}$ is given by

$$\mathbf{V}(\hat{\boldsymbol{\beta}}) = \begin{bmatrix} V(\hat{\beta}_0) & \mathrm{Cov}(\hat{\beta}_0, \hat{\beta}_1) & \cdots & \mathrm{Cov}(\hat{\beta}_0, \hat{\beta}_p) \\ \mathrm{Cov}(\hat{\beta}_0, \hat{\beta}_1) & V(\hat{\beta}_1) & \cdots & \mathrm{Cov}(\hat{\beta}_1, \hat{\beta}_p) \\ \vdots & \vdots & & \vdots \\ \mathrm{Cov}(\hat{\beta}_0, \hat{\beta}_p) & \mathrm{Cov}(\hat{\beta}_1, \hat{\beta}_p) & \cdots & V(\hat{\beta}_p) \end{bmatrix}$$

$$= \sigma^2(\mathbf{X}'\mathbf{X})^{-1} \tag{2.12}$$

This follows from result 1 in Appendix 2, since

$$\mathbf{V}(\hat{\boldsymbol{\beta}}) = \mathbf{V}(\mathbf{A}\mathbf{y}) = \mathbf{A}\mathbf{V}(\mathbf{y})\mathbf{A}' = \mathbf{A}\sigma^2\mathbf{I}\mathbf{A}' = \sigma^2\mathbf{A}\mathbf{A}'$$

$$= \sigma^2(\mathbf{X}'\mathbf{X})^{-1}\mathbf{X}'\mathbf{X}(\mathbf{X}'\mathbf{X})^{-1} = \sigma^2(\mathbf{X}'\mathbf{X})^{-1}$$

where $\mathbf{A} = (\mathbf{X}'\mathbf{X})^{-1}\mathbf{X}'$.

The variance calculations are illustrated on two simple examples. In the *simple linear regression model through the origin*, the variance of the least squares estimate (2.12) is given by

$$V(\hat{\beta}) = \frac{\sigma^2}{\Sigma x_t^2} \tag{2.13}$$

For the *simple linear regression model*, the covariance matrix of the least squares estimates $\hat{\beta}_0$ and $\hat{\beta}_1$ in (2.7) is given by

$$\mathbf{V}(\hat{\boldsymbol{\beta}}) = \begin{bmatrix} V(\hat{\beta}_0) & \operatorname{Cov}(\hat{\beta}_0, \hat{\beta}_1) \\ \operatorname{Cov}(\hat{\beta}_0, \hat{\beta}_1) & V(\hat{\beta}_1) \end{bmatrix} = \sigma^2 \begin{bmatrix} n & \Sigma x_t \\ \Sigma x_t & \Sigma x_t^2 \end{bmatrix}^{-1}$$

$$= \sigma^2 \begin{bmatrix} \dfrac{1}{n} + \dfrac{\bar{x}^2}{\Sigma(x_t - \bar{x})^2} & \dfrac{-\bar{x}}{\Sigma(x_t - \bar{x})^2} \\ \dfrac{-\bar{x}}{\Sigma(x_t - \bar{x})^2} & \dfrac{1}{\Sigma(x_t - \bar{x})^2} \end{bmatrix} \tag{2.14}$$

3. Let us assume that we are interested in estimating a linear combination of the parameters, $\Sigma c_i \beta_i = \mathbf{c}'\boldsymbol{\beta}$. Then it can be shown that among all linear unbiased estimators, the least squares estimator $\mathbf{c}'\hat{\boldsymbol{\beta}}$ will be the one with the smallest variance.

This implies, for example, that the least squares estimator of β_i in $\boldsymbol{\beta} = (\beta_0, \beta_1, \ldots, \beta_p)'$ has the smallest variance among all linear unbiased estimators. In other words, any other linear unbiased estimate will have a larger variance. This property, commonly known under the name *Gauss-Markov theorem*, is not proved here; the interested reader is referred to Rao (1965, p. 179).

4. If it is assumed that the errors in model (2.9) are normally distributed, then

$$\hat{\boldsymbol{\beta}} \sim N_{p+1}\left[\boldsymbol{\beta}, \sigma^2(\mathbf{X}'\mathbf{X})^{-1}\right]$$

where $N_{p+1}(\boldsymbol{\mu}, \boldsymbol{\Sigma})$ is the usual notation for a $(p+1)$-variate normal distribution with mean vector $\boldsymbol{\mu}$ and covariance matrix $\boldsymbol{\Sigma}$.

Therefore, any individual estimate $\hat{\beta}_i$ in $\hat{\boldsymbol{\beta}} = (\hat{\beta}_0, \ldots, \hat{\beta}_p)'$ follows a univariate $N(\beta_i, \sigma^2 c_{ii})$ distribution, where c_{ii} is the corresponding diagonal element in $\mathbf{C} = \{c_{ij}; 0 \leqslant i, j \leqslant p\} = (\mathbf{X}'\mathbf{X})^{-1}$.

5. Substitution of the least squares estimate $\hat{\boldsymbol{\beta}} = (\mathbf{X}'\mathbf{X})^{-1}\mathbf{X}'\mathbf{y}$ in model (2.9) leads to the *fitted values* $\hat{\mathbf{y}} = \mathbf{X}\hat{\boldsymbol{\beta}}$. The differences between the observed and fitted values are called the *residuals* and are given by

$$\mathbf{e} = \mathbf{y} - \hat{\mathbf{y}} = \mathbf{y} - \mathbf{X}(\mathbf{X}'\mathbf{X})^{-1}\mathbf{X}'\mathbf{y} = \left[\mathbf{I} - \mathbf{X}(\mathbf{X}'\mathbf{X})^{-1}\mathbf{X}'\right]\mathbf{y}$$

where $\mathbf{I}$ is the $n \times n$ identity matrix. The fitted values are points on the estimated regression line (surface); the residuals measure the distance between the observations and the fitted line. The residuals e_t can be thought of as estimates of the errors ε_t. Although the errors ε_t are independent by assumption, the residuals e_t are correlated. In fact, there are $p + 1$ linear restrictions among the n residuals, since the $(p+1) \times 1$ vector $\mathbf{X}'\mathbf{e}$ is

$$\begin{aligned} \mathbf{X}'\mathbf{e} &= \mathbf{X}'\left[\mathbf{I} - \mathbf{X}(\mathbf{X}'\mathbf{X})^{-1}\mathbf{X}'\right]\mathbf{y} \\ &= \left[\mathbf{X}' - (\mathbf{X}'\mathbf{X})(\mathbf{X}'\mathbf{X})^{-1}\mathbf{X}'\right]\mathbf{y} = (\mathbf{X}' - \mathbf{X}')\mathbf{y} = \mathbf{0} \end{aligned} \qquad (2.15)$$

For illustration, we consider two simple models. For the *constant mean model*, $y_t = \beta + \varepsilon_t$, we find that the least squares estimate of β is the sample mean, $\hat{\beta} = \bar{y}$. The fitted values and the residuals are given by $\hat{y}_t = \bar{y}$ and $e_t = y_t - \bar{y}$. The single restriction among the residuals is $\mathbf{X}'\mathbf{e} = \Sigma e_t = \Sigma(y_t - \bar{y}) = 0$.

In the *simple linear regression model*, $y_t = \beta_0 + \beta_1 x_t + \varepsilon_t$, the fitted values and the residuals are given by $\hat{y}_t = \hat{\beta}_0 + \hat{\beta}_1 x_t$ and $y_t - \hat{y}_t = y_t - \hat{\beta}_0 - \hat{\beta}_1 x_t$. There are two restrictions among the residuals,

$$\mathbf{X}'\mathbf{e} = \begin{bmatrix} 1 & 1 & \cdots & 1 \\ x_1 & x_2 & \cdots & x_n \end{bmatrix} \mathbf{e} = \begin{bmatrix} \Sigma e_t \\ \Sigma x_t e_t \end{bmatrix} = \begin{bmatrix} 0 \\ 0 \end{bmatrix}$$

6. It is important to learn how much of the variation in the dependent variable Y is accounted for by the variation in the independent variables. Consider the variation of the observations y_t around their mean $\bar{y}$, which is given by $\text{SSTO} = \Sigma(y_t - \bar{y})^2$. We refer to this as the *total sum of squares* (corrected for the mean) and denote it by SSTO. The objective is to partition SSTO into two parts. One portion measures the variation of the fitted values $\hat{y}_t$ around the mean $\bar{y}$. This part is called the *sum of squares due to regression* (SSR), since it measures the variability due to the movement in the variable Y that is explained by the independent variables $X_1, \ldots, X_p$. The other portion measures the variation of the observations around the fitted values. This part is called the *sum of squares due to error (or residual)* (SSE), or the unexplained variation, since it measures the variation that is not explained by the model.

Let us define

$$\text{SSTO} = \Sigma(y_t - \bar{y})^2 = \Sigma y_t^2 - n\bar{y}^2 = \mathbf{y}'\mathbf{y} - n\bar{y}^2$$

$$\text{SSR} = \Sigma(\hat{y}_t - \bar{y})^2 = \Sigma \hat{y}_t^2 - n\bar{y}^2 = \hat{\boldsymbol{\beta}}'\mathbf{X}'\mathbf{y} - n\bar{y}^2$$

since $\hat{\mathbf{y}}'\hat{\mathbf{y}} = \hat{\boldsymbol{\beta}}'\mathbf{X}'\hat{\mathbf{y}} = \hat{\boldsymbol{\beta}}'\mathbf{X}'(\mathbf{y} - \mathbf{e}) = \hat{\boldsymbol{\beta}}'\mathbf{X}'\mathbf{y} - \hat{\boldsymbol{\beta}}'\mathbf{X}'\mathbf{e} = \hat{\boldsymbol{\beta}}'\mathbf{X}'\mathbf{y}$;

$$\text{SSE} = \Sigma(y_t - \hat{y}_t)^2 = \mathbf{e}'\mathbf{e}$$

Then

$$\begin{aligned} \text{SSTO} &= \Sigma(y_t - \bar{y})^2 = \Sigma(y_t - \hat{y}_t + \hat{y}_t - \bar{y})^2 \\ &= \Sigma(\hat{y}_t - \bar{y})^2 + \Sigma(y_t - \hat{y}_t)^2 = \text{SSR} + \text{SSE} \end{aligned} \tag{2.16}$$

since the cross-product term is

$$\begin{aligned} \Sigma(\hat{y}_t - \bar{y})(y_t - \hat{y}_t) &= \Sigma(\hat{y}_t - \bar{y})e_t = \hat{\mathbf{y}}'\mathbf{e} - \bar{\mathbf{y}}\Sigma e_t \\ &= \hat{\boldsymbol{\beta}}'\mathbf{X}'\mathbf{e} - \bar{y}\Sigma e_t = 0 \end{aligned}$$

Here we have used the linear restrictions among the residuals in (2.15); also, as long as an intercept is included in the model, the residuals will sum to zero ($\Sigma e_t = 0$).

In (2.16) we have partitioned the sum of squares into two parts, SSTO = SSR + SSE. The first part describes the variation that is explained by the model; the second part measures the variation that is left unexplained. As a measure of "fit" of the regression model, we consider the *coefficient of determination* R^2, which is defined as the ratio of the sum of squares due to regression and the total sum of squares:

$$R^2 = \frac{\text{SSR}}{\text{SSTO}} = 1 - \frac{\text{SSE}}{\text{SSTO}} \tag{2.17}$$

This coefficient has to be between 0 and 1, since SSR $\leqslant$ SSTO [see Eq. (2.16)]. It measures the proportion of the total variation of the observations around their mean $\bar{y}$ that is explained by the fitted regression model. A large R^2 (near 1) indicates that a large portion of the variation is accounted for by the model. A small value of R^2 (near 0) indicates that only a small fraction of the variation is explained by the regression.

When building forecast models, we always look for independent variables that explain most of the variation in the variable we want to predict. One modeling objective is thus to find models with high explanatory power (high R^2).

The sum of squares decomposition SSTO = SSR + SSE is usually written in the form of a table, commonly referred to as the *analysis of variance (ANOVA) table*:

Source	Sum of Squares	Degrees of Freedom	Mean Square	F Ratio
Regression	$\text{SSR} = \hat{\boldsymbol{\beta}}'\mathbf{X}'\mathbf{y} - n\bar{y}^2$	p	$\text{MSR} = \frac{\text{SSR}}{p}$	$\frac{\text{MSR}}{\text{MSE}}$
Error	$\text{SSE} = \mathbf{e}'\mathbf{e}$	$n - p - 1$	$\text{MSE} = \frac{\text{SSE}}{n - p - 1}$	
Total (corrected for mean)	$\text{SSTO} = \mathbf{y}'\mathbf{y} - n\bar{y}^2$	$n - 1$		

The concepts of degrees of freedom, mean square, and F ratio are discussed below.

7. The variance σ^2 is usually unknown. However, it can be estimated by the mean square error

$$s^2 = \frac{\text{SSE}}{n - p - 1} \tag{2.18}$$

Apart from a different denominator, this is essentially the sample variance of the residuals. For the ordinary sample variance of n observations, we divide the error sum of squares by $n - 1$, since only one parameter, the mean, is estimated. Here we divide by $n - p - 1$, since the regression model contains $p + 1$ estimated parameters. There are $p + 1$ restrictions among the n residuals [see Eq. (2.15)]; thus the degrees of freedom for the error sum of squares stand for the number of independent components in SSE.

A more formal explanation can be based on result $2b$ in Appendix 2, which says that SSE/σ^2 follows a chi-square distribution with $n - p - 1$ degrees of freedom. This result can be used to establish the fact that s^2 in (2.18) is an unbiased estimate of σ^2:

$$E(s^2) = \frac{\sigma^2}{n - p - 1} E\left(\frac{\text{SSE}}{\sigma^2}\right) = \sigma^2 \tag{2.19}$$

since the mean of a chi-square distribution is equal to its degrees of freedom.

8. The estimated covariance matrix of the least squares estimator $\hat{\boldsymbol{\beta}}$ is

$$\hat{\mathbf{V}}(\hat{\boldsymbol{\beta}}) = s^2(\mathbf{X}'\mathbf{X})^{-1} \tag{2.20}$$

The estimated standard error of $\hat{\beta}_i$ is given by $s_{\hat{\beta}_i} = s\sqrt{c_{ii}}$, where c_{ii} is the corresponding diagonal element in $(\mathbf{X}'\mathbf{X})^{-1}$.

9. The least squares estimator $\hat{\boldsymbol{\beta}}$ and s^2 are independent.

2.5. CONFIDENCE INTERVALS AND HYPOTHESIS TESTING

In this section we construct confidence intervals and hypothesis tests for the parameters in $\boldsymbol{\beta}$. We consider the quantity

$$\frac{\hat{\beta}_i - \beta_i}{s\sqrt{c_{ii}}} \qquad i = 0, 1, \ldots, p \tag{2.21}$$

where $\hat{\beta}_i$ is the least squares estimate of β_i and $s_{\hat{\beta}_i} = s\sqrt{c_{ii}}$ is its standard error. This quantity has a t distribution with $n - p - 1$ degrees of freedom (see result $2e$ in Appendix 2). Confidence intervals and hypothesis tests can now be readily derived.

2.5.1. Confidence Intervals

A $100(1 - \alpha)$ percent confidence interval for the unknown parameter β_i is given by

$$\left[\hat{\beta}_i - t_{\alpha/2}(n - p - 1)s\sqrt{c_{ii}},\, \hat{\beta}_i + t_{\alpha/2}(n - p - 1)s\sqrt{c_{ii}}\right] \tag{2.22}$$

where $t_{\alpha/2}(n - p - 1)$ is the $100(1 - \alpha/2)$ percentage point of a t distribution with $n - p - 1$ degrees of freedom. If in repeated samples such confidence intervals were constructed, these random intervals would cover the unknown, but fixed, parameter β_i in $100(1 - \alpha)$ percent of the cases.

2.5.2. Hypothesis Tests for Individual Coefficients

The test statistic for testing the null hypothesis H_0: $\beta_i = \beta_{i0}$ against the alternative H_1: $\beta_i \neq \beta_{i0}$, where β_{i0} is some specified value that is of particular interest, is given by

$$t = \frac{\hat{\beta}_i - \beta_{i0}}{s\sqrt{c_{ii}}} \tag{2.23}$$

If $|t| > t_{\alpha/2}(n - p - 1)$, we reject H_0 in favor of H_1 at significance level α.

If $|t| \leqslant t_{\alpha/2}(n - p - 1)$, there is not enough evidence for rejecting the null hypothesis H_0 in favor of H_1. Thus, loosely speaking, we "accept" H_0.

Of special interest is the case $\beta_{i0} = 0$, since failure to reject H_0: $\beta_i = 0$ indicates that the independent variable X_i has no effect on the dependent variable Y. The test statistic is then given by

$$t_{\hat{\beta}_i} = \frac{\hat{\beta}_i}{s_{\hat{\beta}_i}} = \frac{\hat{\beta}_i}{s\sqrt{c_{ii}}} \tag{2.24}$$

Most regression computer packages report these t statistics as part of their standard output.

2.5.3. A Simultaneous Test for Regression Coefficients

The t statistics discussed above are used to test the significance of each predictor variable individually. A simultaneous test, which considers all predictor variables jointly, tests the null hypothesis H_0: $\beta_1 = \beta_2 = \cdots = \beta_p = 0$ against the alternative H_1 that at least one $\beta_i \neq 0$. Under H_0, none of the predictors has a significant influence on Y. It is thus a test for significance of the regression model, since under H_0 none of the independent variables helps to explain the variation in $y_1, y_2, \ldots, y_n$. If this is the case, the dependent variable is best predicted by its mean $\bar{y}$.

One can show that, given the null hypothesis is true, the F statistic

$$F = \frac{\text{MSR}}{\text{MSE}} \tag{2.25}$$

follows an F distribution with p and $n - p - 1$ degrees of freedom (see result $2h$ in Appendix 2). This statistic is standard output in all regression programs and is usually included in the ANOVA table (see Sec. 2.4). Its significance can be assessed using the critical values $F_\alpha(p, n-p-1)$ given in Table B of the Table Appendix. These values are obtained such that $P[F(p, n-p-1) > F_\alpha(p, n-p-1)] = \alpha$, where $F(p, n-p-1)$ is a random variable that follows an F distribution with p and $n - p - 1$ degrees of freedom.

If $F > F_\alpha(p, n-p-1)$, we reject H_0 in favor of H_1 at level α.
If $F \leqslant F_\alpha(p, n-p-1)$, we do not have enough evidence to reject H_0 in favor of H_1, and thus we accept H_0.

Several computer packages calculate a probability value α_*, which is given by $\alpha_* = P[F(p, n-p-1) > F]$ and where F is the observed F statistic in (2.25). It expresses how "likely" an outcome of magnitude F is, given that H_0 is true. If this probability value is smaller than a specified significance level α, we reject H_0.

2.5.4. General Hypothesis Tests: The Extra Sum of Squares Principle

The question as to whether it is worthwhile to include certain variables in a statistical model arises frequently in empirical model building. The extra regression sum of squares, which can be attributed to these variables, plays an important role in answering this question.

Let us consider the following two models

Model 1: $Y = \beta_0 + \beta_1 X_1 + \cdots + \beta_q X_q + \varepsilon$
Model 2: $Y = \beta_0 + \beta_1 X_1 + \cdots + \beta_q X_q + \cdots + \beta_p X_p + \varepsilon$

Model 2 is the more general model; it includes the $p - q$ extra variables $X_{q+1}, \ldots, X_p$, where $p > q$. A test of whether these extra variables lead to a significant increase in the regression sum of squares, or equivalently in R^2, is now discussed.

The least squares estimates of the $q + 1$ coefficients $\boldsymbol{\beta} = (\beta_0, \beta_1, \ldots, \beta_q)'$ in the first model are given by

$$\hat{\boldsymbol{\beta}}(1) = \left[\hat{\beta}_0(1), \hat{\beta}_1(1), \ldots, \hat{\beta}_q(1)\right]' = (\mathbf{X}_1'\mathbf{X}_1)^{-1}\mathbf{X}_1'\mathbf{y}$$

where $\mathbf{X}_1$ is the $n \times (q+1)$ design matrix obtained from the first q independent variables

$$\mathbf{X}_1 = \begin{bmatrix} 1 & x_{11} & \cdots & x_{1q} \\ 1 & x_{21} & \cdots & x_{2q} \\ \vdots & \vdots & & \vdots \\ 1 & x_{n1} & \cdots & x_{nq} \end{bmatrix}$$

The regression sum of squares, which is explained by the variables $X_1, \ldots, X_q$, is given by

$$\mathrm{SSR}(X_1, \ldots, X_q) = \hat{\boldsymbol{\beta}}(1)'\mathbf{X}_1'\mathbf{y} - n\bar{y}^2 \tag{2.26}$$

For the more general model 2, the least squares estimates of the $p+1$ coefficients $\boldsymbol{\beta} = (\beta_0, \beta_1, \ldots, \beta_q, \ldots, \beta_p)'$ are given by

$$\hat{\boldsymbol{\beta}}(2) = \left[\hat{\beta}_0(2), \hat{\beta}_1(2), \ldots, \hat{\beta}_q(2), \ldots, \hat{\beta}_p(2)\right]' = (\mathbf{X}_2'\mathbf{X}_2)^{-1}\mathbf{X}_2'\mathbf{y}$$

where $\mathbf{X}_2$ is the $n \times (p+1)$ design matrix from all independent variables:

$$\mathbf{X}_2 = \begin{bmatrix} 1 & x_{11} & \cdots & x_{1q} & \cdots & x_{1p} \\ 1 & x_{21} & \cdots & x_{2q} & \cdots & x_{2p} \\ \vdots & \vdots & & \vdots & & \vdots \\ 1 & x_{n1} & \cdots & x_{nq} & \cdots & x_{np} \end{bmatrix}$$

The regression sum of squares, which is explained by all the variables $X_1, \ldots, X_q, \ldots, X_p$, is given by

$$\mathrm{SSR}(X_1, \ldots, X_q, \ldots, X_p) = \hat{\boldsymbol{\beta}}(2)'\mathbf{X}_2'\mathbf{y} - n\bar{y}^2 \tag{2.27}$$

The *extra regression sum of squares*, which can be attributed to the additional variables $X_{q+1}, \ldots, X_p$, is then defined as

$$\begin{aligned} &\mathrm{SSR}(X_{q+1}, \ldots, X_p | X_1, \ldots, X_q) \\ &\quad = \mathrm{SSR}(X_1, \ldots, X_q, \ldots, X_p) - \mathrm{SSR}(X_1, \ldots, X_q) \\ &\quad = \hat{\boldsymbol{\beta}}(2)'\mathbf{X}_2'\mathbf{y} - \hat{\boldsymbol{\beta}}(1)'\mathbf{X}_1'\mathbf{y} \end{aligned} \tag{2.28}$$

To test the hypothesis that none of the additional variables $X_{q+1}, \ldots, X_p$ has an influence on the dependent variable Y (H_0: $\beta_{q+1} = \cdots = \beta_p = 0$), we consider the F statistic

$$F^* = \frac{\text{SSR}(X_{q+1},\ldots, X_p | X_1,\ldots, X_q)/(p-q)}{\text{SSE}(X_1,\ldots, X_p)/(n-p-1)} \tag{2.29}$$

Under H_0, this statistic follows an F distribution with $p-q$ and $n-p-1$ degrees of freedom. Thus the test of H_0: $\beta_{q+1} = \cdots = \beta_p = 0$ against H_1: at least one $\beta_i \neq 0$ (for $q+1 \leqslant i \leqslant p$) is given by

If $F^* > F_\alpha(p-q, n-p-1)$, we reject H_0 in favor of H_1 at level α.

If $F^* \leqslant F_\alpha(p-q, n-p-1)$, we cannot reject H_0, and thus we accept H_0. There is not enough evidence in the data to conclude that the extra variables $X_{q+1},\ldots, X_p$ are important predictor variables.

The calculation of the test statistic F^* in (2.29) requires fitting two regression models, the more general as well as the restricted model. From the entries in the ANOVA tables, the test statistic can be readily computed. This test is useful when comparing two different models. However, one model must be a special case of the other.

2.5.5. Partial and Sequential F Tests

Consider testing the significance of one additional variable, $p = q + 1$. Then the test of H_0: $\beta_p = 0$ against H_1: $\beta_p \neq 0$ is given by

$$F^* = \frac{\text{SSR}(X_p | X_1,\ldots, X_{p-1})}{\text{SSE}(X_1,\ldots, X_p)/(n-p-1)} = \frac{\text{SSR}(X_p | X_1,\ldots, X_{p-1})}{s^2} \tag{2.30}$$

where s^2 is the mean square error from the regression model in which all independent variables $X_1,\ldots, X_p$ are included. It can be shown that

$$\text{SSR}(X_p | X_1,\ldots, X_{p-1}) = \frac{\hat{\beta}_p^2}{c_{pp}}$$

where c_{pp} is the last diagonal element in $(\mathbf{X}'\mathbf{X})^{-1}$. Thus,

$$F^* = \left(\frac{\hat{\beta}_p}{s\sqrt{c_{pp}}} \right)^2 = \left(t_{\hat{\beta}_p} \right)^2 \tag{2.31}$$

where $t_{\hat{\beta}_p}$ is the t statistic given in (2.24). Since it is well known [see Hogg and Craig (1978, Chap. 4)] that $F_\alpha(1, n-p-1) = t^2_{\alpha/2}(n-p-1)$, it follows that in this case the ordinary t test and the extra sum of squares test are equivalent.

This result gives us additional insight into the meaning of t tests. The individual t statistic $t_{\hat{\beta}_i} = \hat{\beta}_i / s\sqrt{c_{ii}}$ actually tests whether the variable X_i leads to a significant reduction in the error sum of squares (or equivalently, a significant increase in the regression sum of squares) after all other variables $X_1, \ldots, X_{i-1}, X_{i+1}, \ldots, X_p$ have been included in the model. Since these tests treat the variable X_i as if it had been included last, these tests are called *partial F* or *partial t tests*.

Since these tests assess the contribution of the predictor X_i as if it were the last variable entered in the model, it is quite possible that whereas the overall regression is significant, all individual t tests are insignificant. For example, consider the case where the dependent variable is strongly correlated with a variable X_1. Now consider a second variable X_2 that is highly correlated with the first one, X_1. Then if both are included in the model, the individual t statistics might well be insignificant, since the extra contribution of X_2 over X_1 (and also X_1 over X_2) is small. This implies that in simplifying the regression model we should never drop all individually insignificant variables at once, but should drop them one at a time.

In *sequential F tests* the variables are added sequentially and $\text{SSR}(X_1), \text{SSR}(X_2|X_1), \text{SSR}(X_3|X_1, X_2), \ldots,$ are calculated. A particular order for including the variables has to be specified. The significance of each added variable can be tested by considering sequential F statistics

$$F^* = \frac{\text{SSR}(X_i | X_{i-1}, \ldots, X_1)}{\text{MSE}(X_1, \ldots, X_i)} \tag{2.32}$$

where $\text{MSE}(X_1, \ldots, X_i)$ is the mean square error from the regression model that includes the variables $X_1, \ldots, X_i$.

2.6. PREDICTION FROM REGRESSION MODELS WITH ESTIMATED COEFFICIENTS

In Section 2.2 we discussed the prediction from regression models when the parameters are known. There it was shown that for given predictor variables $\mathbf{x}_k = (x_{k1}, \ldots, x_{kp})'$ the MMSE prediction of a future y_k from the regression model $y_t = \beta_0 + \beta_1 x_{t1} + \cdots + \beta_p x_{tp} + \varepsilon_t$ is given by

$$y_k^{\text{pred}} = \beta_0 + \beta_1 x_{k1} + \cdots + \beta_p x_{kp} = \mathbf{x}_k' \boldsymbol{\beta}$$

If the parameters $\boldsymbol{\beta} = (\beta_0, \beta_1, \ldots, \beta_p)'$ are unknown, we replace them by their least squares estimates $\hat{\boldsymbol{\beta}} = (\mathbf{X}'\mathbf{X})^{-1}\mathbf{X}'\mathbf{y}$. Then the prediction of y_k is given by

$$\hat{y}_k^{\text{pred}} = \hat{\beta}_0 + \hat{\beta}_1 x_{k1} + \cdots + \hat{\beta}_p x_{kp} = \mathbf{x}'_k \hat{\boldsymbol{\beta}} \tag{2.33}$$

It can be shown quite easily that these predictions have the following properties:

1. The forecast $\hat{y}_k^{\text{pred}}$ is an *unbiased* predictor of y_k, since the expected value of a future forecast error is

$$E\left(y_k - \hat{y}_k^{\text{pred}}\right) = E\left(\mathbf{x}'_k\boldsymbol{\beta} + \varepsilon_k - \mathbf{x}'_k\hat{\boldsymbol{\beta}}\right) = \mathbf{x}'_k\boldsymbol{\beta} - \mathbf{x}'_k\boldsymbol{\beta} = 0$$

2. Furthermore, $\hat{y}_k^{\text{pred}}$ is the *minimum mean square error forecast* among all linear unbiased forecasts. This result follows from the Gauss-Markov theorem (see property 3 in Sec. 2.4).
3. The variance of the forecast error $y_k - \hat{y}_k^{\text{pred}}$ or, equivalently, of the forecast is given by

$$\begin{aligned} V\left(y_k - \hat{y}_k^{\text{pred}}\right) &= V\left[\varepsilon_k + \mathbf{x}'_k(\boldsymbol{\beta} - \hat{\boldsymbol{\beta}})\right] \\ &= \sigma^2 + V\left[\mathbf{x}'_k(\boldsymbol{\beta} - \hat{\boldsymbol{\beta}})\right] = \sigma^2 + \sigma^2\mathbf{x}'_k(\mathbf{X}'\mathbf{X})^{-1}\mathbf{x}_k \\ &= \sigma^2\left[1 + \mathbf{x}'_k(\mathbf{X}'\mathbf{X})^{-1}\mathbf{x}_k\right] \end{aligned} \tag{2.34}$$

   Here we have used the result in Equation (2.12), $\mathbf{V}(\hat{\boldsymbol{\beta}}) = \sigma^2(\mathbf{X}'\mathbf{X})^{-1}$, and have assumed that the future error ε_k is uncorrelated with the errors in the estimation period.
4. The estimated variance of the forecast error is

$$\hat{V}\left(y_k - \hat{y}_k^{\text{pred}}\right) = s^2\left[1 + \mathbf{x}'_k(\mathbf{X}'\mathbf{X})^{-1}\mathbf{x}_k\right] \tag{2.35}$$

   where $s^2 = \text{SSE}/(n - p - 1)$ is the mean square error. A $100(1 - \alpha)$ percent prediction interval for the future y_k is then given by its upper and lower limits:

$$\hat{y}_k^{\text{pred}} \pm t_{\alpha/2}(n - p - 1)s\left[1 + \mathbf{x}'_k(\mathbf{X}'\mathbf{X})^{-1}\mathbf{x}_k\right]^{1/2} \tag{2.36}$$

2.6.1. Examples

To illustrate these results, we consider the same two models we discussed in Section 2.3.

1. In the *simple linear regression model through the origin*, $y_t = \beta x_t + \varepsilon_t$, the forecast of a future y_k is given by $\hat{y}_k^{\text{pred}} = \hat{\beta}x_k$, and from (2.35)

the variance by

$$\hat{V}(y_k - \hat{y}_k^{\text{pred}}) = s^2\left(1 + \frac{x_k^2}{\Sigma x_t^2}\right) \tag{2.37}$$

where $s^2 = \Sigma(y_t - \hat{\beta}x_t)^2/(n-1)$. A $100(1-\alpha)$ percent prediction interval for the future y_k is given by

$$\hat{\beta}x_k \pm t_{\alpha/2}(n-1)s\left(1 + \frac{x_k^2}{\Sigma x_t^2}\right)^{1/2}$$

The variance of the forecast error is smallest if $x_k = 0$. This is to be expected, since the model forces the regression line through the origin; at this point there is no uncertainty due to parameter estimation. For all other x's, however, the prediction depends on the estimated slope $\hat{\beta}$.

2. In the *simple linear regression model*, $y_t = \beta_0 + \beta_1 x_t + \varepsilon_t$, the $2 \times n$ matrix $\mathbf{X}'$ is given by

$$\mathbf{X}' = \begin{bmatrix} 1 & 1 & \cdots & 1 \\ x_1 & x_2 & \cdots & x_n \end{bmatrix}$$

The least squares estimates were derived in (2.7). Substituting $\mathbf{X}$ and the mean square error

$$s^2 = \frac{\Sigma(y_t - \hat{\beta}_0 - \hat{\beta}_1 x_t)^2}{n-2}$$

into Equation (2.35), we find that the error variance of the prediction $\hat{y}_k^{\text{pred}} = \hat{\beta}_0 + \hat{\beta}_1 x_k$ is estimated by

$$\hat{V}(y_k - \hat{y}_k^{\text{pred}}) = s^2\left(1 + \frac{1}{n} + \frac{(x_k - \bar{x})^2}{\Sigma(x_t - \bar{x})^2}\right) \tag{2.38}$$

A $100(1-\alpha)$ prediction interval can be calculated from

$$\hat{\beta}_0 + \hat{\beta}_1 x_k \pm t_{\alpha/2}(n-2)s\left(1 + \frac{1}{n} + \frac{(x_k - \bar{x})^2}{\Sigma(x_t - \bar{x})^2}\right)^{1/2}$$

The variance of the prediction error is smallest if $x_k = \bar{x}$. This means that the uncertainty about a future y value will be smallest if x_k is in the center of the historical data. Again this is plausible, since the fitted line in the simple linear regression model passes through the point $(\bar{x}, \bar{y})$. If we wish to extrapolate away from the experimental region and to predict for values x_k outside this region, this uncertainty increases.

It should be emphasized that the forecasts and their error variances are derived under the assumption that the model is correct. Extrapolations beyond the experimental region should generally be avoided, since the assumed model may not be appropriate outside this region.

2.7. EXAMPLES

Several examples are now considered to illustrate parameter estimation, hypothesis testing, and forecasting in regression models.

Example 2.1: Running Performance

The least squares estimates in the simple linear regression model $y_t = \beta_0 + \beta_1 x_t + \varepsilon_t$, where y_t is the time to run the 10-km race and x_t is the maximal aerobic capacity V_{O_2}, were calculated previously in Section 2.3 and it was found that $\hat{\beta}_0 = 68.494$ and $\hat{\beta}_1 = -0.468$. The fitted values $\hat{y}_t = \hat{\beta}_0 + \hat{\beta}_1 x_t$, the residuals $e_t = y_t - \hat{y}_t$, and the entries in the ANOVA table are easily calculated.

The ANOVA table is

Source	SS	df	MS	F
Regression	48.845	1	48.845	9.249
Error	63.374	12	5.281	
Total	112.219	13		

The coefficient of determination is given by $R^2 = 48.845/112.219 = .435$. This indicates that 43.5 percent of the variation is explained by the regression model. The remaining 56.5 percent of the variation in running time is left unexplained; additional variables should be found to reduce the uncertainty.

The parameter estimate of β_1 is $\hat{\beta}_1 = -0.468$. This means that over the studied range of aerobic capacity an increase in V_{O_2} of one unit (ml kg^{-1}

min^{-1}) will reduce the running time by 0.468 sec. To test whether this effect is significant, we calculate its standard error $s_{\hat{\beta}_1} = s\sqrt{c_{11}}$.

We have shown in (2.14) that for the simple linear regression model the estimated covariance matrix of the least squares estimates $\hat{\beta}_0$ and $\hat{\beta}_1$ is given by

$$\hat{\mathbf{V}}(\hat{\boldsymbol{\beta}}) = s^2(\mathbf{X}'\mathbf{X})^{-1} = s^2 \begin{bmatrix} \dfrac{1}{n} + \dfrac{\bar{x}^2}{\Sigma(x_t - \bar{x})^2} & \dfrac{-\bar{x}}{\Sigma(x_t - \bar{x})^2} \\ \dfrac{-\bar{x}}{\Sigma(x_t - \bar{x})^2} & \dfrac{1}{\Sigma(x_t - \bar{x})^2} \end{bmatrix}$$

$$= \begin{bmatrix} 66.8511 & 1.2544 \\ 1.2544 & 0.0237 \end{bmatrix}$$

The standard errors $s_{\hat{\beta}_i}$ and the t statistics $t_{\hat{\beta}_i} = \hat{\beta}_i / s_{\hat{\beta}_i}$ are given in the following table:

	$\hat{\beta}_i$	$s_{\hat{\beta}_i}$	$t_{\hat{\beta}_i}$
Constant	68.494	8.176	8.38
V_{O_2}	−0.468	0.154	−3.04

In this particular example we are interested in testing whether the running time decreases with increasing maximal aerobic capacity (H_0: $\beta_1 = 0$ against H_1: $\beta_1 < 0$). The critical value for this one-sided test with significance level $\alpha = .05$ is $-t_{.05}(12) = -1.78$. Since the t statistic is smaller than this critical value, we can reject the null hypothesis. In other words, there is evidence for a significant inverse relationship between running time and aerobic capacity.

The F statistic ($F = 9.249$) in the ANOVA table provides a test of H_0: $\beta_1 = 0$ versus H_1: $\beta_1 \neq 0$. Since there is only one predictor variable in our model, this F is the same as $(t_{\hat{\beta}_1})^2 = (-3.04)^2$.

Prediction intervals are readily obtained from Equation (2.38). For example, let us predict the running time for a female athlete with maximal aerobic capacity $x_k = 55$. Note that this value is well within the experimental region (x ranges from 47.17 to 61.32). Then the point forecast is given by

$$\hat{y}_k^{\text{pred}} = \hat{\beta}_0 + \hat{\beta}_1 x_k = 42.75$$

and its standard error by

$$s\left(1 + \frac{1}{n} + \frac{(x_k - \bar{x})^2}{\Sigma(x_t - \bar{x})^2}\right)^{1/2} = 2.40$$

Thus a 95 percent prediction interval for the running time is given by

$$42.75 \pm t_{.025}(12)2.40 = 42.75 \pm 5.23 \qquad \text{or} \qquad (37.52, 47.98)$$

This prediction interval is still quite wide. In fact, all but one of the observed running times fall within this interval. The question that arises next is whether the other variables in Table 2.1 can help to reduce this uncertainty.

To investigate whether these variables (height X_2, weight X_3, skinfold sum X_4, relative body fat X_5) help explain the variation in the data, we fit the model

$$y_t = \beta_0 + \beta_1 x_{t1} + \cdots + \beta_5 x_{t5} + \varepsilon_t \tag{2.39}$$

to the $n = 14$ observations in Table 2.1. Computer programs (for example, regression routines in GLIM, MINITAB [see Ryan et al. (1976)], IDA, SPSS, BMDP, or SAS) can be used to calculate the least squares estimates $\hat{\boldsymbol{\beta}} = (\hat{\beta}_0, \hat{\beta}_1, \ldots, \hat{\beta}_5)'$, the fitted values, the residuals, the entries in the ANOVA table, R^2, the standard errors of the estimates, and the t statistics. We find that the estimates and their standard errors are

	$\hat{\beta}_i$	$s_{\hat{\beta}_i}$	$t_{\hat{\beta}_i}$
Constant	80.921	32.556	2.49
V_{O_2}	−0.479	0.203	−2.36
Height	−0.064	0.262	−0.24
Weight	−0.085	0.285	−0.30
Skinfold sum	0.027	0.079	0.35
Relative body fat	0.047	0.292	0.16

and the ANOVA table

Source	SS	df	MS	F
Regression	57.989	5	11.598	1.71
Error	54.230	8	6.779	
Total	112.219	13		

Note that while the error sum of squares has decreased from 63.374 (for the model including V_{O_2} only) to 54.230, the mean square error has actually increased (from 5.281 to 6.779). Thus, obviously, these additional variables do not help explain the variation in running time.

Examining the t statistics, we find that the estimates $\hat{\beta}_2, \hat{\beta}_3, \hat{\beta}_4, \hat{\beta}_5$ are not statistically significant. It should be pointed out that these t statistics test the individual importance of X_i. They treat the variable X_i as if it were the last variable included in the model. An overall test for the joint significance, H_0: $\beta_2 = \beta_3 = \beta_4 = \beta_5 = 0$ versus H_1: not all $\beta_i = 0$ $(2 \leqslant i \leqslant 5)$, can be derived using the extra sum of squares principle. There the appropriate F statistic is given by

$$F^* = \frac{(\text{SSR}_2 - \text{SSR}_1)/(p - q)}{\text{SSE}_2/(n - p - 1)} = \frac{(57.989 - 48.845)/4}{54.230/8} = 0.34$$

which is insignificant. Here SSR_1 is the regression sum of squares from the restricted model ($y_t = \beta_0 + \beta_1 x_{t1} + \varepsilon_t$), and SSR_2 is the one from the unrestricted model in (2.39). Thus these extra variables do not help in predicting the running time.

Example 2.2: Gas Mileage Data

As another example, consider predicting the gas mileage of an automobile as a function of its size and other engine characteristics. In Table 2.2 we have listed data on 38 automobiles (1978–1979 models). These data, which were originally taken from *Consumer Reports*, were analyzed by Henderson and Velleman (1981). The variables measured include gas mileage in miles per gallon (MPG), number of cylinders, cubic engine displacement, horsepower, weight, acceleration, and engine type [straight(1), V(0)].

The gas consumption (in gallons) should be proportional to the effort (= force × distance) it takes to move the car. Furthermore, since force is proportional to weight, we expect that gas consumption per unit distance is proportional to force and thus proportional to weight. Thus a model of the form

$$y_t = \beta_0 + \beta_1 x_t + \varepsilon_t \qquad (2.40)$$

where $y = \text{MPG}^{-1} = \text{GPM}$ and $x =$ weight should provide a good starting point. Note that $y = \text{GPM}$, or 100 GPM as used in our analysis, measures the gas consumption per 100 traveled miles, which is actually the description of gas efficiency commonly used in Europe.

Table 2.2. Gas Mileage Data on 38 Automobiles (1978–1979 Models)[a]

MPG	X_1	X_2	X_3	X_4	X_5	X_6
16.9	8	350	155	4.360	14.9	1
15.5	8	351	142	4.054	14.3	1
19.2	8	267	125	3.605	15.0	1
18.5	8	360	150	3.940	13.0	1
30.0	4	98	68	2.155	16.5	0
27.5	4	134	95	2.560	14.2	0
27.2	4	119	97	2.300	14.7	0
30.9	4	105	75	2.230	14.5	0
20.3	5	131	103	2.830	15.9	0
17.0	6	163	125	3.140	13.6	0
21.6	4	121	115	2.795	15.7	0
16.2	6	163	133	3.410	15.8	0
20.6	6	231	105	3.380	15.8	0
20.8	6	200	85	3.070	16.7	0
18.6	6	225	110	3.620	18.7	0
18.1	6	258	120	3.410	15.1	0
17.0	8	305	130	3.840	15.4	1
17.6	8	302	129	3.725	13.4	1
16.5	8	351	138	3.955	13.2	1
18.2	8	318	135	3.830	15.2	1
26.5	4	140	88	2.585	14.4	0
21.9	6	171	109	2.910	16.6	1
34.1	4	86	65	1.975	15.2	0
35.1	4	98	80	1.915	14.4	0
27.4	4	121	80	2.670	15.0	0
31.5	4	89	71	1.990	14.9	0
29.5	4	98	68	2.135	16.6	0
28.4	4	151	90	2.670	16.0	0
28.8	6	173	115	2.595	11.3	1
26.8	6	173	115	2.700	12.9	1
33.5	4	151	90	2.556	13.2	0
34.2	4	105	70	2.200	13.2	0
31.8	4	85	65	2.020	19.2	0
37.3	4	91	69	2.130	14.7	0
30.5	4	97	78	2.190	14.1	0
22.0	6	146	97	2.815	14.5	0
21.5	4	121	110	2.600	12.8	0
31.9	4	89	71	1.925	14.0	0

[a] MPG, miles per gallon; X_1, number of cylinders; X_2, engine displacement in cubic inches; X_3, horsepower; X_4, weight in 1000 lb; X_5, acceleration in sec; X_6, engine type [straight(1), V(0)].

Source: Henderson and Velleman (1981).

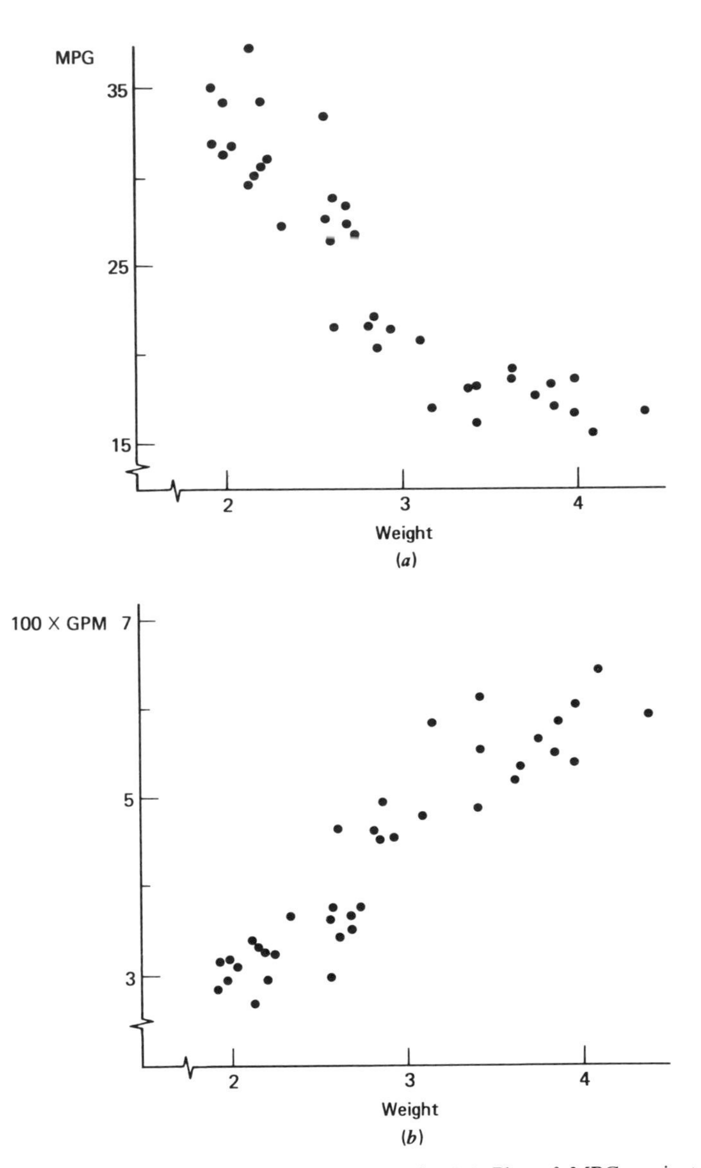

Figure 2.3. Scatter plots for the gas mileage example. (*a*) Plot of MPG against weight. (*b*) Plot of GPM = MPG^{-1} against weight.

Plots of MPG against weight and GPM = MPG^{-1} against weight are given in Figures 2.3*a*, *b*. There we see that while the relationship between MPG and weight appears to be curvilinear, the relationship between the transformed variable GPM and weight is approximately linear. This is a good illustration of a situation where a transformation, in this case the reciprocal, simplifies the model. A more general discussion of transformations is given in a later section (Sec. 2.11). There we illustrate how transformations can be estimated from data.

The parameter estimates of β_0 and β_1 in the model $\text{GPM}_t = \beta_0 + \beta_1(\text{weight}_t) + \varepsilon_t$, their standard errors, the t statistics, and the ANOVA table are given below:

	$\hat{\beta}_i$	$s_{\hat{\beta}_i}$	$t_{\hat{\beta}_i}$
Constant	−0.006	0.303	−0.02
Weight	1.515	0.103	14.75

Source	SS	df	MS	F
Regression	42.424	1	42.424	217.56
Error	7.022	36	0.195	
Total	49.446	37		

The variable weight is a highly significant predictor variable, explaining $100R^2 = 85.8$ percent of the variation in the dependent variable GPM.

To check the model adequacy, we look at the residuals from the fitted model. A detailed discussion of model checking is given in Section 2.11. There we show that a plot of the residuals e_t (or the standardized residuals e_t/s) against fitted values and plots of the residuals against the predictor variables (in our example, weight) are especially useful. In Figure 2.4 we have plotted the standardized residuals against fitted values and against weight.

There is some indication that the linear model in (2.40) overpredicts the gas mileage for very heavy cars; the residuals that come from cars with large weights are mostly negative, while the ones for intermediate weights tend to be positive. This suggests that there could be a quadratic effect.

To check whether a quadratic term $(\text{weight})^2$ is needed, we fit the model

$$y_t = \beta_0 + \beta_1 x_t + \beta_2 x_t^2 + \varepsilon_t \tag{2.41}$$

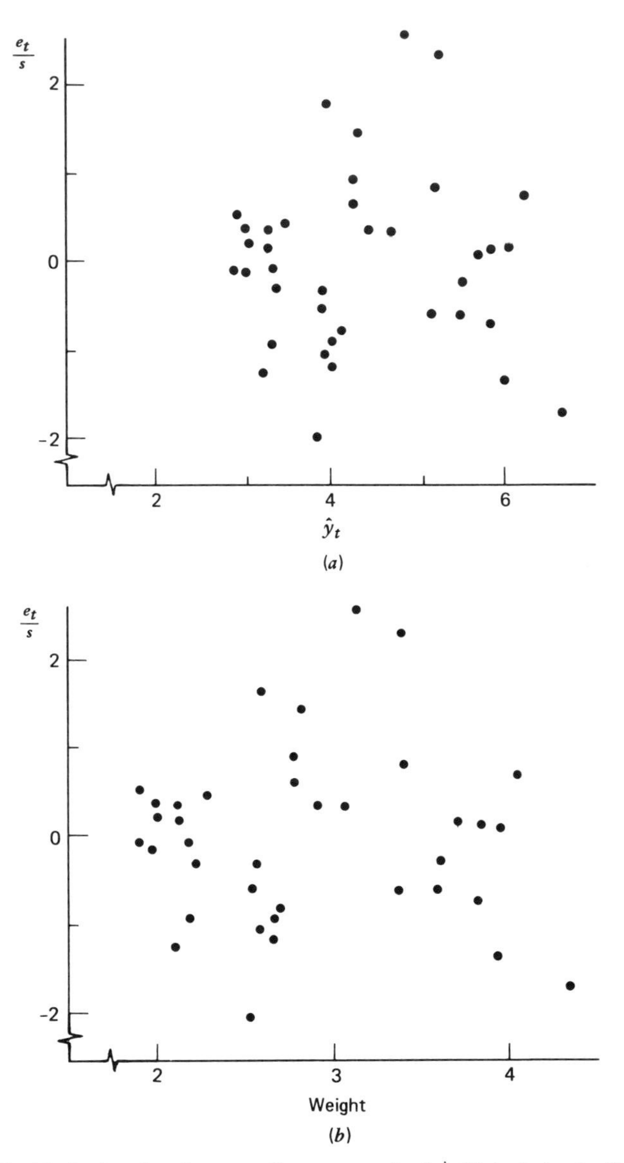

Figure 2.4. Residual plots for the gas mileage example. (*a*) Plot of standardized residuals against fitted values. (*b*) Plot of standardized residuals against weight.

The estimation results are tabulated as follows:

	$\hat{\beta}_i$	$s_{\hat{\beta}_i}$	$t_{\hat{\beta}_i}$
Constant	−2.081	1.414	−1.47
Weight	2.975	0.978	3.04
(Weight)^2	−0.242	0.162	−1.50

Source	SS	df	MS	F
Regression	42.848	2	21.424	113.66
Error	6.597	35	0.189	
Total	49.446	37		

Note that the estimate of β_2 is insignificant, which indicates that the variable (weight)^2 does not lead to a significant reduction in the error sum of squares. Since the predictor variables X and X^2 are related, we notice that the estimate of β_1 in the quadratic model (2.41) is not the same as the estimate of β_1 in the linear model (2.40). In fact, it can be shown that only if the added variable is *orthogonal* to the first one (i.e., $\Sigma x_{t1} x_{t2} = 0$) would these estimates have been the same. We could reparameterize the model (2.41) by transforming the predictor variables such that they are orthogonal. In this special case, we would fit

$$y_t = \beta_0^* + \beta_1^*(x_t - c_1) + \beta_2^*(x_t^2 - c_2 - c_3 x_t) + \varepsilon_t$$

where c_1, c_2, c_3 are chosen such that

$$\Sigma(x_t - c_1) = \Sigma(x_t^2 - c_2 - c_3 x_t) = \Sigma(x_t - c_1)(x_t^2 - c_2 - c_3 x_t) = 0$$

For a detailed discussion of how to transform the design matrix $\mathbf{X}$ to obtain orthogonal columns, see Chapter 5 in Draper and Smith (1981).

2.8. MODEL-SELECTION TECHNIQUES

An important problem in empirical model building is the selection of the independent variables that should be included in the final model. Usually at the beginning of the model-building process, we have an extensive list of possible "predictors", but we are not quite sure how many of these variables, and especially which ones, should be included in the final model.

Obviously, we do not want to omit important explanatory variables. However, on the other hand, we wish to keep the model as simple as possible, since simple models are usually easier to understand and to explain and, more important for forecasting, the estimation of each unnecessary parameter introduces additional uncertainty into the forecasts.

One approach is to estimate all possible regression models. However, for k variables this requires estimating 2^k models. The mean square error s^2, the coefficient of determination R^2, or preferably the adjusted R_a^2,

$$R_a^2 = 1 - \frac{\text{SSE}/(n - p - 1)}{\text{SSTO}/(n - 1)} \tag{2.42}$$

can be calculated for each subset of $p = 1, \ldots, k$ variables. The unadjusted R^2 is nondecreasing as additional variables are included and eventually reaches 1 as the number of parameters equals the number of observations. The adjusted R_a^2 introduces a penalty for each estimated parameter by dividing SSE and SSTO by their degrees of freedom. Thus it can happen that R_a^2 decreases as additional variables are introduced into the model. We could then choose the particular subset regression for which R_a^2 is largest or for which s^2 is smallest. Details can be found in Draper and Smith (1981) and in Neter and Wasserman (1974).

An alternative statistic for model selection is the C_p statistic suggested by Mallows (1973). It measures bias in the regression model and is of the form

$$C_p = \frac{\text{SSE}_p}{s^2} - (n - 2p)$$

In this expression, SSE_p is the residual sum of squares from a model containing p parameters (including intercept β_0; thus $p = 1, \ldots, k + 1$), and s^2 is the mean square error from the model with all k independent variables. Here we assume that s^2 is an unbiased estimate of σ^2. If now a regression with only p parameters is adequate, then $E(\text{SSE}_p) = (n - p)\sigma^2$, and thus, approximately, $E(C_p) = p$. A plot of C_p versus p indicates adequate models as points close to the $C_p = p$ line. We would look for models with a low C_p value that is about equal to p.

There are also several search techniques that help the investigator choose a model without having to look at each possible subset. Three procedures are mentioned here; backward elimination, forward selection, and stepwise regression.

The *backward elimination procedure* starts with the largest possible model and looks at the individual t statistics. If all of them are significant, the model cannot be simplified. If one or more are insignificant (for a chosen significance level α), only the least significant gets dropped from the model. Then the simplified model is reestimated and the procedure is repeated until no variable can be dropped. Most computer packages include programs for backward elimination; usually only the significance level has to be specified.

Another strategy for constructing empirical models is to start with the simplest model and then add variables as necessary. Such an approach is called *forward selection*. There one starts with the variable X_i, which by itself leads to the highest R^2. If its contribution to the regression sum of squares is insignificant [i.e., $F^* = \text{SSR}(X_i)/\text{MSE}(X_i) \leqslant F_\alpha(1, n-2)$ for some specified significance level α; see (2.32)], the variable is not included in the model, and the best model is the mean model $y_t = \beta_0 + \varepsilon_t$. However, if its contribution is significant, X_i is included. The next variable to be included in the model is the one (X_j, $j \neq i$) for which $\text{SSR}(X_j|X_i)$ is largest and also significant [i.e., $F^* = \text{SSR}(X_j|X_i)/\text{MSE}(X_i, X_j) > F_\alpha(1, n-3)$]. This procedure is continued until no significant variables to be entered can be found.

Backward elimination and forward selection have the drawback that once a variable has been eliminated (entered), it never gets entered (eliminated) again. Also, if the independent variables are correlated, it is quite possible that backward elimination and forward selection lead to models that include different explanatory variables and even different numbers of explanatory variables, but which nevertheless have similar R^2.

Stepwise regression is a compromise between backward elimination and forward selection. It starts with forward selection and includes the most significant X_i. At the second stage it includes X_j if its contribution $\text{SSR}(X_j|X_i)$ is significant and the largest among all X_j ($j \neq i$). But now backward elimination is used, and it is checked whether $\text{SSR}(X_i|X_j)$ is significant.

1. If it is not, then X_i is dropped from the model and one looks for the next variable X_k for which $\text{SSR}(X_k|X_j)$ is significant and largest among the remaining variables.
2. If $\text{SSR}(X_i|X_j)$ is significant, then both X_i and X_j are retained in the model, and the search for the next variable to be included is continued.

The iterations stop if no variables can be entered and no variables can be dropped. Usually two significance levels (α to enter and α to remove) have to be specified.

Example 2.2 (Continued)

We use the gas mileage data in Table 2.2 to illustrate the backward elimination and forward selection procedures. After fitting all possible regressions (for example, the procedure RSQUARE in SAS, or ALLS in IDA, can be used), we have listed for each p (number of predictor variables) the model that leads to the largest adjusted R_a^2 (Table 2.3). The results in Table 2.3 show that weight (X_4) by itself explains 85.4 percent of the variation in y = GPM. Adding engine displacement (X_2) leads to additional improvement. Note, however, that the best model with three independent variables is the one with number of cylinders (X_1), horsepower (X_3), and engine type (X_6). Neither weight (X_4) nor engine displacement (X_2) is included.

If the predictor variables themselves are strongly correlated, it can occur that the best model with p predictor variables does not include the variables that have led to the largest R_a^2 among all possible subsets of size $p - 1$. The correlations among the predictor variables are given in Table 2.4. We notice that number of cylinders (X_1), engine displacement (X_2), horsepower (X_3), weight (X_4), and engine type (X_6) are highly correlated and measure similar characteristics. Thus, although the variables X_1, X_3, X_6 achieve the highest R_a^2 among all groups of size $p = 3$, a model with X_1, X_2, X_4 that includes the same variables found for $p = 2$ leads to almost the same R_a^2 (in fact, $R_a^2 = .8943$).

Due to the strong correlations among the predictor variables, the backward elimination and forward selection procedures reach different conclusions (see Table 2.5). Forward selection with significance level $\alpha = .05$ leads to a model with weight (X_4) and engine displacement (X_2). Backward elimination includes *in addition* the number of cylinders (X_1), horsepower (X_3), and engine type (X_6). However, the increase in R_a^2 from .8866 to .9235

Table 2.3. Results from Fitting All Possible Regressions[a]

p			Variables Included				R_a^2
1				X_4			.8540
2		X_2		X_4			.8866
3	X_1		X_3			X_6	.9071
4	X_1		X_3		X_5	X_6	.9231
5	X_1	X_2	X_3	X_4		X_6	.9235
6	X_1	X_2	X_3	X_4	X_5	X_6	.9267

[a] X_1, number of cylinders; X_2, displacement; X_3, horsepower; X_4, weight; X_5, acceleration; X_6, engine type.

Table 2.4. Correlations Among the Predictor Variables[a]

	X_1	X_2	X_3	X_4	X_5	X_6
X_1	1.00					
X_2	.94	1.00				
X_3	.86	.87	1.00			
X_4	.92	.95	.92	1.00		
X_5	−.13	−.14	−.25	−.03	1.00	
X_6	.83	.77	.72	.67	−.31	1.00

[a]Variables as in Table 2.3.

Table 2.5. *t* Statistics from Forward Selection and Backward Elimination Procedures[a]

Step	Constant	X_1	X_2	X_3	X_4	X_5	X_6	R_a^2
				(*a*) Forward Selection				
1	−.02				14.75			.8540
2	−2.78		−3.37		8.39			.8866
3	−3.37	1.88	−3.95		8.08			.8943
				(*b*) Backward Elimination				
1	−3.92	3.62	−1.82	3.50	1.74	1.56	−3.60	.9267
2	−4.29	3.58	−2.54	3.11	2.97		−3.46	.9235

[a]Variables as in Table 2.3.

is rather small, and for simplicity we might want to select a model of the form

$$\text{GPM}_t = \beta_0 + \beta_1(\text{weight}_t) + \beta_2(\text{displacement}_t) + \varepsilon_t$$

or the even simpler model

$$\text{GPM}_t = \beta_0 + \beta_1(\text{weight}_t) + \varepsilon_t$$

2.9. MULTICOLLINEARITY

In many business and economic regression applications, we find that the predictor variables $X_1, \ldots, X_p$ are highly related, making the columns of the

X matrix almost linearly dependent. Such a situation, which is usually described as *multicollinearity* in the predictor variables, is especially common if the models are fitted on observational data. In situations where the investigator can design the experiment (i.e., choose the levels of the independent variables), multicollinearity can be avoided, since one can always choose orthogonal independent variables ($\Sigma x_{ti}x_{tj} = 0$ for $i \neq j$).

When working with observational data, however, the "design matrix" **X** cannot be chosen. Furthermore, quite frequently several predictor variables measure the same concept. Examples are a set of economic indicators measuring the "state of the economy," family income and assets measuring "wealth," store sales and number of employees measuring "size" of a store, and grade-point average and scores on standardized tests measuring "academic potential" of an applicant. All these variables tend to be highly related (correlated). As a consequence, the **X′X** matrix is "ill-conditioned," in the sense that its determinant is very small. This in turn will lead to computational difficulties when this matrix is inverted to get the least squares estimates in (2.11).

To illustrate the consequences of multicollinearity in more detail, let us consider a very simple but hypothetical example. Suppose we study the relationship between the weight of an automobile and its gas mileage (Y). Assume that we have measured the weight in both pounds (X_1) and kilograms (X_2). Obviously, there is no new information in X_2, since pounds and kilograms are exactly proportional. Now suppose we choose to fit a model of the form $Y = \beta_0 + \beta_1 X_1 + \beta_2 X_2 + \varepsilon$ to a sample of size n. We will find that the parameters cannot be estimated, since the 3×3 matrix **X′X** is of rank 2, and its inverse does not exist. If we attempted the estimation on a computer, the program might stop and give us an error message pointing to an overflow in the matrix inversion. It is fairly obvious that the parameters cannot be estimated. If the proportional relationship between X_1 and X_2 (i.e., $X_2 = cX_1$) is incorporated, the model can be written as $Y = \beta_0 + (\beta_1 + \beta_2 c)X_1 + \varepsilon = \beta_0 + \beta_* X_1 + \varepsilon$. The parameter β_* can be estimated. However, there are an infinite number of values for β_1 and β_2 that satisfy $\beta_* = \beta_1 + \beta_2 c$. Thus, individually, the parameters β_1, β_2 are not estimable.

In most cases the linear relations among the independent variables are not exact as assumed in the previous paragraph, but are only approximate. Thus even though the inverse $(\mathbf{X}'\mathbf{X})^{-1}$ exists, the determinant will be very small and the numerical calculation of the inverse will lead to difficulties.

Consequences of an "ill-conditioned" **X′X** matrix are (1) large diagonal elements in $(\mathbf{X}'\mathbf{X})^{-1}$ and therefore large variances for the least squares estimates $\hat{\boldsymbol{\beta}}$, and (2) high correlations among the parameter estimates. Furthermore, due to the large uncertainty, the parameter estimates will be highly unstable. They might even have the wrong sign and be much larger than practical or physical considerations would suggest.

To illustrate these consequences, let us consider the regression model

$$y_t = \beta_0 + \beta_1 x_{t1} + \cdots + \beta_p x_{tp} + \varepsilon_t$$

Equivalently, we can express this model in the form given by

$$y_t = \beta_0^* + \beta_1(x_{t1} - \bar{x}_1) + \cdots + \beta_p(x_{tp} - \bar{x}_p) + \varepsilon_t$$

where $\beta_0^* = \beta_0 + \beta_1\bar{x}_1 + \cdots + \beta_p\bar{x}_p$. The least squares estimate of β_0^* is given by $\bar{y}$, irrespective of the values of x_{ti}, since the new $\mathbf{X}'\mathbf{X}$ matrix can be partitioned into a scalar n and a $p \times p$ matrix. Substituting this estimate, we can write the model as

$$y_t - \bar{y} = \beta_1(x_{t1} - \bar{x}_1) + \cdots + \beta_p(x_{tp} - \bar{x}_p) + \varepsilon_t \qquad (2.43)$$

This amounts to a regression of "mean corrected" variables. Since the variables might also be of different magnitudes, we scale them by

$$s_y\sqrt{n-1} = \left[\Sigma(y_t - \bar{y})^2\right]^{1/2}$$

and

$$s_i\sqrt{n-1} = \left[\Sigma(x_{ti} - \bar{x}_i)^2\right]^{1/2} \qquad \text{for } 1 \leqslant i \leqslant p$$

This leads to the standardized regression model

$$y_t^* = \beta_1^* x_{t1}^* + \cdots + \beta_p^* x_{tp}^* + \varepsilon_t^* \qquad (2.44)$$

where the standardized variables are

$$y_t^* = \frac{y_t - \bar{y}}{s_y\sqrt{n-1}} \qquad x_{ti}^* = \frac{x_{ti} - \bar{x}_i}{s_i\sqrt{n-1}} \qquad 1 \leqslant i \leqslant p$$

The new error terms $\varepsilon_t^* = \varepsilon_t / s_y\sqrt{n-1}$ have variance σ_*^2, and the standardized regression coefficients are given by $\beta_i^* = (s_i/s_y)\beta_i$. For this standardized model, the $\mathbf{X}_*'\mathbf{X}_*$ matrix is given by the correlation matrix

$$\mathbf{X}_*'\mathbf{X}_* = \begin{bmatrix} 1 & r_{12} & \cdots & r_{1p} \\ r_{12} & 1 & \cdots & r_{2p} \\ \vdots & \vdots & & \vdots \\ r_{1p} & r_{2p} & \cdots & 1 \end{bmatrix}$$

where

$$r_{ij} = \frac{\Sigma(x_{ti} - \bar{x}_i)(x_{tj} - \bar{x}_j)}{\left[\Sigma(x_{ti} - \bar{x}_i)^2 \Sigma(x_{tj} - \bar{x}_j)^2\right]^{1/2}}$$

Take the model with $p = 2$, for example. Then from (2.12),

$$V(\hat{\beta}_1^*) = V(\hat{\beta}_2^*) = \sigma_*^2 \frac{1}{1 - r_{12}^2} \qquad \text{and} \qquad \text{Corr}(\hat{\beta}_1^*, \hat{\beta}_2^*) = -r_{12}$$

Thus, as claimed earlier, if r_{12} approaches 1, the variances of the estimates will become very large and the estimates will be highly correlated. The estimates will become very imprecise.

If multicollinearity is present, there is usually not enough information in the data to determine all parameters in the model. Only certain combinations of the parameters can be estimated. This can also be seen from the joint distribution of the parameter estimates. For example, if $p = 2$, the contours of this distribution for an ill-conditioned **X′X** matrix are very long and thin ellipses.

Multicollinearity also makes the interpretation of the individual t statistics more difficult. Assume, for illustration, that X_1 is an important predictor of the dependent variable Y. Suppose X_2 is a variable closely related to X_1 and thus by itself a good predictor of Y. The extra sum of squares contributions $\text{SSR}(X_2 | X_1)$ and $\text{SSR}(X_1 | X_2)$, however, are small and insignificant. Then the individual t statistics, when both variables are included in the model, are insignificant. However, as was pointed out earlier, they should not be dropped from the model at the same time.

Another problem with multicollinearity is that the various selection procedures can lead to different models. However, even though these models will include different predictors, the variation they explain will be very similar.

Multicollinearity causes severe difficulties in assessing the effect of each independent variable. It is usually not so much a problem if the primary purpose of the regression model is to make inferences about the response function or to make predictions of future observations, as long as the inferences are made for independent variables that are within the range of the sample data.

Obviously, multicollinearity raises many difficulties, especially in the interpretation of the model. Remedial measures have to be taken to lessen its impact. Frequent causes for an ill-conditioned **X′X** matrix are the fitting of overparameterized models and trying to infer too much from a data base in which the independent variables have the tendency to change together. If one notices that the parameters are highly correlated, their standard errors are very large, and the estimates are unreasonable, one should check the model specification, impose some prior restrictions on the parameters [as is done in econometric models; see, for example, Theil (1971)], or omit some of the independent variables.

Ill-conditioning of the $\mathbf{X}'\mathbf{X}$ matrix and numerical difficulties in the matrix inversion can also arise if the independent variables are of very different magnitudes. Frequently, reparametcrizing the model, by centering and standardizing the observations as in (2.43) and (2.44), helps to avoid numerical difficulties.

Ridge regression [Hoerl (1962); Hoerl and Kennard (1970a, b)] is another approach to overcome ill-conditioned situations where the near-singularity of the $\mathbf{X}'\mathbf{X}$ matrix leads to unstable and large parameter estimates. Ridge regression estimates are essentially solutions to a constrained regression in which the parameters are spherically restricted ($\boldsymbol{\beta}'\boldsymbol{\beta} \leqslant r^2$). The solutions are evaluated for a range of r values, and the estimates are plotted as a function of r (ridge trace). Various ways of choosing an optimal r are suggested in the literature. Ridge regression estimators are also called *shrinkage estimators*, since they shrink the vector of coefficients toward zero. For a detailed discussion, refer to the original papers by Hoerl (1962), Hoerl and Kennard (1970a, b), and the discussion and many references in Draper and Smith (1981).

2.10. INDICATOR VARIABLES

So far we have assumed that the independent variables are quantitative and are measured on a well-defined scale. Frequently, however, variables are qualitative and have two or more distinct levels. For example, consider predicting the yield from a particular chemical reaction in terms of input concentration, pressure, reaction time (all quantitative variables), and type of catalyst (qualitative: type 1,..., type k). Or consider the effect of graduate grade-point average (quantitative) and sex (male, female) on the starting salary of MBA graduates. Or consider predicting the speed of adoption of an innovation in terms of the size of the firm and type of ownership (public, private). Or consider the factory or machine effect on output, in addition to the effect of other quantitative variables.

In all these examples we observe a qualitative variable at several different levels. In order to model its effect, we have to introduce additional variables. The effect of a qualitative variable that is observed at k different levels (say from k different factories) on the response Y can be represented by

$k - 1$ indicator variables. These *indicator* or *dummy variables* are defined as $\text{IND}_{ti} = 1$ if the observation comes from level i (for $1 \leqslant i \leqslant k - 1$), and 0 otherwise. Combining the effects of these indicators with the effects of p quantitative variables $X_1, \ldots, X_p$, we can write the model as

$$y_t = \beta_0 + \sum_{i=1}^{p} \beta_i x_{ti} + \sum_{i=1}^{k-1} \delta_i \text{IND}_{ti} + \varepsilon_t \tag{2.45}$$

If the observations are from level k, the effect of the quantitative independent variables is given by

$$E(y_t) = \beta_0 + \sum_{i=1}^{p} \beta_i x_{ti}$$

If the observations are from level i ($1 \leqslant i \leqslant k - 1$), their effect is

$$E(y_t) = \beta_0 + \delta_i + \sum_{i=1}^{p} \beta_i x_{ti}$$

The parameter δ_i is thus the effect of level i relative to level k. This effect is additive and does not depend on the other independent variables $X_1, \ldots, X_p$.

In Figure 2.5a we have illustrated the model (2.45) for $p = 1$ and $k = 3$,

$$y_t = \beta_0 + \beta_1 x_t + \delta_1 \text{IND}_{t1} + \delta_2 \text{IND}_{t2} + \varepsilon_t$$

Obviously, the indicator variable representation is not unique. Alternatively, we could have defined $\text{IND}_{ti} = 1$, if the observation is from level $i + 1$ ($1 \leqslant i \leqslant k - 1$), and 0 otherwise. In this case the parameters δ_i represent the level effects as compared to level 1. Or we could have defined k indicators and omitted the intercept β_0. In this case the k parameters δ_i represent the intercepts of k parallel p-dimensional planes.

Whatever representation is adopted, the parameters are easily estimated. The design matrix $\mathbf{X}$ consists of columns corresponding to the independent variables x_{ti} and columns of ones and zeros.

In model (2.45) the levels of the qualitative variable do not interact with the other variables in the model. To model interactions we can introduce cross-product terms $x_{ti}\text{IND}_{tj}$ ($1 \leqslant i \leqslant p$ and $1 \leqslant j \leqslant k - 1$).

For example, for $p = 1$ and $k = 3$, the model involving cross-products is given by

$$y_t = \beta_0 + \beta_1 x_t + \delta_1 \text{IND}_{t1} + \delta_2 \text{IND}_{t2} + \delta_{11} x_t \text{IND}_{t1} + \delta_{22} x_t \text{IND}_{t2} + \varepsilon_t$$

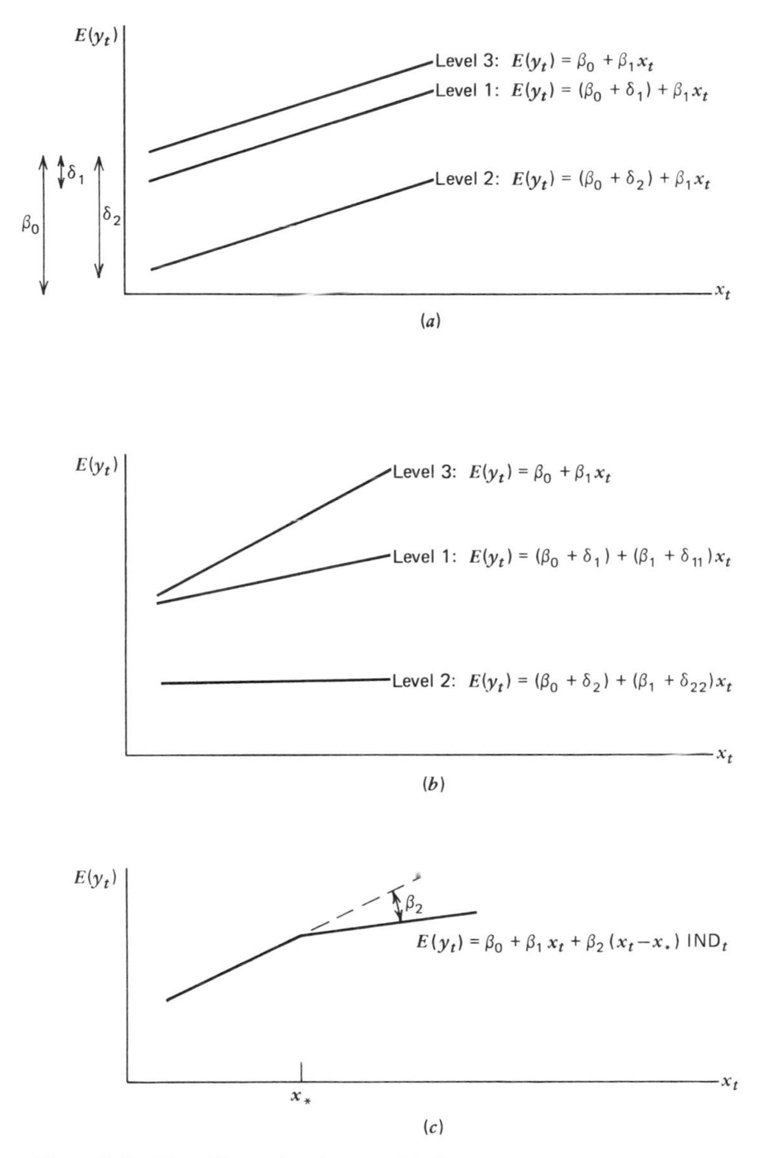

Figure 2.5. Plots illustrating the use of indicator variables in regression.
(a) $y_t = \beta_0 + \beta_1 x_t + \delta_1 \mathrm{IND}_{t1} + \delta_2 \mathrm{IND}_{t2} + \varepsilon_t$.
(b) $y_t = \beta_0 + \beta_1 x_t + \delta_1 \mathrm{IND}_{t1} + \delta_2 \mathrm{IND}_{t2} + \delta_{11} x_t \mathrm{IND}_{t1} + \delta_{22} x_t \mathrm{IND}_{t2} + \varepsilon_t$.
(c) $y_t = \beta_0 + \beta_1 x_t + \beta_2 (x_t - x_*) \mathrm{IND}_t + \varepsilon_t$.

This is illustrated in Figure 2.5*b*. The parameters can be estimated, and it can be tested whether there are any level differences at all ($\delta_1 = \delta_2 = \delta_{11} = \delta_{22} = 0$) and, if such effects exist, whether they are additive ($\delta_{11} = \delta_{22} = 0$). Estimation and testing proceed as if the indicators IND_{ti} were quantitative.

Indicator variables can also be used to estimate piecewise linear regression models. For simplicity, we consider the linear regression model with $p = 1$ predictor variable. Assume, for example, that sales y_t increase linearly with the amount spent on advertising, $y_t = \beta_0 + \beta_1 x_t + \varepsilon_t$. However, suppose that if advertising exceeds a certain amount (say x_*), the effect of each additional dollar spent on advertising will be less than β_1. Such a situation is sketched in Figure 2.5*c*. This piecewise linear regression can be modeled as

$$y_t = \beta_0 + \beta_1 x_t + \beta_2 (x_t - x_*)\text{IND}_t + \varepsilon_t$$

where $\text{IND}_t = 1$ if $x_t \geqslant x_*$, and 0 otherwise. The parameter β_2 measures the change in the slope.

2.11. GENERAL PRINCIPLES OF STATISTICAL MODEL BUILDING

So far we have assumed that the functional form of the regression model is known, and we have discussed the inferences from such a model. We have derived least squares estimates and minimum mean square error forecasts and have discussed their properties. These properties depend on certain model assumptions, such as independence, equal variance, and normality of the error terms.

Statistical inference (model estimation and hypothesis testing) is an important step in data analysis and statistical model building, but it is by no means the only step. Before the parameters can be estimated, the functional relationship among the dependent and independent variables must be specified (*model specification*). Then, after *estimation* of the parameters the adequacy of the fitted model must be evaluated (*model diagnostic checking*). It must be determined whether the assumed functional relationship is adequate and whether the assumptions of normality, equal variance, and independence of the errors are satisfied. If the fitted model is found to be inadequate, a new model must be specified, and the estimation and diagnostic checking cycle must be repeated. Only if the model passes the diagnostic checks should it be used for interpretation and forecasting.

2.11.1. Model Specification

At the first stage of statistical model building, it is necessary to specify a functional form that relates the dependent variable Y to the predictor variables $X_1, \ldots, X_p$. In some cases theory can tell us which functions to consider (for example, economic theory when studying the relationships among economic variables, or theoretical relationships from physics and chemistry in engineering applications). However, often such theory may not exist, and historical data must be used to specify the models. Scatter plots of the dependent variables against the independent variables are usually quite helpful in suggesting classes of models that should be tentatively entertained. Such initial scatter plots of the variables are also helpful in checking (informally) whether theoretical models, if they exist, are confirmed or refuted by the data. However, it must be emphasized that this is only a tentative model specification, since individual scatter plots can be misleading if interaction effects are present.

An important principle in model building is the *principle of parsimony* [see Tukey (1961), Box and Jenkins (1976)]. The principle of parsimony, or *Ockham's razor*, can be paraphrased as, "In a choice among competing hypotheses, other things being equal, the simplest is preferable." Reasons for preferring simple models over models with a large number of parameters are (1) simple models are usually easier to explain and interpret and (2) the estimation of each unnecessary parameter will increase the variance of the prediction error by a factor of $1/n$. To show this result, we note that the covariance matrix of the vector of fitted values $\hat{\mathbf{y}} = \mathbf{X}\hat{\boldsymbol{\beta}} = \mathbf{X}(\mathbf{X}'\mathbf{X})^{-1}\mathbf{X}'\mathbf{y}$ is given by $\mathbf{V}(\hat{\mathbf{y}}) = \sigma^2\mathbf{X}(\mathbf{X}'\mathbf{X})^{-1}\mathbf{X}'$. The average variance of a fitted value is then given by

$$\frac{1}{n}\sum_{t=1}^{n} V(\hat{y}_t) = \frac{\sigma^2}{n}\mathrm{Tr}\left[\mathbf{X}(\mathbf{X}'\mathbf{X})^{-1}\mathbf{X}'\right] = \frac{p\sigma^2}{n}.$$

Note that the trace of a square matrix $\mathbf{A}$, $\mathrm{Tr}(\mathbf{A})$, is the sum of its diagonal elements. The average variance of the forecast error is then given by $\sigma^2(1 + p/n)$.

Even complicated functions of the form $g(\mathbf{x}_t; \boldsymbol{\theta})$ can usually be approximated by linear combinations of integer powers and cross-products of the predictor variables $\mathbf{x}_t$. A first-order approximation (first-order model) is given by

$$y_t = \beta_0 + \sum_{i=1}^{p} \beta_i x_{ti} + \varepsilon_t$$

A second-order approximation (second-order model) can be written as

$$y_t = \beta_0 + \sum_{i=1}^{p} \beta_i x_{ti} + \sum\sum_{i \leqslant j} \beta_{ij} x_{ti} x_{tj} + \varepsilon_t$$

Higher order models will quickly contain a large number of coefficients that have to be estimated from past data. However, in many cases transformations of the dependent and independent variables, such as the natural logarithm or reciprocal transformations, can lead to simpler functional forms.

2.11.2. Model Estimation

Conditional on the specified model, unbiased and efficient estimates of the unknown coefficients (efficient in terms of smallest variances) and optimal forecasts of future observations (optimal in the sense of minimizing the mean square error) can be derived. However, the estimates and forecasts have these optimal properties only if the model assumptions are met. For example, the ordinary least squares estimates do not have the minimum variance property if the errors are correlated or have unequal variance.

2.11.3. Diagnostic Checking

Computers will fit almost any given model, even if the specified model is inadequate. Thus it becomes very important to check the adequacy of the fitted model.

A fitted model can be inadequate for several reasons:

1. The functional form may be incorrect.
2. The error specification may not be adequate. In particular, the errors may not be normally distributed, the error variances may not be constant, or the errors may be correlated.

In diagnostic checking we look at the residuals $e_t = y_t - \hat{y}_t$, since they express the variation that the regression model has not been able to explain. We can think of the residual e_t as an estimate of the error ε_t. Thus if the fitted model is correct, the residuals should confirm the assumptions we have made about the error terms. A histogram (or dot diagram) of the residuals and plots of the residuals e_t against the fitted values $\hat{y}_t$, against each independent variable, and against time (if the observations have been collected in time order) are especially useful at this stage of the analysis.

If the assumptions in the regression model are satisfied, the histogram of the residuals should resemble a normal distribution. The residuals, when plotted against $\hat{y}_t$, against each independent variable, or against time, should vary in a horizontal band around zero (see Fig. 2.6*a*). Any departure from such a horizontal band will be taken as an indication of model inadequacy.

We recommend plotting the residuals e_t against the fitted values $\hat{y}_t$, since the correlation between e and $\hat{y}$ is zero, provided that the fitted model is adequate. This follows from the fact that

$$\Sigma(e_t - \bar{e})(\hat{y}_t - \bar{\hat{y}}) = \Sigma e_t \hat{y}_t = \mathbf{e}'\hat{\mathbf{y}} \qquad \text{since} \quad \Sigma e_t = 0$$

$$= \mathbf{y}'(\mathbf{I} - \mathbf{M})\mathbf{M}\mathbf{y} \qquad \text{where} \quad \mathbf{M} = \mathbf{X}(\mathbf{X}'\mathbf{X})^{-1}\mathbf{X}'$$

$$- \mathbf{y}'(\mathbf{M} - \mathbf{M}^2)\mathbf{y} = 0 \qquad \text{since} \quad \mathbf{M}^2 = \mathbf{M}$$

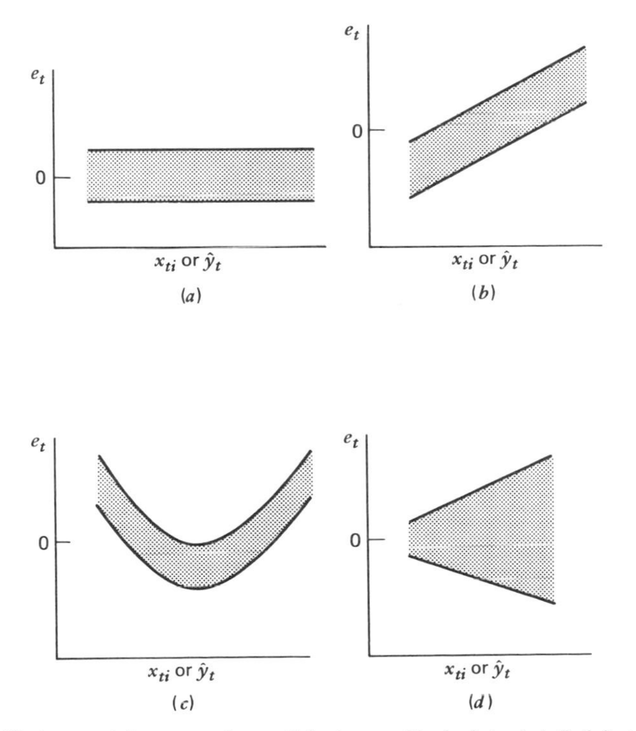

Figure 2.6. Various satisfactory and unsatisfactory residual plots. (*a*) Satisfactory residual plot. (*b*) Incorrect model form (a constant or a linear term should have been included). (*c*) Incorrect model form (a quadratic term should have been included). (*d*) Nonconstant variance.

A plot of e_t against y_t, however, would be meaningless, since the residuals and the dependent variable are always correlated, even if the model is adequate. In fact, it can be easily shown that the sample correlation r_{ey} between e and y is given by $(1 - R^2)^{1/2}$. This follows from

$$\Sigma(e_t - \bar{e})(y_t - \bar{y}) = \Sigma e_t y_t = \mathbf{e}'\mathbf{y} \qquad \text{since} \quad \Sigma e_t = 0$$

$$= \mathbf{e}'\mathbf{e} = \text{SSE} \qquad \text{since} \quad \mathbf{e}'\mathbf{e} = \mathbf{y}'(\mathbf{I} - \mathbf{M})\mathbf{y} = \mathbf{e}'\mathbf{y}$$

Furthermore,

$$\Sigma(e_t - \bar{e})^2 = \Sigma e_t^2 = \mathbf{e}'\mathbf{e} = \text{SSE}$$

$$\Sigma(y_t - \bar{y})^2 = \text{SSTO}$$

Thus

$$r_{ey} = \frac{\text{SSE}}{(\text{SSE} \times \text{SSTO})^{1/2}} = \left(\frac{\text{SSE}}{\text{SSTO}}\right)^{1/2} = (1 - R^2)^{1/2}$$

Incorrect Functional Form

Residual plots will indicate whether the functional form of the regression model is misspecified. For example, if the true model is described by a quadratic relationship but only a linear model is fitted to the data, the residuals e_t, when plotted against $\hat{y}_t$ or the independent variable, will exhibit a curvilinear pattern. Or, if a constant or a linear term in the regression model is incorrectly omitted, the residual plot will show a linear relationship between e_t and $\hat{y}_t$ or the independent variable; see Figures 2.6*b*, *c*.

2.11.4. Lack-of-Fit Tests

Residual plots are important tools in model diagnostic checking. They give us a visual indication of whether the considered model form is adequate and suggest modifications of the model if lack of fit is found. These diagnostic tools are quite general, since they do not assume a specific alternative hypothesis.

A more formal test of lack of fit can be obtained if we have genuine replications at some of the predictor levels. Genuine replications have to be uncorrelated with all other observations. For example, it would not usually be sufficient to take two measurements from the same experiment, since in such a case the measurements would probably be correlated. The replications can be used to partition the error sum of squares into a part that is due to *pure error*, SSPE, and one that is due to *lack of fit*, SSLF. More specifically, let us assume that we have observed the responses at k different settings of the p predictor variables, $\mathbf{x}_1, \mathbf{x}_2, \ldots, \mathbf{x}_k$. At each level $\mathbf{x}_i$ we observe n_i responses $y_1^{(i)}, y_2^{(i)}, \ldots, y_{n_i}^{(i)}$, where $\sum_{i=1}^{k} n_i = n$. Then the SSPE contribution at level $\mathbf{x}_i$ is given by

$$\sum_{t=1}^{n_i} \left(y_t^{(i)} - \bar{y}^{(i)} \right)^2 \qquad \text{where} \quad \bar{y}^{(i)} = \frac{1}{n_i} \sum_{t=1}^{n_i} y_t^{(i)}$$

is the average at level $\mathbf{x}_i$. Since one parameter (the mean) is estimated, this sum of squares contribution has $n_i - 1$ degrees of freedom. Overall, the pure error sum of squares is given by

$$\text{SSPE} = \sum_{i=1}^{k} \sum_{t=1}^{n_i} \left(y_t^{(i)} - \bar{y}^{(i)} \right)^2$$

and has $\sum_{i=1}^{k}(n_i - 1) = n - k$ degrees of freedom. The lack of fit sum of squares is given by SSLF = SSE − SSPE and has $n - p - 1 - (n - k) = k - p - 1$ degrees of freedom. This information is summarized in the following ANOVA table:

Source	SS	df	MS	F
Regression	SSR	p	MSR	
Error	SSE	$n - p - 1$	MSE	
Lack of fit	SSLF	$k - p - 1$	MSLF	$F_{\text{LF}} = \dfrac{\text{MSLF}}{\text{MSPE}}$
Pure error	SSPE	$n - k$	MSPE	
Total (corrected for mean)	SSTO	$n - 1$		

If the model is adequate, MSE, MSLF, and MSPE all estimate the variance σ^2. If there is lack of fit, the mean square error MSE estimates a combination of the variance and lack of fit. To test for lack of fit, we look at the ratio of the lack of fit and the pure error mean square, $F_{\text{LF}} = \text{MSLF}/\text{MSPE}$. If there is no lack of fit, this ratio has an F distribution with $k - p - 1$ and $n - k$ degrees of freedom. In lack-of-fit situations, this ratio will be larger than the values that can be expected from this distribution. Thus lack of fit is indicated if $F_{\text{LF}} > F_\alpha(k - p - 1, n - k)$. In such a case we must modify the original model. If no lack of fit is indicated, we can pool the variance estimates from MSLF and MSPE and use the mean square error MSE in the subsequent significance tests.

2.11.5. Nonconstant Variance and Variance-Stabilizing Transformations

If the scatter plot of e_t against $\hat{y}_t$ does not fall within two horizontal bands around zero but exhibits a "funnel" shape, we can conclude that the equal variance assumption is violated (see Fig. 2.6*d*). In such a case we should use weighted least squares (see Sec. 2.13) or use transformations to stabilize the variance.

To illustrate how such a transformation is chosen, let us consider the general regression model

$$y_t = f(\mathbf{x}_t; \boldsymbol{\beta}) + \varepsilon_t = \eta_t + \varepsilon_t \tag{2.46}$$

where $\eta_t = f(\mathbf{x}_t; \boldsymbol{\beta})$. Furthermore, let us assume that the variance of the errors is functionally related to the mean level η_t according to

$$V(y_t) = V(\varepsilon_t) = h^2(\eta_t)\sigma^2 \tag{2.47}$$

where h is some known function.

Our objective is to find a transformation of the data, $g(y_t)$, that will stabilize the variance. In other words, the variance of the transformed variable $g(y_t)$ should be constant. To achieve this, we expand the function $g(y_t)$ in a first-order Taylor series around η_t:

$$g(y_t) \cong g(\eta_t) + (y_t - \eta_t)g'(\eta_t)$$

where $g'(\eta_t)$ is the first derivative of $g(y_t)$ evaluated at η_t. The variance of the transformed variable can then be approximated as

$$\begin{aligned} V[g(y_t)] &\cong V[g(\eta_t) + (y_t - \eta_t)g'(\eta_t)] = [g'(\eta_t)]^2 V(y_t) \\ &= [g'(\eta_t)]^2[h(\eta_t)]^2\sigma^2 \end{aligned}$$

Thus, in order to stabilize the variance, we have to choose the transformation $g(\cdot)$ such that

$$g'(\eta_t) = \frac{1}{h(\eta_t)} \tag{2.48}$$

Frequently these transformations not only stabilize the variance, but also lead to simplifications in the functional representation of the regression model.

To illustrate the use of variance-stabilizing transformations, let us discuss two special cases, which for practical applications appear to be the most useful ones.

Example 1. Standard deviation is proportional to the level. In this case, $h(\eta_t) = \eta_t$, and the variance-stabilizing transformation $g(\eta_t)$ has to satisfy $g'(\eta_t) = 1/\eta_t$. This implies that $g(\eta_t) = \ln \eta_t$, where ln is the natural logarithm. Thus in cases where the standard deviation of y is proportional to its level, one should consider the logarithmic transformation of the dependent variable y and fit the regression model on $\ln y_t$. A logarithmic transformation is often useful in rate analysis.

Example 2. Variance is proportional to the level. In this case $h(\eta_t) = \eta_t^{1/2}$ and $g'(\eta_t) = \eta_t^{-1/2}$. This implies $g(\eta_t) = 2\eta_t^{1/2}$ and shows that the square root transformation $y_t^{1/2}$ will stabilize the variance. For Poisson counts, such a transformation would be appropriate.

These two examples are special cases from the class of power transformations

$$g(y_t) = \frac{y_t^{\lambda} - 1}{\lambda} \tag{2.49}$$

which is discussed by Box and Cox (1964). If the transformation parameter $\lambda = 1$, we analyze the original observations; if $\lambda = .50$, we analyze $y_t^{1/2}$; if $\lambda = -1$, we analyze the reciprocal $1/y_t$. Since $\lim_{\lambda \to 0}[(y_t^{\lambda} - 1)/\lambda] = \ln y_t$, we analyze $\ln y_t$ if $\lambda = 0$.

Box and Cox treat λ as an additional parameter and discuss how it can be estimated from past data. They show that the maximum likelihood estimate of λ is the one that minimizes SSE(λ), where SSE(λ) is the residual sum of squares from fitting the regression model on

$$y_t^{(\lambda)} = \frac{y_t^{\lambda} - 1}{\lambda \dot{y}^{\lambda - 1}}$$

where

$$\dot{y} = \left(\prod_{t=1}^{n} y_t \right)^{1/n}$$

is the geometric mean. If $\lambda = 0$, we use

$$y_t^{(\lambda = 0)} = \lim_{\lambda \to 0} y_t^{(\lambda)} = \dot{y} \ln y_t$$

Table 2.6. Residual Sum of Squares SSE(λ) from the Model[a, b]

$$y_t^{(\lambda)} = \beta_0 + \beta_1 x_t + \varepsilon_t$$

λ	SSE(λ)
1.0	292.58
.5	257.74
.25	245.47
ln y_t .00	236.34
−.25	230.23
−.50	227.07
−.75	226.86
−1.00	229.67
−1.25	235.63
−1.50	244.98

[a] $y_t^{(\lambda)} = \dfrac{y_t^{\lambda} - 1}{\lambda \dot{y}^{\lambda - 1}}$, where $\dot{y}$ is the geometric mean.

[b] In the gas mileage example, y_t is miles per gallon, x_t is weight.

Standard computer programs can be used to calculate SSE(λ) by repeating the regressions for different values of λ. This has been done for the gas mileage data in Example 2.2. We have estimated the regression parameters and have calculated the error sum of squares for several values of λ in

$$\frac{y_t^{\lambda} - 1}{\lambda \dot{y}^{\lambda - 1}} = \beta_0 + \beta_1 x_t + \varepsilon_t$$

where y = miles per gallon (MPG) and x = weight of the automobile. The error sum of squares is given in Table 2.6 and plotted in Figure 2.7. We find that a value $\hat{\lambda}$ around -0.65 will minimize SSE(λ). Confidence intervals for λ can be obtained [see Box and Cox (1964)]. It can be seen that a value $\hat{\lambda} = -1.0$, which leads to the reciprocal transformation considered earlier in model (2.40), is quite appropriate.

2.12. SERIAL CORRELATION AMONG THE ERRORS

In the ordinary regression model we assume that the errors $\{\ldots, \varepsilon_{t-1}, \varepsilon_t, \varepsilon_{t+1}, \ldots\}$ are uncorrelated. However, if the regression model is fitted on

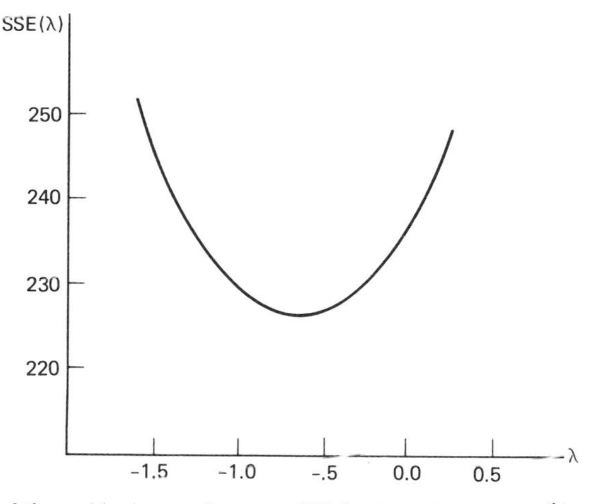

Figure 2.7. Plot of the residual sum of squares SSE(λ) from the model $y_t^{(\lambda)} = \beta_0 + \beta_1 x_t + \varepsilon_t$ for various transformations $y_t^{(\lambda)} = (y_t^\lambda - 1)/(\lambda \dot{y}^{\lambda-1})$.

time series data, it is quite likely that the errors are serially correlated. Thus diagnostic checks that test for correlation among the errors are of particular importance.

The consequences of ignoring correlated errors in regression models can be quite serious. For example, if there is positive serial correlation among the errors, the true standard errors of the regression coefficients can be considerably underestimated by the usual standard errors $s_{\hat{\beta}_i} = s\sqrt{c_{ii}}$. This means that if the usual least squares procedures are employed in the presence of serially correlated errors, the parameter estimates may appear significantly different from zero when in fact they are not. This phenomenon is called *spurious regression* [Box and Newbold (1971), Granger and Newbold (1974)].

2.12.1. Serial Correlation in a Time Series

Serial correlation among a time-ordered sequence of random variables $\{\ldots, z_{t-1}, z_t, z_{t+1}, \ldots\}$ indicates that the random variables at different time periods are correlated. Serial correlation is measured by autocovariances γ_k or autocorrelations ρ_k. The lag k *autocovariance* is defined by

$$\gamma_k = \text{Cov}(z_t, z_{t-k}) = E(z_t - \mu)(z_{t-k} - \mu) \qquad k = 0, 1, 2, \ldots \tag{2.50}$$

Here we have assumed that the mean $\mu = E(z_t)$ is constant over time and furthermore that the autocovariance depends only on the time difference k, but not on time t. This assumption, which requires that the first two moments (mean μ, variance γ_0, and autocovariance γ_k) are invariant with respect to changes along the time axis, is called the *stationarity condition*. A detailed explanation will be given in Chapter 5. From the definition in (2.50), and from the assumption of stationarity, it follows that

$$\gamma_{-k} = E(z_t - \mu)(z_{t+k} - \mu) = E(z_{t+k} - \mu)(z_t - \mu) = \gamma_k$$

The *autocorrelations* ρ_k are defined as

$$\rho_k = \frac{\text{Cov}(z_t, z_{t-k})}{[V(z_t)V(z_{t-k})]^{1/2}} = \frac{\gamma_k}{\gamma_0} \qquad k = 0, 1, 2, \ldots \tag{2.51}$$

The set of autocorrelations ρ_k, considered as a function of the lag k, is called the correlogram or the *autocorrelation function*. Since $\rho_{-k} = \rho_k$, only nonnegative k have to be considered. Furthermore, $\rho_0 = 1$.

Estimates of the autocovariances γ_k from a sample series $(z_1, z_2, \ldots, z_n)$ are given by

$$c_k = \frac{1}{n} \sum_{t=k+1}^{n} (z_t - \bar{z})(z_{t-k} - \bar{z}) \qquad k = 0, 1, 2, \ldots \tag{2.52}$$

where $\bar{z}$ is the sample mean $\bar{z} = (1/n)\sum_{t=1}^{n} z_t$.

Estimates of the autocorrelations ρ_k are given by the sample autocorrelations

$$r_k = \frac{c_k}{c_0} = \frac{\sum_{t=k+1}^{n} (z_t - \bar{z})(z_{t-k} - \bar{z})}{\sum_{t=1}^{n} (z_t - \bar{z})^2} \qquad k = 1, 2, \ldots \tag{2.53}$$

Basic results for the distribution theory of sample autocorrelations were derived by Bartlett (1946). He showed, among other results, that if there is no correlation among observations that are more than q steps apart ($\rho_k = 0$ for $k > q$), the variance of r_k can be approximated by

$$V(r_k) \cong \frac{1}{n}\left(1 + 2\sum_{k=1}^{q} \rho_k^2\right) \qquad \text{for } k > q \tag{2.54}$$

In the special case when all observations are uncorrelated ($\rho_k = 0$ for $k > 0$), this equation reduces to

$$V(r_k) \cong n^{-1} \qquad \text{for } k > 0 \tag{2.55}$$

Furthermore, for large n and $\rho_k = 0$, the distribution of r_k will be approximately normal. Therefore, one can test the null hypothesis (H_0: $\rho_k = 0$) by comparing r_k with its standard error $n^{-1/2}$ and reject H_0 at the commonly used significance level $\alpha = .05$ if

$$\frac{|r_k|}{n^{-1/2}} = \sqrt{n}\,|r_k| > 1.96 \tag{2.56}$$

It should be emphasized that we always look at several r_k simultaneously. Bartlett (1946) has shown that under the null hypothesis (H_0: $\rho_k = 0$ for all k), the correlation between r_k and r_{k-s} ($s \neq 0$) is negligible. Nevertheless, we should be aware that, while the significance level can be controlled for each separate test at $\alpha = .05$, the probability of getting at least one r_k from many such tests outside $\pm 1.96 n^{-1/2}$ is quite large. For example, for 20 such correlations, this probability is given by $1 - .95^{20} = .64$.

2.12.2. Detection of Serial Correlation Among the Errors in the Regression Model

It has been discussed previously that the residuals e_t from the fitted model are estimates of the errors ε_t that are assumed to be uncorrelated. Thus, the residuals could be expected to exhibit no or little correlation. However, we also know that the least squares estimation introduces $p + 1$ restrictions among the n residuals [see Eq. (2.15)]. Thus, obviously, the residuals cannot be uncorrelated, even if the errors are. However, this correlation will become negligible if the sample size n is large compared with the number of estimated parameters.

To check whether the errors are correlated, we calculate r_k, the lag k sample autocorrelation among the residuals. Since $\Sigma e_t = 0$, provided an intercept has been included in the regression model, these autocorrelations are simply given by

$$r_k = \frac{\sum_{t=k+1}^{n} e_t e_{t-k}}{\sum_{t=1}^{n} e_t^2} \tag{2.57}$$

As a rule of thumb, we will question the assumption of uncorrelated errors whenever the sample autocorrelation of the residuals r_k exceeds twice its standard error, that is, whenever $|r_k| > 2n^{-1/2}$. This test, however, is only approximate, since we have to use the residuals as estimates of the unobserved errors. The estimation of the regression parameters will itself introduce additional variation and make the approximation to the limiting distribution of r_k somewhat worse. In addition, it must be remembered that we always look at several r_k's jointly and that, even if the errors are uncorrelated at all lags, it is quite likely that at least one of the r_k's exceeds twice its standard error.

The advantage of this diagnostic check is that it is very general and does not require us to specify a particular alternative hypothesis when we are testing for the absence of serial correlation. In this sense this test can be considered nonparametric.

A more specific test of zero correlation against a particular alternative hypothesis was developed by Durbin and Watson (1950, 1951, 1971). There one tests whether the parameter ϕ in the error process

$$\varepsilon_t = \phi\varepsilon_{t-1} + a_t \tag{2.58}$$

where now the a_t are uncorrelated random variables, is significant. Such a model is referred to as a first-order autoregressive model and is discussed in Chapter 5. Obviously, if $\phi = 0$, the errors ε_t in (2.58) are uncorrelated. Furthermore, it will be shown in Chapter 5 that the autocorrelations ρ_k that are implied by this process decrease geometrically with the time lag k ($\rho_k = \phi^k$ for $k = 1, 2, \ldots$), and thus in particular $\rho_1 = \phi$. To test H_0: $\phi = 0$, Durbin and Watson consider the statistic

$$D = \frac{\sum_{t=2}^{n} (e_t - e_{t-1})^2}{\sum_{t=1}^{n} e_t^2} = \frac{\sum_{t=2}^{n} e_t^2 + \sum_{t=2}^{n} e_{t-1}^2 - 2\sum_{t=2}^{n} e_t e_{t-1}}{\sum_{t=1}^{n} e_t^2} \cong 2(1 - r_1) \tag{2.59}$$

This test statistic, which is commonly known as the *Durbin-Watson statistic*, is standard output in most regression computer packages.

Since the lag 1 autocorrelation r_1 has to be between -1 and $+1$, this test statistic has to be between 0 and 4. For positive correlation ($r_1 > 0$), D falls into the range from 0 to 2; for negative correlation, D is between 2 and 4.

Although the exact significance points for the D statistic vary with the design matrix $\mathbf{X}$, they are subject to lower and upper bounds that depend only on the number of independent variables. Durbin and Watson have calculated these bounds, d_L and d_U. Values of D that are either smaller than d_L or larger than d_U lead to definite conclusions. However, for test statistics that fall between these two bounds, the test is inconclusive.

More specifically, when testing H_0: $\phi = 0$ against H_1: $\phi > 0$ (positive autocorrelation), the Durbin-Watson test specifies the following decision rule:

If $D > d_U$, conclude D is not significant at level α; do not reject H_0.
If $D < d_L$, conclude D is significant; reject H_0 in favor of H_1.
If $d_L \leqslant D \leqslant d_U$, the test is inconclusive.

To test H_0: $\phi = 0$ against H_1: $\phi < 0$ (negative autocorrelation), the decision rule is given by:

If $4 - D > d_U$, conclude D is not significant at level α; do not reject H_0.
If $4 - D < d_L$, conclude D is significant; reject H_0 in favor of H_1.
If $d_L \leqslant 4 - D \leqslant d_U$, the test is inconclusive.

To test H_0: $\phi = 0$ against H_1: $\phi \neq 0$, the decision rule is given by:

If $D > d_U$ and $4 - D > d_U$, conclude D is not significant at level 2α; do not reject H_0.
If $D < d_L$ or $4 - D < d_L$, conclude D is significant; reject H_0 in favor of H_1.
Otherwise the test is inconclusive.

The lower and upper bounds d_L and d_U, which were tabulated by Durbin and Watson, are given in the Table Appendix, Table D. They depend on the significance level of the test, the number of observations n, and the number of predictor variables p.

If it is found that the test is inconclusive, one could use the data and calculate the exact significance point numerically, using methods described by Durbin and Watson (1971). Alternatively, one could also treat the inconclusive area as part of the rejection region. For economic data, which are typically slowly changing, this approach is usually quite good, since in this case the exact percentage point in tests for positive (negative) autocorrelation is close to d_U (d_L).

Compared with our earlier test, which checked whether the residual autocorrelations r_k exceed twice their standard error $n^{-1/2}$, the Durbin-Watson test is certainly more efficient when the null hypothesis of zero correlation is tested against the alternative of a first-order autoregressive model. However, if the alternative hypothesis is of a different structure, the Durbin-Watson test can be quite inefficient. For example, let us suppose that the observations are recorded quarterly and that the errors exhibit a seasonal correlation structure. Let us assume that a large value of r_4 is found, but that all other sample autocorrelations (including r_1) are of small magnitude. Since the Durbin-Watson test statistic is essentially a function of r_1, we would conclude incorrectly from this test that the errors are uncorrelated. For this reason we recommend that the autocorrelations at lags other than lag 1 also be investigated.

2.12.3. Regression Models with Correlated Errors

If serial correlation is found among the errors, the regression model must be generalized. In Chapter 5 we discuss a class of models that can be used to describe a variety of different autocorrelation structures. One such model was already introduced in Equation (2.58), when we discussed the Durbin-Watson test statistic. In the first-order autoregressive model, $\varepsilon_t = \phi\varepsilon_{t-1} + a_t$, the error at time t is a function of the previous error ε_{t-1} and an uncorrelated random variable a_t. If we assume this particular error structure, we can write the linear regression model as

$$y_t = \beta_0 + \sum_{i=1}^{p} \beta_i x_{ti} + \varepsilon_t \tag{2.60}$$

where $\varepsilon_t = \phi\varepsilon_{t-1} + a_t$.

Since at time $t - 1$,

$$y_{t-1} = \beta_0 + \sum_{i=1}^{p} \beta_i x_{t-1,i} + \varepsilon_{t-1}$$

we can transform the model in (2.60) into one with uncorrelated errors by multiplying y_{t-1} by ϕ and subtracting it from (2.60). This leads to

$$y_t = \phi y_{t-1} + \beta_0(1 - \phi) + \sum_{i=1}^{p} \beta_i(x_{ti} - \phi x_{t-1,i}) + a_t \tag{2.61}$$

Least squares estimates of the parameters ϕ, β_0, $\beta_1, \ldots, \beta_p$ and their standard errors can now be found. However, since the new model is nonlinear in the parameters, we must use nonlinear least squares procedures to minimize the error sum of squares:

$$\mathrm{SSE}(\phi, \beta_0, \beta_1, \ldots, \beta_p)$$

$$= \sum_{t=2}^{n} \left[y_t - \phi y_{t-1} - \beta_0(1 - \phi) - \sum_{i=1}^{p} \beta_i (x_{ti} - \phi x_{t-1,i}) \right]^2 \tag{2.62}$$

Here the summation starts with $t = 2$, since a lagged variable is included in the model. Nonlinear minimization procedures are part of most statistical computer packages.

If $\phi = 1$ in (2.58), the model

$$\varepsilon_t = \varepsilon_{t-1} + a_t \tag{2.63}$$

is called a *random walk model*. In this special case the transformed regression model (2.61) simplifies to

$$y_t - y_{t-1} = \sum_{i=1}^{p} \beta_i (x_{ti} - x_{t-1,i}) + a_t \tag{2.64}$$

Estimates of the regression coefficients $\beta_1, \ldots, \beta_p$ can be derived by regressing the first differences of y_t on the first differences of the predictor variables. Note that in this case the intercept β_0 cannot be estimated.

In our discussion we have described just one possible model for the correlation among the errors. As pointed out before, this model implies that the autocorrelations ρ_k decrease geometrically ($\rho_k = \phi^k$). Thus if we observe that the sample autocorrelations r_k of the residuals from the ordinary regression model assuming uncorrelated errors exhibit such a pattern, we should transform the model into (2.61) and find the estimates by minimizing the error sum of squares in (2.62). If the autocorrelations decrease very slowly with increasing k, we should consider a regression model of the differenced series.

Example 2.3: Advertising-Sales Relationship

In this example we study the relationship between advertising and sales, and consider the time series data reported in Blattberg and Jeuland (1981). These data consist of 36 consecutive monthly sales (y_t) and advertising

Table 2.7. Sales and Advertising for $n = 36$ Consecutive Months[a]

Sales	Advertising	Sales	Advertising
12.0	15.0	30.5	33.0
20.5	16.0	28.0	62.0
21.0	18.0	26.0	22.0
15.5	27.0	21.5	12.0
15.3	21.0	19.7	24.0
23.5	49.0	19.0	3.0
24.5	21.0	16.0	5.0
21.3	22.0	20.7	14.0
23.5	28.0	26.5	36.0
28.0	36.0	30.6	40.0
24.0	40.0	32.3	49.0
15.5	3.0	29.5	7.0
17.3	21.0	28.3	52.0
25.3	29.0	31.3	65.0
25.0	62.0	32.2	17.0
36.5	65.0	26.4	5.0
36.5	46.0	23.4	17.0
29.6	44.0	16.4	1.0

[a]Read downwards, left to right.

Source: Blattberg and Jeuland (1981).

expenditures (x_t) of a dietary weight control product (see Table 2.7). The data were originally analyzed by Bass and Clarke (1972), who also give a description of the series.

Researchers in marketing have realized that the effects of advertising may not totally dissipate in the period in which the advertisement is seen. Thus, models of the form

$$y_t = \beta_0 + \alpha_0 x_t + \varepsilon_t \tag{2.65}$$

may not be adequate. This is, in fact, confirmed by the results in Table 2.8, where we list the parameter estimates and the autocorrelations of the residuals. The Durbin-Watson test statistic is DW = 1.25. From Table D in the Table Appendix we find that for $n = 36$ and $p = 1$ the 5 percent lower and upper significance bounds are given by $d_L = 1.41$ and $d_U = 1.52$. Since DW $< d_L$, we can conclude that there is significant positive correlation among the errors. Also the lag 1 residual autocorrelation $r_1 = .31$ is quite large compared to its approximate standard error $n^{-1/2} = .17$.

Figure 2.8 also shows that the residuals e_t are still correlated with the lagged advertising expenditures x_{t-1}. This confirms that advertising effects

Table 2.8. Advertising-Sales Relationship ($n = 36$). Estimation Results from Fitting the Model

$$y_t = \beta_0 + \alpha_0 x_t + \varepsilon_t$$

	$\hat{\beta}_i$	$s_{\hat{\beta}_i}$	$t_{\hat{\beta}_i}$
Constant	18.32	1.49	12.31
α_0	.208	.044	4.75

$R^2 = .399 \qquad s = 4.863$

Autocorrelations of Residuals

k	1	2	3	4	5	6
r_k	.32	.16	.20	.10	.09	$-.13$

DW = 1.25

do not dissipate in the period in which advertising is seen but are distributed over the next periods. Thus models of the form

$$y_t = \beta_0 + \sum_{k \geqslant 0} \alpha_k x_{t-k} + \varepsilon_t \tag{2.66}$$

are commonly considered in the marketing literature. Since the effect of the

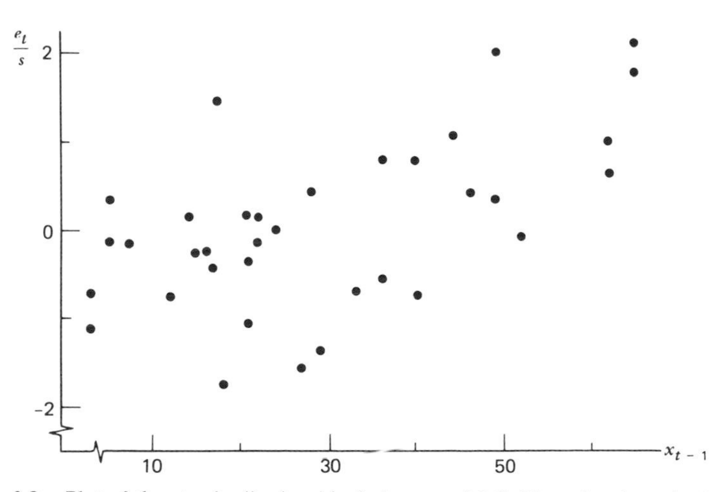

Figure 2.8. Plot of the standardized residuals from model (2.65) against lagged advertising.

independent variable is distributed over several lags, these models are known as *distributed lag models*.

Since we have only limited data to estimate the coefficients, we have to make simplifying assumptions. Frequently it is assumed that the weights decrease exponentially, $\alpha_k = \alpha_0 \lambda^k$, where $|\lambda| < 1$. Then the model can be written as

$$\begin{aligned} y_t &= \beta_0 + \alpha_0 \sum_{k \geq 0} \lambda^k x_{t-k} + \varepsilon_t \\ &= \beta_0 + \alpha_0 x_t + \lambda \alpha_0 \sum_{k \geq 0} \lambda^k x_{t-1-k} + \varepsilon_t \end{aligned} \tag{2.67}$$

Substituting $\alpha_0 \Sigma_{k \geq 0} \lambda^k x_{t-1-k} = y_{t-1} - \beta_0 - \varepsilon_{t-1}$, we find that

$$\begin{aligned} y_t &= \beta_0(1 - \lambda) + \lambda y_{t-1} + \alpha_0 x_t + \varepsilon_t - \lambda \varepsilon_{t-1} \\ &= \beta_0^* + \lambda y_{t-1} + \alpha_0 x_t + \varepsilon_t^* \end{aligned} \tag{2.68}$$

Note that even if the errors ε_t in model (2.67) are uncorrelated, the errors $\varepsilon_t^* = \varepsilon_t - \lambda \varepsilon_{t-1}$ are correlated. However, in marketing studies it is usually assumed that the errors ε_t^* are uncorrelated, and ordinary least squares estimates for the parameters in (2.68) are calculated. This strategy is also adopted here. However, if a residual analysis would show that the errors in (2.68) are still correlated, the error model would be changed accordingly.

The estimation results are shown in Table 2.9. The residuals from this model show no correlation. All estimates are significant.

Table 2.9. Advertising-Sales Relationship ($n = 36$). Estimation Results from Fitting the Model

$$y_t = \beta_0 + \lambda y_{t-1} + \alpha_0 x_t + \varepsilon_t$$

	$\hat{\beta}_i$	$s_{\hat{\beta}_i}$	$t_{\hat{\beta}_i}$
Constant	7.453	2.467	3.02
λ	.528	.102	5.17
α_0	.147	.033	4.43

$R^2 = .672 \qquad s = 3.480$

Autocorrelations of Residuals

k	1	2	3	4	5	6
r_k	−.05	−.14	.06	−.09	−.02	−.19

DW = 2.03

It can be seen that a one-time unit change in advertising leads eventually to an increase in sales of $\alpha_0(1 + \lambda + \lambda^2 + \cdots) = \alpha_0/(1 - \lambda)$. After only k time periods the effect is

$$\alpha_0(1 + \lambda + \cdots + \lambda^k) = \frac{\alpha_0(1 - \lambda^{k+1})}{1 - \lambda}$$

Thus at time k, $100(1 - \lambda^{k+1})$ percent of the total advertising effect has been realized. If we want to find the time period at which $100p$ percent of the effect has been observed, we can solve for k. This leads to

$$k = \frac{\log(1 - p)}{\log \lambda} - 1 \tag{2.69}$$

For example, for $p = 0.95$ and $\lambda = 0.528$ (as in Table 2.9), we find that $k = 3.69$. This implies that after four periods more than 95 percent of the advertising effect has been realized.

Example 2.4: Lydia Pinkham Data

As another example to illustrate the modeling of sales-advertising relationships, we consider the Lydia E. Pinkham data set. This data base has been studied extensively in the marketing literature [Palda (1964); see also Clarke (1976), Helmer and Johansson (1977), Houston and Weiss (1975), Pollay (1979), Weiss, Houston, and Windal (1978)]. In Table 2.10 we have listed annual sales and annual advertising expenditures for the period 1907–1960. The data were taken from Erickson (1981). Over the 54-year history, four different advertising strategies were used (1907–1914, 1915–1925, 1926–1940, 1941–1960). Indicator variables are introduced to model the effects of these changes. We define

$$\text{IND}_{t1} = \begin{cases} 1 & \text{if } 1907 \leqslant t \leqslant 1914 \\ 0 & \text{otherwise} \end{cases}$$

$$\text{IND}_{t2} = \begin{cases} 1 & \text{if } 1915 \leqslant t \leqslant 1925 \\ 0 & \text{otherwise} \end{cases}$$

$$\text{IND}_{t3} = \begin{cases} 1 & \text{if } 1926 \leqslant t \leqslant 1940 \\ 0 & \text{otherwise} \end{cases}$$

Several models for capturing the lagged advertising effects are possible. We could include past advertising effects and estimate a model of the form

$$y_t = \beta_0 + \sum_{i=0}^{m} \alpha_i x_{t-i} + \sum_{i=1}^{3} \delta_i \text{IND}_{ti} + \varepsilon_t \tag{2.70}$$

Table 2.10. The Lydia Pinkham Data Base; Annual Data (in thousands of dollars)

Year	Sales	Advertising	Year	Sales	Advertising
1907	1016	608	1934	1770	1504
1908	921	451	1935	1518	807
1909	934	529	1936	1103	339
1910	976	543	1937	1266	562
1911	930	525	1938	1473	745
1912	1052	549	1939	1423	749
1913	1184	525	1940	1767	862
1914	1089	578	1941	2161	1034
1915	1087	609	1942	2336	1054
1916	1154	504	1943	2602	1164
1917	1330	752	1944	2518	1102
1918	1980	613	1945	2637	1145
1919	2223	862	1946	2177	1012
1920	2203	866	1947	1920	836
1921	2514	1016	1948	1910	941
1922	2726	1360	1949	1984	981
1923	3185	1482	1950	1787	974
1924	3351	1608	1951	1689	766
1925	3438	1800	1952	1866	920
1926	2917	1941	1953	1896	964
1927	2359	1229	1954	1684	811
1928	2240	1373	1955	1633	789
1929	2196	1611	1956	1657	802
1930	2111	1568	1957	1569	770
1931	1806	983	1958	1390	639
1932	1644	1046	1959	1387	644
1933	1814	1453	1960	1289	564

Or we could consider a distributed lag model as in (2.68):

$$y_t = \beta_0 + \lambda y_{t-1} + \alpha_0 x_t + \sum_{i=1}^{3} \delta_i \text{IND}_{ti} + \varepsilon_t \tag{2.71}$$

The estimation results for the model in (2.71) are given in Table 2.11. We notice that this model is not adequate, since the residuals are still autocorrelated. The Durbin-Watson test statistic DW = 1.20 is significant (for $n = 55$, $p = 5$, and $\alpha = .05$: $d_L = 1.38$, $d_U = 1.77$). Also the lag 1 sample autocorrelation r_1 is almost three times as large as its standard error.

Table 2.11. Lydia Pinkham Annual Data (n = 54). Estimation Results from Fitting the Model

$$y_t = \beta_0 + \lambda y_{t-1} + \alpha_0 x_t + \sum_{i=1}^{3} \delta_i \text{IND}_{ti} + \varepsilon_t$$

	$\hat{\beta}_i$	$s_{\hat{\beta}_i}$	$t_{\hat{\beta}_i}$
Constant	255	96	2.64
λ	0.61	0.081	7.46
α_0	0.53	0.136	3.93
δ_1	−133	89	−1.50
δ_2	217	67	3.23
δ_3	−203	67	−3.02

$R^2 = .929$ $s = 176.2$

Autocorrelations of Residuals

k	1	2	3	4	5	6	7	8	9	10
r_k	.39	.20	.09	.10	.13	−.12	−.15	−.20	−.18	−.08

DW = 1.20

The autocorrelations in Table 2.11 exhibit a pattern similar to that of ϕ^k (exponential decay). It was pointed out above, and it will be explained in more detail in Chapter 5, that such an autocorrelation pattern can be represented by a first-order autoregressive model. Combining the models in (2.71) and (2.58) leads to

$$y_t = \beta_0 + \lambda y_{t-1} + \alpha_0 x_t + \sum_{i=1}^{3} \delta_i \text{IND}_{ti} + \varepsilon_t \tag{2.72}$$

where

$$\varepsilon_t = \phi \varepsilon_{t-1} + a_t$$

Using the same approach as in Equation (2.61), we can transform this model into one with uncorrelated errors and determine the parameter estimates of β_0, λ, α_0, δ_1, δ_2, δ_3, and ϕ by minimizing the sum of squares:

$$\text{SSE}(\beta_0, \lambda, \alpha_0, \delta_1, \delta_2, \delta_3, \phi) = \sum_{t=3}^{n} [(y_t - f_t) - \phi(y_{t-1} - f_{t-1})]^2 \tag{2.73}$$

Table 2.12. Lydia Pinkham Annual Data ($n = 54$). Estimation Results from Fitting the Model

$$y_t = \beta_0 + \lambda y_{t-1} + \alpha_0 x_t + \sum_{i=1}^{3} \delta_i \text{IND}_{ti} + \varepsilon_t;\ \varepsilon_t = \phi \varepsilon_{t-1} + a_t$$

	$\hat{\beta}_i$	$s_{\hat{\beta}_i}$	$t_{\hat{\beta}_i}$
Constant	371	185	2.01
λ	0.51	0.11	4.87
α_0	0.59	0.13	4.53
δ_1	−49	153	−0.32
δ_2	231	118	1.97
δ_3	−212	105	−2.02
ϕ	0.51	0.18	2.83

$R^2 = .942$ $s = 162.7$

Autocorrelations of Residuals

k	1	2	3	4	5	6	7	8	9	10
r_k	−.08	.08	.02	.03	.24	−.14	−.06	−.12	−.14	.04

DW = 2.15

where

$$f_t = f_t(\beta_0, \lambda, \alpha_0, \delta_1, \delta_2, \delta_3)$$

$$= \beta_0 + \lambda y_{t-1} + \alpha_0 x_t + \sum_{i=1}^{3} \delta_i \text{IND}_{ti}$$

Nonlinear least squares procedures can be used to estimate the parameters. The parameter estimates and their standard errors, which are obtained by linearizing the model [for a detailed discussion, see Draper and Smith (1981), p. 463], are given in Table 2.12. The autocorrelations show that for this model the residuals are uncorrelated.

2.13. WEIGHTED LEAST SQUARES

The two major assumptions about the error terms ε_t in the regression model are that (1) the errors are uncorrelated and (2) the errors have equal variances.

The equal-variance assumption implies that each observation is measured with the same precision or reliability. However, it sometimes happens that some of the observations in the regression model are measured with more precision (i.e., have smaller variance) than others. If the variance is related to the level of the observations, one can transform the dependent variable to stabilize the variance. Such variance-stabilizing transformations were discussed in Section 2.11.

However, situations can occur in which the observations have different precision, but the precision is not a function of the level of the series. In such instances the Gauss-Markov theorem does not apply for the ordinary least squares estimator $\hat{\boldsymbol{\beta}} = (\mathbf{X}'\mathbf{X})^{-1}\mathbf{X}'\mathbf{y}$. This implies that other linear unbiased estimators can be found that have smaller variances than the ordinary least squares estimator. The one with the smallest variance, which is called the *weighted least squares estimator*, is now derived.

Let us write the model in vector form as

$$\mathbf{y} = \mathbf{X}\boldsymbol{\beta} + \boldsymbol{\varepsilon} \tag{2.74}$$

where $E(\boldsymbol{\varepsilon}) = \mathbf{0}$ and

$$\mathbf{V}(\boldsymbol{\varepsilon}) = \sigma^2\boldsymbol{\Omega}^{-1} = \sigma^2 \begin{bmatrix} \omega_1^{-1} & & & \\ & \omega_2^{-1} & & \mathbf{0} \\ \mathbf{0} & & \ddots & \\ & & & \omega_n^{-1} \end{bmatrix}$$

Note that here we have assumed that the covariance matrix is a diagonal matrix in which the diagonal elements (variances) are not equal. The diagonal matrix $\sigma^{-2}\boldsymbol{\Omega}$ is called the precision matrix, since it expresses how reliable (or precise) the observations are. The next step is to transform this model into a regression model in which the variances are equal. This can be achieved by multiplying the model in (2.74) by

$$\boldsymbol{\Omega}^{1/2} = \begin{bmatrix} \sqrt{\omega_1} & & & \\ & \sqrt{\omega_2} & & \mathbf{0} \\ \mathbf{0} & & \ddots & \\ & & & \sqrt{\omega_n} \end{bmatrix}$$

which leads to

$$\boldsymbol{\Omega}^{1/2}\mathbf{y} = \boldsymbol{\Omega}^{1/2}\mathbf{X}\boldsymbol{\beta} + \boldsymbol{\varepsilon}_* \tag{2.75}$$

where $\boldsymbol{\varepsilon}_* = \boldsymbol{\Omega}^{1/2}\boldsymbol{\varepsilon}$. The error in the transformed model (2.75) satisfies the usual regression assumptions $E(\boldsymbol{\varepsilon}_*) = \mathbf{0}$, and $\mathbf{V}(\boldsymbol{\varepsilon}_*) = \mathbf{V}(\boldsymbol{\Omega}^{1/2}\boldsymbol{\varepsilon}) = \sigma^2\boldsymbol{\Omega}^{1/2}\boldsymbol{\Omega}^{-1}\boldsymbol{\Omega}^{1/2} = \sigma^2\mathbf{I}$.

The least squares estimates in the transformed model (2.75) can be calculated using the result in (2.11):

$$\hat{\boldsymbol{\beta}}_* = (\mathbf{X}'\boldsymbol{\Omega}^{1/2}\boldsymbol{\Omega}^{1/2}\mathbf{X})^{-1}\mathbf{X}'\boldsymbol{\Omega}^{1/2}\boldsymbol{\Omega}^{1/2}\mathbf{y} = (\mathbf{X}'\boldsymbol{\Omega}\mathbf{X})^{-1}\mathbf{X}'\boldsymbol{\Omega}\mathbf{y} \qquad (2.76)$$

Since these estimates minimize the weighted sum of squares

$$S(\boldsymbol{\beta}) = (\mathbf{y} - \mathbf{X}\boldsymbol{\beta})'\boldsymbol{\Omega}(\mathbf{y} - \mathbf{X}\boldsymbol{\beta}) = \sum_{t=1}^{n} \omega_t(y_t - \mathbf{x}_t'\boldsymbol{\beta})^2$$

they are called the *weighted least squares estimates*. In this sum of squares the observations are weighted in proportion to the reciprocal of their variances. The weighted least squares estimates are unbiased, since

$$\begin{aligned} E(\hat{\boldsymbol{\beta}}_*) &= E\left[(\mathbf{X}'\boldsymbol{\Omega}\mathbf{X})^{-1}\mathbf{X}'\boldsymbol{\Omega}\mathbf{y}\right] \\ &= (\mathbf{X}'\boldsymbol{\Omega}\mathbf{X})^{-1}\mathbf{X}'\boldsymbol{\Omega}E(\mathbf{y}) = (\mathbf{X}'\boldsymbol{\Omega}\mathbf{X})^{-1}\mathbf{X}'\boldsymbol{\Omega}\mathbf{X}\boldsymbol{\beta} = \boldsymbol{\beta} \end{aligned}$$

Furthermore, the covariance matrix is given by

$$\begin{aligned} \mathbf{V}(\hat{\boldsymbol{\beta}}_*) &= \mathbf{V}(\mathbf{A}\mathbf{y}) = \mathbf{A}\mathbf{V}(\mathbf{y})\mathbf{A}' \\ &= \sigma^2\mathbf{A}\boldsymbol{\Omega}^{-1}\mathbf{A}' = \sigma^2(\mathbf{X}'\boldsymbol{\Omega}\mathbf{X})^{-1}\mathbf{X}'\boldsymbol{\Omega}\boldsymbol{\Omega}^{-1}\boldsymbol{\Omega}\mathbf{X}(\mathbf{X}'\boldsymbol{\Omega}\mathbf{X})^{-1} \\ &= \sigma^2(\mathbf{X}'\boldsymbol{\Omega}\mathbf{X})^{-1} \end{aligned} \qquad (2.77)$$

where $\mathbf{A} = (\mathbf{X}'\boldsymbol{\Omega}\mathbf{X})^{-1}\mathbf{X}'\boldsymbol{\Omega}$. Comparing $\mathbf{V}(\hat{\boldsymbol{\beta}}_*)$ in (2.77) with the covariance matrix of the ordinary least squares estimates, which for unequal variances is given by

$$\begin{aligned} \mathbf{V}(\hat{\boldsymbol{\beta}}) &= \mathbf{V}\left[(\mathbf{X}'\mathbf{X})^{-1}\mathbf{X}'\mathbf{y}\right] = (\mathbf{X}'\mathbf{X})^{-1}\mathbf{X}'\mathbf{V}(\mathbf{y})\mathbf{X}(\mathbf{X}'\mathbf{X})^{-1} \\ &= \sigma^2(\mathbf{X}'\mathbf{X})^{-1}\mathbf{X}'\boldsymbol{\Omega}^{-1}\mathbf{X}(\mathbf{X}'\mathbf{X})^{-1} \end{aligned} \qquad (2.78)$$

it can be shown that weighted least squares will always lead to smaller (or at worst, the same) variances than ordinary least squares, for individual coefficients as well as for linear combinations of the coefficients.

To model nonconstant variances among uncorrelated errors, we have assumed a diagonal matrix $\boldsymbol{\Omega}$. However, note that the matrix $\boldsymbol{\Omega}$ in weighted least squares does not have to be diagonal; it can be any positive definite symmetric matrix.

To illustrate weighted least squares, we consider the *simple linear regression through the origin*, $y_t = \beta x_t + \varepsilon_t$, as an example. Let us assume that the variance of the errors ε_t is not constant but is given by $V(\varepsilon_t) = \sigma^2/\omega_t$.

Then the weighted least squares estimate of β is the one that minimizes $\Sigma\omega_t(y_t - \beta x_t)^2$. According to (2.76) it is given by

$$\hat{\beta}_* = (\mathbf{X}'\boldsymbol{\Omega}\mathbf{X})^{-1}\mathbf{X}'\boldsymbol{\Omega}\mathbf{y} = \frac{\Sigma\omega_t x_t y_t}{\Sigma\omega_t x_t^2} \tag{2.79}$$

It is unbiased, and it follows from (2.77) that its variance is given by

$$V(\hat{\beta}_*) = \frac{\sigma^2}{\Sigma\omega_t x_t^2} \tag{2.80}$$

In the special case when the variance is proportional to the squared predictor variable [i.e., $V(\varepsilon_t) = \sigma^2 x_t^2$ or $\omega_t = 1/x_t^2$], the weighted least squares estimate and its variance are

$$\hat{\beta}_* = \frac{1}{n}\Sigma\left(\frac{y_t}{x_t}\right) \quad \text{and} \quad V(\hat{\beta}_*) = \frac{\sigma^2}{n} \tag{2.81}$$

The weighted least squares estimate is the average of the ratios y_t/x_t.

In the case when the variance is proportional to $x_t > 0$ [i.e., $V(\varepsilon_t) = \sigma^2 x_t$, or $\omega_t = 1/x_t$], the weighted least squares estimate and its variance are given by

$$\hat{\beta}_* = \frac{\Sigma y_t}{\Sigma x_t} = \frac{\bar{y}}{\bar{x}} \quad \text{and} \quad V(\hat{\beta}_*) = \frac{\sigma^2}{\Sigma x_t} \tag{2.82}$$

Here the weighted least squares estimate is the ratio of the averages.

A further application of weighted least squares is given in Exercise 2.22. It arises in the context of fitting regressions on grouped data with unequal group sizes.

APPENDIX 2: SUMMARY OF DISTRIBUTION THEORY RESULTS

Result 1

Suppose $\mathbf{y} = (y_1, \ldots, y_n)'$ is a random vector with mean $E(\mathbf{y})$ and covariance matrix $\mathbf{V}(\mathbf{y})$. Consider the linear transformation $\mathbf{z} = \mathbf{A}\mathbf{y}$, where $\mathbf{z} =$

$(z_1, \ldots, z_m)'$ and $\mathbf{A}$ is an $m \times n$ matrix of constants. Then:

(*a*) $E(\mathbf{z}) = \mathbf{A}E(\mathbf{y})$ and $\mathbf{V}(\mathbf{z}) = \mathbf{A}\mathbf{V}(\mathbf{y})\mathbf{A}'$.
(*b*) If $\mathbf{y}$ has a multivariate normal distribution, so does $\mathbf{z}$.

Result 2

Consider the regression model $\mathbf{y} = \mathbf{X}\boldsymbol{\beta} + \boldsymbol{\varepsilon}$ in Eq. (2.9). Then it can be shown that:

(*a*) $(\hat{\beta}_i - \beta_i)/\sigma\sqrt{c_{ii}} \sim N(0, 1)$
(*b*) $\mathrm{SSE}/\sigma^2 = (\mathbf{y} - \mathbf{X}\hat{\boldsymbol{\beta}})'(\mathbf{y} - \mathbf{X}\hat{\boldsymbol{\beta}})/\sigma^2 \sim \chi^2_{n-p-1}$
(*c*) $\hat{\boldsymbol{\beta}}$ and s^2 are statistically independent.
(*d*) If $Z \sim N(0, 1)$, $W \sim \chi^2_r$, and Z and W are independent, then $Z/\sqrt{W/r} \sim t_r$.
(*e*) From (*a*)–(*d*) it follows that

$$\frac{(\hat{\beta}_i - \beta_i)/\sigma\sqrt{c_{ii}}}{\left[\mathrm{SSE}/\sigma^2(n-p-1)\right]^{1/2}} = \frac{\hat{\beta}_i - \beta_i}{s\sqrt{c_{ii}}} \sim t_{n-p-1}$$

(*f*) If $\beta_1 = \cdots = \beta_p = 0$, then $\mathrm{SSR}/\sigma^2 \sim \chi^2_p$.
(*g*) If $Z \sim \chi^2_{r_1}$, $W \sim \chi^2_{r_2}$, and Z and W are independent, then $(Z/r_1)/(W/r_2) \sim F_{r_1, r_2}$.
(*h*) From (*f*) and (*g*) it follows that $F = \mathrm{MSR}/\mathrm{MSE} \sim F_{p, n-p-1}$.

CHAPTER 3

Regression and Exponential Smoothing Methods to Forecast Nonseasonal Time Series

In Chapter 2 we have discussed the regression model, which relates a dependent variable Y to a set of explanatory or predictor variables $X_1, \ldots, X_p$. With such a model we can predict the dependent variable Y, provided we can specify future values of the explanatory variables.

Frequently, however, the forecaster has observations on only a single series and has to develop forecasts without being able to include other explanatory variables. In such a case only the past values of this single variable are available for modeling and forecasting.

3.1. FORECASTING A SINGLE TIME SERIES

Two main approaches are traditionally used to model a single time series $z_1, z_2, \ldots, z_n$. One approach, and this is the one discussed in Chapters 3 and 4, models the observation z_t as a function of time. In general, such models can be written as $z_t = f(t; \boldsymbol{\beta}) + \varepsilon_t$, where $f(t; \boldsymbol{\beta})$ is a function of time t and unknown coefficients $\boldsymbol{\beta}$, and ε_t are uncorrelated errors. Through the selection of appropriate fitting functions $f(t; \boldsymbol{\beta})$ one can represent a variety of nonseasonal and seasonal series. Models for nonseasonal series are discussed in this chapter. We consider polynomial trend models, which include the constant mean model $z_t = \beta + \varepsilon_t$ and the linear trend model $z_t = \beta_0 + \beta_1 t + \varepsilon_t$ as special cases. Models for seasonal series are discussed in Chapter 4. Seasonality is modeled by either seasonal indicators or trigonometric functions of time. For example, a model of the form $z_t = \beta_0 +$

$\beta_1 \sin(2\pi/12)t + \beta_2 \cos(2\pi/12)t + \varepsilon_t$ describes monthly observations with a very simple seasonal pattern.

The parameters $\boldsymbol{\beta}$ in these models are estimated either by *ordinary least squares*, which weights all observations equally, or by *weighted (discounted) least squares*, where the weights in $\Sigma_{t=1}^{n} w_t[z_t - f(t; \boldsymbol{\beta})]^2$ decrease geometrically with the age of the observations. The discount factor ω in the weights $w_t = \omega^{n-t}$ $(t = 1, 2, \ldots, n)$ determines how fast the information from previous observations is discounted. Usually this coefficient is chosen somewhere around .9. This estimation and forecasting approach is known under the name of *discounted least squares* or *general exponential smoothing* [see Brown (1962)]. Special cases of these models lead to single, double, and triple exponential smoothing procedures; these are discussed by Holt (1957), Muth (1960), Brown and Meyer (1961), and Brown (1962).

The second approach is a time series modeling approach. There the observation at time t is modeled as a linear combination of previous observations, $z_t = \Sigma_{j \geqslant 1} \pi_j z_{t-j} + \varepsilon_t$. Such a representation is called an *autoregressive model*, since the series at time t is regressed on itself at lagged time periods. Depending on the "memory" of the series, such an autoregressive representation can lead to models with many parameters that may be difficult to interpret. However, to achieve more parsimonious representations, we can approximate the autoregressive models by *autoregressive moving average models* of the form $z_t = \phi_1 z_{t-1} + \cdots + \phi_p z_{t-p} + \varepsilon_t - \theta_1 \varepsilon_{t-1} - \cdots - \theta_q \varepsilon_{t-q}$. In these models the observation z_t is written as a linear combination of past observations and past errors. Usually, p and q, the orders of the autoregression and of the moving average, are small.

The form of the autoregressive moving average model (i.e., values of p and q) depends on the autocorrelation structure of the observations. Box and Jenkins (1970) have developed an empirical model-building approach in which the appropriate model is determined from past data. Once the appropriate model has been found, it is easily used for forecasting. In fact, the π weights in the corresponding autoregressive representation determine the forecast weights that are applied to the past observations. A detailed description of the time series modeling approach for nonseasonal and seasonal series is given in Chapters 5 and 6.

The general exponential smoothing and the time series approach for forecasting single series are introduced separately. Relationships between these two approaches, however, exist. These are explored in Chapter 7. We show that exponential smoothing forecast procedures arise as very special cases of the more general time series modeling approach. In fact, the exponential smoothing forecast techniques are appropriate only if the observations follow particular restricted time series models. Implications of these results and recommendations for forecast practitioners are discussed in Chapter 7.

3.2. CONSTANT MEAN MODEL

To introduce regression and smoothing methods for the prediction of nonseasonal series, we consider a very special case of the general model $z_t = f(t; \boldsymbol{\beta}) + \varepsilon_t$. We assume that the observations z_t are generated from

$$z_t = \beta + \varepsilon_t \tag{3.1}$$

where β is a constant mean level and ε_t is a sequence of uncorrelated errors with constant variance σ^2. Series that follow this model are characterized by random variation around a constant mean; thus (3.1) is referred to as the *constant mean model.*

If the parameter β is *known*, the minimum mean square error forecast of a future observation at time $n + l$, $z_{n+l} = \beta + \varepsilon_{n+l}$, is given by

$$z_n(l) = \beta \tag{3.2}$$

This forecast is unbiased, in the sense that the forecast error $z_{n+l} - z_n(l) = \varepsilon_{n+l}$ has expectation $E[z_{n+l} - z_n(l)] = 0$. Its mean square error is given by $E[z_{n+l} - z_n(l)]^2 = E(\varepsilon_{n+l}^2) = \sigma^2$. If in addition we assume that the errors are normally distributed, $100(1 - \lambda)$ percent prediction intervals for a future realization are given by $[\beta - u_{\lambda/2}\sigma; \beta + u_{\lambda/2}\sigma]$, where $u_{\lambda/2}$ is the $100(1 - \lambda/2)$ percentage point of the standard normal distribution. In order to avoid confusion with the smoothing constant that is introduced in Section 3.3, we denote the significance level by λ.

If the parameter β is *unknown*, we estimate it from past data $(z_1, z_2, \ldots, z_n)$ and replace β in (3.2) by its least squares estimate. The least squares estimate is given by the sample mean

$$\hat{\beta} = \bar{z} = \frac{1}{n} \sum_{t=1}^{n} z_t$$

and the l-step-ahead forecast of z_{n+l} from time origin n by

$$\hat{z}_n(l) = \bar{z} \tag{3.3}$$

The caret ("hat") on the z in (3.3) reflects the fact that we replaced the model parameter β by its estimate.

The forecasts are the same for all l. It also follows from Section 2.6 that these forecasts are unbiased, with mean square error

$$E[z_{n+l} - \hat{z}_n(l)]^2 = \sigma^2\left(1 + \frac{1}{n}\right) \tag{3.4}$$

An unbiased estimate of σ^2 can be calculated from

$$\hat{\sigma}^2 = \frac{1}{n-1}\sum_{t=1}^{n}(z_t - \bar{z})^2 \tag{3.5}$$

A $100(1-\lambda)$ percent prediction interval for a future realization at time $n + l$ is given by

$$\left[\bar{z} - t_{\lambda/2}(n-1)\hat{\sigma}\left(1 + \frac{1}{n}\right)^{1/2};\ \bar{z} + t_{\lambda/2}(n-1)\hat{\sigma}\left(1 + \frac{1}{n}\right)^{1/2}\right]$$

where $t_{\lambda/2}(n-1)$ is the $100(1-\lambda/2)$ percentage point of a t distribution with $n-1$ degrees of freedom.

3.2.1. Updating Forecasts

The forecast at time origin $n + 1$ can be written as

$$\hat{z}_{n+1}(1) = \frac{1}{n+1}(z_1 + \cdots + z_n + z_{n+1}) = \frac{1}{n+1}[z_{n+1} + n\hat{z}_n(1)]$$

$$= \frac{n}{n+1}\hat{z}_n(1) + \frac{1}{n+1}z_{n+1} \tag{3.6}$$

or

$$\hat{z}_{n+1}(1) = \frac{1}{n+1}[z_{n+1} - \hat{z}_n(1) + (n+1)\hat{z}_n(1)]$$

$$= \hat{z}_n(1) + \frac{1}{n+1}[z_{n+1} - \hat{z}_n(1)] \tag{3.7}$$

Equation (3.6) indicates how forecasts from time origin $n + 1$ can be expressed as a linear combination of the forecast from origin n and the most recent observation. Since the mean β in the model (3.1) is assumed constant, each observation contributes equally to the forecast. Alternatively, Equation (3.7) expresses the new forecast as the previous forecast, corrected by a fixed fraction of the most recent forecast error. For the computation of the new forecast, it is important that only the last observation and the most recent forecast error have to be stored.

3.2.2. Checking the Adequacy of the Model

In the constant mean model it is assumed that the observations vary independently around a constant level. To investigate whether this model describes past data adequately, one should always calculate the sample autocorrelations of the residuals. For the constant mean model the residuals are $z_t - \bar{z}$. The sample autocorrelations (see Sec. 2.12) are given by

$$r_k = \frac{\sum_{t=k+1}^{n} (z_t - \bar{z})(z_{t-k} - \bar{z})}{\sum_{t=1}^{n} (z_t - \bar{z})^2} \qquad k = 1, 2, \ldots$$

To judge their significance, one should compare the estimated autocorrelations with their approximate standard error $n^{-1/2}$. If sample autocorrelations exceed twice their standard error (i.e., $|r_k| > 2n^{-1/2}$), one can conclude that the observations are likely to be correlated and that a constant mean model may not be appropriate.

Example 3.1: Annual U.S. Lumber Production

As an example we consider the annual U.S. lumber production from 1947 through 1976. The data were obtained from the U.S. Department of Commerce *Survey of Current Business*. The 30 observations are listed in Table 3.1.

Table 3.1. Annual Total U.S. Lumber Production (Millions of Board Feet), 1947–1976[a]

35,404	36,762	32,901	38,902	37,515
37,462	36,742	36,356	37,858	38,629
32,901	33,385	37,166	32,926	32,019
33,178	34,171	35,733	35,697	35,710
34,449	36,124	35,791	34,548	36,693
38,044	38,658	34,592	32,087	37,153

[a] Table reads from left to right.

The plot of the data in Figure 3.1 shows no trend and thus suggests that a constant mean model may provide an adequate representation. The sample mean and the sample standard deviation are given by

$$\bar{z} = 35{,}652 \qquad \hat{\sigma} = [\tfrac{1}{29}\Sigma(z_t - \bar{z})^2]^{1/2} = 2037$$

Before accepting this model for forecasting purposes, we have to check its validity. Only if the model describes historic data adequately and if the structure of the model does not change in the future will forecasts derived from this model lead to the smallest possible forecast errors. In the constant mean model we assume that the observations are uncorrelated. To check this assumption we calculate the sample autocorrelations of the observations. These are listed below:

Lag k	1	2	3	4	5	6
Sample autocorrelation r_k	.20	−.05	.13	.14	.04	−.17

Comparing the sample autocorrelations with their standard error $1/\sqrt{30} = .18$, we cannot find enough evidence to reject the assumption of uncorre-

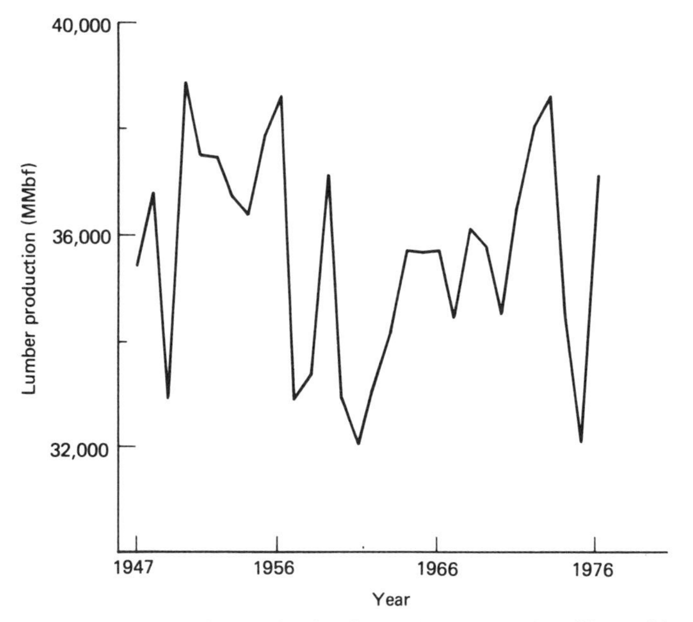

Figure 3.1. Annual U.S. lumber production from 1947 to 1976 (in millions of board feet).

lated error terms. (It should be kept in mind, however, that this standard error is a large sample approximation, which may not be suitable for small n.) We conclude that the constant mean model gives an adequate description of the historic data.

Conditional on the assumption that the structure of the model does not change in the near future, we can compute forecasts for future lumber production. The forecasts from the constant mean model are the same for all forecast lead times and are given by $\hat{z}_{1976}(l) = \bar{z} = 35{,}652$. The standard error of these forecasts is given by $\hat{\sigma}\sqrt{1 + 1/n} = 2071$; a 95 percent prediction interval by $[35{,}652 \pm (2.045)(2071)]$, or $[31{,}417; 39{,}887]$. This prediction interval is quite large, since there is considerable variability in the historic data.

If new observations become available, the forecasts are easily updated. Lumber production in 1977, for example, was 37,520 million board feet. Then the revised forecasts are given by

$$\begin{aligned} \hat{z}_{1977}(l) &= \hat{z}_{1976}(1) + \frac{1}{n+1}\left[z_{1977} - \hat{z}_{1976}(1)\right] \\ &= 35{,}652 + \tfrac{1}{31}[37{,}520 - 35{,}652] \\ &= 35{,}712 \end{aligned}$$

3.3. LOCALLY CONSTANT MEAN MODEL AND SIMPLE EXPONENTIAL SMOOTHING

The model in Equation (3.1) assumes that the mean is constant over all time periods. As a consequence, in the forecast computations each observation carries the same weight.

In many instances, however, the assumption of a time constant mean is restrictive, and it would be more reasonable to allow for a mean that moves slowly over time. Heuristically, in such a case it would be reasonable to give more weight to the most recent observations and less to the observations in the distant past.

If one chooses weights that decrease geometrically with the age of the observations, the forecast of the future observation at time $n + l$ can be calculated from

$$\hat{z}_n(l) = c\sum_{t=0}^{n-1} \omega^t z_{n-t} = c\left[z_n + \omega z_{n-1} + \cdots + \omega^{n-1} z_1\right] \tag{3.8}$$

The constant ω ($|\omega| < 1$) is a *discount coefficient*. This coefficient, which should depend on how fast the mean level changes, is usually chosen between .7 and .95; in many applications a value of .9 is suggested [see Brown (1962)]. The factor $c = (1 - \omega)/(1 - \omega^n)$ is needed to normalize the sum of the weights to 1. Since

$$\sum_{t=0}^{n-1} \omega^t = \frac{1 - \omega^n}{1 - \omega}$$

it follows that

$$c \sum_{t=0}^{n-1} \omega^t = 1$$

If n is large, then the term ω^n in the normalizing constant c goes to zero, and exponentially weighted forecasts can be written as

$$\begin{aligned} \hat{z}_n(l) &= (1 - \omega) \sum_{j \geq 0} \omega^j z_{n-j} \\ &= (1 - \omega)\left[z_n + \omega z_{n-1} + \omega^2 z_{n-2} + \cdots\right] \end{aligned} \tag{3.9}$$

The forecasts are the same for all lead times l. The coefficient $\alpha = 1 - \omega$ is called the *smoothing constant* and is usually chosen between .05 and .30. The expression

$$\begin{aligned} S_n = S_n^{[1]} &= (1 - \omega)\left[z_n + \omega z_{n-1} + \omega^2 z_{n-2} + \cdots\right] \\ &= \alpha\left[z_n + (1 - \alpha) z_{n-1} + (1 - \alpha)^2 z_{n-2} + \cdots\right] \end{aligned} \tag{3.10}$$

is called the *smoothed statistic* or the smoothed value. The last available smoothed statistic S_n serves as forecast for all future observations, $\hat{z}_n(l) = S_n$. Since it is an exponentially weighted average of previous observations, this method is called *simple exponential smoothing*.

3.3.1. Updating Forecasts

The forecast in (3.9), or equivalently the smoothed statistic in (3.10), can be updated in several alternative ways. By simple substitution it can be shown that

$$\begin{aligned} S_n &= (1 - \omega) z_n + \omega S_{n-1} \\ \hat{z}_n(1) &= (1 - \omega) z_n + \omega \hat{z}_{n-1}(1) \end{aligned} \tag{3.11}$$

or

$$S_n = S_{n-1} + (1 - \omega)[z_n - S_{n-1}]$$

$$\hat{z}_n(1) = \hat{z}_{n-1}(1) + (1 - \omega)[z_n - \hat{z}_{n-1}(1)] \tag{3.12}$$

Expressions (3.11) and (3.12) show how the forecasts can be updated after a new observation has become available. Equation (3.11) expresses the new forecast as a combination of the old forecast and the most recent observation. If ω is small, more weight is given to the last observation and the information from previous periods is heavily discounted. If ω is close to 1, a new observation will change the old forecast only very little. Equation (3.12) expresses the new forecast as the previous forecast corrected by a fraction ($\alpha = 1 - \omega$) of the last forecast error.

3.3.2. Actual Implementation of Simple Exponential Smoothing

The recurrence equation in (3.11) can be used to update the smoothed statistics at any time period t. In practice, one starts the recursion with the first observation z_1 and calculates $S_1 = (1 - \omega)z_1 + \omega S_0$. This is then substituted into (3.11) to calculate $S_2 = (1 - \omega)z_2 + \omega S_1$, and the smoothing is continued until S_n is reached. This procedure is somewhat simpler than the one in (3.8) and is usually adopted in practice. To carry out these operations we need to know (1) a starting value S_0, and (2) a smoothing constant $\alpha = 1 - \omega$.

Initial Value for S_0

Through repeated application of Equation (3.11), it can be shown that

$$S_n = (1 - \omega)[z_n + \omega z_{n-1} + \cdots + \omega^{n-1}z_1] + \omega^n S_0$$

Thus the influence of S_0 on S_n is negligible, provided n is moderately large and ω smaller than 1. For example, if $n = 30$ and $\omega = .9$ (or $\alpha = .1$), the weight given to the initial smoothed value S_0 (i.e., $\omega^n = .042$) is very small compared with the combined weight that is given to $z_1, \ldots, z_{30}$ (i.e., $1 - \omega^n = .958$).

We take the simple arithmetic average of the available historical data $(z_1, z_2, \ldots, z_n)$ as the initial estimate of S_0. Such a choice has also been suggested by Brown (1962) and Montgomery and Johnson (1976). The arithmetic average will perform well, provided that the mean level changes only slowly (small α, or ω close to 1).

Alternative solutions to choosing the initial value S_0 have been suggested in the literature. Makridakis and Wheelwright (1978), for example, use the first observation as initial smoothed statistic; $S_0 = z_1$, which implies that $S_1 = (1 - \omega)z_1 + \omega S_0 = z_1$. Such a choice will be preferable if the level changes rapidly (α close to 1, or ω close to 0).

Another slightly different solution to the initialization in exponential smoothing is to choose S_0 as the "backforecast" value of z_0. This is achieved by reversing the time order and smoothing backwards; $S_j^* = (1 - \omega)z_j + \omega S_{j+1}^*$, where $S_{n+1}^* = z_n$. The prediction of z_0, which is given by S_1^*, can then be used as initial value S_0.

Choice of the Smoothing Constant

The smoothing constant $\alpha = 1 - \omega$ determines the extent to which past observations influence the forecast. A small α results in a slow response to changes in the level; a large α results in a rapid response, which, however, will also make the forecast respond to irregular movements in the time series.

The smoothing constant α is frequently determined by simulation. Forecasts are generated for various α's (usually over the range .05 to .30) and are then compared to the actual observations $z_1, z_2, \ldots, z_n$. For each α, the one-step-ahead forecast errors

$$e_{t-1}(1) = z_t - \hat{z}_{t-1}(1) = z_t - S_{t-1}$$

and the sum of the squared one-step-ahead forecast errors

$$\mathrm{SSE}(\alpha) = \sum_{t=1}^{n} e_{t-1}^2(1)$$

are calculated. The smoothing constant, which minimizes the sum of the squared forecast errors, is then used as smoothing constant in the derivation of future forecasts.

The notation $e_{t-1}(1)$ expresses the fact that it is the one-step-ahead forecast error of the forecast that is calculated from past data up to and including time $t - 1$. In general, $e_t(l) = z_{t+l} - \hat{z}_t(l)$ is the l-step-ahead forecast error corresponding to the l-step-ahead forecast made at time t.

The smoothing constant that is obtained by simulation depends on the value of S_0. A poorly chosen starting value S_0 will lead to an increase in the smoothing constant. Ideally, since the choice of α depends on S_0, one should choose α and S_0 jointly. For example, if $\alpha = 0$, one should choose the sample mean as starting value; if $\alpha = 1$, one should choose $S_0 = z_1$. If

$0 < \alpha < 1$, one could choose S_0 as the "backforecast" value; $S_0 = \alpha[z_1 + (1-\alpha)z_2 + \cdots + (1-\alpha)^{n-2}z_{n-1}] + (1-\alpha)^{n-1}z_n$. Further discussion of this point can be found in Ledolter and Abraham (1983).

Once the smoothing constant has been determined from past data, the forecasts are easily updated as each new observation becomes available:

$$S_{n+1} = \alpha z_{n+1} + (1-\alpha)S_n \qquad \hat{z}_{n+1}(l) = S_{n+1}$$

It is usually assumed that the smoothing constant α stays fixed over time. Thus it is not necessary to reestimate the value of α as each new observation becomes available.

How to detect whether the smoothing constant (or in general, any parameter of a model) has changed is discussed in Chapter 8. There we discuss tracking signals and also adaptive smoothing methods, in which the smoothing parameter adapts itself to changes in the underlying time series.

3.3.3. Additional Comments and Example

Forecasts from simple exponential smoothing were introduced without reference to a particular model. In Chapter 7 we discuss the model under which these forecasts are minimum mean square error (MMSE) forecasts. There it will be shown that simple exponential smoothing leads to optimal (MMSE) forecasts, if the mean β_t in $z_t = \beta_t + \varepsilon_t$ changes according to a random walk model $\beta_t = \beta_{t-1} + a_t$ or, equivalently, if the observations are generated from a particular time series model.

The main reason for the widespread use of simple exponential smoothing comes from the updating equations (3.11), since they make the calculation of new forecasts computationally very convenient. Only the previous forecast and the most recent observation have to be stored when updating the forecast. This is especially important if a large number of different items have to be predicted.

Another reason exponential smoothing techniques have received broad attention in the business literature is that they are fully automatic. Once a computer program has been written and a discount coefficient ω (or equivalently a smoothing constant $\alpha = 1 - \omega$) has been chosen, forecasts for any time series can be derived without manual intervention of the forecaster. The fact that they are fully automatic has been put forward as an advantage of the scheme. However, it can equally well be argued that this is a great disadvantage, since every time series is treated identically. Thus it is very important to perform diagnostic checks to see whether this forecasting technique is in fact adequate.

Example 3.2: Quarterly Iowa Nonfarm Income

As an example we consider the quarterly Iowa nonfarm income for 1948–1979. The 128 observations are listed in series 2 of the Data Appendix; a plot of the data is given in Figure 3.2.

The data exhibit exponential growth, a pattern typical of many economic series of this time period. Instead of analyzing and forecasting the original series, we first model the quarterly growth rates of nonfarm income. The quarterly growth rates or percentage changes

$$z_t = \frac{I_{t+1} - I_t}{I_t} 100 \qquad t = 1, 2, \ldots, 127$$

are also given in the Data Appendix and are plotted in Figure 3.3.

Analyzing percentage changes is approximately equivalent to analyzing successive differences of logarithmically transformed data, since

$$\log I_{t+1} - \log I_t - \log \frac{I_{t+1}}{I_t} = \log \frac{(1 + z_t^*) I_t}{I_t}$$

$$= \log(1 + z_t^*) \cong z_t^* \qquad \text{for small } z_t^* = z_t/100$$

The plot of the growth rates indicates that the mean level is not constant but changes slowly over time. Especially in the middle part of the series

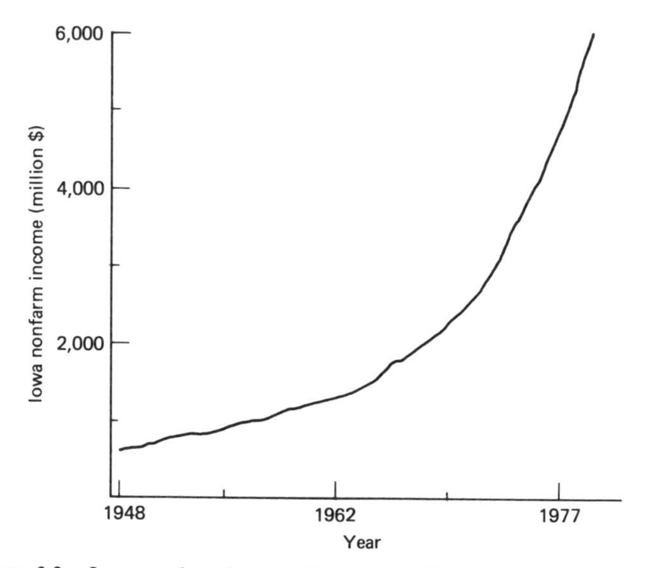

Figure 3.2. Iowa nonfarm income, first quarter 1948 to fourth quarter 1979.

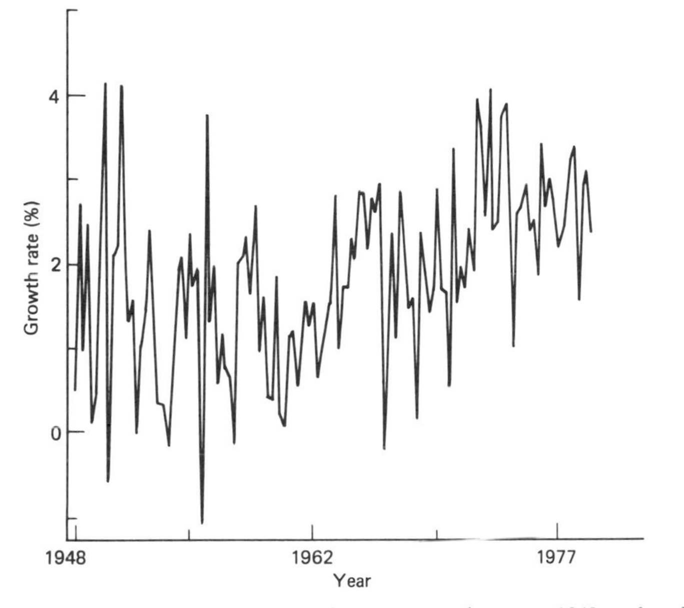

Figure 3.3. Growth rates of Iowa nonfarm income, second quarter 1948 to fourth quarter 1979.

(during the 1960s), the mean level increases toward a slightly higher new level. However, it does not appear to grow as a linear function of time, nor do we expect such a pattern in the near future.

The constant mean model would be clearly inappropriate. This can also be seen from the sample autocorrelations of the growth rates, which are given in Table 3.2. Compared with their standard errors $1/\sqrt{127} = .089$, most autocorrelations are significantly different from zero.

Since the mean is slowly changing, simple exponential smoothing appears to be an appropriate method. Smoothing constants somewhere between .05 and .30 are usually recommended. To illustrate the calculations we use a smoothing constant $\alpha = .11$. This particular choice will be justified in the next section.

The average of all 127 observations, $\bar{z} = 1.829$, is used as initial smoothed statistic S_0. We could also have chosen the average of a subset of the observations [let's say the first 6 or the first 10 observations; see Bowerman and O'Connell (1979)]. The value $S_0 = 1.829$ is used to predict the first observation. With $z_1 = .50$, the first one-step-ahead forecast error is given by $e_0(1) = z_1 - S_0 = .50 - 1.829 = -1.329$. The updated smoothed statistic $S_1 = \alpha z_1 + (1 - \alpha)S_0 = (.11)(.50) + (.89)(1.829) = 1.683$ is used to pre-

Table 3.2. Sample Autocorrelations r_k of Growth Rates of Iowa Nonfarm Income ($n = 127$)

Lag k	1	2	3	4	5	6
Sample autocorrelation r_k	.25	.32	.18	.35	.18	.22

Table 3.3. Simple Exponential Smoothing—Growth Rates of Iowa Nonfarm Income

		$\alpha = .11$		$\alpha = .40$	
Time t	Observation z_t	Smoothed Statistic S_t	One-Step-Ahead Forecast Error $e_{t-1}(1) = z_t - S_{t-1}$	Smoothed Statistic S_t	One-Step-Ahead Forecast Error $e_{t-1}(1) = z_t - S_{t-1}$
0		1.829		1.829	
1	0.50	1.683	−1.329	1.297	−1.329
2	2.65	1.789	0.967	1.838	1.353
3	0.97	1.699	−0.819	1.491	−0.868
4	2.40	1.776	0.701	1.855	0.909
5	0.16	1.598	−1.616	1.177	−1.695
6	0.47	1.474	−1.128	0.894	−0.707
	⋮	⋮	⋮	⋮	⋮
123	3.38	2.736	0.723	3.032	0.579
124	1.55	2.606	−1.186	2.439	−1.482
125	2.93	2.642	0.324	2.636	0.491
126	3.10	2.692	0.458	2.821	0.464
127	2.35	2.654	−0.342	2.633	−0.471
		SSE(.11) = 118.19		SSE(.40) = 132.56	

dict z_2. With $z_2 = 2.65$, the one-step-ahead forecast error is $e_1(1) = z_2 - S_1 = 2.65 - 1.683 = .967$. The next smoothed statistic $S_2 = \alpha z_2 + (1 - \alpha)S_1 = (.11)(2.65) + (.89)(1.683) = 1.789$ is the prediction for z_3; since $z_3 = .97$, the one-step-ahead forecast error is $e_2(1) = z_3 - S_2 = -.819$. From z_3 we can calculate $S_3 = 1.699$; $e_3(1) = .701$, etc. The observations z_t, the smoothed statistics S_t, and the one-step-ahead forecast errors are given in Table 3.3.

Through repeated updating, $S_t = (.11)z_t + (.89)S_{t-1}$, we eventually find that $S_{127} = (.11)(2.35) + (.89)(2.692) = 2.654$. The last smoothed statistic is then used to predict all future growth rates $\hat{z}_{127}(l) = 2.654$. This implies that our prediction of the Iowa nonfarm income for the first quarter of 1980 is given by $\hat{I}_{128}(1) = (1.02654)I_{128} = (1.02654)(5965) = 6123$, for the second quarter by $\hat{I}_{128}(2) = (1.02654)^2(5965) = 6286$, and in general $\hat{I}_{128}(l) = (1.02654)^l(5965)$.

In our example the sum of the squared one-step-ahead forecast errors for the smoothing constant $\alpha = .11$ is given by

$$\text{SSE}(.11) = \sum_{t=1}^{n} e_{t-1}^2(1)$$

$$= (-1.329)^2 + (.967)^2 + \cdots + (.458)^2 + (-.342)^2 = 118.19$$

For illustration we have also calculated the smoothed statistics S_t and the one-step-ahead forecast errors $e_{t-1}(1) = z_t - S_{t-1}$ for the smoothing constant $\alpha = .40$. The results are given in the last two columns of Table 3.3. There it is found that $\text{SSE}(.40) = (-1.329)^2 + (1.353)^2 + \cdots + (.464)^2 + (-.471)^2 = 132.56$. Changing the smoothing constant from .01 to .30 in increments of .01 leads to the sums of squared one-step-ahead forecast errors in Table 3.4; they are plotted in Figure 3.4. The minimum is achieved for $\alpha = .11$, which explains our previous choice of α.

Before the model is used for forecasting, we must investigate whether it gives an adequate description of the historical data. If the forecast model is appropriate, then the one-step-ahead forecast errors should be uncorrelated. Correlation among the one-step-ahead forecast errors would imply that the current forecast error can be used to improve the next forecast. In such a case, we would incorporate this information into the next forecast and would thus use a different forecast model.

Table 3.4. Sums of Squared One-Step-Ahead Forecast Errors for Different Values of α; Simple Exponential Smoothing—Growth Rates of Iowa Nonfarm Income

α	SSE(α)	α	SSE(α)	α	SSE(α)
.01	140.78	.11	118.19	.21	120.95
.02	134.42	.12	118.24	.22	121.39
.03	128.84	.13	118.38	.23	121.86
.04	124.86	.14	118.57	.24	122.35
.05	122.22	.15	118.81	.25	122.86
.06	120.51	.16	119.09	.26	123.38
.07	119.44	.17	119.41	.27	123.93
.08	118.78	.18	119.75	.28	124.49
.09	118.41	.19	120.13	.29	125.07
.10	118.23	.20	120.53	.30	125.67

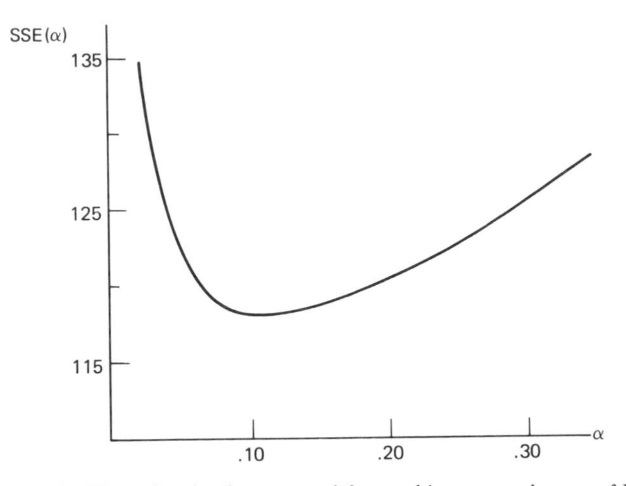

Figure 3.4. Plot of SSE(α), for simple exponential smoothing—growth rates of Iowa nonfarm income.

As a diagnostic check, we calculate the sample autocorrelations of the one-step-ahead forecast errors

$$r_k = \frac{\sum_{t=k}^{n-1} [e_t(1) - \bar{e}][e_{t-k}(1) - \bar{e}]}{\sum_{t=0}^{n-1} [e_t(1) - \bar{e}]^2} \qquad \bar{e} = \frac{1}{n} \sum_{t=0}^{n-1} e_t(1)$$

If the forecast errors are uncorrelated, the sample autocorrelations should vary around mean zero with standard error $1/n^{1/2}$.

The sample autocorrelations of the one-step-ahead forecast errors are given in Table 3.5. The autocorrelations of the forecast errors, when using the optimal smoothing constant .11, are compared with those for a smoothing constant of .40. Comparing the autocorrelations in the first column ($\alpha = .11$) with their standard error $1/\sqrt{127} = .089$, we find that they are within ± 2 standard errors. For this smoothing constant, successive one-step-ahead forecast errors are uncorrelated. On the other hand, one-step-ahead forecast errors for exponential smoothing with $\alpha = .40$ are correlated. The significant lag 1 correlation indicates that forecast errors one step apart are still correlated. This shows that for this data set the smoothing constant $\alpha = .40$ is inappropriate.

In addition to checking for correlation among the forecast errors, one should also check for possible bias in the forecasts. A mean of the forecast

Table 3.5. Means, Standard Errors, and Sample Autocorrelations of the One-Step-Ahead Forecast Errors from Exponential Smoothing (with $\alpha = .11$ and $\alpha = .40$)—Growth Rates of Iowa Nonfarm Income

Lag k	Sample Autocorrelations of One-Step-Ahead Forecast Errors $\alpha = .11$	$\alpha = .40$
1	−.02	.22
2	.08	−.02
3	−.09	−.19
4	.14	.13
5	−.07	−.11
6	.00	.00
Mean of historical forecast errors	.059	.016
Standard error of mean	.086	.091

errors that is significantly larger (smaller) than zero indicates that the forecast procedure underpredicts (overpredicts) the observations. To assess the significance of the mean of the forecast errors, we compare it with its standard error $s/n^{1/2}$, where

$$s^2 = \frac{1}{n} \sum_{t=0}^{n-1} [e_t(1) - \bar{e}]^2$$

Means and standard errors of the one-step-ahead forecast errors are also given in Table 3.5. Since the mean .059 lies within ± 2 standard errors, we conclude that the mean of the forecast errors is not significantly different from zero. Thus, exponential smoothing with a smoothing constant $\alpha = .11$ leads to unbiased forecasts.

3.4. REGRESSION MODELS WITH TIME AS INDEPENDENT VARIABLE

We now extend the constant mean model to the class of regression models in which certain functions of time are taken as independent variables. We consider models of the form

$$z_{n+j} = \sum_{i=1}^{m} \beta_i f_i(j) + \varepsilon_{n+j} = \mathbf{f}'(j)\boldsymbol{\beta} + \varepsilon_{n+j} \tag{3.13}$$

where $\boldsymbol{\beta} = (\beta_1, \beta_2, \ldots, \beta_m)'$ is a vector of parameters and $\mathbf{f}(j) = [f_1(j), \ldots, f_m(j)]'$ is a vector of specified fitting or forecast functions; $\{\varepsilon_{n+j}, j = 0, \pm 1, \pm 2, \ldots\}$ is a sequence of independent $N(0, \sigma^2)$ random variables.

Furthermore, we assume that the fitting functions $\mathbf{f}(j)$ satisfy the difference equation

$$\mathbf{f}(j + 1) = \mathbf{L}\mathbf{f}(j) \tag{3.14}$$

where $\mathbf{L}$ is an $m \times m$ nonsingular fixed transition matrix

$$\mathbf{L} = \begin{bmatrix} l_{11} & \cdots & l_{1m} \\ \vdots & & \vdots \\ l_{m1} & \cdots & l_{mm} \end{bmatrix}$$

For example, exponentials, polynomials, sines, cosines, and linear combinations of these functions can be shown to satisfy these equations.

In Equation (3.13) the m fitting or forecast functions $f_i(j)$ $(i = 1, \ldots, m)$ are defined relative to time origin n. Equivalently, we could express the model as an ordinary regression in which the forecast functions are defined in relation to the time origin zero:

$$z_{n+j} = \sum_{i=1}^{m} \beta_i^* f_i(n + j) + \varepsilon_{n+j} = \mathbf{f}'(n + j)\boldsymbol{\beta}^* + \varepsilon_{n+j} \tag{3.15}$$

Since we assume that the fitting functions satisfy (3.14), we can write $\mathbf{f}(n + j) = \mathbf{L}\mathbf{f}(n + j - 1) = \cdots = \mathbf{L}^n\mathbf{f}(j)$ and express (3.15) as

$$z_{n+j} = [\mathbf{L}^n\mathbf{f}(j)]'\boldsymbol{\beta}^* + \varepsilon_{n+j} = \mathbf{f}'(j)\boldsymbol{\beta} + \varepsilon_{n+j}$$

The coefficients in this representation are then given by $\boldsymbol{\beta} = (\mathbf{L}^n)'\boldsymbol{\beta}^*$.

The parameterizations in Equations (3.13) and (3.15) lead to equivalent representations. However, when updating the parameter estimates and forecasts, the parameterization (3.13) turns out to be more convenient.

3.4.1. Examples

To familiarize ourselves with this class of forecast models, we consider several important special cases. To be consistent with the regression notation in Chapter 2, we have used β_0 as the parameter corresponding to a constant fitting function. This amounts to a relabeling of the parameters in model (3.13).

1. **Constant mean model:**

$$z_{n+j} = \beta_0 + \varepsilon_{n+j}$$

This model is obtained by choosing a single constant fitting function $f_1(j) = 1$. In this case, $L = f(0) = 1$.

2. **Linear trend model:**

$$z_{n+j} = \beta_0 + \beta_1 j + \varepsilon_{n+j}$$

Here the two fitting functions are $f_1(j) = 1$ and $f_2(j) = j$. The transition matrix in (3.14) is given by

$$\mathbf{L} = \begin{bmatrix} 1 & 0 \\ 1 & 1 \end{bmatrix} \quad \text{and} \quad \mathbf{f}(0) = \begin{bmatrix} 1 \\ 0 \end{bmatrix}$$

3. **Quadratic trend model:**

$$z_{n+j} = \beta_0 + \beta_1 j + \beta_2 \frac{j^2}{2} + \varepsilon_{n+j}$$

In this case we have three fitting functions $f_1(j) = 1$, $f_2(j) = j$, and $f_3(j) = j^2/2$. Choosing $j^2/2$ instead of the quadratic term j^2 simplifies the difference equation in (3.14). The transition matrix is

$$\mathbf{L} = \begin{bmatrix} 1 & 0 & 0 \\ 1 & 1 & 0 \\ \frac{1}{2} & 1 & 1 \end{bmatrix}$$

Furthermore,

$$\mathbf{f}(0) = \begin{bmatrix} 1 \\ 0 \\ 0 \end{bmatrix}$$

4. **General kth-order polynomial model:**

$$z_{n+j} = \beta_0 + \beta_1 j + \beta_2 \frac{j^2}{2} + \cdots + \beta_k \frac{j^k}{k!} + \varepsilon_{n+j}$$

$$= \sum_{i=0}^{k} \beta_i \frac{j^i}{i!} + \varepsilon_{n+j}$$

This is obtained by specifying $k + 1$ fitting functions

$$f_{i+1}(j) = \frac{j^i}{i!} \qquad i = 0, 1, \ldots, k$$

The transition matrix and the initial vector $\mathbf{f}(0)$ are given by

$$\mathbf{L} = \begin{bmatrix} 1 & & & & \\ 1 & 1 & & \mathbf{0} & \\ 1/2 & 1 & \ddots & & \\ \vdots & \vdots & & 1 & \\ 1/k! & 1/(k-1)! & \ldots & 1 & 1 \end{bmatrix} \qquad \mathbf{f}(0) = \begin{bmatrix} 1 \\ 0 \\ \vdots \\ 0 \end{bmatrix}$$

$\mathbf{L}$ is a lower triangular matrix with elements $l_{ij} = 1/(i-j)!$ for $i \geqslant j$.

5. 12-point sinusoidal model:

$$z_{n+j} = \beta_0 + \beta_1 \sin\frac{2\pi}{12}j + \beta_2 \cos\frac{2\pi}{12}j + \varepsilon_{n+j}$$

In this case the fitting functions are given by

$$f_1(j) = 1, \qquad f_2(j) = \sin\frac{2\pi}{12}j, \qquad f_3(j) = \cos\frac{2\pi}{12}j$$

Furthermore,

$$\mathbf{L} = \begin{bmatrix} 1 & 0 & 0 \\ 0 & \sqrt{3}/2 & 1/2 \\ 0 & -1/2 & \sqrt{3}/2 \end{bmatrix} \qquad \mathbf{f}(0) = \begin{bmatrix} 1 \\ 0 \\ 1 \end{bmatrix}$$

The sinusoidal model can be used to represent seasonal time series. Additional seasonal models are introduced in Chapter 4.

3.4.2. Forecasts

If the parameters $\boldsymbol{\beta}$ in model (3.13) are known, the forecast of a future observation at time $n + l$ is given by

$$z_n(l) = \sum_{i=1}^{m} \beta_i f_i(l) = \mathbf{f}'(l)\boldsymbol{\beta} \tag{3.16}$$

The forecast is unbiased, with mean square error σ^2. A $100(1-\lambda)$ percent prediction interval for the future observation z_{n+l} is given by $[z_n(l) - u_{\lambda/2}\sigma;\ z_n(l) + u_{\lambda/2}\sigma]$.

If the coefficients $\boldsymbol{\beta}$ in (3.13) are unknown, we have to estimate them from past data $z_1, z_2, \ldots, z_n$. The least squares estimates $\hat{\boldsymbol{\beta}}$ minimize

$$\sum_{j=1}^{n} \left[z_j - \mathbf{f}'(j-n)\boldsymbol{\beta}\right]^2 = \sum_{j=0}^{n-1} \left[z_{n-j} - \mathbf{f}'(-j)\boldsymbol{\beta}\right]^2 \tag{3.17}$$

In the general least squares formulation of Section 2.3,

$$\mathbf{y}' = (z_1, \ldots, z_n)$$

$$\mathbf{X}' = [\mathbf{f}(-n+1), \ldots, \mathbf{f}(0)]$$

$$\mathbf{X}'\mathbf{X} = \mathbf{f}(0)\mathbf{f}'(0) + \cdots + \mathbf{f}(-n+1)\mathbf{f}'(-n+1)$$

$$= \sum_{j=0}^{n-1} \mathbf{f}(-j)\mathbf{f}'(-j)$$

and

$$\mathbf{X}'\mathbf{y} = \sum_{j=0}^{n-1} \mathbf{f}(-j) z_{n-j}$$

The least squares estimates are given by

$$\hat{\boldsymbol{\beta}}_n = \mathbf{F}_n^{-1}\mathbf{h}_n \tag{3.18}$$

where

$$\mathbf{F}_n = \sum_{j=0}^{n-1} \mathbf{f}(-j)\mathbf{f}'(-j) \qquad \mathbf{h}_n = \sum_{j=0}^{n-1} \mathbf{f}(-j) z_{n-j}$$

The subscript n in the above notation expresses the fact that the estimates are calculated from observations up to and including time n.

Substituting these estimates into Equation (3.16), we forecast a future observation z_{n+l} from

$$\hat{z}_n(l) = \mathbf{f}'(l)\hat{\boldsymbol{\beta}}_n \tag{3.19}$$

It follows from results in Section 2.6 that $\hat{z}_n(l)$ is an unbiased forecast. The variance of the forecast error $e_n(l) = z_{n+l} - \hat{z}_n(l)$ is given by

$$V[e_n(l)] = \sigma^2[1 + \mathbf{f}'(l)\mathbf{F}_n^{-1}\mathbf{f}(l)] \tag{3.20}$$

The variance σ^2 can be estimated from

$$\hat{\sigma}^2 = \frac{1}{n-m}\sum_{j=0}^{n-1}\left(z_{n-j} - \mathbf{f}'(-j)\hat{\boldsymbol{\beta}}_n\right)^2 \tag{3.21}$$

and a $100(1-\lambda)$ percent prediction interval for a future realization is given by

$$\hat{z}_n(l) \pm t_{\lambda/2}(n-m)\hat{\sigma}[1 + \mathbf{f}'(l)\mathbf{F}_n^{-1}\mathbf{f}(l)]^{1/2} \tag{3.22}$$

3.4.3. Updating Parameter Estimates and Forecasts

If an additional observation z_{n+1} becomes available, the estimate $\hat{\boldsymbol{\beta}}_{n+1}$ can be written as

$$\hat{\boldsymbol{\beta}}_{n+1} = \mathbf{F}_{n+1}^{-1}\mathbf{h}_{n+1} \tag{3.23}$$

where

$$\mathbf{F}_{n+1} = \sum_{j=0}^{n}\mathbf{f}(-j)\mathbf{f}'(-j) = \mathbf{F}_n + \mathbf{f}(-n)\mathbf{f}'(-n)$$

and

$$\mathbf{h}_{n+1} = \sum_{j=0}^{n}\mathbf{f}(-j)z_{n+1-j} = \mathbf{f}(0)z_{n+1} + \sum_{j=0}^{n-1}\mathbf{L}^{-1}\mathbf{f}(-j)z_{n-j}$$

$$= \mathbf{f}(0)z_{n+1} + \mathbf{L}^{-1}\mathbf{h}_n$$

In the above derivation we used the fact that $\mathbf{f}(-j) = \mathbf{L}\mathbf{f}(-j-1)$; see (3.14). The new forecast is given by

$$\hat{z}_{n+1}(l) = \mathbf{f}'(l)\hat{\boldsymbol{\beta}}_{n+1}$$

The result in (3.23) can be used to update the parameter estimate $\hat{\boldsymbol{\beta}}_{n+1}$. For the recursive calculation, only $\mathbf{F}_n$ and $\mathbf{h}_n$ have to be stored.

3.5. DISCOUNTED LEAST SQUARES AND GENERAL EXPONENTIAL SMOOTHING

In model (3.13) we have assumed that the parameters $\boldsymbol{\beta}$ are constant. Estimating the coefficients by ordinary least squares implies that each observation (recent or past) has the same importance in determining the estimates.

If we assume that the model in (3.13) is only locally constant and if we wished to guard against possible parameter changes, we could give more weight to recent observations and discount past observations. This reasoning motivated Brown (1962) to consider a discount factor in the least squares criterion. In *discounted least squares* or *general exponential smoothing*, the parameter estimates are determined by minimizing

$$\sum_{j=0}^{n-1} \omega^j \left[z_{n-j} - \mathbf{f}'(-j)\boldsymbol{\beta} \right]^2 \tag{3.24}$$

The constant ω ($|\omega| < 1$) is a discount factor that discounts past observations exponentially. Brown suggests choosing ω such that $.70 < \omega^m < .95$, where m is the number of estimated parameters in the model.

It should be emphasized that at this point no model for possible changes in the coefficients is assumed, and exponential discounting is introduced as an ad hoc procedure. In a later chapter (Chap. 7) we will show that this procedure leads to minimum mean square error forecasts, but only if the underlying process generating the data is of a particular form.

Weighted least squares procedures can be used to find the solution to the minimization problem in (3.24). In the notation of Section 2.13 we can write the model as $\mathbf{y} = \mathbf{X}\boldsymbol{\beta} + \boldsymbol{\varepsilon}$, where $\mathbf{y}' = (z_1, \ldots, z_{n-1}, z_n)$; $\mathbf{X}' = [\mathbf{f}(-n+1), \ldots, \mathbf{f}(-1), \mathbf{f}(0)]$; $\mathbf{V}(\boldsymbol{\varepsilon}) = \sigma^2 \boldsymbol{\Omega}^{-1}$, where

$$\boldsymbol{\Omega} = \begin{bmatrix} \omega^{n-1} & & & \\ & \ddots & & \mathbf{0} \\ & \mathbf{0} & \omega & \\ & & & 1 \end{bmatrix};$$

and

$$\mathbf{X}'\boldsymbol{\Omega}\mathbf{X} = \sum_{j=0}^{n-1} \omega^j \mathbf{f}(-j)\mathbf{f}'(-j) \qquad \mathbf{X}'\boldsymbol{\Omega}\mathbf{y} = \sum_{j=0}^{n-1} \omega^j \mathbf{f}(-j) z_{n-j}$$

The estimate of $\boldsymbol{\beta}$ is thus given by

$$\hat{\boldsymbol{\beta}}_n = \mathbf{F}_n^{-1}\mathbf{h}_n \tag{3.25}$$

where now

$$\mathbf{F}_n = \sum_{j=0}^{n-1} \omega^j \mathbf{f}(-j)\mathbf{f}'(-j) \qquad \text{and} \qquad \mathbf{h}_n = \sum_{j=0}^{n-1} \omega^j \mathbf{f}(-j) z_{n-j}$$

The forecasts are given by

$$\hat{z}_n(l) = \mathbf{f}'(l)\hat{\boldsymbol{\beta}}_n \tag{3.26}$$

3.5.1. Updating Parameter Estimates and Forecasts

Using the observations up to period $n + 1$, the estimated coefficients $\hat{\boldsymbol{\beta}}_{n+1}$ are given by

$$\hat{\boldsymbol{\beta}}_{n+1} = \mathbf{F}_{n+1}^{-1}\mathbf{h}_{n+1}$$

where

$$\begin{aligned}
\mathbf{h}_{n+1} &= \sum_{j=0}^{n} \omega^j \mathbf{f}(-j) z_{n+1-j} = \mathbf{f}(0) z_{n+1} + \omega \sum_{j=1}^{n} \omega^{j-1} \mathbf{f}(-j) z_{n+1-j} \\
&= \mathbf{f}(0) z_{n+1} + \omega \sum_{j=1}^{n} \omega^{j-1} \mathbf{L}^{-1} \mathbf{f}(-j+1) z_{n+1-j} \\
&= \mathbf{f}(0) z_{n+1} + \omega \mathbf{L}^{-1} \mathbf{h}_n
\end{aligned} \tag{3.27}$$

In the derivation of (3.27) we have used the fact that $\mathbf{f}(j+1) = \mathbf{L}\mathbf{f}(j)$. Furthermore,

$$\mathbf{F}_{n+1} = \sum_{j=0}^{n} \omega^j \mathbf{f}(-j)\mathbf{f}'(-j) = \omega^n \mathbf{f}(-n)\mathbf{f}'(-n) + \mathbf{F}_n$$

For polynomial and sinusoidal fitting functions, the elements of $\omega^n \mathbf{f}(-n)\mathbf{f}'(-n)$ go to zero as $n \to \infty$, since ω^n tends to zero faster than $\mathbf{f}(-n)$ can grow. For example, in the linear trend model with $f_1(j) = 1$ and $f_2(j) = j$, the elements of the matrix

$$\omega^n \mathbf{f}(-n)\mathbf{f}'(-n) = \omega^n \begin{bmatrix} 1 \\ -n \end{bmatrix} [1 \quad -n] = \begin{bmatrix} \omega^n & -n\omega^n \\ -n\omega^n & n^2\omega^n \end{bmatrix}$$

go to zero provided $|\omega| < 1$. For a decreasing exponential fitting function $f(j) = \phi^j$ ($|\phi| < 1$), we have to assume that $|\omega| < \phi^2$; in this case, $\omega^n \phi^{-2n} = (\omega\phi^{-2})^n$ goes to zero.

Under the above assumption, the elements of the matrix $\omega^n \mathbf{f}(-n)\mathbf{f}'(-n)$ go to zero, and $\mathbf{F}_{n+1}$ reaches a limit (or steady state) that we denote by $\mathbf{F}$;

$$\lim_{n \to \infty} \mathbf{F}_{n+1} = \mathbf{F} = \sum_{j \geqslant 0} \omega^j \mathbf{f}(-j)\mathbf{f}'(-j) \tag{3.28}$$

Then the steady-state solution can be written as

$$\begin{aligned}
\hat{\boldsymbol{\beta}}_{n+1} &= \mathbf{F}^{-1}\mathbf{h}_{n+1} = \mathbf{F}^{-1}\left[\mathbf{f}(0) z_{n+1} + \omega \mathbf{L}^{-1}\mathbf{h}_n\right] \\
&= \mathbf{F}^{-1}\mathbf{f}(0) z_{n+1} + \omega \mathbf{F}^{-1}\mathbf{L}^{-1}\mathbf{h}_n \\
&= \mathbf{F}^{-1}\mathbf{f}(0) z_{n+1} + \omega \mathbf{F}^{-1}\mathbf{L}^{-1}\mathbf{F}\hat{\boldsymbol{\beta}}_n
\end{aligned}$$

Furthermore,

$$\begin{aligned}
\omega \mathbf{F}^{-1}\mathbf{L}^{-1}\mathbf{F} &= \omega \mathbf{F}^{-1}\mathbf{L}^{-1}\mathbf{F}\mathbf{L}'^{-1}\mathbf{L}' = \omega \mathbf{F}^{-1}\mathbf{L}^{-1}\left[\sum_{j \geqslant 0} \omega^j \mathbf{f}(-j)\mathbf{f}'(-j)\right]\mathbf{L}'^{-1}\mathbf{L}' \\
&= \omega \mathbf{F}^{-1}\left\{\sum_{j \geqslant 0} \omega^j \left[\mathbf{L}^{-1}\mathbf{f}(-j)\right]\left[\mathbf{L}^{-1}\mathbf{f}(-j)\right]'\right\}\mathbf{L}' \\
&= \omega \mathbf{F}^{-1}\left[\sum_{j \geqslant 0} \omega^j \mathbf{f}(-j-1)\mathbf{f}'(-j-1)\right]\mathbf{L}' \\
&= \mathbf{F}^{-1}\left[\mathbf{F} - \mathbf{f}(0)\mathbf{f}'(0)\right]\mathbf{L}' = \mathbf{L}' - \mathbf{F}^{-1}\mathbf{f}(0)\mathbf{f}'(0)\mathbf{L}'
\end{aligned}$$

Therefore,

$$\begin{aligned}
\hat{\boldsymbol{\beta}}_{n+1} &= \mathbf{F}^{-1}\mathbf{f}(0) z_{n+1} + \left[\mathbf{L}' - \mathbf{F}^{-1}\mathbf{f}(0)\mathbf{f}'(0)\mathbf{L}'\right]\hat{\boldsymbol{\beta}}_n \\
&= \mathbf{L}'\hat{\boldsymbol{\beta}}_n + \mathbf{F}^{-1}\mathbf{f}(0) z_{n+1} - \mathbf{F}^{-1}\mathbf{f}(0)\mathbf{f}'(1)\hat{\boldsymbol{\beta}}_n \\
&= \mathbf{L}'\hat{\boldsymbol{\beta}}_n + \mathbf{F}^{-1}\mathbf{f}(0)\left[z_{n+1} - \hat{z}_n(1)\right] \qquad (3.29)
\end{aligned}$$

and

$$\hat{z}_{n+1}(l) = \mathbf{f}'(l)\hat{\boldsymbol{\beta}}_{n+1}$$

The new parameter estimate in Equation (3.29) is a linear combination of the previous parameter estimate and the one-step-ahead forecast error. Since $\mathbf{L}$, $\mathbf{F}$, and $\mathbf{f}(0)$ are fixed for each model and each discount coefficient ω, only the last parameter estimate and the last forecast error have to be stored for parameter and forecast updating.

We now consider several special cases. One special case, the *locally constant mean model or simple exponential smoothing*, was discussed previously (Sec. 3.3). In terms of the general notation (3.13), the mean model is described by a single constant fitting function, $f_1(j) = 1$.

It follows from Equation (3.25) that for this special case

$$\hat{z}_n(l) = \hat{\beta}_n = \frac{1-\omega}{1-\omega^n} \sum_{j=0}^{n-1} \omega^j z_{n-j} \tag{3.30}$$

For fixed n, as $\omega \to 1$, we find from L'Hospital's rule on the evaluation of limits that Equation (3.30) reduces to

$$\hat{z}_n(l) = \frac{1}{n} \sum_{j=0}^{n-1} z_{n-j}$$

This is the minimum mean square error forecast for the globally constant mean model (see Sec. 3.2).

In the steady-state case ($n \to \infty$) and for fixed ω, the forecast is given by

$$\hat{z}_n(l) = (1-\omega) \sum_{j \geqslant 0} \omega^j z_{n-j} = S_n \tag{3.31}$$

This corresponds to simple exponential smoothing; the forecast of a future value is given by the last smoothed statistic S_n. Equation (3.29) simplifies in this special case to

$$\hat{z}_{n+1}(1) = \hat{z}_n(1) + (1-\omega)[z_{n+1} - \hat{z}_n(1)] \tag{3.32}$$

Other special cases of discounted least squares are discussed in the next two sections. In Section 3.6 we discuss the locally constant trend model or double exponential smoothing. In Section 3.7 we discuss triple exponential smoothing.

3.6. LOCALLY CONSTANT LINEAR TREND MODEL AND DOUBLE EXPONENTIAL SMOOTHING

The linear trend model $z_{n+j} = \beta_0 + \beta_1 j + \varepsilon_{n+j}$ is described by $m = 2$ fitting functions $f_1(j) = 1$, $f_2(j) = j$, transition matrix

$$\mathbf{L} = \begin{bmatrix} 1 & 0 \\ 1 & 1 \end{bmatrix}$$

and initial vector

$$\mathbf{f}(0) = \begin{bmatrix} 1 \\ 0 \end{bmatrix}$$

The expressions in Equation (3.25) simplify to

$$\mathbf{F}_n = \begin{bmatrix} \Sigma\omega^j & -\Sigma j\omega^j \\ -\Sigma j\omega^j & \Sigma j^2\omega^j \end{bmatrix} \qquad \mathbf{h}_n = \begin{bmatrix} \Sigma\omega^j z_{n-j} \\ -\Sigma j\omega^j z_{n-j} \end{bmatrix}$$

Furthermore, since

$$\lim_{n\to\infty} \sum_{j=0}^{n-1} \omega^j = \frac{1}{1-\omega}$$

$$\lim_{n\to\infty} \sum_{j=0}^{n-1} j\omega^j = \frac{\omega}{(1-\omega)^2}$$

$$\lim_{n\to\infty} \sum_{j=0}^{n-1} j^2\omega^j = \frac{\omega(1+\omega)}{(1-\omega)^3}$$

the steady-state value of $\mathbf{F}_n$ is given by

$$\mathbf{F} = \begin{bmatrix} \dfrac{1}{1-\omega} & \dfrac{-\omega}{(1-\omega)^2} \\ \dfrac{-\omega}{(1-\omega)^2} & \dfrac{\omega(1+\omega)}{(1-\omega)^3} \end{bmatrix}$$

Thus, for large n, the estimates $\hat{\boldsymbol{\beta}}_n = (\hat{\beta}_{0,n}, \hat{\beta}_{1,n})'$ are given by

$$\hat{\boldsymbol{\beta}}_n = \mathbf{F}^{-1}\mathbf{h}_n = \begin{bmatrix} 1-\omega^2 & (1-\omega)^2 \\ (1-\omega)^2 & \dfrac{(1-\omega)^3}{\omega} \end{bmatrix} \begin{bmatrix} \Sigma\omega^j z_{n-j} \\ -\Sigma j\omega^j z_{n-j} \end{bmatrix}$$

or

$$\hat{\beta}_{0,n} = (1-\omega^2)\Sigma\omega^j z_{n-j} - (1-\omega)^2\Sigma j\omega^j z_{n-j}$$

$$\hat{\beta}_{1,n} = (1-\omega)^2\Sigma\omega^j z_{n-j} - \frac{(1-\omega)^3}{\omega}\Sigma j\omega^j z_{n-j} \tag{3.33}$$

We now relate these estimates to statistics that can be obtained from an extension of simple exponential smoothing. For this purpose, we introduce smoothed statistics of higher orders. The single smoothed statistic $S_n^{[1]} = S_n$ was defined previously,

$$S_n^{[1]} = (1 - \omega) z_n + \omega S_{n-1}^{[1]} = (1 - \omega)\Sigma\omega^j z_{n-j} \tag{3.34}$$

We now define the double smoothed statistic $S_n^{[2]}$ as

$$S_n^{[2]} = (1 - \omega) S_n^{[1]} + \omega S_{n-1}^{[2]} = (1 - \omega)^2 \Sigma(j + 1)\omega^j z_{n-j} \tag{3.35}$$

and in general the smoothed statistic of order k,

$$S_n^{[k]} = (1 - \omega) S_n^{[k-1]} + \omega S_{n-1}^{[k]} \tag{3.36}$$

The constant ω is the discount coefficient; $\alpha = 1 - \omega$ is called the smoothing constant. $S_n^{[2]}$ is called the double smoothed statistic because it is derived by smoothing the single smoothed statistic $S_n^{[1]}$. Similarly, $S_n^{[k]}$ is derived by k successive smoothing operations.

By simple substitution it can be shown that the estimates $\hat{\beta}_{0,n}$ and $\hat{\beta}_{1,n}$ in (3.33) can be expressed as a function of the single and double smoothed statistics:

$$\hat{\beta}_{0,n} = 2S_n^{[1]} - S_n^{[2]} \qquad \hat{\beta}_{1,n} = \frac{1 - \omega}{\omega}\left(S_n^{[1]} - S_n^{[2]}\right) \tag{3.37}$$

The forecasts can then be computed from

$$\begin{aligned}
\hat{z}_n(l) &= \hat{\beta}_{0,n} + \hat{\beta}_{1,n} l \\
&= 2S_n^{[1]} - S_n^{[2]} + l\frac{1 - \omega}{\omega}\left(S_n^{[1]} - S_n^{[2]}\right) \\
&= \left(2 + \frac{1 - \omega}{\omega} l\right) S_n^{[1]} - \left(1 + \frac{1 - \omega}{\omega} l\right) S_n^{[2]}
\end{aligned} \tag{3.38}$$

Or, in terms of the smoothing constant α,

$$\hat{z}_n(l) = \left(2 + \frac{\alpha}{1 - \alpha} l\right) S_n^{[1]} - \left(1 + \frac{\alpha}{1 - \alpha} l\right) S_n^{[2]}$$

Since the forecasts can be expressed as a function of the single and double smoothed statistics, this forecast procedure is known as *double exponential smoothing*.

3.6.1. Updating Coefficient Estimates

Using the updating relation in Equation (3.29), we can furthermore write

$$\hat{\beta}_{0,n+1} = \hat{\beta}_{0,n} + \hat{\beta}_{1,n} + (1-\omega^2)[z_{n+1} - \hat{z}_n(1)]$$

$$\hat{\beta}_{1,n+1} = \hat{\beta}_{1,n} + (1-\omega)^2[z_{n+1} - \hat{z}_n(1)] \tag{3.39}$$

By substitution it can be shown that

$$\hat{\beta}_{0,n+1} = (1-\omega^2)z_{n+1} + \omega^2(\hat{\beta}_{0,n} + \hat{\beta}_{1,n})$$

$$\hat{\beta}_{1,n+1} = \frac{1-\omega}{1+\omega}(\hat{\beta}_{0,n+1} - \hat{\beta}_{0,n}) + \frac{2\omega}{1+\omega}\hat{\beta}_{1,n} \tag{3.40}$$

The first equation follows by substituting $\hat{z}_n(1) = \hat{\beta}_{0,n} + \hat{\beta}_{1,n}$. The second equation is derived by substituting

$$z_{n+1} - \hat{z}_n(1) = \frac{1}{1-\omega^2}(\hat{\beta}_{0,n+1} - \hat{\beta}_{0,n} - \hat{\beta}_{1,n})$$

into the second equation of (3.39).

3.6.2. Another Interpretation of Double Exponential Smoothing

A slightly different approach will aid in the interpretation of the updating equations in (3.40). We follow an approach originally used by Holt (1957).

We assume a linear trend model for the mean μ_t and write it in slightly different form: $\mu_t = \mu_n + (t-n)\beta$. In this representation, β is the slope parameter and μ_n is the level at time n. Then the mean at time $n+1$ is defined as $\mu_{n+1} = \mu_n + \beta$. An estimate of this mean can be found from two different sources: (1) from z_{n+1}, which represents the present estimate of μ_{n+1} and (2) from $\hat{\mu}_n + \hat{\beta}_n$, which is the estimate of μ_{n+1} from observations up to and including time n. Note that $\hat{\beta}_n$ is the estimate of the slope at time n. Holt considers a linear combination of these estimates,

$$\hat{\mu}_{n+1} = (1-\omega_1)z_{n+1} + \omega_1(\hat{\mu}_n + \hat{\beta}_n)$$

where $\alpha_1 = 1 - \omega_1$ $(0 < \alpha_1 < 1)$ is a smoothing constant that determines how quickly past information is discounted.

Similarly, information about the slope comes from two sources: (1) from the difference of the mean estimates $\hat{\mu}_{n+1} - \hat{\mu}_n$ (i.e., the most recent estimate of β_{n+1}) and (2) from the previous estimate of the slope $\hat{\beta}_n$. Again, these estimates are linearly weighted to give

$$\hat{\beta}_{n+1} = (1 - \omega_2)(\hat{\mu}_{n+1} - \hat{\mu}_n) + \omega_2 \hat{\beta}_n$$

The coefficient $\alpha_2 = 1 - \omega_2$ $(0 < \alpha_2 < 1)$ is another smoothing constant. Thus, the parameter estimates are updated according to

$$\hat{\mu}_{n+1} = (1 - \omega_1) z_{n+1} + \omega_1 (\hat{\mu}_n + \hat{\beta}_n)$$

$$\hat{\beta}_{n+1} = (1 - \omega_2)(\hat{\mu}_{n+1} - \hat{\mu}_n) + \omega_2 \hat{\beta}_n \tag{3.41}$$

and the forecasts of future observations are given by

$$\hat{z}_{n+1}(l) = \hat{\mu}_{n+1} + \hat{\beta}_{n+1} l \tag{3.42}$$

Equations (3.41) are more general than those of double exponential smoothing in (3.40), since two different discount coefficients are used. However, if $\omega_1 = \omega^2$ and $\omega_2 = 2\omega/(1 + \omega)$, Holt's procedure is equivalent to double exponential smoothing with discount coefficient ω (or smoothing constant $\alpha = 1 - \omega$).

3.6.3. Actual Implementation of Double Exponential Smoothing

The forecasts in double exponential smoothing can be computed from Equation (3.38), which implies that it is not necessary to calculate the discounted least squares estimates $\hat{\beta}_{0,n}$ and $\hat{\beta}_{1,n}$ directly. However, we need to know (1) the smoothing constant $\alpha = 1 - \omega$ and (2) the most recent smoothed statistics $S_n^{[1]}$ and $S_n^{[2]}$. To calculate these smoothed statistics, we can start the recursions (3.34) and (3.35) from the first observation. For this, we need to specify initial values $S_0^{[1]}$ and $S_0^{[2]}$. Equations (3.34) and (3.35) can then be used to update the smoothed statistics until $S_n^{[1]}$ and $S_n^{[2]}$ are reached.

Initial Values for $S_0^{[1]}$ and $S_0^{[2]}$

Equations (3.37) express the estimates of the model coefficients as a function of the smoothed statistics. These equations can be solved for $S_n^{[1]}$ and $S_n^{[2]}$. For $n = 0$, it follows that

$$S_0^{[1]} = \hat{\beta}_{0,0} - \frac{\omega}{1 - \omega} \hat{\beta}_{1,0}$$

$$S_0^{[2]} = \hat{\beta}_{0,0} - 2\frac{\omega}{1 - \omega} \hat{\beta}_{1,0} \tag{3.43}$$

The estimates of the coefficients at time zero, $\hat{\beta}_{0,0}$ and $\hat{\beta}_{1,0}$, are usually found by fitting the constant linear trend model $z_t = \beta_0 + \beta_1 t + \varepsilon_t$ either to all available observations $z_1, z_2, \ldots, z_n$, or to a subset consisting of the first n_1 observations. This approach is similar to the one in simple exponential smoothing where the estimate of the constant mean model (i.e., the sample mean) is used as starting value for $S_0^{[1]}$. Ordinary least squares [see Eq. (2.7)] leads to the following estimates:

$$\hat{\beta}_1 = \frac{\sum_{t=1}^{n}\left(t - \frac{n+1}{2}\right) z_t}{\sum_{t=1}^{n}\left(t - \frac{n+1}{2}\right)^2} = \frac{12 \sum_{t=1}^{n}\left(t - \frac{n+1}{2}\right) z_t}{n^3 - n}$$

$$\hat{\beta}_0 = \bar{z} - \hat{\beta}_1 \frac{n+1}{2} \tag{3.44}$$

These estimates are used for $\hat{\beta}_{0,0}$ and $\hat{\beta}_{1,0}$ (i.e., $\hat{\beta}_{0,0} = \hat{\beta}_0$ and $\hat{\beta}_{1,0} = \hat{\beta}_1$), and the initial values $S_0^{[1]}$ and $S_0^{[2]}$ are derived from Equations (3.43).

Choice of the Smoothing Constant

As in simple exponential smoothing, the discount coefficient ω (or equivalently the smoothing coefficient $\alpha = 1 - \omega$) is chosen by generating one-step-ahead forecast errors for several values of the smoothing constant α. The sum of the squared one-step-ahead forecast errors is calculated for each chosen α. The smoothing constant that minimizes

$$\begin{aligned} \text{SSE}(\alpha) &= \Sigma[z_t - \hat{z}_{t-1}(1)]^2 \\ &= \Sigma\left[z_t - \left(2 + \frac{\alpha}{1-\alpha}\right) S_{t-1}^{[1]} + \left(1 + \frac{\alpha}{1-\alpha}\right) S_{t-1}^{[2]}\right]^2 \end{aligned} \tag{3.45}$$

is the coefficient that is used for future forecasting.

In the forecasting literature it is usually suggested that the discount coefficient ω be chosen close to 1 (or equivalently the smoothing constant $\alpha = 1 - \omega$ close to zero). Brown (1962), for example, suggests that the discount factor ω in double exponential smoothing be chosen between $\sqrt{.70} = .84$ and $\sqrt{.95} = .97$.

However, in many applications of simple and double exponential smoothing it is found that ω lies outside the suggested range. The closer ω is to zero, the less likely it is that the data are described by a locally constant mean model (as in simple exponential smoothing) or a locally constant linear trend model (as in double exponential smoothing).

For example, let us consider the extreme case when $\omega \to 0$ (or equivalently $\alpha \to 1$). In simple exponential smoothing the smoothed value becomes $S_n = (1)z_n + (0)S_{n-1} = z_n$. Thus the last recorded value is the best forecast of all future observations; $\hat{z}_n(l) = z_n$. Information from all other past observations is ignored.

In double exponential smoothing, the smoothed statistics in (3.34) and (3.35) are given by $S_n^{[1]} = S_n^{[2]} = z_n$. From (3.38) it follows that the forecasts, for example for $l = 1$, are

$$\hat{z}_n(1) = 2S_n^{[1]} - S_n^{[2]} + \frac{1-\omega}{\omega}\left(S_n^{[1]} - S_n^{[2]}\right)$$

$$= 2(1-\omega)\sum_{j\geq 0} \omega^j z_{n-j} - (1-\omega)^2 \sum_{j\geq 0} (j+1)\omega^j z_{n-j}$$

$$+ \frac{1-\omega}{\omega}\left[(1-\omega)\sum_{j\geq 0} \omega^j z_{n-j} - (1-\omega)^2 \sum_{j\geq 0} (j+1)\omega^j z_{n-j}\right]$$

Now, as $\omega \to 0$,

$$\lim_{\omega \to 0} \hat{z}_n(1) = 2z_n - z_n + \lim_{\omega \to 0} \frac{1-\omega}{\omega}\Big\{\omega(1-\omega)z_n$$

$$+\omega(1-\omega)\sum_{j\geq 1} \omega^{j-1}[\omega - (1-\omega)j]z_{n-j}\Big\}$$

$$= 2z_n - z_n + z_n - z_{n-1} = 2z_n - z_{n-1}$$

This result implies that double exponential smoothing with a discount factor $\omega \to 0$ corresponds to fitting a linear trend model to only the last two observations. No other observations contribute to the forecast. This choice of ω is very different from the case $\omega = 1$. There the best forecast is achieved by fitting a constant linear trend model to all available observations and giving each observation the same weight.

3.6.4. Examples

Example 3.3: Weekly Thermostat Sales

As an example for double exponential smoothing, we analyze a sequence of 52 weekly sales observations. The data, which are listed in Table 3.6 and plotted in Figure 3.5, were originally analyzed by Brown (1962). The plot of

Table 3.6. Weekly Thermostat Sales, 52 Observations[a]

206	189	172	255
245	244	210	303
185	209	205	282
169	207	244	291
162	211	218	280
177	210	182	255
207	173	206	312
216	194	211	296
193	234	273	307
230	156	248	281
212	206	262	308
192	188	258	280
162	162	233	345

[a]Read downwards, left to right.

Source: Reprinted by permission of Prentice-Hall, Inc. from R. G. Brown (1962), *Smoothing, Forecasting and Prediction of Discrete Time Series*, p. 431.

the data indicates an upward trend in the thermostat sales. This trend, however, does not appear to be constant but seems to change over time. A constant linear trend model would therefore not be appropriate.

Nevertheless, we first fit a constant linear trend model $z_t = \beta_0 + \beta_1 t + \varepsilon_t$ to all 52 observations. The time origin in our constant linear trend model is at time zero. This representation is somewhat different from the models in Section 3.4, where the time origin is always at time period n. However, we choose the current representation because we need the estimates of β_0 and β_1 to calculate the initial values for $S_0^{[1]}$ and $S_0^{[2]}$. Another reason for fitting a constant linear trend model is to illustrate how a residual analysis can detect model inadequacies.

The least squares estimates of β_0 and β_1 are calculated from (3.44):

$$\hat{\beta}_1 = \frac{\sum_{t=1}^{52} (t - 26.5) z_t}{\sum_{t=1}^{52} (t - 26.5)^2} = 2.325$$

$$\hat{\beta}_0 = \bar{z} - \hat{\beta}_1 (26.5) = 166.40$$

If the constant linear trend model were appropriate, the residuals

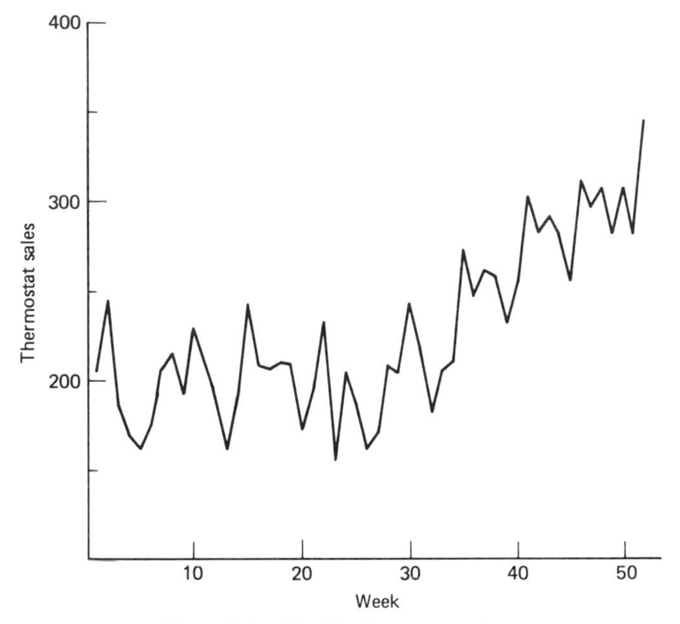

Figure 3.5. Weekly thermostat sales.

$e_t = z_t - (\hat{\beta}_0 + \hat{\beta}_1 t)$ for $t = 1, 2, \ldots, 52$ should be uncorrelated. The sample autocorrelations of the residuals,

$$r_k = \frac{\sum_{t=k+1}^{52} e_t e_{t-k}}{\sum_{t=1}^{52} e_t^2}$$

are given in Table 3.7. The mean $\bar{e}$ can be omitted in the calculation of r_k, since the least squares fit introduces restrictions among the residuals [see Eq. (2.15)]. Since an intercept is included in the model, one of the restrictions forces the residuals to add to zero.

Comparing the sample autocorrelations with their standard error $1/\sqrt{52} = .14$, we find that the residuals are correlated and that the constant linear trend model is certainly not appropriate for this particular data set.

To model the time-changing trend, we now consider double exponential smoothing. We choose a smoothing constant $\alpha = .14$ ($\omega = .86$). Later we will explain why we choose this particular value. To derive the initial values

Table 3.7. Sample Autocorrelations of the Residuals from the Constant Linear Trend Model—Thermostat Sales

Lag k	1	2	3	4	5	6
Autocorrelation r_k	.41	.26	.18	.19	.27	.40

$S_0^{[1]}$ and $S_0^{[2]}$, we substitute the least squares estimates $\hat{\beta}_0$ and $\hat{\beta}_1$ into Equations (3.43).

$$S_0^{[1]} = 166.40 - \frac{.86}{.14}2.325 = 152.12$$

$$S_0^{[2]} = 166.40 - 2\left(\frac{.86}{.14}\right)(2.325) = 137.84$$

From these initial smoothed statistics we can calculate the forecasts $\hat{z}_0(l)$. For example, for $l = 1$, it follows from (3.38) that

$$\hat{z}_0(1) = \left(2 + \frac{.14}{.86}\right)S_0^{[1]} - \left(1 + \frac{.14}{.86}\right)S_0^{[2]} = 168.72$$

Similarly, for $l = 2$ we find

$$\hat{z}_0(2) = \left(2 + \frac{.14}{.86}2\right)S_0^{[1]} - \left(1 + \frac{.14}{.86}2\right)S_0^{[2]} = 171.05$$

Since $z_1 = 206$, we can calculate the first one-step-ahead forecast error:

$$e_0(1) = z_1 - \hat{z}_0(1) = 206 - 168.72 = 37.28$$

The smoothed statistics can be updated [Eqs. (3.34), (3.35)]:

$$S_1^{[1]} = (1 - \omega)z_1 + \omega S_0^{[1]}$$

$$= (.14)(206) + (.86)(152.12) = 159.66$$

$$S_1^{[2]} = (1 - \omega)S_1^{[1]} + \omega S_0^{[2]}$$

$$= (.14)(159.66) + (.86)(137.84) = 140.89$$

The forecasts $\hat{z}_1(l)$ can then be calculated from (3.38):

$$\hat{z}_1(1) = \left(2 + \frac{.14}{.86}\right)(159.66) - \left(1 + \frac{.14}{.86}\right)(140.89) = 181.48$$

and the next one-step-ahead forecast error is given by

$$e_1(1) = z_2 - \hat{z}_1(1) = 245 - 181.48 = 63.52$$

With the new observation z_2 we can update the smoothed statistics and calculate $S_2^{[1]}$ and $S_2^{[2]}$. The observations z_t; the smoothed statistics $S_t^{[1]}$, $S_t^{[2]}$; the forecasts $\hat{z}_t(1)$, $\hat{z}_t(2)$; and the one-step-ahead forecast errors $e_{t-1}(1) = z_t - \hat{z}_{t-1}(1)$ are listed in Table 3.8. Eventually, through repeated application of (3.34) and (3.35), we find $S_{52}^{[1]} = 289.12$, $S_{52}^{[2]} = 263.18$. Forecasts for the next l periods are then given by

$$\hat{z}_{52}(l) = \left(2 + \frac{.14}{.86}l\right)289.12 - \left(1 + \frac{.14}{.86}l\right)263.18$$

For example, the forecast for the next period ($l = 1$) is given by 319.29; for two periods ahead it is given by $\hat{z}_{52}(2) = 323.51$. All other forecasts lie on a straight line determined by $\hat{z}_{52}(1)$ and $\hat{z}_{52}(2)$.

Choice of the smoothing constant. For the smoothing constant $\alpha = .14$, the sum of the squared one-step-ahead forecast errors is given by

$$\text{SSE}(.14) = (37.28)^2 + (63.52)^2 + \cdots + (-28.52)^2 + (40.48)^2 = 41{,}469$$

Table 3.8. Double Exponential Smoothing with Smoothing Constant $\alpha = .14$—Thermostat Sales

t	z_t	$S_t^{[1]}$	$S_t^{[2]}$	$\hat{z}_t(1)$	$\hat{z}_t(2)$	$e_{t-1}(1)$ $[= z_t - \hat{z}_{t-1}(1)]$
0		152.12	137.84	168.72	171.05	
1	206	159.66	140.89	181.48	184.54	37.28
2	245	171.61	145.19	202.32	206.62	63.52
3	185	173.48	149.15	201.77	205.73	−17.32
4	169	172.86	152.47	196.55	199.88	−32.77
5	162	171.34	155.11			−34.56
⋮	⋮	⋮	⋮	⋮	⋮	⋮
48	307	274.58	247.64	305.90	310.29	7.42
49	281	275.48	251.54	303.31	307.21	−24.90
50	308	280.03	255.53	308.52	312.51	4.69
51	280	280.03	258.96	304.53	307.96	−28.52
52	345	289.12	263.18	319.29	323.51	40.48
						SSE(.14) = 41,469

Table 3.9. Sums of Squared One-Step-Ahead Forecast Errors for Different Values of α; Double Exponential Smoothing—Thermostat Sales

α	SSE(α)	α	SSE(α)	α	SSE(α)
0.02	49,305	0.11	42,018	0.21	43,132
0.03	48,935	0.12	41,707	0.22	43,558
0.04	48,149	0.13	41,530	0.23	44,014
0.05	47,108	0.14	41,469	0.24	44,496
0.06	45,979	0.15	41,507	0.25	45,001
0.07	44,888	0.16	41,630	0.26	45,526
0.08	43,920	0.17	41,824	0.27	46,070
0.09	43,114	0.18	42,079	0.28	46,629
0.10	42,482	0.19	42,387	0.29	47,203
		0.20	42,740	0.30	47,790

Changing the smoothing constant from .02 to .30 in increments of .01 leads to the sums of squared one-step-ahead forecast errors that are listed in Table 3.9 and plotted in Figure 3.6. It is found that the minimum of SSE(α) occurs when $\alpha = .14$, which explains our previous choice. As in simple exponential smoothing, the choice of $S_0^{[1]}$ and $S_0^{[2]}$ will influence the value of α obtained by simulation. Ideally, one should choose the starting values as a function of α as well as the data $z_1, \ldots, z_n$.

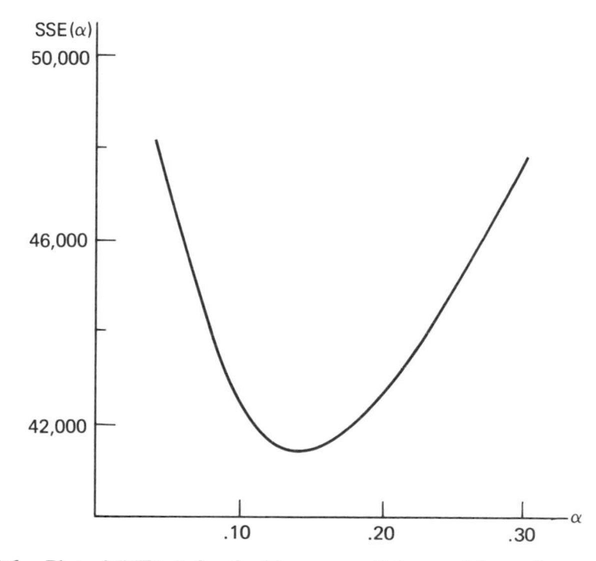

Figure 3.6. Plot of SSE(α), for double exponential smoothing—thermostat sales.

Table 3.10. Mean, Standard Error, and Sample Autocorrelations of the One-Step-Ahead Forecast Errors from Double Exponential Smoothing ($\alpha = .14$)—Thermostat Sales

Lag k	Sample Autocorrelations of One-Step-Ahead Forecast Errors r_k
1	.13
2	−.09
3	−.16
4	−.13
5	.05
Mean of historical forecast errors	1.86
Standard error of mean	3.95

Checking the adequacy of double exponential smoothing. To check whether double exponential smoothing is an appropriate forecast procedure for the thermostat sales data, we calculate the mean, the standard error, and the autocorrelations of the one-step-ahead forecast errors $e_{t-1}(1) = z_t - \hat{z}_{t-1}(1)$. The results in Table 3.10 indicate that double exponential smoothing is an appropriate forecasting procedure for the thermostat sales data, since (1) it leads to unbiased forecasts and (2) the forecast errors are uncorrelated. The standard error of r_k is $1/\sqrt{52} = .14$; all sample autocorrelations are well within two standard errors.

Example 3.4: University of Iowa Student Enrollments

As another example, we consider the annual student enrollments (fall and spring semesters combined) at the University of Iowa. Observations for the

Table 3.11. Total Annual Student Enrollment at the University of Iowa, 1951/52 to 1979/80[a]

14,348	14,307	15,197	16,715	18,476	19,404
20,173	20,645	20,937	21,501	22,788	23,579
25,319	28,250	32,191	34,584	36,366	37,865
39,173	40,119	39,626	39,107	39,796	41,567
43,646	43,534	44,157	44,551	45,572	

[a]Read across, top to bottom.

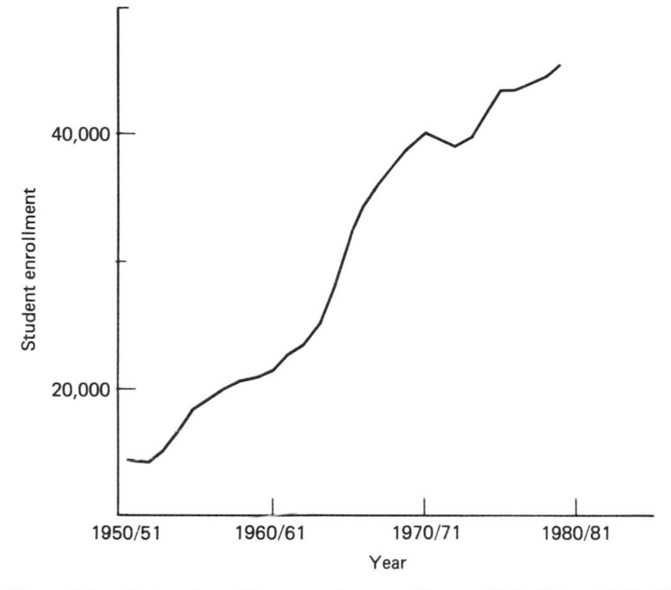

Figure 3.7. University of Iowa student enrollment, 1951/52 to 1979/80.

last 29 years (1951/52 through 1979/80) are summarized in Table 3.11. A plot of the observations is given in Figure 3.7.

Because of the growing trend pattern of the series, we decided to use double exponential smoothing. To decide on the smoothing constant α, we simulated the forecast errors for several different smoothing coefficients. Sums of the squared one-step-ahead forecast errors,

$$\begin{aligned} \text{SSE}(\alpha) &= \sum_{t=1}^{n} [z_t - \hat{z}_{t-1}(1)]^2 \\ &= \sum_{t=1}^{n} \left[z_t - \left(2 + \frac{\alpha}{1-\alpha}\right) S_{t-1}^{[1]} + \left(1 + \frac{\alpha}{1-\alpha}\right) S_{t-1}^{[2]} \right]^2 \end{aligned}$$

are given in Table 3.12 and plotted in Figure 3.8.

The optimal smoothing constant is given by $\alpha = .87$. We notice that it is not in the range usually suggested for exponential smoothing. The function SSE(α) is very flat from .80 to 1.00, indicating that the smoothing constant α is close to 1. A smoothing constant near 1 implies that the linear trend depends mostly on the last two observations and can change very quickly. The trend is not of a deterministic nature; it is stochastic, changing rapidly with each new observation.

Table 3.12. Sums of Squared One-Step-Ahead Forecast Errors for Different Values of α; Double Exponential Smoothing—University of Iowa Student Enrollment

Smoothing Constant α	Sum of Squared One-Step-Ahead Forecast Errors (in 1000) SSE(α)
.10	120,717
.20	94,041
.30	68,669
.40	51,814
.50	41,171
.60	34,657
.70	30,906
.80	29,066
.81	28,968
.84	28,760
.87	28,679
.90	28,727
.93	28,908
.96	29,228
.99	29,700

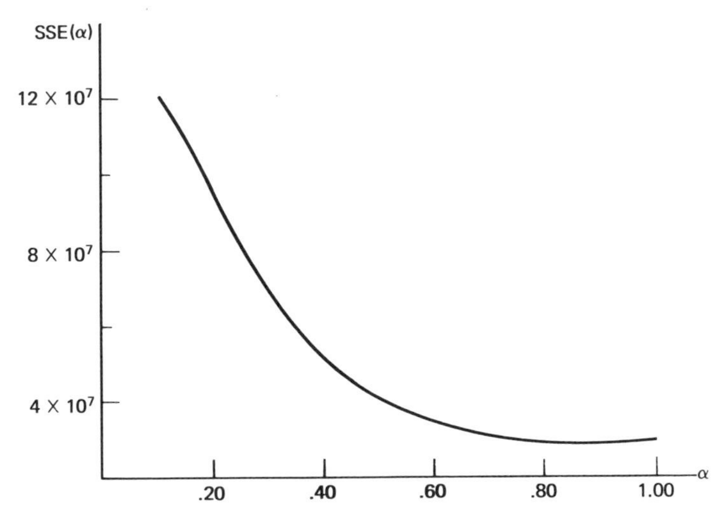

Figure 3.8. Plot of SSE(α), for double exponential smoothing—University of Iowa student enrollment.

The observations, the first- and second-order smoothed statistics, and the one-step-ahead forecast errors for the smoothing constant $\alpha = .87$ are given in Table 3.13. The initial smoothed statistics are calculated from (3.43) and (3.44). All 29 observations are used to determine the estimates $\hat{\beta}_0$ and $\hat{\beta}_1$. Substituting the last smoothed statistics $S_{29}^{[1]} = 45{,}431$ and $S_{29}^{[2]} = 45{,}300$ in the forecast equation (3.38), we can calculate the enrollments for future

Table 3.13. Double Exponential Smoothing ($\alpha = .87$)—University of Iowa Student Enrollment

t	z_t	$S_t^{[1]}$	$S_t^{[2]}$	$\hat{z}_t(1)$	$\hat{z}_t(2)$	$e_{t-1}(1)$ $[= z_t - \hat{z}_{t-1}(1)]$
0		11,418	11,230	12,863	14,121	
1	14,348	13,967	13,611	16,704	19,085	1,485
2	14,307	14,263	14,178	14,914	15,481	−2,397
3	15,197	15,076	14,959	15,973	16,754	283
4	16,715	16,502	16,301	18,045	19,387	742
5	18,476	18,219	17,970	20,137	21,806	431
6	19,404	19,250	19,084	20,530	21,644	−733
7	20,173	20,053	19,927	21,022	21,866	−357
8	20,645	20,568	20,485	21,209	21,767	−377
9	20,937	20,889	20,836	21,293	21,645	−272
10	21,501	21,421	21,345	22,006	22,515	208
11	22,788	22,610	22,446	23,875	24,976	782
12	23,579	23,453	23,322	24,460	25,337	−296
13	25,319	25,076	24,848	26,831	28,357	859
14	28,250	27,837	27,449	30,827	33,427	1,419
15	32,191	31,625	31,082	35,801	39,435	1,364
16	34,584	34,199	33,794	37,317	40,029	−1,217
17	36,366	36,084	35,787	38,375	40,367	−951
18	37,865	37,634	37,393	39,480	41,087	−510
19	39,173	38,973	38,768	40,552	41,927	−307
20	40,119	39,970	39,814	41,173	42,219	−433
21	39,626	39,671	39,689	39,528	39,404	−1,547
22	39,107	39,180	39,246	38,671	38,229	−421
23	39,796	39,716	39,655	40,185	40,594	1,125
24	41,567	41,326	41,109	42,998	44,452	1,382
25	43,646	43,344	43,054	45,580	47,525	648
26	43,534	43,509	43,450	43,965	44,361	−2,046
27	44,157	44,073	43,992	44,696	45,237	192
28	44,551	44,489	44,424	44,986	45,418	−145
29	45,572	45,431	45,300	46,438	47,314	586
						SSE(.87) = 28,679,000

Table 3.14. Mean, Standard Error, and Sample Autocorrelations of the One-Step-Ahead Forecast Errors from Double Exponential Smoothing (α = .87)—University of Iowa Student Enrollment

Lag k	Sample Autocorrelations of One-Step-Ahead Forecast Errors r_k
1	.03
2	−.20
3	−.26
4	.09
5	.04
Mean of historical forecast errors	−17.4
Standard error of mean	187.9

academic years. The forecast for 1980/81 ($l = 1$ year ahead) is given by $\hat{z}_{29}(1) = 46{,}438$; for 1981/82 we predict an enrollment of $\hat{z}_{29}(2) = 47{,}314$; etc.

The diagnostics in Table 3.14 indicate that double exponential smoothing with a smoothing constant $\alpha = .87$ leads to unbiased forecasts and uncorrelated forecast errors. The standard error of r_k is $1/\sqrt{29} = .19$.

It should be emphasized that a smoothing constant of $\alpha = .87$ implies a rapidly changing linear trend model. As we have shown earlier, for $\omega \to 0$ (or $\alpha \to 1$), all forecasts fall on a straight line that is determined by the last two observations. The optimal forecast system for University of Iowa student enrollments essentially remembers only the last two observations; all previous observations are ignored.

3.7. LOCALLY QUADRATIC TREND MODEL AND TRIPLE EXPONENTIAL SMOOTHING

The quadratic model

$$z_{n+j} = \beta_0 + \beta_1 j + \beta_2 \frac{j^2}{2} + \varepsilon_{n+j}$$

is characterized by $m = 3$ fitting functions $f_1(j) = 1, f_2(j) = j, f_3(j) = j^2/2$. The vector of fitting functions $\mathbf{f}(j) = [f_1(j), f_2(j), f_3(j)]'$ satisfies the difference equation $\mathbf{f}(j+1) = \mathbf{L}\mathbf{f}(j)$ [see Eq. (3.14)], where the transition matrix is

$$\mathbf{L} = \begin{bmatrix} 1 & 0 & 0 \\ 1 & 1 & 0 \\ \frac{1}{2} & 1 & 1 \end{bmatrix}$$

and

$$\mathbf{f}(0) = \begin{bmatrix} 1 \\ 0 \\ 0 \end{bmatrix}$$

The matrix $\mathbf{F}_n = \Sigma_{j=0}^{n-1} \omega^j \mathbf{f}(-j)\mathbf{f}'(-j)$ in (3.25) is given by

$$\mathbf{F}_n = \begin{bmatrix} \Sigma\omega^j & -\Sigma j\omega^j & \frac{1}{2}\Sigma j^2\omega^j \\ -\Sigma j\omega^j & \Sigma j^2\omega^j & -\frac{1}{2}\Sigma j^3\omega^j \\ \frac{1}{2}\Sigma j^2\omega^j & -\frac{1}{2}\Sigma j^3\omega^j & \frac{1}{4}\Sigma j^4\omega^j \end{bmatrix}$$

and the vector $\mathbf{h}_n$ by

$$\mathbf{h}_n = \begin{bmatrix} \Sigma\omega^j z_{n-j} \\ -\Sigma j\omega^j z_{n-j} \\ \frac{1}{2}\Sigma j^2\omega^j z_{n-j} \end{bmatrix}$$

For large n, it can be shown [see Brown (1962), p. 135] that

$$\lim_{n\to\infty} \sum_{j=0}^{n-1} j^3\omega^j = \frac{\omega(1 + 4\omega + \omega^2)}{(1-\omega)^4}$$

$$\lim_{n\to\infty} \sum_{j=0}^{n-1} j^4\omega^j = \frac{\omega(1 + 11\omega + 11\omega^2 + \omega^3)}{(1-\omega)^5}$$

The limits of $\Sigma\omega^j$, $\Sigma j\omega^j$, and $\Sigma j^2\omega^j$ were already discussed in Section 3.6. Thus the steady-state value of $\mathbf{F}_n$ is given by

$$\mathbf{F} = \begin{bmatrix} \dfrac{1}{1-\omega} & \dfrac{-\omega}{(1-\omega)^2} & \dfrac{\omega(1+\omega)}{2(1-\omega)^3} \\ \dfrac{-\omega}{(1-\omega)^2} & \dfrac{\omega(1+\omega)}{(1-\omega)^3} & \dfrac{-\omega(1+4\omega+\omega^2)}{2(1-\omega)^4} \\ \dfrac{\omega(1+\omega)}{2(1-\omega)^3} & \dfrac{-\omega(1+4\omega+\omega^2)}{2(1-\omega)^4} & \dfrac{\omega(1+11\omega+11\omega^2+\omega^3)}{4(1-\omega)^5} \end{bmatrix}$$

For large n, the estimates $\hat{\boldsymbol{\beta}}_n = (\hat{\beta}_{0,n}, \hat{\beta}_{1,n}, \hat{\beta}_{2,n})'$ are then given by

$$\hat{\boldsymbol{\beta}}_n = \mathbf{F}^{-1}\mathbf{h}_n$$

$$= \begin{bmatrix} (1-\omega)(1+\omega+\omega^2) & \frac{3}{2}(1-\omega)^2(1+\omega) & (1-\omega)^3 \\ \frac{3}{2}(1-\omega)^2(1+\omega) & \dfrac{(1-\omega)^3(1+10\omega+9\omega^2)}{4\omega^2} & \dfrac{(1-\omega)^4(1+3\omega)}{2\omega^2} \\ (1-\omega)^3 & \dfrac{(1-\omega)^4(1+3\omega)}{2\omega^2} & \dfrac{(1-\omega)^5}{\omega^2} \end{bmatrix} \begin{bmatrix} \Sigma\omega^j z_{n-j} \\ -\Sigma j\omega^j z_{n-j} \\ \frac{1}{2}\Sigma j^2\omega^j z_{n-j} \end{bmatrix} \tag{3.46}$$

The elements of $\mathbf{h}_n$, and thus the estimates $\hat{\boldsymbol{\beta}}_n$, can be expressed in terms of the first three smoothed statistics:

$$S_n^{[1]} = (1-\omega)z_n + \omega S_{n-1}^{[1]}$$

$$S_n^{[2]} = (1-\omega)S_n^{[1]} + \omega S_{n-1}^{[2]}$$

$$S_n^{[3]} = (1-\omega)S_n^{[2]} + \omega S_{n-1}^{[3]}$$

We have shown earlier that

$$S_n^{[1]} = (1-\omega)\sum_{j\geq 0}\omega^j z_{n-j} \qquad S_n^{[2]} = (1-\omega)^2\sum_{j\geq 0}(j+1)\omega^j z_{n-j}$$

By substitution it can be shown that

$$S_n^{[3]} = \frac{(1-\omega)^3}{2}\sum_{j\geq 0}(j+1)(j+2)\omega^j z_{n-j}$$

and in general for $k \geqslant 2$,

$$S_n^{[k]} = \frac{(1-\omega)^k}{(k-1)!}\sum_{j\geq 0}\left[\prod_{i=1}^{k-1}(j+i)\right]\omega^j z_{n-j}$$

Thus the elements in $\mathbf{h}_n$ can be written as functions of the first three smoothed statistics.

The coefficient estimates $\hat{\boldsymbol{\beta}}_n$, expressed in terms of the smoothed statistics, are given by

$$\hat{\beta}_{0,n} = 3S_n^{[1]} - 3S_n^{[2]} + S_n^{[3]}$$

$$\hat{\beta}_{1,n} = \frac{\alpha}{2\omega^2}\left[(6-5\alpha)S_n^{[1]} - 2(5-4\alpha)S_n^{[2]} + (4-3\alpha)S_n^{[3]}\right]$$

$$\hat{\beta}_{2,n} = \frac{\alpha^2}{\omega^2}\left[S_n^{[1]} - 2S_n^{[2]} + S_n^{[3]}\right] \tag{3.47}$$

where $\alpha = 1 - \omega$.

The l-step-ahead forecast can be calculated from

$$\hat{z}_n(l) = \hat{\beta}_{0,n} + \hat{\beta}_{1,n}l + \hat{\beta}_{2,n}\frac{l^2}{2}$$

$$= \left[3 + \frac{\alpha(6-5\alpha)}{2\omega^2}l + \frac{\alpha^2}{2\omega^2}l^2\right]S_n^{[1]} - \left[3 + \frac{\alpha(5-4\alpha)}{\omega^2}l + \frac{\alpha^2}{\omega^2}l^2\right]S_n^{[2]}$$

$$+ \left[1 + \frac{\alpha(4-3\alpha)}{2\omega^2}l + \frac{\alpha^2}{2\omega^2}l^2\right]S_n^{[3]} \tag{3.48}$$

Since the forecasts can be expressed as a function of the first three smoothed statistics, this procedure is known under *triple exponential smoothing*.

3.7.1. Implementation of Triple Exponential Smoothing

Equation (3.48) implies that the forecasts from triple exponential smoothing can be obtained from the smoothed statistics $S_n^{[1]}$, $S_n^{[2]}$, and $S_n^{[3]}$. Thus it is not necessary to compute the discounted least squares estimates $\hat{\beta}_{0,n}, \hat{\beta}_{1,n}, \hat{\beta}_{2,n}$.

The smoothed statistics can be calculated recursively from period 1 onwards. This requires initial values $S_0^{[1]}$, $S_0^{[2]}$, and $S_0^{[3]}$.

Initial Values for the Smoothed Statistics

Initial values for the smoothed statistics can be found by first fitting a constant quadratic model

$$z_t = \beta_0 + \beta_1 t + \beta_2\frac{t^2}{2} + \varepsilon_t$$

to all past data $(z_1, z_2, \ldots, z_n)$. The quadratic model is parameterized such that $\boldsymbol{\beta}_0 = (\beta_0, \beta_1, \beta_2)'$ represents the coefficients at time origin zero. The least squares estimates are given by $\hat{\boldsymbol{\beta}}_0 = (\mathbf{X}'\mathbf{X})^{-1}\mathbf{X}'\mathbf{z}$, where the $3 \times n$ matrix $\mathbf{X}' = [\mathbf{f}(1), \ldots, \mathbf{f}(n)]$ and $\mathbf{z}' = (z_1, \ldots, z_n)$.

The least squares estimates $\hat{\boldsymbol{\beta}}_0 = (\hat{\beta}_0, \hat{\beta}_1, \hat{\beta}_2)'$ can then be substituted into Equations (3.47) and solved for $S_0^{[1]}$, $S_0^{[2]}$, and $S_0^{[3]}$. The solutions are given by

$$S_0^{[1]} = \hat{\beta}_0 - \frac{\omega}{\alpha}\hat{\beta}_1 + \frac{\omega(2-\alpha)}{2\alpha^2}\hat{\beta}_2$$

$$S_0^{[2]} = \hat{\beta}_0 - \frac{2\omega}{\alpha}\hat{\beta}_1 + \frac{2\omega(3-2\alpha)}{2\alpha^2}\hat{\beta}_2$$

$$S_0^{[3]} = \hat{\beta}_0 - \frac{3\omega}{\alpha}\hat{\beta}_1 + \frac{3\omega(4-3\alpha)}{2\alpha^2}\hat{\beta}_2$$

Choice of the Smoothing Constant

Brown (1962) suggests choosing ω such that $.70 < \omega^3 < .95$, which implies a value of ω between .89 and .98 (or $\alpha = 1 - \omega$ between .02 and .11). If past data are available, the smoothing constant can be found by simulation. The sum of squared one-step-ahead forecast errors can be calculated for various values of α (or $\omega = 1 - \alpha$). The value that minimizes this sum is used for future smoothing and forecasting.

There are very few genuine data sets that are best forecast by smoothing methods of orders higher than 2. The reason for this will become clear when we investigate the models that imply these procedures (Chap. 7). We will find that time series data do not usually follow these models.

Model Checking

Again, to check the adequacy of triple exponential smoothing, we recommend calculating past one-step-ahead forecast errors. If triple exponential smoothing is adequate, the forecast errors should have mean zero and furthermore be uncorrelated.

3.7.2. Extension to the General Polynomial Model and Higher Order Exponential Smoothing

Exponential smoothing in the context of the general polynomial model

$$z_{n+j} = \beta_0 + \beta_1 j + \beta_2 \frac{j^2}{2} + \cdots + \beta_k \frac{j^k}{k!} + \varepsilon_{n+j}$$

is discussed by Brown and Meyer (1961) and Brown (1962). They show that discounted least squares leads to coefficient estimates and forecasts that can be expressed as linear combinations of the first $k + 1$ smoothed statistics $S_n^{[1]}, \ldots, S_n^{[k+1]}$. Explicit expressions for $\hat{\boldsymbol{\beta}}_n$ and the forecasts, however, quickly become cumbersome and are not pursued further.

3.8. PREDICTION INTERVALS FOR FUTURE VALUES

We have shown in Section 3.5 that the discounted least squares estimate of $\boldsymbol{\beta}$ in the general model $z_{n+j} = \mathbf{f}'(j)\boldsymbol{\beta} + \varepsilon_{n+j}$ is given by

$$\hat{\boldsymbol{\beta}}_n = \mathbf{F}_n^{-1}\mathbf{h}_n$$

where

$$\mathbf{F}_n = \sum_{j=0}^{n-1} \omega^j \mathbf{f}(-j)\mathbf{f}'(-j) \qquad \text{and} \qquad \mathbf{h}_n = \sum_{j=0}^{n-1} \omega^j \mathbf{f}(-j) z_{n-j}$$

[see Eq. (3.25)]. From this estimate we can calculate the l-step-ahead forecast of a future observation z_{n+l},

$$\hat{z}_n(l) = \mathbf{f}'(l)\hat{\boldsymbol{\beta}}_n$$

To assess the uncertainty associated with this forecast, we have to derive the variance of the l-step-ahead forecast error

$$e_n(l) = z_{n+l} - \hat{z}_n(l) \tag{3.49}$$

The variance is given by

$$\begin{aligned} V[e_n(l)] &= V[z_{n+l} - \hat{z}_n(l)] = V(z_{n+l}) + V[\hat{z}_n(l)] \\ &= \sigma^2 + \mathbf{f}'(l)\mathbf{V}(\hat{\boldsymbol{\beta}}_n)\mathbf{f}(l) \end{aligned} \tag{3.50}$$

since $\hat{z}_n(l)$ is a function of the random variables $z_1, \ldots, z_n$, which in our model are assumed uncorrelated with z_{n+l}. To calculate this variance we have to derive the variance of the discounted least squares estimate $\hat{\boldsymbol{\beta}}_n$. The $m \times m$ covariance matrix of $\hat{\boldsymbol{\beta}}_n$ is given by

$$\mathbf{V}(\hat{\boldsymbol{\beta}}_n) = \mathbf{V}(\mathbf{F}_n^{-1}\mathbf{h}_n) = \mathbf{F}_n^{-1}\mathbf{V}(\mathbf{h}_n)\mathbf{F}_n^{-1} \tag{3.51}$$

In the variance derivations given in the literature [Brown (1962), Montgomery and Johnson (1976)], it is assumed that the observations z_t are uncorrelated and have the same variance σ^2 (see the comment given below). Then it follows that

$$\mathbf{V}(\mathbf{h}_n) = \mathbf{V}\left(\sum_{j=0}^{n-1} \omega^j \mathbf{f}(-j) z_{n-j}\right) = \sigma^2 \sum_{j=0}^{n-1} \omega^{2j} \mathbf{f}(-j)\mathbf{f}'(-j)$$

and

$$\mathbf{V}(\hat{\boldsymbol{\beta}}_n) = \sigma^2 \mathbf{F}_n^{-1}\left(\sum_{j=0}^{n-1} \omega^{2j} \mathbf{f}(-j)\mathbf{f}'(-j)\right)\mathbf{F}_n^{-1} \tag{3.52}$$

For large n, $\mathbf{F}_n$ and $\sum_{j=0}^{n-1}\omega^{2j}\mathbf{f}(-j)\mathbf{f}'(-j)$ approach steady-state values $\mathbf{F}$ and $\mathbf{F}_*$, respectively, and

$$\mathbf{V}(\hat{\boldsymbol{\beta}}_n) = \sigma^2 \mathbf{F}^{-1}\mathbf{F}_*\mathbf{F}^{-1} \tag{3.53}$$

Substitution of (3.53) into (3.50) leads to the variance of the l-step-ahead forecast error

$$V[e_n(l)] = \sigma^2 c_l^2 \tag{3.54}$$

where $c_l^2 = 1 + \mathbf{f}'(l)\mathbf{F}^{-1}\mathbf{F}_*\mathbf{F}^{-1}\mathbf{f}(l)$. Then a $100(1-\lambda)$ percent prediction interval for the future observation z_{n+l} is

$$\hat{z}_n(l) \pm u_{\lambda/2}\sigma c_l \tag{3.55}$$

where $u_{\lambda/2}$ is the $100(1-\lambda/2)$ percentage point of the standard normal distribution.

Comment

If we assume a model in which all observations have equal variance, we should calculate ordinary, and not discounted, least squares estimates. In (3.54) we have derived the variance of the discounted least squares estimates under the assumption that the observations have equal variance. Such an approach leads to correct standard errors only if the model (mean, trend, etc.) stays constant over time. For a large smoothing constant α (or small ω), which indicates rapid changes in the model, these variance approximations will be rather poor.

3.8.1. Prediction Intervals for Sums of Future Observations

Sometimes the forecaster is interested not only in a forecast of one single observation but also in forecasts of a sum of K future realizations. For example, if sales are recorded monthly, the forecaster might be interested in the forecast of next year's total sales.

The forecast of a sum of K future observations, $\sum_{l=1}^{K} z_{n+l}$, is given by the sum of their respective forecasts:

$$\sum_{l=1}^{K} \hat{z}_n(l) = \sum_{l=1}^{K} \mathbf{f}'(l)\hat{\boldsymbol{\beta}}_n = \left(\sum_{l=1}^{K} \mathbf{f}'(l)\right)\hat{\boldsymbol{\beta}}_n \tag{3.56}$$

The variance of the cumulative forecast error, $\sum_{l=1}^{K} z_{n+l} - \sum_{l=1}^{K} \hat{z}_n(l)$, can be written as

$$\begin{aligned} V\left(\sum_{l=1}^{K} z_{n+l} - \sum_{l=1}^{K} \hat{z}_n(l)\right) &= V\left(\sum_{l=1}^{K} z_{n+l}\right) + V\left(\sum_{l=1}^{K} \hat{z}_n(l)\right) \\ &= K\sigma^2 + V\left[\left(\sum_{l=1}^{K} \mathbf{f}'(l)\right)\hat{\boldsymbol{\beta}}_n\right] \\ &= K\sigma^2 + \left(\sum_{l=1}^{K} \mathbf{f}'(l)\right)\mathbf{V}(\hat{\boldsymbol{\beta}}_n)\left(\sum_{l=1}^{K} \mathbf{f}(l)\right) \\ &= \sigma^2\left[K + \left(\sum_{l=1}^{K} \mathbf{f}'(l)\right)\mathbf{F}^{-1}\mathbf{F}_*\mathbf{F}^{-1}\left(\sum_{l=1}^{K} \mathbf{f}(l)\right)\right] \end{aligned} \tag{3.57}$$

A $100(1 - \lambda)$ percent prediction interval for the sum of K future observations, $\sum_{l=1}^{K} z_{n+l}$, can be calculated from

$$\sum_{l=1}^{K} \mathbf{f}'(l)\hat{\boldsymbol{\beta}}_n \pm u_{\lambda/2}\sigma\left[K + \left(\sum_{l=1}^{K} f'(l)\right)\mathbf{F}^{-1}\mathbf{F}_*\mathbf{F}^{-1}\left(\sum_{l=1}^{K} f(l)\right)\right]^{1/2} \tag{3.58}$$

3.8.2. Examples

To illustrate these general expressions we consider now several special cases.

1. **Locally constant mean model:**

$$z_{n+j} = \beta + \varepsilon_{n+j}$$

In this model we have only one fitting function; $f_1(j) = 1$ for all j. This fitting function implies $F^{-1} = 1 - \omega$ and $F_* = \Sigma\omega^{2j} = 1/(1 - \omega^2)$. Substituting these expressions into Equation (3.53), we find that the variance of $\hat{\beta}_n = S_n = (1 - \omega)\Sigma\omega^j z_{n-j}$ is

$$V(\hat{\beta}_n) = \frac{(1 - \omega)^2}{1 - \omega^2}\sigma^2 = \frac{\alpha^2}{1 - \omega^2}\sigma^2$$

From (3.54) it follows that the variance of the l-step-ahead forecast error is given by

$$V[e_n(l)] = \sigma^2\left(1 + \frac{\alpha^2}{1 - \omega^2}\right) = \sigma^2\frac{2\alpha}{1 - \omega^2} \tag{3.59}$$

which is the same for all l. A $100(1 - \lambda)$ percent prediction interval for a future observation z_{n+l} is given by

$$S_n \pm u_{\lambda/2}\sigma\sqrt{\frac{2\alpha}{1 - \omega^2}} \tag{3.60}$$

The prediction intervals are constant for all l. This implies, for example, that the uncertainty associated with a 10-step-ahead forecast is the same as that of a one-step-ahead forecast. For models with rapidly changing mean (or ω close to zero), this will be a very poor approximation. In this case the observations are described by a random walk $z_n = z_{n-1} + \varepsilon_n$, and the l-step-ahead forecast is $\hat{z}_n(l) = z_n$. Then the variance of the l-step-ahead forecast error $z_{n+l} - \hat{z}_n(l) = z_{n+l} - z_n = \varepsilon_{n+l} + \cdots + \varepsilon_{n+1}$ is given by $V[z_{n+l} - \hat{z}_n(l)] = l\sigma^2$. The forecast error variance increases as a linear function of the forecast lead time l. A detailed discussion of forecast error variances will be given in Chapter 5.

Similarly, it can be shown from Equation (3.58) that a $100(1 - \lambda)$ prediction interval for the sum of K future observations is given by

$$KS_n \pm u_{\lambda/2}\sigma\left[K\left(1 + K\frac{\alpha^2}{1 - \omega^2}\right)\right]^{1/2} \tag{3.61}$$

2. Locally constant linear trend model:

$$z_{n+j} = \beta_0 + \beta_1 j + \varepsilon_{n+j}$$

In this model we have $m = 2$ fitting functions, $f_1(j) = 1$ and $f_2(j) = j$. It was shown earlier (see Sec. 3.6) that

$$\mathbf{F}^{-1} = \begin{bmatrix} 1-\omega^2 & (1-\omega)^2 \\ (1-\omega)^2 & \dfrac{(1-\omega)^3}{\omega} \end{bmatrix}$$

Furthermore,

$$\mathbf{F}_* = \sum_{j \geq 0} \omega^{2j} \begin{bmatrix} 1 & -j \\ -j & j^2 \end{bmatrix} = \begin{bmatrix} \dfrac{1}{1-\omega^2} & \dfrac{-\omega^2}{(1-\omega^2)^2} \\ \dfrac{-\omega^2}{(1-\omega^2)^2} & \dfrac{\omega^2(1+\omega^2)}{(1-\omega^2)^3} \end{bmatrix}$$

After some algebra it can be shown that the 2×2 covariance matrix of $\hat{\boldsymbol{\beta}}_n = (\hat{\beta}_{0,n}, \hat{\beta}_{1,n})'$ reduces to

$$\begin{aligned} \mathbf{V}(\hat{\boldsymbol{\beta}}_n) &= \sigma^2 \mathbf{F}^{-1} \mathbf{F}_* \mathbf{F}^{-1} \\ &= \sigma^2 \frac{1-\omega}{(1+\omega)^3} \begin{bmatrix} 1+4\omega+5\omega^2 & (1-\omega)(1+3\omega) \\ (1-\omega)(1+3\omega) & 2(1-\omega)^2 \end{bmatrix} \end{aligned}$$

With substitution of the above result into Equation (3.54), the variance of the l-step-ahead forecast error for the locally constant linear trend model simplifies to

$$V[e_n(l)] = \sigma^2 c_l^2 \tag{3.62}$$

where

$$c_l^2 = 1 + \frac{1-\omega}{(1+\omega)^3} \Big[(1+4\omega+5\omega^2) + 2l(1-\omega)(1+3\omega) + 2l^2(1-\omega)^2 \Big]$$

A $100(1-\lambda)$ percent prediction interval for z_{n+l} is therefore given by

$$\left[\left(2 + \frac{1-\omega}{\omega} l\right) S_n^{[1]} - \left(1 + \frac{1-\omega}{\omega} l\right) S_n^{[2]} \right] \pm u_{\lambda/2} \sigma c_l \tag{3.63}$$

3.8.3. Estimation of the Variance

The variance of the forecast errors in (3.54) and the prediction intervals (3.55) include σ^2, the unknown population variance of the errors ε_t. This unknown variance can be estimated from the sample variance of the one-step-ahead forecast errors.

It was shown earlier [Eq. (3.54)] that the variance of the ($l = 1$)-step-ahead forecast error is given by

$$\sigma_e^2 = V\left[e_n(1)\right] = \sigma^2\left[1 + \mathbf{f}'(1)\mathbf{F}^{-1}\mathbf{F}_*\mathbf{F}^{-1}\mathbf{f}(1)\right] = \sigma^2 c_1^2$$

Then it follows that

$$\sigma^2 = \frac{\sigma_e^2}{c_1^2} \tag{3.64}$$

where $c_1^2 = [1 + \mathbf{f}'(1)\mathbf{F}^{-1}\mathbf{F}_*\mathbf{F}^{-1}\mathbf{f}(1)]$.

The observed forecast errors $e_{t-1}(1) = z_t - \hat{z}_{t-1}(1)$ ($t = 1, 2, \ldots, n$) can be used to estimate the variance of the one-step-ahead forecast errors. If the model is correct, these errors have mean zero, and hence the variance estimate is given by

$$\hat{\sigma}_e^2 = \frac{\sum_{t=1}^{n}\left[z_t - \hat{z}_{t-1}(1)\right]^2}{n} \tag{3.65}$$

A mean correction in (3.65) is not needed, since a correct model will lead to unbiased forecasts. This expression is substituted into Equation (3.64) to get an estimate of σ^2:

$$\hat{\sigma}^2 = \frac{\hat{\sigma}_e^2}{c_1^2} \tag{3.66}$$

Substituting (3.66) into (3.55) we find that an estimated $100(1 - \lambda)$ percent prediction interval for a future observation z_{n+l} is given by

$$\hat{z}_n(l) \pm u_{\lambda/2}\frac{\hat{\sigma}_e}{c_1}\left[1 + \mathbf{f}'(l)\mathbf{F}^{-1}\mathbf{F}_*\mathbf{F}^{-1}\mathbf{f}(l)\right]^{1/2} \tag{3.67}$$

Similarly, it follows from (3.58) that the prediction interval for a sum of future observations $\sum_{l=1}^{K} z_{n+l}$ is

$$\sum_{l=1}^{K} \hat{z}_n(l) \pm u_{\lambda/2}\frac{\hat{\sigma}_e}{c_1}\left[K + \left(\sum_{l=1}^{K}\mathbf{f}'(l)\right)\mathbf{F}^{-1}\mathbf{F}_*\mathbf{F}^{-1}\left(\sum_{l=1}^{K}\mathbf{f}(l)\right)\right]^{1/2} \tag{3.68}$$

Example 1: Simple Exponential Smoothing

In the case of simple exponential smoothing, it follows from (3.59) that $c_l^2 = 2\alpha/(1 - \omega^2)$. Thus, for any l the estimated prediction interval for z_{n+l} is

$$S_n \pm u_{\lambda/2}\hat{\sigma}_e \tag{3.69}$$

From Section 3.3, we recall that the Iowa nonfarm income growth rates were best predicted by simple exponential smoothing with a smoothing constant $\alpha = .11$. The prediction for all future growth rates was $\hat{z}_{127}(l) = 2.654$. The sum of the squared one-step-ahead forecast errors was

$$\sum_{t=1}^{127} (z_t - S_{t-1})^2 = 118.19.$$

Thus $\hat{\sigma}_e^2 = 118.19/127 = .9306$, and $\hat{\sigma}_e = .965$. A 95 percent prediction interval for all future growth rates z_{127+l} is given by $2.654 \pm (1.96)(.965)$ or $(.763, 4.545)$.

Example 2: Double Exponential Smoothing

For double exponential smoothing, the estimated prediction interval for z_{n+l} is given by

$$\left[\left(2 + \frac{1-\omega}{\omega} l\right) S_n^{[1]} - \left(1 + \frac{1-\omega}{\omega} l\right) S_n^{[2]}\right] \pm u_{\lambda/2}\frac{\hat{\sigma}_e}{c_1} c_l \tag{3.70}$$

where

$$c_l^2 = 1 + \frac{1-\omega}{(1+\omega)^3}\Big[(1 + 4\omega + 5\omega^2)$$

$$+ 2l(1-\omega)(1+3\omega) + 2l^2(1-\omega)^2\Big]$$

We illustrate the calculation of the prediction intervals using the thermostat sales series that was analyzed in Section 3.6. With a smoothing constant $\alpha = .14$, the double exponential smoothing forecasts were $\hat{z}_{52}(1) = 319.29$, $\hat{z}_{52}(2) = 323.51$, $\hat{z}_{52}(3) = 327.73$, and so on. Furthermore, the sum of the squared one-step-ahead forecast errors was 41,469. Therefore, $\hat{\sigma}_e^2 = 41{,}469/52 = 797.48$ and $\hat{\sigma}_e = 28.24$. Evaluating c_l for $l = 1, 2, 3$ and $\alpha = .14$ $(\omega = .86)$ leads to $c_1 = 1.095$, $c_2 = 1.106$, $c_3 = 1.118$. Thus the 95 percent prediction intervals are:

One step ahead (z_{53}):

$$319.29 \pm (1.96)(28.24) \quad \text{or} \quad (263.94, 374.64)$$

Two steps ahead (z_{54}):

$$323.51 \pm (1.96)(28.24)(1.106/1.095) \quad \text{or} \quad (267.60, 379.42)$$

Three steps ahead (z_{55}):

$$327.73 \pm (1.96)(28.24)(1.118/1.095) \quad \text{or} \quad (271.23, 384.23)$$

Updating Probability Intervals

The probability intervals for future observations are easily updated as each new observation becomes available. The estimated variance of the one-step-ahead forecast errors at time $n + 1$, $\hat{\sigma}_e^2(n + 1)$, can be expressed as a function of the previous variance estimate and the most recent one-step-ahead forecast error:

$$\hat{\sigma}_e^2(n+1) = \frac{\sum_{t=1}^{n+1} [z_t - \hat{z}_{t-1}(1)]^2}{n+1} = \frac{n\hat{\sigma}_e^2(n) + [z_{n+1} - \hat{z}_n(1)]^2}{n+1}$$

3.8.4. An Alternative Variance Estimate

Equation (3.65) is used to estimate the variance of the one-step-ahead forecast errors from the corresponding sample variance. An alternative estimate that is more convenient for updating can be calculated from the mean absolute deviation of the one-step-ahead forecast errors [see Montgomery and Johnson (1976)]. It is given by

$$\hat{\sigma}_e = 1.25\hat{\Delta}_e \tag{3.71}$$

where

$$\hat{\Delta}_e = \frac{\sum_{t=1}^{n} |e_{t-1}(1)|}{n} = \frac{\sum_{t=1}^{n} |z_t - \hat{z}_{t-1}(1)|}{n}$$

Substituting this estimate into Equations (3.55) and (3.58) leads to alternative prediction intervals.

The prediction intervals are easily updated as each new observation becomes available, since the mean absolute deviation at time $n + 1$, $\hat{\Delta}_e(n + 1)$, can be expressed in terms of the previous estimate and the most recent forecast error:

$$\hat{\Delta}_e(n+1) = \frac{\sum_{t=1}^{n+1} |z_t - \hat{z}_{t-1}(1)|}{n+1} = \frac{n\hat{\Delta}_e(n) + |z_{n+1} - \hat{z}_n(1)|}{n+1}$$

The justification for this alternative variance estimate is given below. Let us assume that a random variable X follows a normal distribution with mean μ and variance σ^2. Then the expected value of the absolute deviation (mean absolute deviation) is given by

$$\Delta = E|X - \mu| = \int_{-\infty}^{+\infty} |x - \mu| f(x)\, dx$$

$$= \int_{-\infty}^{+\infty} |x - \mu| \frac{1}{\sigma\sqrt{2\pi}} \exp\left[-\frac{1}{2\sigma^2}(x-\mu)^2\right] dx$$

$$= \frac{1}{\sigma}\sqrt{\frac{2}{\pi}} \int_0^{\infty} y \exp\left(-\frac{1}{2\sigma^2} y^2\right) dy = \sigma\sqrt{\frac{2}{\pi}}$$

If we assume that the one-step-ahead forecast errors are normally distributed, then it follows that

$$\Delta_e = \sigma_e\sqrt{\frac{2}{\pi}} \qquad \text{or} \qquad \sigma_e = \Delta_e\sqrt{\frac{\pi}{2}} \simeq 1.25\Delta_e$$

This is a reasonable approximation, even if the errors are nonnormal.

3.9. FURTHER COMMENTS

Exponential smoothing procedures have received considerable attention in the business forecasting literature. Two main reasons for the popularity of these techniques among routine business forecasters are:

1. Easy updating relationships allow the forecaster to update the forecasts without storing all past observations; only the most recent smoothed statistics have to be stored.
2. These procedures are said to be automatic and easy to use.

The claim that exponential smoothing procedures are automatic is true only if one has already decided on the order of smoothing and on the value of the smoothing constant. The order of smoothing is usually decided after an ad hoc visual inspection of the data. For observations with slowly changing mean, first-order (or simple) exponential smoothing is suggested. Observations that increase (decrease) linearly over time are usually predicted by double exponential smoothing. A value of α between .05 and .30 is usually suggested for the smoothing constant. A value of .1 seems to be preferred in many forecasting textbooks.

It should be pointed out that a visual inspection of the data alone can lead to incorrect conclusions about the order of exponential smoothing (or equivalently the order of the locally constant polynomial trend model). For example, stock price data, which are known to follow random walks ($z_t = z_{t-1} + \varepsilon_t$), can sometimes give the impression of local linear and quadratic trends.

Furthermore, it should be emphasized that a smoothing constant $\alpha = .1$ (or discount coefficient $\omega = .9$) will not always lead to good forecasts. For the University of Iowa student enrollments, for example, it was found that the estimated ω in double exponential smoothing is close to zero. This essentially implies a stochastic trend model that remembers only the two most recent observations.

In Chapter 7 we take another look at exponential smoothing methods and gain additional insights into why these procedures work well in some instances but perform poorly in other cases. There we show that it depends on the underlying stochastic process whether exponential smoothing will lead to good forecasts. For exponential smoothing to perform well, the stochastic process has to be from a special, particularly restricted, subclass of the stochastic models, which are discussed in Chapter 5. After learning more about this subclass, we can ask whether these restricted models are more likely to occur than others. We will learn that there are no good reasons why real series should follow these restricted models. In particular, models that imply higher order exponential smoothing are rarely found in practice.

CHAPTER 4

Regression and Exponential Smoothing Methods to Forecast Seasonal Time Series

Many time series exhibit a cyclical pattern that has the tendency to repeat itself over a certain fixed period of time. We refer to such a tendency as *seasonality* and to the length of the cycle as seasonal period s. For example, for monthly series with a yearly seasonal pattern, the seasonal period is $s = 12$; for quarterly series it is $s = 4$.

4.1. SEASONAL SERIES

Series that contain seasonal components are quite common, especially in economics, business, and the natural sciences. Seasonal economic series, for example, are those concerned with production, sales, inventories, employment, government receipts and expenditures, and personal income. For illustration we have plotted five seasonal series: monthly electricity demand in Iowa City (Fig. 4.1), monthly traffic fatalities in Ontario (Fig. 4.2), monthly U.S. single-family housing starts (Fig. 4.3), monthly new car sales in Quebec (Fig. 4.4), and quarterly expenditures for new plant and equipment in U.S. industries (Fig. 4.5). Additional examples will be given in Chapter 6.

Much of the seasonality can be explained on the basis of physical reasons. The earth's rotation around the sun, for example, introduces a yearly seasonal pattern into many of the meteorological variables. This in

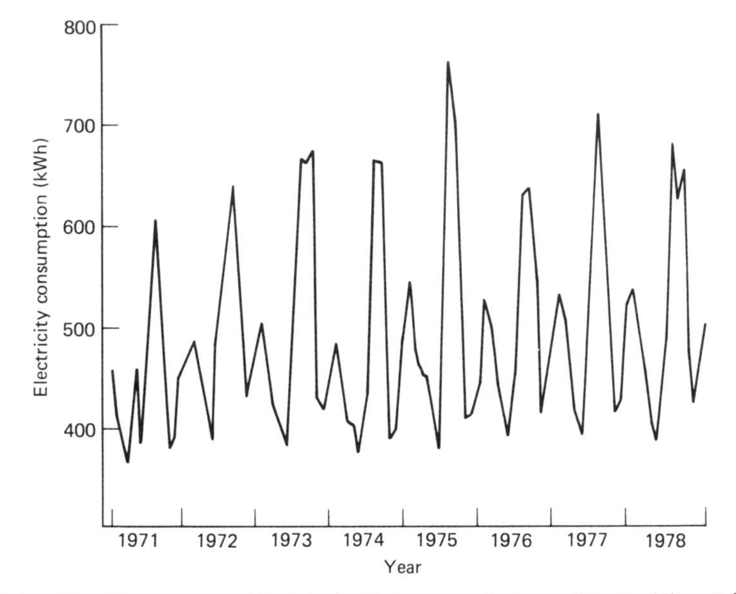

Figure 4.1. Monthly average residential electricity usage in Iowa City (in kilowatt-hours), January 1971 to December 1978.

turn leads to seasonality in series such as electricity demand, which is high during the summer period in Iowa, owing to air-conditioning. Housing starts are lower during the winter season, since adverse meteorological conditions limit winter construction in many parts of the country. Meteorological conditions also affect the demand for many seasonal apparel items. Seasonality in traffic volume and thus, indirectly, traffic fatalities; seasonality in car sales; and seasonality in expenditures for new plant and equipment can also be explained from institutional time tables, such as the timing of holidays and vacations, the timing of the introduction of new car models, and tax-filing deadlines.

The seasonal pattern in certain variables, such as the one in meteorological variables, is usually quite stable and deterministic and repeats itself year after year. The seasonal pattern in business and economic series, however, is frequently stochastic and changes with time. The stochastic nature of the seasonality in economic and business series can be explained by the fact that these series are influenced by many factors that do not repeat themselves exactly every season.

Apart from a seasonal component, we observe in many series an additional trend component. In some of the examples (Fig. 4.4, for example; see Example 4.1) the trend is fairly constant; in others (Figs. 4.3, 4.5), however,

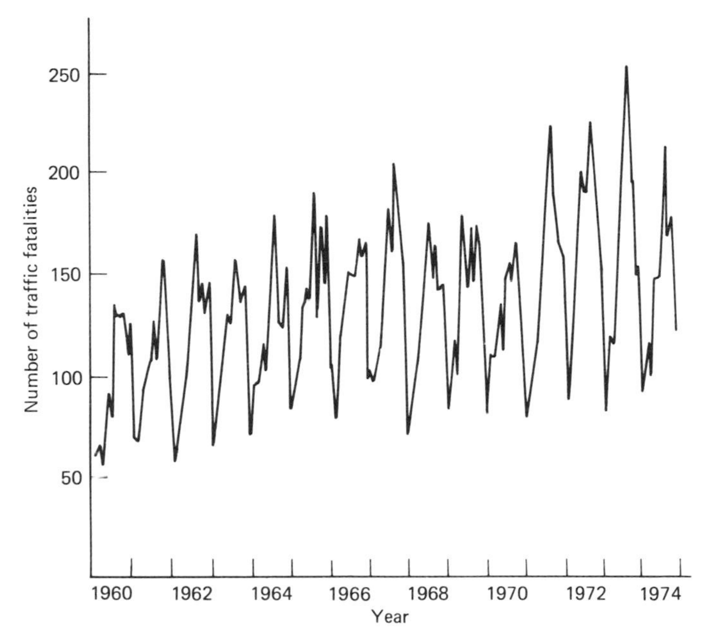

Figure 4.2. Monthly traffic fatalities in Ontario, January 1960 to December 1974.

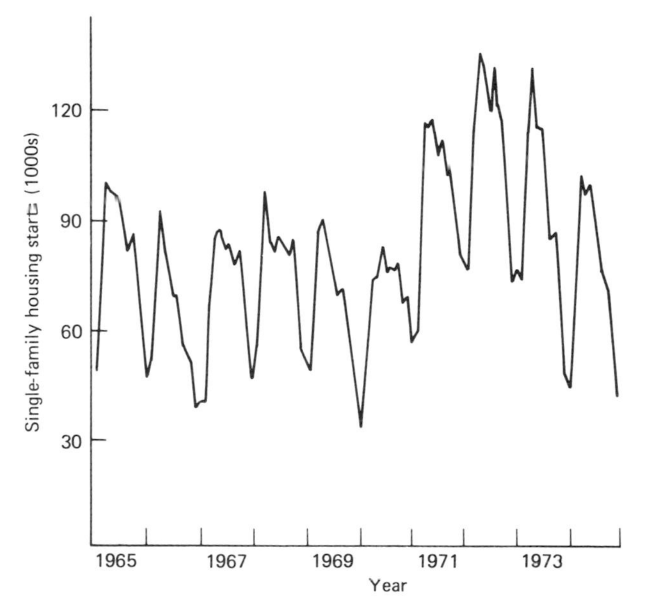

Figure 4.3. Monthly housing starts of privately owned single-family structures, January 1965 to December 1974.

the trend changes with time. In several of the series (Figs. 4.2, 4.5), the seasonal swings increase with the level of the series; in others (Figs. 4.3, 4.4) the magnitude of the seasonal swings appears to be unaffected by the level.

The traditional approach to modeling seasonal data is to decompose the series into three components: a trend T_t, a seasonal component S_t, and an irregular (or error) component ε_t.

The *additive decomposition* approach models the seasonal series as a sum of these three components,

$$z_t = T_t + S_t + \varepsilon_t \tag{4.1}$$

The trend component is frequently modeled by low-order polynomials of time t. The seasonal component is modeled by either seasonal indicators or combinations of trigonometric functions. The irregular components ε_t are usually assumed to be uncorrelated. Additive decomposition is appropriate if the seasonal swings and the variation in the errors do not depend on the level of the series.

The *multiplicative decomposition* approach specifies the model to be of the form

$$z_t = T_t \times S_t \times \varepsilon_t \tag{4.2}$$

In this model the seasonal swings depend on the level of the series; furthermore, the irregular component constitutes a multiplicative, or percentage, error. This model, however, is readily transformed into an additive model by considering the logarithmic transformation

$$\log z_t = T_t^* + S_t^* + \varepsilon_t^*$$

The model in (4.2) is not the only possible multiplicative model. One could also consider decompositions of the form

$$z_t = T_t \times S_t + \varepsilon_t \tag{4.3}$$

In this formulation the seasonal component is proportional to the level; the error component, however, is additive. Such a decomposition is considered in Section 4.4.

In Section 4.2 (globally constant seasonal models) we discuss the forecasts from seasonal decompositions in which the parameters are assumed to be constant. In Section 4.3 (locally constant seasonal models) we relax this assumption and allow for changes in the parameters. We adopt a discounted least squares or general exponential smoothing approach to estimate the parameters. In Section 4.4 we discuss Winters' forecasting methods for additive and multiplicative seasonal models. In Section 4.5 an elementary introduction to seasonal adjustment procedures is given. The Census X-11 program, which is the official U.S. seasonal adjustment method, is described.

4.2. GLOBALLY CONSTANT SEASONAL MODELS

We start our discussion with the traditional additive decomposition of the series into trend T_t, seasonal S_t, and irregular ε_t components:

$$z_t = T_t + S_t + \varepsilon_t \tag{4.4}$$

It was pointed out in Section 4.1 that also the multiplicative model in (4.2) can be transformed into such a representation.

Traditionally the *trend component* T_t is modeled by low-order polynomials of time t:

$$T_t = \beta_0 + \sum_{i=1}^{k} \beta_i \frac{t^i}{i!} \tag{4.5}$$

Usually the order of the polynomial, k, is small; either $k = 1$ (linear trend: $T_t = \beta_0 + \beta_1 t$) or $k = 2$ [quadratic trend: $T_t = \beta_0 + \beta_1 t + \beta_2(t^2/2)$]. The reason for parameterizing the trend model in terms of $\beta_i/i!$ was explained in Chapter 3. There it was shown that this particular parameterization simplifies the difference equation representation of the fitting functions.

The *seasonal component* S_t can be described by seasonal indicators

$$S_t = \sum_{i=1}^{s} \delta_i \text{IND}_{ti} \tag{4.6}$$

where $\text{IND}_{ti} = 1$ if t corresponds to the seasonal period i, and 0 otherwise, or by trigonometric functions

$$S_t = \sum_{i=1}^{m} A_i \sin\left(\frac{2\pi i}{s}t + \phi_i\right) \tag{4.7}$$

where A_i and ϕ_i are the amplitude and the phase shift of the sine function with frequency $f_i = 2\pi i/s$. In some situations, the parameterization in (4.7) may be more parsimonious, since it can involve fewer parameters than the s coefficients in the seasonal indicator representation.

The *error* components $\{\varepsilon_t\}$ are usually assumed to be uncorrelated and identically distributed with mean zero and constant variance σ^2. In addition, it is usually assumed that they are from a normal distribution.

Combining these three components leads to seasonal models of the form

$$z_t = \beta_0 + \sum_{i=1}^{k} \beta_i \frac{t^i}{i!} + \sum_{i=1}^{s} \delta_i \,\mathrm{IND}_{ti} + \varepsilon_t \tag{4.8}$$

or

$$z_t = \beta_0 + \sum_{i=1}^{k} \beta_i \frac{t^i}{i!} + \sum_{i=1}^{m} A_i \sin\left(\frac{2\pi i}{s} t + \phi_i\right) + \varepsilon_t \tag{4.9}$$

In the following sections we study these models in more detail.

4.2.1. Modeling the Seasonality with Seasonal Indicators

In model (4.8) we have introduced an intercept β_0 and s seasonal indicators. Such a representation, however, leads to an indeterminacy, since it uses $s + 1$ parameters to model s seasonal intercepts. Restrictions have to be imposed before the parameters can be estimated. Several equivalent parameterizations are possible:

1. Omit the intercept; $\beta_0 = 0$.
2. Restrict $\sum_{i=1}^{s} \delta_i = 0$.
3. Set one of the δ's equal to zero; for example, $\delta_s = 0$.

If the intercept β_0 is omitted, the δ_i represent the seasonal intercepts of the s seasonal parallel trend lines. If $\sum_{i=1}^{s} \delta_i = 0$, the parameters δ_i represent the seasonal effects as compared to an average trend line. If $\delta_s = 0$, the coefficient δ_i $(i \neq s)$ represents the seasonal effect of the ith seasonal period as compared to the seasonal period s.

The three parameterizations are mathematically equivalent, and it is easy to go from one to another. For example, the coefficients under parameterization 2, (β^*, δ^*), can be obtained from the ones in parameterization 1, (β, δ), by

$$\beta_0^* = \frac{\sum_{i=1}^{s} \delta_i}{s} \qquad \beta_i^* = \beta_i \qquad (1 \leqslant i \leqslant k)$$

$$\delta_i^* = \delta_i - \beta_0^* \qquad (1 \leqslant i \leqslant s)$$

From parameterization 3, $(\beta^{**}, \delta^{**})$, they can be obtained by:

$$\beta_0^* = \beta_0^{**} + \frac{\sum_{i=1}^{s-1} \delta_i^{**}}{s} \qquad \beta_i^* = \beta_i^{**} \qquad (1 \leqslant i \leqslant k)$$

$$\delta_s^* = -\frac{\sum_{i=1}^{s-1} \delta_i^{**}}{s} \qquad \delta_i^* = \delta_i^{**} - \frac{\sum_{j=1}^{s-1} \delta_j^{**}}{s} \qquad (1 \leqslant i \leqslant s-1)$$

In the following discussion we adopt parameterization 3 and omit δ_s in the model (4.8). Assuming that the parameters in the model

$$z_t = \beta_0 + \sum_{i=1}^{k} \beta_i \frac{t^i}{i!} + \sum_{i=1}^{s-1} \delta_i \, \mathrm{IND}_{ti} + \varepsilon_t \tag{4.10}$$

are constant, we can use ordinary least squares to estimate the parameters and to calculate the forecasts and the prediction intervals. It follows from Section 2.3 that the least squares estimates of $\boldsymbol{\beta} = (\beta_0, \beta_1, \ldots, \beta_k, \delta_1, \ldots, \delta_{s-1})'$ are given by

$$\hat{\boldsymbol{\beta}} = (\mathbf{X}'\mathbf{X})^{-1}\mathbf{X}'\mathbf{z} \tag{4.11}$$

where $\mathbf{z} = (z_1, z_2, \ldots, z_n)'$ is the vector of the available observations and $\mathbf{X}$ is an $n \times (k+s)$ matrix with the tth row given by

$$\mathbf{f}'(t) = \left(1, t, \frac{t^2}{2}, \ldots, \frac{t^k}{k!}, \mathrm{IND}_{t1}, \ldots, \mathrm{IND}_{t,s-1}\right)$$

The minimum mean square error forecast of z_{n+l} can be calculated from

$$\hat{z}_n(l) = \mathbf{f}'(n+l)\hat{\boldsymbol{\beta}} \tag{4.12}$$

A $100(1-\lambda)$ percent prediction interval is given by

$$\hat{z}_n(l) \pm t_{\lambda/2}(n-k-s)\hat{\sigma}\left[1 + \mathbf{f}'(n+l)(\mathbf{X}'\mathbf{X})^{-1}\mathbf{f}(n+l)\right]^{1/2} \tag{4.13}$$

where

$$\hat{\sigma}^2 = \frac{1}{n-k-s} \sum_{t=1}^{n} \left[z_t - \mathbf{f}'(t)\hat{\boldsymbol{\beta}}\right]^2$$

is the regression mean square error.

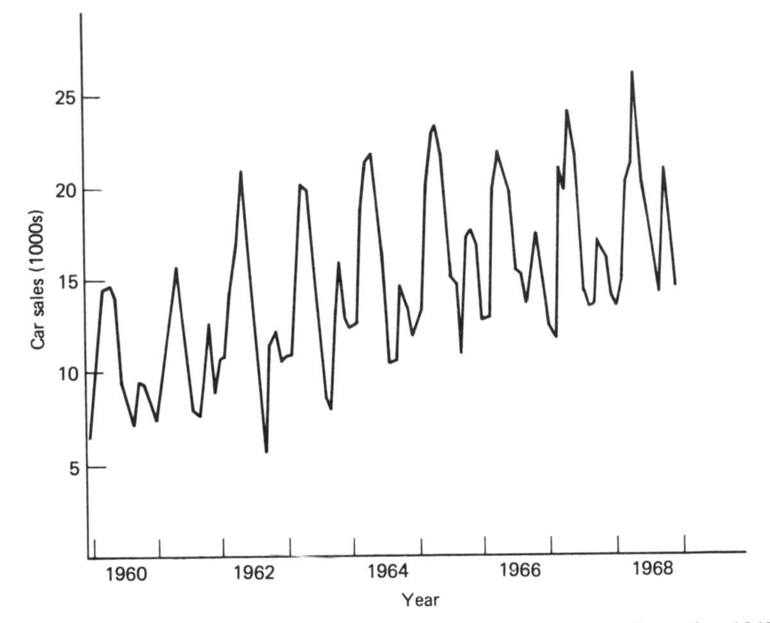

Figure 4.4. Monthly car sales in Quebec, Canada; January 1960 to December 1968.

Alternatively, models under parameterization 1 or 2 can be fitted. Then the vector of fitting functions $\mathbf{f}(t)$ is given by

$$\mathbf{f}(t) = \left(t, \frac{t^2}{2}, \ldots, \frac{t^k}{k!}, \mathrm{IND}_{t1}, \ldots, \mathrm{IND}_{ts}\right)'$$

for parameterization 1, and

$$\mathbf{f}(t) = \left(1, t, \frac{t^2}{2}, \ldots, \frac{t^k}{k!}, \mathrm{IND}^*_{t1}, \ldots, \mathrm{IND}^*_{t,s-1}\right)'$$

for parameterization 2, where $\mathrm{IND}^*_{ti} = \mathrm{IND}_{ti} - \mathrm{IND}_{ts}$ for $1 \leqslant i \leqslant s-1$.

Example 4.1: Car Sales

To illustrate the seasonal indicator model, we consider the monthly car sales in Quebec from January 1960 through December 1967 ($n = 96$ observations). The remaining 12 observations (1968) are used as a holdout period to evaluate the forecast performance. The data are listed as Series 4 in the Data Appendix. An initial inspection of the series in Figure 4.4 shows that the data may be described by an additive model with a linear trend ($k = 1$) and a yearly seasonal pattern; the trend and the seasonal components appear fairly constant.

The regression output (including parameter estimates, their standard errors, the ANOVA table, and the autocorrelation function of the residuals) for the model,

$$z_t = \beta_0 + \beta_1 t + \sum_{i=1}^{11} \delta_i \mathrm{IND}_{ti} + \varepsilon_t \tag{4.14}$$

is given in Table 4.1. In the same table we list the forecasts and the forecast errors for the next 12 periods ($t = 97, \ldots, 108$).

To investigate whether model assumptions (i.e., the functional form of the model, the constancy of the trend and the seasonal component, uncorrelated errors) are violated, we focus on the residuals $e_t = z_t - \mathbf{f}'(t)\hat{\boldsymbol{\beta}}$ and calculate their sample autocorrelations. These are also given in Table 4.1. Approximate standard errors for these autocorrelations are $n^{-1/2} = 96^{-1/2} = .10$. We find that the lag 1 autocorrelation is significant. This puts the adequacy of the globally constant model in question, since the model fails to extract all information from the data. Thus the forecaster should look for a different method to predict these data. We will return to this data set in Section 4.3 when we discuss locally constant seasonal models and seasonal exponential smoothing.

Example 4.2: New Plant and Equipment Expenditures

As an additional example we consider quarterly new plant and equipment expenditures for the first quarter of 1964 through the fourth quarter of 1974 ($n = 44$). The time series plot in Figure 4.5 indicates that the size of the seasonal swings increases with the level of the series; hence a logarithmic transformation must be considered. Figure 4.6 shows that this transformation has stabilized the variance.

An additive seasonal model of the form

$$z_t = \ln y_t = \beta_0 + \beta_1 t + \sum_{i=1}^{3} \delta_i \mathrm{IND}_{ti} + \varepsilon_t$$

is fitted to the transformed data. Figure 4.6 indicates that a globally constant linear trend model will certainly not be appropriate for this series. The trend component is not fixed, but changes over time. Nevertheless, we proceed with this model and estimate the parameters by ordinary least squares, mainly to illustrate how a careful residual analysis can point to violations in the model assumptions.

Table 4.1. Least Squares Estimation and Forecasting Results for Quebec Monthly Car Sales (1960–1967; $n = 96$)

$$z_t = \beta_0 + \beta_1 t + \sum_{i=1}^{11} \delta_i \mathrm{IND}_{ti} + \varepsilon_t$$

Coefficient	Estimate	Standard Error	t Ratio
β_0	7.40	.59	12.62
β_1	.088	.0054	16.45
δ_1	−.60	.72	−.84
δ_2	−.05	.72	−.07
δ_3	5.34	.72	7.38
δ_4	7.52	.72	10.40
δ_5	8.69	.72	12.03
δ_6	6.31	.72	8.73
δ_7	1.41	.72	1.95
δ_8	−.87	.72	−1.21
δ_9	−2.28	.72	−3.16
δ_{10}	1.79	.72	2.48
δ_{11}	2.36	.72	3.27

ANOVA Table

Source	SS	df	MS
Regression	1671.8	12	139.32
Error	172.9	83	2.08
Total (corrected for mean)	1844.7	95	

Autocorrelations of the Residuals

Lag k	r_k	Lag k	r_k
1	.29	7	−.02
2	.20	8	−.02
3	.12	9	.04
4	.11	10	.02
5	.08	11	.17
6	.14	12	.09

Forecasts

Lead Time l

l	1	2	3	4	5	6
z_{96+l}	13.21	14.25	20.14	21.73	26.10	21.08
$\hat{z}_{96}(l)$	15.34	15.99	21.46	23.73	24.99	22.70
$z_{96+l} - \hat{z}_{96}(l)$	−2.13	−1.74	−1.32	−2.00	1.11	−1.62

l	7	8	9	10	11	12
z_{96+l}	18.02	16.72	14.39	21.34	17.18	14.58
$\hat{z}_{96}(l)$	17.89	15.69	14.37	18.53	19.19	16.92
$z_{96+l} - \hat{z}_{96}(l)$	.13	1.03	.02	2.81	−2.01	−2.34

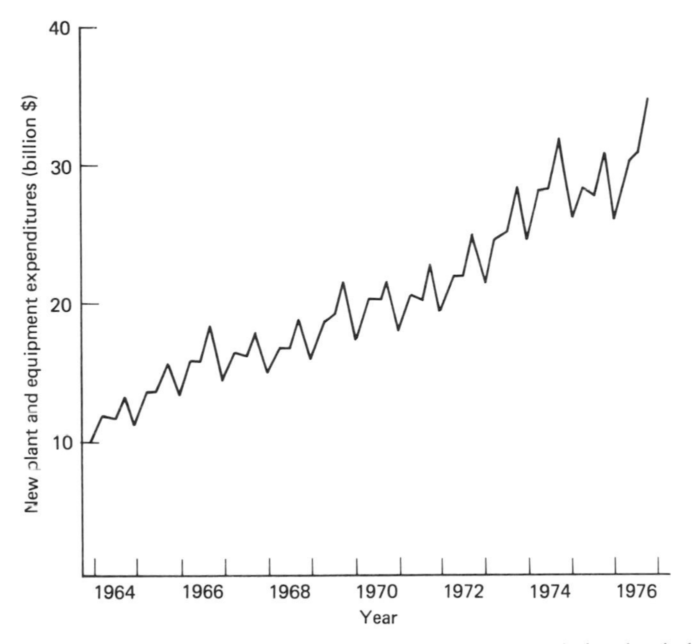

Figure 4.5. Quarterly new plant and equipment expenditures in U.S. industries (in billions of dollars), first quarter 1964 to fourth quarter 1976.

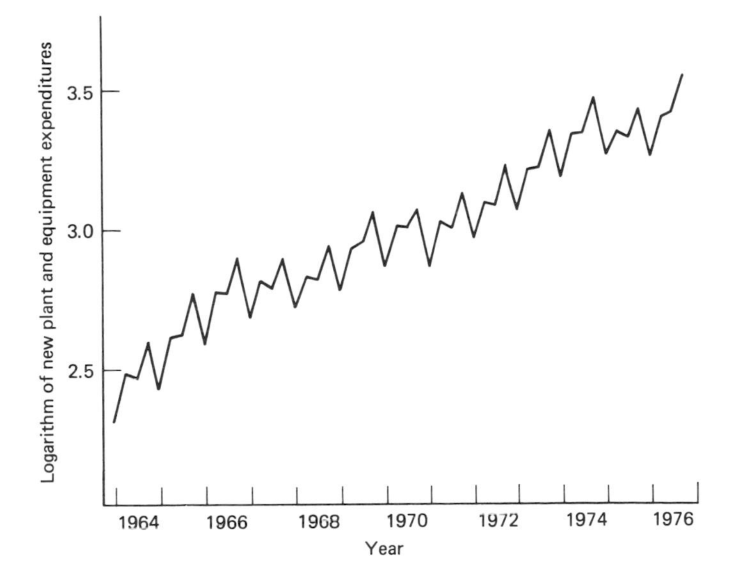

Figure 4.6. Logarithm of quarterly new plant and equipment expenditures, first quarter 1964 to fourth quarter 1976.

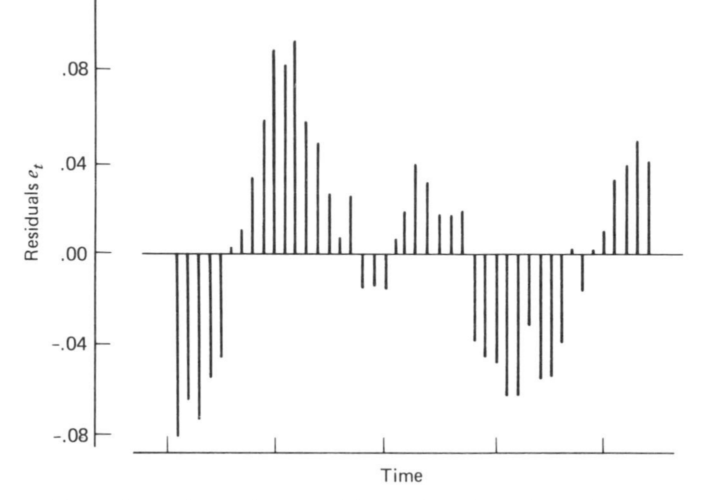

Figure 4.7. Time series plot of the residuals from the model

$$\ln y_t = \beta_0 + \beta_1 t + \sum_{i=1}^{3} \delta_i \mathrm{IND}_{ti} + \varepsilon_t$$

New plant and equipment expenditures.

The fitting results are summarized in Table 4.2. The time series plot of the residuals in Figure 4.7 indicates that the model overpredicts at the beginning and toward the end and underpredicts in the middle of the series. The sample autocorrelations of the residuals are large and highly significant, and illustrate the inappropriateness of the globally constant linear trend model. A model with a quadratic trend component $\beta_0 + \beta_1 t + \beta_2 t^2/2$ was also tried. The least squares estimate of β_2, however, was insignificant, and the residuals and their autocorrelations were similar to the ones reported in Table 4.2.

Globally constant models are not appropriate for this data series. The trend parameter β_1 especially appears to be changing over time. Thus the assumption of globally constant parameters should be relaxed, and smoothing or exponentially discounted least squares methods should be considered.

Change of Time Origin in the Seasonal Indicator Model

The trend coefficients in model (4.10) are expressed in relation to time origin zero. As in Section 3.4, we can shift the time origin to n, where n is the time period of the last available observation. Then the model can be

Table 4.2. Least Squares Estimation Results—New Plant and Equipment Expenditures (1964–1974; $n = 44$)

$$\ln y_t = \beta_0 + \beta_1 t + \sum_{i=1}^{3} \delta_i \,\mathrm{IND}_{ti} + \varepsilon_t$$

Coefficient	Estimate	Standard Error	t Ratio
β_0	2.580	.020	128.82
β_1	.019	.00057	33.20
δ_1	−.216	.021	−10.45
δ_2	−.081	.021	−3.93
δ_3	−.104	.021	−5.07

ANOVA Table

Source	SS	df	MS
Regression	2.973	4	.74
Error	.091	39	.0023
Total (corrected for mean)	3.063	43	

Autocorrelations of the Residuals

Lag k	r_k	Lag k	r_k
1	.84	5	−.02
2	.68	6	−.18
3	.47	7	−.33
4	.24	8	−.42

written as

$$z_{n+j} = \beta_0 + \sum_{i=1}^{k} \beta_i \frac{j^i}{i!} + \sum_{i=1}^{s-1} \delta_i \,\mathrm{IND}_{ji} + \varepsilon_{n+j} \tag{4.15}$$

where the fitting function is

$$\mathbf{f}(j) = \left(1, j, \ldots, \frac{j^k}{k!}, \mathrm{IND}_{j1}, \ldots, \mathrm{IND}_{j,s-1}\right)'$$

It should be emphasized that the coefficients $\boldsymbol{\beta} = (\beta_0, \beta_1, \ldots, \beta_k, \delta_1, \ldots, \delta_{s-1})'$ in this model are not the same as those in model (4.10). They are functionally related, however, and in order to avoid the introduction of additional symbols, we have used the same notation for these coefficients.

From the result in (3.18) we find that at time n the least squares estimate of $\boldsymbol{\beta}$ is given by

$$\hat{\boldsymbol{\beta}}_n = \mathbf{F}_n^{-1}\mathbf{h}_n \tag{4.16}$$

where

$$\mathbf{F}_n = \sum_{j=0}^{n-1} \mathbf{f}(-j)\mathbf{f}'(-j) \qquad \mathbf{h}_n = \sum_{j=0}^{n-1} \mathbf{f}(-j)z_{n-j}$$

The forecast of z_{n+l} is then given by

$$\hat{z}_n(l) = \mathbf{f}'(l)\hat{\boldsymbol{\beta}}_n \tag{4.17}$$

and $100(1-\lambda)$ percent prediction intervals can be calculated from

$$\hat{z}_n(l) \pm t_{\lambda/2}(n-k-s)\hat{\sigma}\left[1 + \mathbf{f}'(l)\mathbf{F}_n^{-1}\mathbf{f}(l)\right]^{1/2} \tag{4.18}$$

It can be shown that the forecast or fitting functions in (4.15) follow the difference equation $\mathbf{f}(j) = \mathbf{L}\mathbf{f}(j-1)$, where $\mathbf{L}$ is a $(k+s)\times(k+s)$ transition matrix

$$\mathbf{L} = \left[\begin{array}{c|c} \mathbf{L}_{11} & \mathbf{0} \\ \hline \mathbf{L}_{21} & \mathbf{L}_{22} \end{array}\right]$$

The $(k+1)\times(k+1)$ matrix $\mathbf{L}_{11}$ is the transition matrix of the polynomial trend model discussed in Section 3.4, with elements $l_{ij} = 1/(i-j)!$ for $i \geqslant j$, and zero otherwise. The elements of the $(s-1)\times(k+1)$ matrix $\mathbf{L}_{21}$ are $l_{11} = 1$, and zero otherwise. The elements of the $(s-1)\times(s-1)$ matrix $\mathbf{L}_{22}$ are $l_{1j} = -1$ for $1 \leqslant j \leqslant s-1$, $l_{ij} = 1$ for $i \geqslant 2$ and $j = i-1$, and zero otherwise. The vector of the initial fitting function $\mathbf{f}(0)$ is a $(k+s)\times 1$ vector with a 1 in the first row, and zero otherwise. This initialization implies that the last available observation at time n corresponds to the last seasonal period s.

As an illustration, let us consider a model with quadratic trend and seasonal period $s = 4$:

$$z_{n+j} = \beta_0 + \beta_1 j + \beta_2\frac{j^2}{2} + \sum_{i=1}^{3}\delta_i \,\mathrm{IND}_{ji} + \varepsilon_{n+j}$$

Then the transition matrix **L** and the initial vector **f**(0) are given by

$$\mathbf{L} = \left[\begin{array}{ccc|ccc} 1 & 0 & 0 & 0 & 0 & 0 \\ 1 & 1 & 0 & 0 & 0 & 0 \\ 1/2 & 1 & 1 & 0 & 0 & 0 \\ \hline 1 & 0 & 0 & -1 & -1 & -1 \\ 0 & 0 & 0 & 1 & 0 & 0 \\ 0 & 0 & 0 & 0 & 1 & 0 \end{array}\right] \qquad \mathbf{f}(0) = \begin{bmatrix} 1 \\ 0 \\ 0 \\ 0 \\ 0 \\ 0 \end{bmatrix}$$

Successive application of the difference equation $\mathbf{f}(j) = \mathbf{L}\mathbf{f}(j-1)$ leads to $\mathbf{f}(j) = (1, j, j^2/2, \text{IND}_{j1}, \text{IND}_{j2}, \text{IND}_{j3})'$.

The fact that the fitting functions follow such a difference equation will become important when the parameters are estimated by discounted least squares, since then the estimates can be updated via easy recursive algorithms.

4.2.2. Modeling the Seasonality with Trigonometric Functions

In Equation (4.7) the seasonal component S_t is modeled as a linear combination of trigonometric functions. In the characterization

$$S_t = \sum_{i=1}^{m} A_i \sin\left(\frac{2\pi i}{s} t + \phi_i\right) \tag{4.19}$$

the seasonal component is a linear combination of m sine functions (or harmonics) with frequencies $f_i = 2\pi i/s$ and phase shifts ϕ_i. The amplitudes A_i correspond to the weights in the above linear combination. For example, for monthly data with yearly seasonal pattern ($s = 12$), the first ($i = 1$) harmonic, $\sin[(2\pi/12)t + \phi_1]$, describes a sine wave that completes its cycle in 12 time periods. The second ($i = 2$) harmonic completes the cycle in six periods. In general, it takes s/i time periods until the ith harmonic has gone through a complete cycle.

In discrete time series (t integer valued), we can consider at most $m = s/2$ harmonics, since the period corresponding to the $(s/2)$th harmonic is 2, which is the shortest possible cycle length. Frequently, not all $s/2$ harmonics are needed; usually the first few are capable of generating complex seasonal patterns. This is illustrated in Figure 4.8, where we plot

$$E(z_t) = \sum_{i=1}^{2} A_i \sin\left(\frac{2\pi i}{12} t + \phi_i\right)$$

for $A_1 = 1, \phi_1 = 0, A_2 = -.70, \phi_2 = .6944\pi$ (corresponding to 125°).

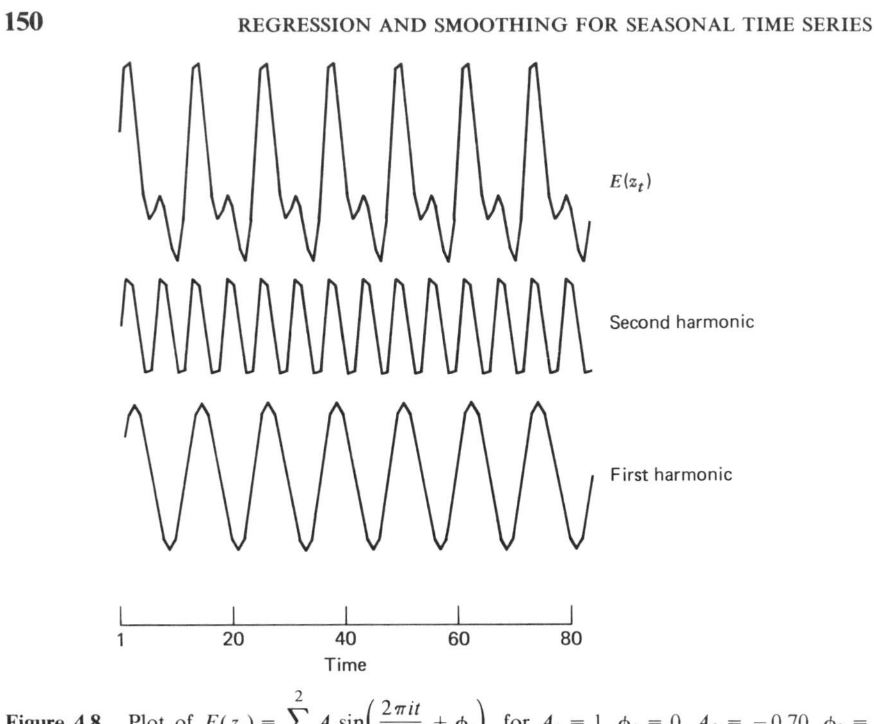

Figure 4.8. Plot of $E(z_t) = \sum_{i=1}^{2} A_i \sin\left(\frac{2\pi i t}{12} + \phi_i\right)$, for $A_1 = 1$, $\phi_1 = 0$, $A_2 = -0.70$, $\phi_2 = (0.6944)\pi$ (or 125°).

Combining the model for S_t in (4.19) with the trend model in (4.5) leads us to the model

$$z_t = \beta_0 + \sum_{i=1}^{k} \beta_i \frac{t^i}{i!} + \sum_{i=1}^{m} A_i \sin(f_i t + \phi_i) + \varepsilon_t \tag{4.20}$$

where $f_i = 2\pi i/s$ is the frequency. This should not be confused with the fitting function.

The model in (4.20) incorporates the parameters in a nonlinear fashion. However, we can use the trigonometric identity $\sin(x + y) = \sin x \cos y + \sin y \cos x$ and express (4.20) as

$$\begin{aligned} z_t &= \beta_0 + \sum_{i=1}^{k} \beta_i \frac{t^i}{i!} + \sum_{i=1}^{m} (\beta_{1i} \sin f_i t + \beta_{2i} \cos f_i t) + \varepsilon_t \\ &= \mathbf{f}'(t)\boldsymbol{\beta} + \varepsilon_t \end{aligned} \tag{4.21}$$

where $\mathbf{f}(t) = (1, t, \dots, t^k/k!, \sin f_1 t, \cos f_1 t, \dots, \sin f_m t, \cos f_m t)'$ is the vector of fitting functions, and $\boldsymbol{\beta} = (\beta_0, \beta_1, \ldots, \beta_k, \beta_{11}, \beta_{21}, \ldots, \beta_{1m}, \beta_{2m})'$ is the vector of coefficients; these coefficients are functionally related to the ones in model (4.20). Since an intercept β_0 is included in the model, we can fit at most $m = s/2 - 1$ full harmonics if s is even. The coefficient $\beta_{1,s/2}$ of the $(s/2)$th harmonic has to be set to zero. If s is odd, we can fit $(s-1)/2$ full harmonics.

The least squares estimates $\hat{\boldsymbol{\beta}}$, the forecasts $\hat{z}_n(l)$, and the prediction intervals are easily calculated using the general expressions in Chapter 2.

Example 4.3: Another Look at Car Sales

We consider the same data as in Example 4.1 and fit the model

$$z_t = \beta_0 + \beta_1 t + \sum_{i=1}^{5}\left(\beta_{1i}\sin\frac{2\pi i}{12}t + \beta_{2i}\cos\frac{2\pi i}{12}t\right) + \beta_{26}\cos\pi t + \varepsilon_t$$

The ANOVA table for this model is given in Table 4.3. The regression sum of squares is further partitioned into the individual regression contributions of the various harmonics, which are included in sequential order from the lowest to the highest frequency. This is a natural sequential ordering, since one would want to include low frequencies, which are the ones leading to long periods, first. Also note that in general the inclusion of the trend component in the model will lead to a nonorthogonal **X** matrix; thus the regression contribution of the ith harmonic will depend on the other harmonics included in the model.

Table 4.3. Analysis of Variance Table—Car Sales (1960–1967; $n = 96$)

$$z_t = \beta_0 + \beta_1 t + \sum_{i=1}^{5}\left(\beta_{1i}\sin\frac{2\pi i}{12}t + \beta_{2i}\cos\frac{2\pi i}{12}t\right) + \beta_{26}\cos\pi t + \varepsilon_t$$

Source	SS		df
Regression	1671.8		12
trend		517.8	1
1st harmonic		659.0	2
2nd harmonic		451.4	2
3rd harmonic		17.1	2
4th harmonic		0.8	2
5th harmonic		25.7	2
6th harmonic		0.0	1
Error	172.9		83
Total (corrected for mean)	1844.7		95

Table 4.4. Least Squares Estimation and Forecasting Results—Car Sales (1960–1967; $n = 96$)

$$z_t = \beta_0 + \beta_1 t + \sum_{i=1}^{2} \left(\beta_{1i} \sin \frac{2\pi i}{12} t + \beta_{2i} \cos \frac{2\pi i}{12} t \right) + \varepsilon_t$$

Coefficient	Estimate	Standard Error	t Ratio
β_0	9.88	.32	30.80
β_1	0.088	.0057	15.30
β_{11}	2.57	.22	11.45
β_{21}	−2.66	.22	−11.90
β_{12}	−2.96	.22	−13.19
β_{22}	0.83	.22	3.71

Autocorrelations of the Residuals

Lag k	r_k	Lag k	r_k
1	.13	7	.09
2	.13	8	.02
3	.11	9	.02
4	.10	10	−.01
5	.16	11	.07
6	−.08	12	.25

Forecasts

Lead Time l						
l	1	2	3	4	5	6
z_{96+l}	13.21	14.25	20.14	21.73	26.10	21.08
$\hat{z}_{96}(l)$	15.24	16.42	20.32	24.38	25.33	22.34
$z_{96+l} - \hat{z}_{96}(l)$	−2.03	−2.17	−.18	−2.65	.77	−1.26
l	7	8	9	10	11	12
z_{96+l}	18.02	16.72	14.39	21.34	17.18	14.58
$\hat{z}_{96}(l)$	17.81	15.15	15.70	17.78	18.66	17.54
$z_{96+l} - \hat{z}_{96}(l)$	.21	1.57	−1.31	3.56	−1.48	−2.96

We find from Table 4.3 that the first two harmonics explain most of the seasonal variation. Detailed estimation results for the model with $m = 2$ are given in Table 4.4.

The sample autocorrelation function of the residuals in Table 4.4 shows that the residuals 12 steps apart are still correlated. This indicates that the seasonal model can be further improved by adding terms that include higher frequencies. According to Table 4.3, the third and fifth harmonic are, although small, still significant. For example, the F statistic for testing H_0: $\beta_{15} = \beta_{25} = 0$, which is given by $F = (25.7/2)/(172.9/83) = 6.17$, is significant compared to the .01 critical value from the F distribution with 2 and 83 degrees of freedom.

The forecasts and the forecast errors for the next 12 periods are also given in Table 4.4. We can make a rough comparison between the forecasts from this model and those from the seasonal indicator model by calculating the average of the squared forecast errors,

$$\frac{1}{12}\sum_{l=1}^{12}\left[z_{96+l} - \hat{z}_{96}(l)\right]^2.$$

For the current model this measure is 3.83. This is somewhat larger than the one from the seasonal indicator model in Table 4.1, which is 2.97. In this sense the seasonal indicator model leads to a better description of the seasonal component than the model with the first two harmonics.

Change of Time Origin in the Seasonal Trigonometric Model

The coefficients in model (4.21) are expressed relative to time origin zero. Shifting the time origin to n leads to the model

$$z_{n+j} = \mathbf{f}'(j)\boldsymbol{\beta} + \varepsilon_{n+j} \tag{4.22}$$

It follows from the properties of polynomials and trigonometric functions

$$\sin(x + y) = \sin x \cos y + \sin y \cos x$$

$$\cos(x + y) = \cos x \cos y - \sin x \sin y$$

that the vector of fitting or forecast functions

$$\mathbf{f}(j) = \left(1, j, \ldots, \frac{j^k}{k!}, \sin f_1 j, \cos f_1 j, \ldots, \sin f_m j, \cos f_m j\right)'$$

follows the difference equation $\mathbf{f}(j) = \mathbf{L}\mathbf{f}(j-1)$, where the $(k+1+2m) \times (k+1+2m)$ transition matrix is given by

$$\mathbf{L} = \left[\begin{array}{c|c} \mathbf{L}_1 & \mathbf{0} \\ \hline \mathbf{0} & \mathbf{L}_2 \end{array}\right]$$

The $(k+1) \times (k+1)$ matrix $\mathbf{L}_1$ is the transition matrix of the polynomial trend model in Section 3.4 [$l_{ij} = 1/(i-j)!$ for $i \geqslant j$, and 0 otherwise]; the

$2m \times 2m$ matrix $\mathbf{L}_2$ is a block diagonal matrix

$$\mathbf{L}_2 = \begin{bmatrix} \mathbf{L}_{21} & & & \\ & \mathbf{L}_{22} & & \\ & & \ddots & \\ & & & \mathbf{L}_{2m} \end{bmatrix}$$

where $\mathbf{L}_{2i}(1 \leqslant i \leqslant m)$ is a 2×2 matrix such that

$$\mathbf{L}_{2i} = \begin{bmatrix} \cos f_i & \sin f_i \\ -\sin f_i & \cos f_i \end{bmatrix}$$

The $(k + 1 + 2m)$-dimensional initial vector $\mathbf{f}(0)$ is given by $\mathbf{f}(0) = (1, \mathbf{0}', 0, 1, \ldots, 0, 1)'$, where $\mathbf{0}$ is a $k \times 1$ vector of zeros. For illustration we consider two special cases.

Example 1. 12-point sinusoidal model ($k = 0$, $s = 12$, $m = 1$):

$$z_{n+j} = \beta_0 + \beta_{11} \sin \frac{2\pi j}{12} + \beta_{21} \cos \frac{2\pi j}{12} + \varepsilon_{n+j}$$

In this case,

$$\mathbf{L} = \begin{bmatrix} 1 & 0 & 0 \\ 0 & \sqrt{3}/2 & 1/2 \\ 0 & -1/2 & \sqrt{3}/2 \end{bmatrix} \qquad \mathbf{f}(0) = \begin{bmatrix} 1 \\ 0 \\ 1 \end{bmatrix}$$

Example 2. Linear trend model with two superimposed harmonics ($k = 1$, $s = 12$, $m = 2$):

$$z_{n+j} = \beta_0 + \beta_1 j + \beta_{11} \sin \frac{2\pi j}{12} + \beta_{21} \cos \frac{2\pi j}{12} + \beta_{12} \sin \frac{4\pi j}{12} + \beta_{22} \cos \frac{4\pi j}{12} + \varepsilon_{n+j}$$

In this case,

$$\mathbf{L} = \begin{bmatrix} 1 & 0 & 0 & 0 & 0 & 0 \\ 1 & 1 & 0 & 0 & 0 & 0 \\ 0 & 0 & \sqrt{3}/2 & 1/2 & 0 & 0 \\ 0 & 0 & -1/2 & \sqrt{3}/2 & 0 & 0 \\ 0 & 0 & 0 & 0 & 1/2 & \sqrt{3}/2 \\ 0 & 0 & 0 & 0 & -\sqrt{3}/2 & 1/2 \end{bmatrix} \qquad \mathbf{f}(0) = \begin{bmatrix} 1 \\ 0 \\ 0 \\ 1 \\ 0 \\ 1 \end{bmatrix}$$

4.3. LOCALLY CONSTANT SEASONAL MODELS

The globally constant models discussed in Section 4.2 assume that (1) the parameters in the trend component T_t and in the seasonal component S_t are fixed constants that do not change over time and (2) the errors $\{\varepsilon_t\}$ are uncorrelated. These models are capable of tracing trends and seasonal patterns with fixed amplitudes and phases and work well for stable phenomena. For economic and business data, however, it is more reasonable to allow for adaptive, time-changing trend and seasonal components. While the assumption of fixed seasonals is occasionally appropriate, it is in most cases unrealistic to suppose that the trend stays fixed and that the error terms are uncorrelated. This fact was illustrated by the two examples in the last section.

We must relax the assumption of globally constant parameters. In the following sections we adopt a heuristic approach and allow for shifting trends and seasonals. In Chapter 6 we discuss more general models, which incorporate adaptive trends and seasonals, as well as correlated errors.

If one expects that the parameters change over time and are constant only locally, one can discount past observations in deriving their estimates and use the discounted least squares (or general exponential smoothing) approach. General expressions for the discounted least squares estimates $\hat{\boldsymbol{\beta}}_n$ of the coefficients $\boldsymbol{\beta}$ in

$$z_{n+j} = \mathbf{f}'(j)\boldsymbol{\beta} + \varepsilon_{n+j} \tag{4.23}$$

are given in Chapter 3 [Eq. (3.25)]. There the estimates are chosen to minimize

$$\sum_{j=0}^{n-1} \omega^j \left[z_{n-j} - \mathbf{f}'(-j)\boldsymbol{\beta} \right]^2$$

where $\omega = 1 - \alpha$ is a discount coefficient. The forecast of z_{n+l} is then given by $\hat{z}_n(l) = \mathbf{f}'(l)\hat{\boldsymbol{\beta}}_n$. If a new observation z_{n+1} becomes available, the parameter estimates can be updated according to Equation (3.29) in Chapter 3,

$$\hat{\boldsymbol{\beta}}_{n+1} = \mathbf{L}'\hat{\boldsymbol{\beta}}_n + \mathbf{F}^{-1}\mathbf{f}(0)\left[z_{n+1} - \hat{z}_n(1) \right] \tag{4.24}$$

where $\mathbf{L}$ and $\mathbf{f}(0)$ are the transition matrix and the initial vector, which generate the forecast functions $\mathbf{f}(l)$ from the difference equation $\mathbf{f}(l) = \mathbf{L}\mathbf{f}(l-1)$. The choice of $\mathbf{L}$ and $\mathbf{f}(0)$ for forecast functions with seasonal indicators and trigonometric functions was discussed in the previous sections.

For seasonal models the steady-state matrix $\mathbf{F} = \Sigma_{j \geq 0} \omega^j \mathbf{f}(-j)\mathbf{f}'(-j)$ in (4.24) consists of infinite sums of exponentially discounted polynomials, sines, cosines, and their combinations, such as

$$\Sigma \omega^j j^k \qquad \Sigma \omega^j j^k \sin fj \qquad \Sigma \omega^j j^k \cos fj$$

$$\Sigma \omega^j j^k \sin f_1 j \sin f_2 j \qquad \Sigma \omega^j j^k \cos f_1 j \cos f_2 j \qquad \Sigma \omega^j j^k \sin f_1 j \cos f_2 j$$

It can be shown that

$$\Sigma \omega^j = \frac{1}{1 - \omega}$$

$$\Sigma \omega^j \sin fj = \frac{\omega \sin f}{1 - 2\omega \cos f + \omega^2}$$

$$\Sigma \omega^j \cos fj = \frac{1 - \omega \cos f}{1 - 2\omega \cos f + \omega^2}$$

All other required sums can be derived by differentiation and application of basic properties of trigonometric functions

$$\sin f_1 \sin f_2 = \tfrac{1}{2}\left[\cos(f_1 - f_2) - \cos(f_1 + f_2)\right]$$

$$\cos f_1 \cos f_2 = \tfrac{1}{2}\left[\cos(f_1 + f_2) + \cos(f_1 - f_2)\right]$$

$$\cos f_1 \sin f_2 = \tfrac{1}{2}\left[\sin(f_1 + f_2) - \sin(f_1 - f_2)\right]$$

[See *Standard Mathematical Tables* (Selby 1965, p. 503).] For example, $\Sigma \omega^j j \cos fj$ can be derived from

$$\Sigma \omega^j j \cos fj = \omega \frac{\partial}{\partial \omega} \Sigma \omega^j \cos fj = \omega \frac{\partial}{\partial \omega}\left[\frac{1 - \omega \cos f}{1 - 2\omega \cos f + \omega^2}\right]$$

$$= \frac{\omega(1 + \omega^2)\cos f - 2\omega^2}{(1 - 2\omega \cos f + \omega^2)^2}$$

A collection of infinite sums needed to calculate $\mathbf{F}$ for seasonal models is given in Table 4.5. Apart from two corrections, this table is similar to the one in Montgomery and Johnson (1976, p. 91).

Equation (4.24) gives a procedure for updating or revising the estimates as new observations become available; to start the updating process, one has

Table 4.5. Infinite Sums Needed in General Exponential Smoothing

$$\Sigma\omega^j = \frac{1}{1-\omega} \qquad \Sigma\omega^j j = \frac{\omega}{(1-\omega)^2} \qquad \Sigma\omega^j j^2 = \frac{\omega(1+\omega)}{(1-\omega)^3}$$

$$\Sigma\omega^j j^3 = \frac{\omega(1+4\omega+\omega^2)}{(1-\omega)^4} \qquad \Sigma\omega^j j^4 = \frac{\omega(1+11\omega+11\omega^2+\omega^3)}{(1-\omega)^5}$$

$$\Sigma\omega^j \sin fj = \frac{\omega \sin f}{g_1} \qquad \Sigma\omega^j \cos fj = \frac{1-\omega\cos f}{g_1}$$

$$\Sigma\omega^j j \sin fj = \frac{\omega(1-\omega^2)\sin f}{g_1^2} \qquad \Sigma\omega^j j \cos fj = \frac{\omega(1+\omega^2)\cos f - 2\omega^2}{g_1^2}$$

$$\Sigma\omega^j \sin f_1 j \sin f_2 j = \frac{1}{2}\left[\frac{1-\omega\cos(f_1-f_2)}{g_2} - \frac{1-\omega\cos(f_1+f_2)}{g_3}\right]$$

$$\Sigma\omega^j \sin f_1 j \cos f_2 j = \frac{1}{2}\left[\frac{\omega\sin(f_1-f_2)}{g_2} + \frac{\omega\sin(f_1+f_2)}{g_3}\right]$$

$$\Sigma\omega^j \cos f_1 j \cos f_2 j = \frac{1}{2}\left[\frac{1-\omega\cos(f_1-f_2)}{g_2} + \frac{1-\omega\cos(f_1+f_2)}{g_3}\right]$$

where

$$g_1 = 1 - 2\omega\cos f + \omega^2$$

$$g_2 = 1 - 2\omega\cos(f_1 - f_2) + \omega^2$$

$$g_3 = 1 - 2\omega\cos(f_1 + f_2) + \omega^2$$

to specify a starting value $\hat{\boldsymbol{\beta}}_0$. The elements of the matrix $\mathbf{F}$ are functions of the discount coefficient $\omega = 1 - \alpha$; a smoothing constant α must be chosen.

It is usually suggested that the least squares estimate of $\boldsymbol{\beta}$ in the regression model

$$z_t = \mathbf{f}'(t)\boldsymbol{\beta} + \varepsilon_t$$

be taken as initial vector $\hat{\boldsymbol{\beta}}_0$. One suggestion, the one adopted here, is to calculate this estimate from all available observations $(z_1, z_2, \ldots, z_n)$. Such a choice will be appropriate if the trend and seasonal components are fairly stable. Other choices are possible. For example, one can estimate $\boldsymbol{\beta}$ from the first g observations, where g is the number of parameters. If there is considerable variation in the parameters of the model (4.23), such a starting value will be preferable. Or, one can use a "backforecasting" approach, which for the special case of simple exponential smoothing was described in Section 3.3. A discussion of this approach can be found in Ledolter and Abraham (1983).

To update the estimates in (4.24), a smoothing constant must be determined. Brown (1962) suggests that the value of ω should lie between $(.70)^{1/g}$ and $(.95)^{1/g}$, where g is the number of parameters. Or, if sufficient historical data are available, one can estimate $\omega = 1 - \alpha$ by simulation and choose the smoothing constant that minimizes the sum of the squared one-step-ahead forecast errors

$$\mathrm{SSE}(\alpha) = \sum_{t=1}^{n} [z_t - \hat{z}_{t-1}(1)]^2$$

After estimating the smoothing constant α, one should always check the adequacy of the model. The sample autocorrelation function of the one-step-ahead forecast errors should be calculated. Significant autocorrelations indicate that the particular forecast model is not appropriate.

4.3.1. Locally Constant Seasonal Models Using Seasonal Indicators

For illustration, we consider the logarithm of the quarterly expenditures for new plant and equipment that we analyzed in Example 4.2. There we found that the globally constant linear trend and seasonal indicator model ($s = 4$)

$$z_t = \ln y_t = \beta_0 + \beta_1 t + \sum_{i=1}^{3} \delta_i \mathrm{IND}_{ti} + \varepsilon_t \tag{4.25}$$

was inappropriate; it could not adequately capture the time-changing trend.

Shifting the time origin to n and introducing the additional parameter $\omega = 1 - \alpha$ to discount past observations, we can write the model in (4.25) as

$$z_{n+j} = \mathbf{f}'(j)\boldsymbol{\beta} + \varepsilon_{n+j} \tag{4.26}$$

where the fitting functions $\mathbf{f}(j) = (1, j, \mathrm{IND}_{j1}, \mathrm{IND}_{j2}, \mathrm{IND}_{j3})'$ follow the difference equation $\mathbf{f}(j) = \mathbf{L}\mathbf{f}(j-1)$, with

$$\mathbf{L} = \begin{bmatrix} 1 & 0 & 0 & 0 & 0 \\ 1 & 1 & 0 & 0 & 0 \\ 1 & 0 & -1 & -1 & -1 \\ 0 & 0 & 1 & 0 & 0 \\ 0 & 0 & 0 & 1 & 0 \end{bmatrix} \qquad \mathbf{f}(0) = \begin{bmatrix} 1 \\ 0 \\ 0 \\ 0 \\ 0 \end{bmatrix}$$

In this formulation, the last available observation z_n is the one corresponding to the last seasonal period $s = 4$.

The updating weights in $\hat{\boldsymbol{\beta}}_{n+1} = \mathbf{L}'\hat{\boldsymbol{\beta}}_n + \mathbf{F}^{-1}\mathbf{f}(0)[z_{n+1} - \hat{z}_n(1)]$ can be calculated from $\mathbf{f}(0)$ and the symmetric matrix

$$\mathbf{F} = \begin{bmatrix} \frac{1}{1-\omega} & \frac{-\omega}{(1-\omega)^2} & \frac{\omega^3}{1-\omega^4} & \frac{\omega^2}{1-\omega^4} & \frac{\omega}{1-\omega^4} \\ & \frac{\omega(1+\omega)}{(1-\omega)^3} & \frac{-\omega^3(3+\omega^4)}{(1-\omega^4)^2} & \frac{-\omega^2(2+2\omega^4)}{(1-\omega^4)^2} & \frac{-\omega(1+3\omega^4)}{(1-\omega^4)^2} \\ & & \frac{\omega^3}{1-\omega^4} & 0 & 0 \\ & \text{symmetric} & & \frac{\omega^2}{1-\omega^4} & 0 \\ & & & & \frac{\omega}{1-\omega^4} \end{bmatrix} \tag{4.27}$$

For general s, it is easy to show that the nonzero elements of the $(s+1) \times (s+1)$ matrix $\mathbf{F}$ are given by

$$f_{11} = \frac{1}{1-\omega} \qquad f_{22} = \frac{\omega(1+\omega)}{(1-\omega)^3} \qquad f_{12} = f_{21} = \frac{-\omega}{(1-\omega)^2}$$

$$f_{1j} = f_{j1} = f_{jj} = \frac{\omega^{s+2-j}}{1-\omega^s} \qquad \text{for } 3 \leqslant j \leqslant s+1$$

$$f_{2j} = f_{j2} = \frac{-\omega^{s+2-j}[(s+2-j) + (j-2)\omega^s]}{(1-\omega^s)^2} \qquad \text{for } 3 \leqslant j \leqslant s+1$$

To start the updating recursions (4.24), we have to calculate $\hat{\boldsymbol{\beta}}_0$. The ordinary least squares estimates of the parameters in model (4.25) are used as starting values. These estimates were calculated previously in Table 4.2, where it was found that $\hat{\boldsymbol{\beta}}_0 = (2.58, .019, -.216, -.081, -.104)'$. To update

Table 4.6. Sums of the Squared One-Step-Ahead Forecast Errors from the Model

$$z_{n+j} = \ln y_{n+j} = \beta_0 + \beta_1 j + \sum_{i=1}^{3} \delta_i \mathrm{IND}_{ji} + \varepsilon_{n+j}$$

for Various Values of the Smoothing Constant α.—New Plant and Equipment Expenditures ($n = 44$)

α	SSE(α)
.10	.1114
.30	.0904
.50	.0691
.70	.0540
.90	.0455
1.10	.0421
1.20	.0418
1.30	.0422
1.50	.0445

the parameters we have to invert the matrix **F** in (4.27). We have written a computer program EXPSIND, given in Appendix 4, to calculate $\mathbf{F}^{-1}\mathbf{f}(0)$, to update the coefficients $\hat{\boldsymbol{\beta}}_t$, and to derive the forecasts $\hat{z}_n(l) = \mathbf{f}'(l)\hat{\boldsymbol{\beta}}_n$ for the model with forecast function $\mathbf{f}(l) = (1, l, \mathrm{IND}_{l1}, \ldots, \mathrm{IND}_{l,s-1})'$.

The first 44 one-step-ahead forecast errors $z_t - \mathbf{f}'(1)\hat{\boldsymbol{\beta}}_{t-1}$ are calculated for various smoothing constants α. The corresponding sums of the squared one-step-ahead forecast errors, $\mathrm{SSE}(\alpha) = \sum_{t=1}^{n}[z_t - \mathbf{f}'(1)\hat{\boldsymbol{\beta}}_{t-1}]^2$, are given in Table 4.6 and are plotted in Figure 4.9. We find that the optimal smoothing constant is $\alpha = 1.20$. However, SSE(α), as a function of α, is quite flat over the range 0.90 to 1.50; the estimate of α is essentially 1, which implies a discount coefficient $\omega = 0$. This value is not at all in the range $(.70)^{1/5} = .93 \leqslant \omega \leqslant (.95)^{1/5} = .99$ that is usually suggested in the traditional forecasting literature. This indicates that in many genuine (i.e., not generated) data sets, the optimal discount coefficient may not fall within the suggested range.

The estimate $\alpha = 1.20$ implies a negative discount coefficient $\omega = 1 - \alpha = -.20$. Such a discount coefficient might appear unusual to readers familiar with the traditional forecasting literature, since it discounts past deviations $z_{n-j} - \mathbf{f}'(-j)\boldsymbol{\beta}$ in $\sum_{j=0}^{n-1}\omega^j[z_{n-j} - \mathbf{f}'(-j)\boldsymbol{\beta}]^2$ with weights that alternate in sign. A negative discount coefficient, however, does not cause any difficulties in the numerical calculation of the coefficient estimates or in updating the forecasts, since $|\omega| < 1$ is the only condition for the convergence of the infinite sums in the matrix **F**.

In Table 4.7 we use the estimated smoothing constant $\alpha = 1.20$ to update the coefficients $\hat{\boldsymbol{\beta}}_t$, the one-step-ahead forecasts, and the one-step-ahead

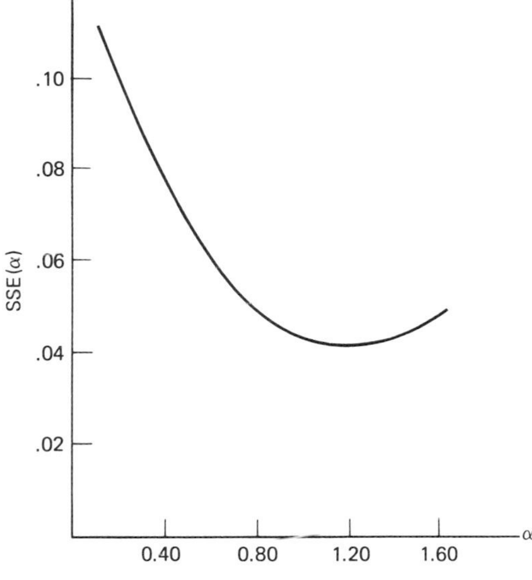

Figure 4.9. Plot of the sum of the squared one-step-ahead forecast errors from the model

$$z_{n+j} = \ln y_{n+j} = \beta_0 + \beta_1 j + \sum_{i=1}^{3} \delta_i \text{IND}_{ji} + \varepsilon_{n+j}$$

New plant and equipment expenditures.

forecast errors. For this choice of α, the smoothing vector $\mathbf{F}^{-1}\mathbf{f}(0)$ is given by $(1.00032, .29952, -.09984, -.39936, -.69888)'$. We notice that the estimate of the intercept $\hat{\beta}_{0,t+1}$ is very close to z_{t+1}. This follows from the fact that for $\alpha = 1.20$ the updating equation for the intercept in the linear trend model with seasonal indicators [see Eq. (4.24)] is given by

$$\hat{\beta}_{0,t+1} = \hat{\beta}_{0,t} + \hat{\beta}_{1,t} + \hat{\delta}_{1,t} + 1.00032\left(z_{t+1} - \hat{\beta}_{0,t} - \hat{\beta}_{1,t} - \hat{\delta}_{1,t}\right) \cong z_{t+1}.$$

The sample autocorrelations of the 44 one-step-ahead forecast errors, $z_1 - \hat{z}_0(1) = -.081$, $z_2 - \hat{z}_1(1) = .032, \ldots, z_{44} - \hat{z}_{43}(1) = -.003$, are given in Table 4.8. The autocorrelations at low lags are all insignificant compared with their standard error $n^{-1/2} = .15$. At seasonal lags, especially at lags 4 and 8, we find significant correlations; this could be due to the fact that the single smoothing constant α tries to balance the changes in the trend and the seasonal component. In the current example, the trend exhibits more variability than the seasonal component, which appears fairly stable. A procedure that would use two smoothing constants, one for the trend and one for the seasonal component, would probably perform better.

Table 4.7. General Exponential smoothing (α = 1.20) for the Model[a]

$$z_{n+j} = \ln y_{n+j} = \beta_0 + \beta_1 j + \sum_{i=1}^{3} \delta_i \mathrm{IND}_{ji} + \varepsilon_{n+j}$$

New Plant and Equipment Expenditures

Time	z_t	$\hat{\beta}_0$	$\hat{\beta}_1$	$\hat{\delta}_1$	$\hat{\delta}_2$	$\hat{\delta}_3$	$\hat{z}_t(1)$	$e_{t-1}(1)$ $[= z_t - \hat{z}_{t-1}(1)]$
0		2.580	.019	−.216	−.081	−.104	2.384	
1	2.303	2.303	−.005	.143	.144	.272	2.440	−.081
2	2.472	2.472	.004	−.002	.117	−.165	2.474	.032
3	2.460	2.460	.000	.120	−.157	.013	2.580	−.015
⋮								
43	3.340	3.340	.029	.097	−.100	.027	3.466	−.013
44	3.463	3.463	.028	−.196	−.069	−.095	3.295	−.003
45	3.251	3.251	.015	.132	.119	.227	3.398	−.044
46	3.347	3.347	.000	−.008	.115	−.097	3.340	−.050
47	3.325	3.325	−.005	.125	−.083	.018	3.444	−.015
48	3.426	3.426	−.010	−.206	−.099	−.111	3.210	−.019
49	3.253	3.253	.003	.102	.077	.175	3.358	.043
50	3.391	3.391	.013	−.029	.060	−.125	3.375	.033
51	3.415	3.415	.025	.084	−.113	.001	3.524	.040
52	3.542	3.541	.030	−.199	−.091	−.097	3.373	.018

[a]Smoothing constant α and initial value $\hat{\boldsymbol{\beta}}_0$ are determined from the first 44 observations.

The transformed forecasts $\hat{y}_t(l) = e^{\hat{z}_t(l)}$ of the next eight observations are given in Table 4.9.

Implications of the Smoothing Constant α = 1

In the previous example we found that the smoothing constant α is essentially 1. If $\alpha = 1$ or $\omega = 0$, the matrix $\mathbf{F}$ in (4.27) is singular, and its inverse cannot be calculated. However, taking the limit of $\mathbf{F}^{-1}\mathbf{f}(0)$ as $\omega \to 0$, we can show after substantial algebra that

$$\mathbf{f}_* = \lim_{\omega \to 0} \mathbf{F}^{-1}\mathbf{f}(0) = \left(1, \frac{1}{s}, -\frac{1}{s}, -\frac{2}{s}, \ldots, -\frac{s-1}{s}\right)' \qquad (4.28)$$

Substituting this limit into the updating equations $\hat{\boldsymbol{\beta}}_{n+1} = \mathbf{L}'\hat{\boldsymbol{\beta}}_n + \mathbf{f}_*[z_{n+1} - \hat{z}_n(1)]$, it can be shown (again, after some algebra) that the forecasts $\hat{z}_n(l) = \mathbf{f}'(l)\hat{\boldsymbol{\beta}}_n$ follow the difference equation given by

$$\hat{z}_n(1) = z_n + z_{n+1-s} - z_{n-s} = z_{n+1-s} + (z_n - z_{n-s}) \qquad (4.29)$$

Table 4.8. Sample Autocorrelations of the One-Step-Ahead Forecast Errors from the Model

$$z_{n+j} = \ln y_{n+j} = \beta_0 + \beta_1 j + \sum_{i=1}^{3} \delta_i \, \mathrm{IND}_{ji} + \varepsilon_{n+j}$$

with Smoothing Constant $\alpha = 1.20$.—New Plant and Equipment Expenditures ($n = 44$)

Lag k	r_k	Lag k	r_k
1	−.02	5	−.18
2	.07	6	−.08
3	.22	7	−.17
4	−.27	8	−.39

and for general l,

$$\hat{z}_n(l) = \hat{z}_n(l-1) + \hat{z}_n(l-s) - \hat{z}_n(l-s-1)$$

where $\hat{z}_n(j) = z_{n+j}$ for $j \leqslant 0$. This result implies that only the last $s+1$ observations $(z_n, z_{n-1}, \ldots, z_{n-s})$ are used in the forecast calculations. The forecast of the observation z_{n+1}, let's say the first quarter (if $s = 4$), is the sum of (1) the observation in the same quarter from the previous year (z_{n+1-s}) and (2) the yearly change in the preceding fourth quarter $(z_n - z_{n-s})$.

Table 4.9. Forecasts $\hat{y}_t(l) = \exp[\hat{z}_t(l)]$ of Original Data y_{t+l}.—New Plant and Equipment Expenditures

Time[a] t	$\hat{y}_t(1)$	$\hat{y}_t(2)$	$\hat{y}_t(3)$	$\hat{y}_t(4)$	y_t
44	27.00	31.53	31.63	35.77	31.92
45	29.90	29.96	33.89	27.41	25.82
46	28.22	31.91	25.82	28.42	28.43
47	31.31	25.33	27.91	27.28	27.79
48	24.78	27.28	26.66	29.49	30.74
49	28.73	28.07	31.06	26.15	25.87
50	29.22	32.33	27.19	31.22	29.70
51	33.92	28.53	32.75	33.55	30.41
52	29.17	33.48	34.26	38.90	34.52

[a] $t = 44$ corresponds to the fourth quarter of 1964.

4.3.2. Locally Constant Seasonal Models Using Trigonometric Functions

Monthly Quebec car sales, which were previously analyzed in Section 4.2, are used as an additional illustrative example for general exponential smoothing. In Example 4.3 we considered a globally constant model with a linear trend and the first two harmonics. After fitting this globally constant model, we found the residuals still slightly correlated, especially at the seasonal lag 12.

We now allow for the possibility of time-changing coefficients and consider the locally constant model

$$z_{n+j} = \mathbf{f}'(j)\boldsymbol{\beta} + \varepsilon_{n+j}$$

where

$$\mathbf{f}(j) = (1, j, \sin f_1 j, \cos f_1 j, \sin f_2 j, \cos f_2 j)' \qquad \text{and} \qquad f_i = \frac{2\pi i}{12}.$$

These fitting functions follow the difference equation $\mathbf{f}(j) = \mathbf{L}\mathbf{f}(j-1)$, where the transition matrix $\mathbf{L}$ and the initial vector $\mathbf{f}(0)$ are given in one of the special cases in Section 4.2.2.

To update the coefficient vectors $\hat{\boldsymbol{\beta}}_t$, we have to calculate the smoothing vector $\mathbf{F}^{-1}\mathbf{f}(0)$. For the fitting functions $\mathbf{f}(j) = (1, j, \sin f_1 j, \cos f_1 j, \ldots, \sin f_m j, \cos f_m j)'$, the elements of the symmetric matrix $\mathbf{F} = \sum_{j \geqslant 0} \omega^j \mathbf{f}(-j)\mathbf{f}'(-j)$ are given by

$$f_{11} = \Sigma\omega^j \quad f_{12} = -\Sigma\omega^j j \quad f_{22} = \Sigma\omega^j j^2$$

$$f_{1,1+2i} = -\Sigma\omega^j \sin f_i j \qquad f_{1,2+2i} = \Sigma\omega^j \cos f_i j$$

$$f_{2,1+2i} = \Sigma\omega^j j \sin f_i j \qquad f_{2,2+2i} = -\Sigma\omega^j j \cos f_i j$$

$$f_{1+2i,1+2k} = \Sigma\omega^j \sin f_i j \sin f_k j \qquad f_{2+2i,2+2k} = \Sigma\omega^j \cos f_i j \cos f_k j$$

$$f_{1+2i,2+2k} = -\Sigma\omega^j \sin f_i j \cos f_k j$$

where $1 \leqslant i, k \leqslant m$. The results in Table 4.5 can be used to evaluate these infinite sums. We have written a computer program EXPHARM, given in Appendix 4, to evaluate $\mathbf{F}^{-1}\mathbf{f}(0)$, update the coefficient estimates $\hat{\boldsymbol{\beta}}_t$, and compute the forecasts $\hat{z}_n(l) = \mathbf{f}'(l)\hat{\boldsymbol{\beta}}_n$.

As starting value for the updating equation (4.24) we use the ordinary least squares estimate of $\boldsymbol{\beta}$ in the globally constant model $z_t = \mathbf{f}'(t)\boldsymbol{\beta} + \varepsilon_t$.

Table 4.10. Sums of the Squared One-Step-Ahead Forecast Errors from the Model

$$z_{n+j} = \beta_0 + \beta_1 j + \sum_{i=1}^{2} (\beta_{1i} \sin f_i j + \beta_{2i} \cos f_i j) + \varepsilon_{n+j}$$

for Various Values of the Smoothing Constant—Car Sales (n = 96)

α	SSE(α)
.03	233.4
.04	233.8
.05	234.4
.06	235.9
.07	238.7
.08	242.7
.09	247.9
.10	254.3
.15	301.4
.20	374.7

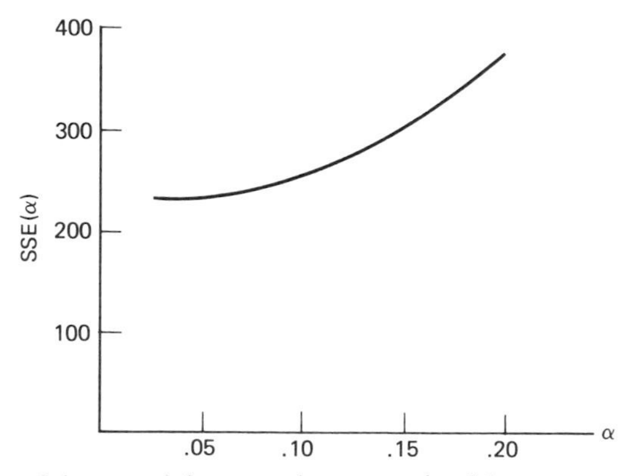

Figure 4.10. Plot of the sum of the squared one-step-ahead forecast errors from the model

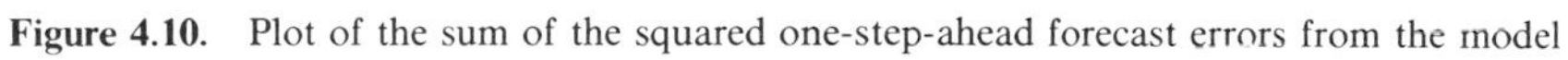

$$z_{n+j} = \beta_0 + \beta_1 j + \sum_{i=1}^{2} (\beta_{1i} \sin f_i j + \beta_{2i} \cos f_i j) + \varepsilon_{n+j}$$

Car sales.

Table 4.11. General Exponential Smoothing (α = .03) for the Car Sales Model

$$z_{n+j} = \beta_0 + \beta_1 j + \sum_{i=1}^{2} (\beta_{1i} \sin f_i j + \beta_{2i} \cos f_i j) + \varepsilon_{n+j}$$

Time	z_t	$\hat{\beta}_0$	$\hat{\beta}_1$	$\hat{\beta}_{11}$	$\hat{\beta}_{21}$	$\hat{\beta}_{12}$	$\hat{\beta}_{22}$	$\hat{z}_t(1)$	$e_{t-1}(1)$ $[= z_t - \hat{z}_{t-1}(1)]$
0		9.877	.088	2.575	−2.665	−2.956	.832	6.800	
1	6.550	9.950	.088	3.561	−1.034	−2.200	−2.158	7.939	−.250
2	8.728	10.082	.088	3.606	.929	.775	−2.940	11.978	.789
3	12.026	10.174	.088	2.658	2.610	2.934	−.797	15.994	.048
⋮									
95	16.119	17.939	.081	.764	−3.554	−.902	3.194	16.140	−1.753
96	13.713	17.884	.079	2.425	−2.831	−3.235	.681	14.263	−2.427
97	13.210	17.904	.078	3.510	−1.297	−2.216	−2.519	15.435	−1.053
⋮									
107	17.180	18.767	.080	.352	−3.551	−1.099	3.143	16.568	−1.417
108	14.577	18.735	.078	2.070	−3.010	−3.287	.510	14.650	−1.991

This estimate was calculated previously and was given in Table 4.4; $\hat{\boldsymbol{\beta}}_0 = (9.88, 0.088, 2.57, -2.66, -2.96, 0.83)'$. The parameter estimates and the forecasts are updated for several values of the smoothing constant α. Sums of the first $n = 96$ squared one-step-ahead forecast errors are given in Table 4.10 and are plotted in Figure 4.10. We find that the optimal smoothing constant is close to zero, which indicates that the trend and the seasonal components are quite stable.

The parameter estimates, the forecasts, and the one-step-ahead forecast errors are updated with the smoothing constant $\alpha = .03$. For this smoothing constant, the vector $\mathbf{F}^{-1}\mathbf{f}(0)$ is given by $(0.05610, 0.00085, 0.00549, 0.05566, 0.00742, 0.05527)'$. The coefficient estimates, forecasts, and one-step-ahead forecast errors are given in Table 4.11.

Other Possible Seasonal Forecast Functions

Brown (1962) considers models in which the seasonal swings grow with the level of the series; for example,

$$z_{n+j} = \beta_0 + \beta_1 j + (\beta_{11} + \beta_{12} j) \sin fj + (\beta_{21} + \beta_{22} j) \cos fj + \varepsilon_{n+j} \tag{4.30}$$

In this case the vector of fitting functions is given by

$$\mathbf{f}(j) = (1, j, \sin fj, \cos fj, j\sin fj, j\cos fj)'$$

and the transition matrix is

$$\mathbf{L} = \begin{bmatrix} 1 & 0 & 0 & 0 & 0 & 0 \\ 1 & 1 & 0 & 0 & 0 & 0 \\ 0 & 0 & \cos f & \sin f & 0 & 0 \\ 0 & 0 & -\sin f & \cos f & 0 & 0 \\ 0 & 0 & \cos f & \sin f & \cos f & \sin f \\ 0 & 0 & -\sin f & \cos f & -\sin f & \cos f \end{bmatrix}.$$

However, these models are not considered here.

4.4. WINTERS' SEASONAL FORECAST PROCEDURES

The discounted least squares approach uses only one parameter (the smoothing constant α) to take account of the time-changing coefficients; it is smoothing the trend and the seasonal components equally. Frequently, however, the seasonal component is more stable than the trend, and a single smoothing parameter will be just a bad compromise that discounts the seasonal component too much and the trend component too little.

Winters (1960) addresses this criticism and considers updating equations with several smoothing constants. He considers the linear trend model with seasonal indicators. The seasonal and the trend components can be either additive ($z_{n+j} = T_{n+j} + S_{n+j} + \varepsilon_{n+j}$) or multiplicative ($z_{n+j} = T_{n+j} \times S_{n+j} + \varepsilon_{n+j}$). Instead of using only one smoothing constant as in general exponential smoothing, Winters' forecast procedure incorporates three possibly different smoothing constants: one to update the level, one for the slope, and one for the seasonal components. It is an extension of Holt's two-parameter version of double exponential smoothing that was discussed in Section 3.6.

4.4.1. Winters' Additive Seasonal Forecast Procedure

Let us assume that the data come from the additive model

$$z_{n+j} = T_{n+j} + S_{n+j} + \varepsilon_{n+j} \tag{4.31}$$

where the trend component is linear, $T_{n+j} = \mu_{n+j} = \mu_n + \beta j$, and where the

s seasonal factors $S_i = S_{i+s} = S_{i+2s} = \cdots$ (for $i = 1, \ldots, s$), are restricted to add to zero, $\Sigma_{i=1}^{s} S_i = 0$.

An estimate of the level of the series at time $n + 1$ ($\hat{\mu}_{n+1}$) can be found from two sources:

1. The most recent observation, z_{n+1}, adjusted by its seasonal factor $\hat{S}_{n+1-s}$, which is available before z_{n+1} becomes known.
2. The trend estimate $\hat{\mu}_n + \hat{\beta}_n$, which uses the observations available at time n; $\hat{\beta}_n$ is the estimate of the slope at time n.

Winters considers a weighted average of these two estimates and updates $\hat{\mu}_{n+1}$ according to

$$\hat{\mu}_{n+1} = \alpha_1(z_{n+1} - \hat{S}_{n+1-s}) + (1 - \alpha_1)(\hat{\mu}_n + \hat{\beta}_n) \tag{4.32}$$

The coefficient α_1 is a smoothing constant; if α_1 is large, more weight is given to the most recent observation; this allows the level to update more quickly.

Similarly, the slope estimate $\hat{\beta}_{n+1}$ is taken as a weighted average of two estimates:

1. The most recent estimate of the slope, $\hat{\mu}_{n+1} - \hat{\mu}_n$.
2. The previous estimate of the slope $\hat{\beta}_n$.

Winters updates the slope estimate according to

$$\hat{\beta}_{n+1} = \alpha_2(\hat{\mu}_{n+1} - \hat{\mu}_n) + (1 - \alpha_2)\hat{\beta}_n \tag{4.33}$$

After adjusting for the seasonal factors, the updating equations (4.32) and (4.33) are identical to Holt's equations (3.41).

The estimate of the seasonal coefficient $\hat{S}_{n+1}$ is a weighted average of two estimates:

1. The most recent estimate of the seasonal factor, which is $z_{n+1} - \hat{\mu}_{n+1}$.
2. The previous estimate $\hat{S}_{n+1-s}$.

Winters revises the seasonal coefficient according to

$$\hat{S}_{n+1} = \alpha_3(z_{n+1} - \hat{\mu}_{n+1}) + (1 - \alpha_3)\hat{S}_{n+1-s} \tag{4.34}$$

The seasonal coefficients are updated only once every full season; the new seasonal estimate $\hat{S}_{n+1}$ is updated from the previous one s steps ago. Note that as the estimates of the seasonal factors are updated, they may not add to zero. One could modify this procedure by normalizing the seasonal factors at the end of each season.

After obtaining estimates for the trend, the slope, and the seasonal components, we can calculate the forecast of the future value z_{n+l} from time origin n as

$$\hat{z}_n(l) = \hat{\mu}_n + \hat{\beta}_n l + \hat{S}_{n+l-s} \qquad \text{for } l = 1, 2, \ldots, s$$

$$\hat{z}_n(l) = \hat{\mu}_n + \hat{\beta}_n l + \hat{S}_{n+l-2s} \qquad \text{for } l = s+1, \ldots, 2s \tag{4.35}$$

and so on.

The updating equations (4.32) to (4.34) depend on three smoothing constants $\alpha_1, \alpha_2, \alpha_3$, which lie somewhere between 0 and 1. If the components change rapidly, large smoothing constants should be considered; for stable components, the smoothing constants should be close to zero. If sufficient historical data are available, we can determine the smoothing constants by simulation, minimizing $\text{SSE} = \sum_{t=1}^{n}[z_t - \hat{z}_{t-1}(1)]^2$. Since three parameters are involved, this minimization will be more difficult and time-consuming than the minimization in general exponential smoothing with just one smoothing constant.

In order to start the updating equations (4.32) to (4.34), we must specify initial values for $\hat{\mu}_0$, $\hat{\beta}_0$, and $\hat{S}_{j-s}$ $(j = 1, 2, \ldots, s)$. We can obtain these starting values from the ordinary least squares estimates in the regression model

$$z_t = \mu + \beta t + \sum_{i=1}^{s-1} \delta_i (\text{IND}_{ti} - \text{IND}_{ts}) + \varepsilon_t \tag{4.36}$$

where IND_{ti} are seasonal indicators (1 if t is in seasonal period i, and 0 otherwise). At least the first $s + 1$ observations are needed to derive these estimates. If the trend and the seasonal pattern are stable, we suggest using all available observations $(z_1, \ldots, z_n)$. From the least squares estimates $\hat{\mu}$, $\hat{\beta}$, and $\hat{\delta}_j$ $(j = 1, \ldots, s-1)$ we obtain the initial values as

$$\hat{\mu}_0 = \hat{\mu} \qquad \hat{\beta}_0 = \hat{\beta}$$

$$\hat{S}_{j-s} = \hat{\delta}_j \qquad \text{for } 1 \leqslant j \leqslant s-1$$

$$\hat{S}_0 = -\sum_{j=1}^{s-1} \hat{\delta}_j \tag{4.37}$$

Example 4.4: Car Sales

As an example we consider the monthly Quebec car sales previously analyzed in Examples 4.1 and 4.3 (globally constant models), and in Section 4.3.2 (exponential smoothing for linear trend models with added harmonics). The smoothing constants $\alpha_1, \alpha_2, \alpha_3$ are determined from the first 96 observations. The computer program WINTERS1 in Appendix 4 is used to update the coefficients, determine the forecasts, and calculate the sum of the squared one-step-ahead forecast errors. We find that this sum is minimized if $\alpha_1 = .17$ and $\alpha_2 = \alpha_3 = .01$. This indicates that the seasonal factors are quite stable; however, the level exhibits slight changes. The autocorrelations of the historical one-step-ahead forecast errors are also calculated; the correlations are quite small and insignificant.

4.4.2. Winters' Multiplicative Seasonal Forecast Procedure

If the seasonal swings and the variation in the errors are proportional to the level, we should transform the data and consider the additive model for the logarithmically transformed observations. However, if the seasonal swings are proportional to the level, but the errors are not, we can consider Winters' multiplicative version of seasonal exponential smoothing. There it is assumed that the model is of the form

$$z_{n+j} = (\mu_n + \beta j) S_{n+j} + \varepsilon_{n+j} \tag{4.38}$$

where the s multiplicative seasonal factors $S_i = S_{i+s} = S_{i+2s} = \cdots$ (for $i = 1, \ldots, s$) are restricted to add to s; $\Sigma_{i=1}^{s} S_i = s$. The updating equations for $\hat{\mu}_n$, $\hat{\beta}_n$, and $\hat{S}_n$ are similar to Equations (4.32) to (4.34), except that in the multiplicative model the "seasonally adjusted" observation is given by the ratio $z_{n+1}/\hat{S}_{n+1-s}$, compared to the difference $z_{n+1} - \hat{S}_{n+1-s}$ in the additive version:

$$\hat{\mu}_{n+1} = \alpha_1 \frac{z_{n+1}}{\hat{S}_{n+1-s}} + (1 - \alpha_1)(\hat{\mu}_n + \hat{\beta}_n) \tag{4.39}$$

$$\hat{\beta}_{n+1} = \alpha_2(\hat{\mu}_{n+1} - \hat{\mu}_n) + (1 - \alpha_2)\hat{\beta}_n \tag{4.40}$$

$$\hat{S}_{n+1} = \alpha_3 \frac{z_{n+1}}{\hat{\mu}_{n+1}} + (1 - \alpha_3)\hat{S}_{n+1-s} \tag{4.41}$$

The smoothing constants $\alpha_1, \alpha_2, \alpha_3$ are supposed to be between 0 and 1. The forecasts are calculated from

$$\hat{z}_n(l) = (\hat{\mu}_n + \hat{\beta}_n l)\hat{S}_{n+l-s} \qquad l = 1, 2, \ldots, s$$
$$= (\hat{\mu}_n + \hat{\beta}_n l)\hat{S}_{n+l-2s} \qquad l = s+1, \ldots, 2s \tag{4.42}$$

and so on. Initial values for $\hat{\mu}_0$, $\hat{\beta}_0$, and the seasonal coefficients $\hat{S}_{j-s}$ ($j = 1, 2, \ldots, s$) can be obtained as follows [see Montgomery and Johnson (1976, p. 102)]. Let us denote the averages of the first k complete seasons by $\bar{z}_1, \bar{z}_2, \ldots, \bar{z}_k$. Then the initial values can be calculated from

$$\hat{\beta}_0 = \frac{\bar{z}_k - \bar{z}_1}{(k-1)s} \qquad \hat{\mu}_0 = \bar{z}_1 - \frac{s}{2}\hat{\beta}_0 \qquad \hat{S}_{j-s} = \frac{s\hat{R}_j}{\sum_{i=1}^{s} \hat{R}_i} \qquad j = 1, 2, \ldots, s \tag{4.43}$$

where the raw unstandardized seasonal coefficients $\hat{R}_j$ are calculated as

$$\hat{R}_j = \frac{1}{k}\sum_{t=1}^{k} \frac{z_{(t-1)s+j}}{\bar{z}_t - \left(\dfrac{s+1}{2} - j\right)\hat{\beta}_0} \tag{4.44}$$

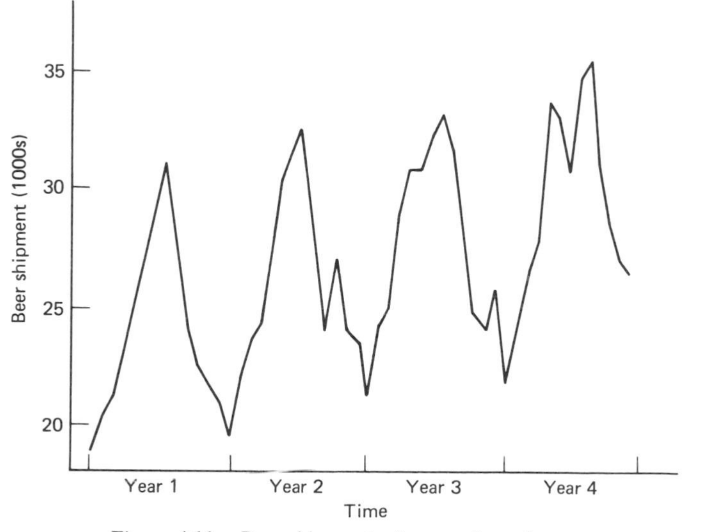

Figure 4.11. Beer shipments, four-week totals.

Table 4.12. Winters' Multiplicative Seasonal Forecasting Procedure—Beer Shipments ($s = 13$; $\alpha_1 = \alpha_2 = \alpha_3 = .05$)

Time	z_t	$\hat{\mu}_t$	$\hat{\beta}_t$	$\hat{S}_t$	$\hat{z}_t(1)$	$e_{t-1}(1)$ [$= z_t - \hat{z}_{t-1}(1)$]
−12				0.787		
−11				0.865		
−10				0.904		
−9				0.977		
−8				1.084		
−7				1.146		
−6				1.218		
−5				1.248		
−4				1.145		
−3				0.957		
−2				0.935		
−1				0.867		
0		22.958	0.150	0.867	18.188	
1	18.705	23.141	0.152	0.788	20.144	0.517
2	20.232	23.298	0.152	0.865	21.200	0.088
3	20.467	23.410	0.150	0.903	23.013	−0.733
4	22.123	23.514	0.148	0.975	25.656	−0.890
5	25.036	23.633	0.146	1.083	27.247	−0.620
6	26.839	23.762	0.145	1.145	29.119	−0.408
7	29.640	23.929	0.147	1.219	30.048	0.521
8	30.935	24.111	0.148	1.250	27.771	0.887
9	28.278	24.281	0.149	1.146	23.386	0.507
10	24.235	24.475	0.152	0.959	23.022	0.849
11	22.370	24.592	0.150	0.934	21.461	−0.652
12	21.224	24.728	0.149	0.867	21.563	−0.237
13	21.061	24.848	0.148	0.866	19.700	−0.502
⋮						
39	25.368	28.849	0.149	0.867	22.821	
40	21.260	28.899	0.144	0.784	25.118	−1.561
41	24.109	28.985	0.142	0.863	26.333	−1.009
42	26.320	29.126	0.142	0.904	28.594	−0.016
43	27.701	29.222	0.139	0.976	31.838	−0.893
44	34.502	29.484	0.145	1.089	33.950	2.664
45	33.297	29.601	0.144	1.145	36.225	−0.653
46	31.252	29.541	0.134	1.210	37.027	−4.973
47	35.173	29.600	0.130	1.245	34.030	−1.856
48	36.207	29.825	0.135	1.148	28.676	2.177
49	31.511	30.108	0.142	0.962	28.277	2.835
50	28.560	30.265	0.143	0.935	26.375	0.283
51	26.828	30.434	0.144	0.868	26.506	0.453
52	26.660	30.588	0.145	0.867	24.095	0.154

Estimates of the smoothing constants can be obtained by simulation. Frequently, the estimates of α_i ($i = 1, 2, 3$) will be correlated; an increase in α_1 will often lead to decreases in α_2 and α_3.

Example 4.5: Beer Shipment Data

To illustrate Winters' multiplicative seasonal forecast procedure, we consider shipment data from a major beer producer. The observations are listed as series 8 in the Data Appendix and are plotted in Figure 4.11. These data, which are total shipments in consecutive four-week periods, exhibit a strong seasonal pattern with seasonal period $s = 13$. The first 3 years ($n = 39$) are used to calculate the starting values $\hat{\mu}_0$, $\hat{\beta}_0$, and $\hat{S}_{j-s}$ ($j = 1, 2, \ldots, 13$). The updated estimates and the one-step-ahead forecasts for the following season ($t = 40, \ldots, 52$) are given in Table 4.12. The computer program WINTERS2, which is listed in Appendix 4, can be used to update the estimates, calculate the forecasts, and compute the sum of the squared one-step-ahead forecast errors.

It is not quite clear whether an additive or a multiplicative model should be considered here. For illustration, we have used the multiplicative version with smoothing constants $\alpha_1 = \alpha_2 = \alpha_3 = .05$. As more data points become available, one should check whether the amplitude of the seasonal component grows with the level. If it does not, then the additive model should be used.

4.5. SEASONAL ADJUSTMENT

Many economic and social time series exhibit seasonal variation, which makes trend assessment difficult. Thus there is great interest in government and industry to seasonally adjust the data and remove seasonal fluctuations. However, empirical comparisons indicate that forecasts obtained through seasonal adjustment procedures are no better than those generated from the unadjusted data [Abraham and Chatterjee (1982), Plosser (1979)]. Thus this topic is discussed only briefly here.

Seasonal adjustment consists of estimating and removing an unobservable seasonal component S_t from an observable series z_t. In the additive model (which is the only one considered here),

$$z_t = T_t + S_t + \varepsilon_t \tag{4.45}$$

the seasonally adjusted series is given by

$$z_t^{(a)} = z_t - \hat{S}_t \tag{4.46}$$

where $\hat{S}_t$ is an estimate of the seasonal component series. In general, there are three different approaches for estimating S_t: (1) regression, (2) smoothing, and (3) signal extraction. We consider the first two here and the third in Chapter 6 after our discussion of seasonal time series models.

4.5.1. Regression Approach

In this approach S_t may be modeled as a linear combination of seasonal indicator variables as in (4.6) or as a linear combination of trigonometric functions of different frequencies as in (4.7). Regression of z_t on these indicator variables or on these trigonometric functions leads to a set of residuals that can be taken as a seasonally adjusted series. This is roughly equivalent to subtracting the seasonal means from the original series. If the time series exhibits stable seasonal patterns that can be represented by deterministic functions of time, then the regression approach will produce a reasonable estimate of S_t and hence a "good" seasonally adjusted series. However, if the seasonal pattern is not stable, this procedure may not correctly adjust for seasonality [see Lovell (1963), Shiskin et al. (1967)].

4.5.2. Smoothing Approach

It is well known that moving averages (or linear smoothing filters) can smooth irregular fluctuations in the data. Hence, most seasonal adjustment procedures use various moving average techniques to smooth seasonal data.

Moving Averages

From a set of observations $z_1, \ldots, z_t, \ldots, z_n$ we define a symmetric $(2m + 1)$-term moving average (MA) as

$$\bar{z}_t(m) = \sum_{j=-m}^{m} \frac{z_{t+j}}{2m+1} \tag{4.47}$$

and a symmetric $(2m + 1)$-term weighted moving average (WMA) as

$$\bar{z}_{t,\omega}(m) = \sum_{j=-m}^{m} \omega_j z_{t+j} \tag{4.48}$$

where $\sum_{j=-m}^{m} \omega_j = 1$, and $\omega_j = \omega_{-j}$. Thus we can generate sets of averages moving with time. It should be noted that for n observations we will be left with only $n - 2m$ MA's (or WMA's). We lose m averages at each end of the series.

The computation of some special MA's and WMA's is illustrated here, using a set of artificial numbers. The results are presented in Table 4.13.

Table 4.13. Calculation of Some Special Moving Averages

(1) t	(2) z_t	(3) three-term MA	(4) 3 × 3-term MA	(5) 3 × 5-term MA	(6) 12-term MA	(7) 12-term centered MA	(8) Henderson's 9-term MA
1	7						
2	8	9.000					
3	12	11.000	11.333				
4	13	14.000	13.333	12.600			
5	17	15.000	14.333	13.000			15.011
6	15	14.000	13.333	12.400	10.750		13.617
7	10	11.000	11.000	11.400	11.000	10.875	10.975
8	8	8.000	9.333	10.400	11.000	11.000	9.062
9	6	9.000	9.000	9.800	11.250	11.125	8.739
10	13	10.000	10.000	9.600	11.500	11.375	9.709
11	11	11.000	10.333	9.800	11.250	11.375	10.223
12	9	10.000	10.000	10.200	11.000	11.125	9.882
13	10	9.000	10.000	10.800	11.000	11.000	9.823
14	8	11.000	11.000	11.600	11.000	11.000	11.015
15	15	13.000	13.000	12.400			13.143
16	16	15.000	14.000	13.000			14.506
17	14	14.000	13.667	12.800			
18	12	12.000	12.000				
19	10	10.000					
20	8						

Three-term MA

$$\bar{z}_t(3) = \frac{\sum_{j=-1}^{1} z_{t+j}}{3}$$

These are shown in column (3) of Table 4.13 for the set of numbers given in column (2). For example,

$$\bar{z}_2(3) = (7 + 8 + 12)/3 = 9$$

$$\bar{z}_3(3) = (8 + 12 + 13)/3 = 11$$

$$\vdots$$

$$\bar{z}_{19}(3) = (12 + 10 + 8)/3 = 10$$

The end values $\bar{z}_1(3)$ and $\bar{z}_{20}(3)$ are not available.

Table 4.14. **Weights ω_j Used in Henderson's Moving Averages**

j	0	± 1	± 2	± 3	± 4	± 5	± 6	± 7	± 8	± 9	± 10	± 11
5-term ($m = 2$)	.558	.294	$-.073$									
9-term ($m = 4$)	.330	.267	.119	$-.010$	$-.041$							
13-term ($m = 6$)	.240	.214	.147	.066	.000	$-.028$	$-.019$					
23-term ($m = 11$)	.148	.138	.122	.097	.068	.039	.013	$-.005$	$-.015$	$-.016$	$-.011$	$-.004$

(3 × 5)-term MA

$$\bar{z}_t(3 \times 5) = \frac{\sum_{j=-2}^{2} \bar{z}_{t+j}(3)}{5}$$

Here we take a five-term MA of the three-term MA's already generated. Column (5) lists the five-term MA's of the three-term MA's in column (3). For example,

$$\bar{z}_4(3 \times 5) = (9 + 11 + 14 + 15 + 14)/5 = 12.6$$

$$\bar{z}_5(3 \times 5) = (11 + 14 + 15 + 14 + 11)/5 = 13.0$$

$$\vdots$$

$$\bar{z}_{17}(3 \times 5) = (13 + 15 + 14 + 12 + 10)/5 = 12.8$$

The averages $\bar{z}_t(3 \times 5)$ (for $t = 1, 2, 3, 18, 19, 20$) are not available.

If a three-term instead of the five-term MA is used, then we get the double MA's presented in column (4).

12-term MA

$$\bar{z}_{t+}(12) = \frac{\sum_{j=-5}^{6} z_{t+j}}{12} \tag{4.49}$$

Since the number of observations in the average is even, there is no unique middle term; hence we use the modified notation in (4.49). The computations are exactly the same as those for the three-term MA's except that there are 12 observations to consider here. These are shown in column (6). For example,

$$\bar{z}_{6+}(12) = (7 + 8 + \cdots + 11 + 9)/12 = 10.75$$

$$\bar{z}_{7+}(12) = (8 + 12 + \cdots + 9 + 10)/12 = 11.00$$

$$\vdots$$

$$\bar{z}_{14+}(12) = (6 + 13 + \cdots + 10 + 8)/12 = 11.00$$

Centered 12-term MA

$$\bar{z}_t(12 \times 2) = \frac{\bar{z}_{t-1+}(12) + \bar{z}_{t+}(12)}{2}$$

These are obtained by taking a two-term MA of the 12-term MA's already generated. Each average corresponds to a specific time point in the sequence; this was not possible in the 12-term MA. Centered 12-term MA's are shown in column (7). For example,

$$\bar{z}_7(12 \times 2) = (10.75 + 11.00)/2 = 10.875$$

$$\bar{z}_8(12 \times 2) = (11.00 + 11.00)/2 = 11.00$$

$$\vdots$$

$$\bar{z}_{14}(12 \times 2) = (11.00 + 11.00)/2 = 11.00$$

Here we lose the first six and the last six averages.

Henderson's WMA

$$\bar{z}_{t,H}(2m+1) = \sum_{j=-m}^{m} \omega_j z_{t+j} \qquad \sum_{j=-m}^{m} \omega_j = 1 \tag{4.50}$$

where the ω_j's ($\omega_j = \omega_{-j}$) are specific weights corresponding to each m. When $m = 2$, the WMA is referred to as Henderson's five-term WMA; $m = 4$ corresponds to Henderson's nine-term WMA; $m = 6$ corresponds to Henderson's 13-term WMA; and $m = 11$ implies a 23-term WMA. The weights corresponding to 5-, 9-, 13-, and 23-term WMA's are shown in Table 4.14. Henderson's nine-term WMA is calculated in column (8) of Table 4.13. For example,

$$\begin{aligned}\bar{z}_{4,H}(9) = {} & (-.041 \times 7) + (-.010 \times 8) + (.119 \times 12) + (.267 \times 13) \\ & + (.330 \times 17) + (.267 \times 15) + (.119 \times 10) + (-.010 \times 8) \\ & + (-.041 \times 6) = 15.011\end{aligned}$$

$$\begin{aligned}\bar{z}_{5,H}(9) = {} & (-.041 \times 8) + (-.010 \times 12) + (.119 \times 13) + (.267 \times 17) \\ & + (.330 \times 15) + (.267 \times 10) + (.119 \times 8) + (-.010 \times 6) \\ & + (-.041 \times 13) = 13.617\end{aligned}$$

$$\begin{aligned}\bar{z}_{16,H}(9) = {} & (-.041 \times 9) + (-.010 \times 10) + (.119 \times 8) + (.267 \times 15) \\ & + (.330 \times 16) + (.267 \times 14) + (.119 \times 12) + (-.010 \times 10) \\ & + (-.041 \times 8) = 14.506\end{aligned}$$

4.5.3. Seasonal Adjustment Procedures

Several procedures for seasonal adjustment are currently available; five of these are:

1. The X-11 program developed at the Census Bureau in Washington, D.C. [see Shiskin et al. (1967)].
2. The BE program used by the Bank of England [see Burman (1965, 1967)].
3. The SABL program developed at Bell Laboratories [see Cleveland et al. (1979)].
4. The SEABIRD method (or the replacement DAINTIES) used by the Statistical Office of the European Economic Community [see Bongard (1960), Mesnage (1968)].
5. The X-11-ARIMA program developed at Statistics Canada [see Dagum (1975)].

Since X-11 is the most commonly used program, we give a brief outline of this method.

X-11 Method

The basic ideas behind the Census Method II-X-11 variant program can be summarized in the following 12 steps. In our discussion we assume (1) an additive model and (2) monthly data with yearly seasonal pattern.

Step 1. Calculate a centered 12-term MA. An initial estimate of the seasonal and irregular component series is obtained by subtracting these MA's from the original series.

Step 2. Apply a weighted five-term [or a (3×3)-term] MA to each month separately to obtain an estimate of the seasonal factors.

Step 3. Compute a centered 12-term MA of the seasonal factors from step 2 for the entire series. Fill in the six missing values at either end of this average. [One method used is to repeat the first (last) available MA six times.] Adjust the seasonal factors from step 2 by subtracting the centered 12-term MA's; the adjusted seasonal factors sum to zero (approximately) over any 12-month period.

Step 4. Obtain an estimate of the irregular component series by subtracting the seasonal factor estimates from the seasonal and irregular component series obtained in step 1. This component is needed for outlier adjustment.

Step 5. Treatment of outliers: Compute a moving five-year standard deviation s_* of the estimates of the irregular component. Assign a zero weight to irregulars beyond $2.5s_*$ and a weight of 1 to irregulars within $1.5s_*$. Assign a linearly graduated weight between 0 and 1 to irregulars between $2.5s_*$ and $1.5s_*$. Use this weight function to adjust the seasonal and irregular component series from step 1 for outliers.

Step 6. Apply a weighted seven-term MA to this adjusted series for each month separately to estimate preliminary seasonal factors.

Step 7. Repeat step 3 to standardize the seasonal factors.

Step 8. Obtain a preliminary seasonally adjusted series by subtracting the series obtained in step 7 from the original series.

Step 9. To obtain a trend estimate, apply a 9 (or 13 or 23)-term Henderson WMA to the seasonally adjusted series. Subtract this estimate from the original data to get a second estimate of the seasonal and irregular component series.

Step 10. Apply a weighted seven-term [or (3×5)-term] MA to each month separately to obtain a second estimate of the seasonal component.

Step 11. Repeat step 3 to standardize the seasonal factors.

Step 12. Subtract these final estimates of the seasonal component from the original series to obtain the final seasonally adjusted series.

Modifications for quarterly data and for multiplicative models are straightforward and are implemented in the X-11 program. In addition, various options for "trading-day adjustments" are possible, which adjust for the variation in the number of working days per month.

Although this multistep procedure is complex, the resulting seasonally adjusted series can be approximated [Wallis (1974)] by the output from a linear filter of the form

$$z_t^{(a)} = \sum_{j=-m}^{m} \alpha_j z_{t+j} \tag{4.51}$$

The "half-length" m is the sum of the half-lengths of the component moving averages and depends mainly on the choice in step 9. This depends on the relative contributions of the trend cycle and the irregular components to the variability of the preliminary seasonally adjusted series in step 8. The

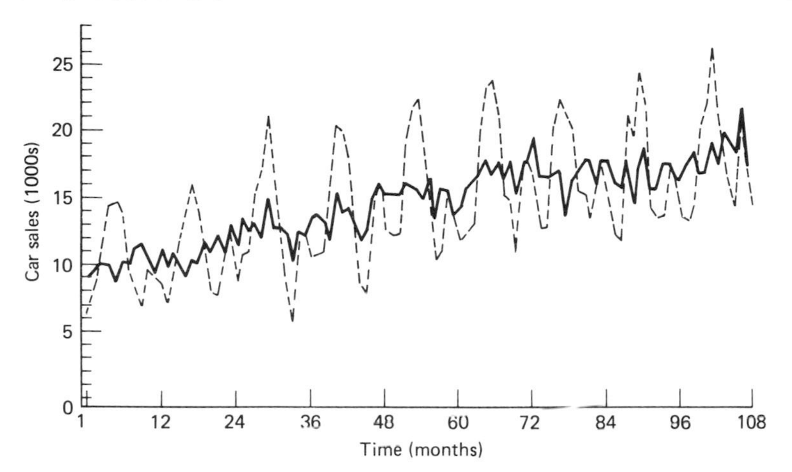

Figure 4.12. Original (broken line) and seasonally adjusted (solid line)—car sales.

greater the irregular contribution, the longer the half-length of the moving average.

We consider the monthly car sales in Figure 4.4 to illustrate the additive version of the X-11 program. The original series and the seasonally adjusted series are presented in Figure 4.12. The trend behavior is clearly visible from the seasonally adjusted time series.

Comments on X-11 and Other Seasonal Adjustment Procedures

The seasonal adjustment procedures imply that the seasonality is stochastic rather than deterministic. Thus they are quite appropriate for most economic series, because their seasonal movements are usually not quite stable. Another feature of these procedures is that they are fully automatic and can be used to seasonally adjust a large number of series within a short time. Although this saves effort it can be a weakness, because the same "model" is used to isolate seasonality from different types of series. The question is whether a single seasonal adjustment program, however flexible and comprehensive, can serve all the various needs.

Another problem in X-11 and other related procedures is that the symmetric moving averages cannot be applied to the initial and last part of the series, since the observations z_t ($t \leqslant 0$ and $t > n$) are not available. Hence asymmetric moving averages are used. This means that usually, when new data become available, large revisions must be made. This problem is addressed by the X-11-ARIMA method, which replaces the necessary initial and final unobserved values by their forecasts. These forecasts are obtained from corresponding ARIMA time series models (see Chap. 6).

APPENDIX 4: COMPUTER PROGRAMS FOR SEASONAL EXPONENTIAL SMOOTHING

In this appendix we list computer programs for general exponential smoothing with seasonal indicators (EXPSIND), general exponential smoothing with trigonometric forecast functions (EXPHARM), and Winters' additive (WINTERS1) and Winters' multiplicative (WINTERS2) forecast procedures.

```
C**** PROGRAM    E X P S I N D
C**** GENERAL EXPONENTIAL SMOOTHING
C**** LINEAR TREND MODEL WITH IS-1 SEASONAL INDICATORS
C**** TOTAL NUMBER OF PARAMETERS IS IDEM=IS+1
C**** IS IS SEASONAL PERIOD
C**** NOBS IS THE TOTAL NUMBER OF OBSERVATIONS
C**** NOBS1  NUMBER OF OBSERVATIONS TO DETERMINE THE INITIAL VALUES
C**** AL IS THE SMOOTHING CONSTANT
C**** LEAD IS THE FORECAST LEAD TIME
      IMPLICIT REAL*8(A-H,O-Z)
      DIMENSION Z(200),ER(200),X(200,15),WK1(400),WK2(400),C(15),
     1FO(200,26),T(15,15),FM(15,15),FINV(15,15),XXI(15,15),
     2F(15),H(15),XX(15,15),B(200,15)
      READ(10,800)IS,NOBS,NOBS1,LEAD,AL
  800 FORMAT(4I5,F5.0)
      READ(10,801)(Z(I),I=1,NOBS)
  801 FORMAT(      )
      IDEM=IS+1
C**** T IS THE TRANSITION MATRIX, FM IS THE MATRIX F IN THE UPDATING EQUATIONS
      OM=1.-AL
      OM1=OM**IS
      OM2=1.-OM1
      OM3=OM2*OM2
      DO 100 I=1,IDEM
      DO 100 J=1,IDEM
      T(I,J)=0.
      FM(I,J)=0.
  100 CONTINUE
C**** INITIALIZATION OF THE TRANSITION MATRIX T
      T(1,1)=1.
      T(2,1)=1.
      T(2,2)=1.
      T(3,1)=1.
      DO 101 I=3,IDEM
  101 T(3,I)=-1.
      DO 102 I=4,IDEM
  102 T(I,I-1)=1.
C**** INITIALIZATION OF THE MATRIX FM
      FM(1,1)=1./AL
      FM(2,2)=OM*(1.+OM)/(AL**3)
      FM(1,2)=-OM/(AL**2)
      FM(2,1)=FM(1,2)
      DO 110 I=2,IS
      FM(2,I+1)=(((IDEM-I)*(OM**(IDEM-I)))+((I-1)*(OM**(2*IS+1-I))))/OM3
      FM(2,I+1)=-FM(2,I+1)
      FM(I+1,2)=FM(2,I+1)
  110 CONTINUE
```

```
      DO 111 I=2,IS
      FM(1,I+1)=(OM**(IDEM-I))/OM2
      FM(I+1,1)=FM(1,I+1)
  111 FM(I+1,I+1)=FM(1,I+1)
C**** LINV2F IS THE IMSLS MATRIX INVERSION ROUTINE
C**** IF NOT AVAILABLE, REPLACE BY ANY OTHER INVERSION PROGRAM
      CALL LINV2F(FM,IDEM,15,FINV,6,WK1,IER)
      WRITE(1,900)IER
  900 FORMAT(' DIAGNOSTIC ERROR FROM THE IMSL MATRIX INVERSION',I5)
      DO 113 I=1,IDEM
  113 F(I)=0.
      F(1)=1.
      DO 115 I=1,IDEM
      H(I)=0.
      DO 115 J=1,IDEM
  115 H(I)=H(I)+FINV(I,J)*F(J)
      WRITE(1,901)
  901 FORMAT(' GENERAL EXPONENTIAL SMOOTHING')
      WRITE(1,902)
  902 FORMAT(' LINEAR TREND MODEL WITH SEASONAL INDICATORS')
      WRITE(1,903) AL
  903 FORMAT(' SMOOTHING CONSTANT',F6.3)
      WRITE(1,904)
  904 FORMAT(' SMOOTHING VECTOR  FINV*F(0)')
      WRITE(1,905)(H(I),I=1,IDEM)
  905 FORMAT(10F10.5)
C**** CALCULATION OF THE STARTING VALUES
      KK=(NOBS1/IS)+1
      DO 400 I=1,NOBS1
      DO 400 J=1,IDEM
  400 X(I,J)=0.
      DO 401 I=1,NOBS1
      X(I,1)=1.
  401 X(I,2)=I
      DO 402 I=1,KK
      DO 402 J=1,IS
      I1=IS*(I-1)+J
      I2=J+2
      IF(J.LT.IS) X(I1,I2)=1.
  402 CONTINUE
      DO 403 I=1,IDEM
      DO 403 J=1,IDEM
      XX(I,J)=0.
      DO 403 K=1,NOBS1
  403 XX(I,J)=XX(I,J)+X(K,I)*X(K,J)
      CALL LINV2F(XX,IDEM,15,XXI,4,WK2,IER)
      WRITE(1,900)IER
      DO 409 J=1,IDEM
      C(J)=0.
      DO 409 K=1,NOBS1
  409 C(J)=C(J)+X(K,J)*Z(K)
      DO 410 J=1,IDEM
      B(1,J)=0.
      DO 410 K=1,IDEM
  410 B(1,J)=B(1,J)+XXI(J,K)*C(K)
C**** UPDATING OF THE PARAMETER ESTIMATES AND CALCULATION OF THE FORECASTS
      N11=NOBS1+1
      DO 500 I=1,N11
      DO 510 L=1,LEAD
      DO 511 JJ=1,IDEM
  511 F(JJ)=0.
```

```
      F(1)=1.
      F(2)=L
      K1=(L-1)/IS
      K2=L-(K1*IS)
      IF(K2.LT.IS) F(K2+2)=1.
      FO(I,L)=0.
      DO 512 K=1,IDEM
 512  FO(I,L)=FO(I,L)+B(I,K)*F(K)
 510  CONTINUE
      IF(I.EQ.N11) GO TO 500
      DO 322 J=1,IDEM
      B(I+1,J)=0.
      DO 321 K=1,IDEM
 321  B(I+1,J)=B(I+1,J)+T(K,J)*B(I,K)
 322  B(I+1,J)=B(I+1,J)+H(J)*(Z(I)-FO(I,1))
 500  CONTINUE
C**** OUTPUT OF RESULTS
      IND=0
      WRITE(1,910)
 910  FORMAT('    T      Z(T)     Z(T-1,1)    E(T-1,1)   PAR.  ESTIMATES')
      WRITE(1,911) IND,(B(1,J),J=1,IDEM)
 911  FORMAT(I5,32X,15F6.3)
      DO 570 I=1,NOBS
      ER(I)=Z(I)-FO(I,1)
 570  WRITE(1,912)I,Z(I),FO(I,1),ER(I),(B(I+1,J),J=1,IDEM)
 912  FORMAT(I5,3F10.3,2X,15F6.3)
      ERR=0.
      DO 580 I=1,NOBS
 580  ERR=ERR+ER(I)*ER(I)
      WRITE(1,915) ERR
 915  FORMAT(' SSE(ALPHA) = ', F10.3)
      WRITE(1,916) NOBS
 916  FORMAT(' L STEP AHEAD FORECASTS FROM TIME ORIGIN',I10)
      DO 590 I=1,LEAD
 590  WRITE(1,917) I,FO(N11,I)
 917  FORMAT(I10,F10.3)
      STOP
      END
```

```
C**** PROGRAM    E X P H A R M
C**** GENERAL EXPONENTIAL SMOOTHING
C**** LINEAR TREND MODEL WITH M HARMONICS
C**** TOTAL NUMBER OF PARAMETERS IS IDEM=2*(M+1)
C**** IS IS SEASONAL PERIOD
C**** NOBS IS THE TOTAL NUMBER OF OBSERVATIONS
C**** NOBS1  NUMBER OF OBSERVATIONS TO DETERMINE THE INITIAL VALUES
C**** AL IS THE SMOOTHING CONSTANT
C**** LEAD IS THE FORECAST LEAD TIME
      IMPLICIT REAL*8(A-H,O-Z)
      DIMENSION Z(200),ER(200),X(200,15),WK1(400),WK2(400),C(15),
     1FO(200,26),T(15,15),FM(15,15),FINV(15,15),XXI(15,15),
     2B(200,15),XX(15,15),F(15),H(15)
      READ(5,800)IS,NOBS,NOBS1,M,LEAD,AL
 800  FORMAT(5I5,F5.0)
      READ(5,801)(Z(I),I=1,NOBS)
 801  FORMAT(12F6.3)
      PI=3.141592653589793
      ANG=2.*PI/IS
      IDEM=2+2*M
      OM=1.-AL
C**** T IS THE TRANSITION MATRIX, FM IS THE MATRIX F IN THE UPDATING EQUATIONS
      DO 100 I=1,IDEM
      DO 100 J=1,IDEM
      T(I,J)=0.
 100  FM(I,J)=0.
C**** INITIALIZATION OF THE TRANSITION MATRIX T
      T(1,1)=1.
      T(2,1)=1.
      T(2,2)=1.
      DO 101 I=1,M
      AA=ANG*I
      I1=1+2*I
      I2=2+2*I
      T(I1,I1)=DCOS(AA)
      T(I1,I2)=DSIN(AA)
      T(I2,I1)=-DSIN(AA)
      T(I2,I2)=DCOS(AA)
 101  CONTINUE
C**** INITIALIZATION OF THE MATRIX F, STORED IN FM
      OM1=1-OM
      FM(1,1)=1./OM1
      FM(2,2)=OM*(1.+OM)/(OM1**3)
      FM(1,2)=-OM/(OM1**2)
      FM(2,1)=FM(1,2)
      DO 110 I=1,M
      AA=-ANG*I
      I1=1+2*I
      I2=2+2*I
      OM2=1.-(2.*OM*DCOS(AA))+(OM*OM)
      FM(1,I1)=(OM*DSIN(AA))/OM2
      FM(1,I2)=(1.-OM*DCOS(AA))/OM2
      FM(2,I1)=-(OM*(1.-OM*OM)*DSIN(AA))/(OM2*OM2)
      FM(2,I2)=(2.*OM*OM-(OM*(1.+OM*OM)*DCOS(AA)))/(OM2*OM2)
      FM(I1,1)=FM(1,I1)
      FM(I2,1)=FM(1,I2)
      FM(I1,2)=FM(2,I1)
      FM(I2,2)=FM(2,I2)
 110  CONTINUE
```

```
      DO 111 I=1,M
      I1=1+2*I
      I2=2+2*I
      A1=-ANG*I
      DO 111 J=1,M
      A2=-J*ANG
      J1=1+2*J
      J2=2+2*J
      OM2=1.-(2.*OM*DCOS(A1+A2))+(OM*OM)
      OM3=1.-(2.*OM*DCOS(A1-A2))+(OM*OM)
      OM4=1.-(OM*DCOS(A1+A2))
      OM5=1.-(OM*DCOS(A1-A2))
      FM(I1,J1)=-.5*((OM4/OM2)-(OM5/OM3))
      FM(I2,J2)=.5*((OM4/OM2)+(OM5/OM3))
      FM(I1,J2)=.5*((OM*(DSIN(A1+A2))/OM2)+(OM*(SIN(A1-A2))/OM3))
      FM(J2,I1)=FM(I1,J2)
  111 CONTINUE
C**** LINV2F IS THE IMSL MATRIX INVERSION ROUTINE
C**** IF NOT AVAILABLE, REPLACE BY ANY OTHER INVERSION PROGRAM
      CALL LINV2F(FM,IDEM,15,FINV,6,WK1,IER)
      WRITE(6,900)IER
  900 FORMAT(' DIAGNOSTIC ERROR FROM THE IMSL MATRIX INVERSION ',I5)
      DO 113 I=1,IDEM
  113 F(I)=0.
      F(1)=1.
      DO 114 I=1,M
      I1=2+2*I
  114 F(I1)=1.
      DO 115 I=1,IDEM
      H(I)=0.
      DO 115 J=1,IDEM
  115 H(I)=H(I)+FINV(I,J)*F(J)
      WRITE(6,901)
  901 FORMAT(' GENERAL EXPONENTIAL SMOOTHING')
      WRITE(6,902) M
  902 FORMAT(' LINEAR TREND MODEL WITH',I3,' ADDED HARMONICS')
      WRITE(6,903) AL
  903 FORMAT(' SMOOTHING CONSTANT',F6.3)
      WRITE(6,904)
  904 FORMAT(' SMOOTHING VECTOR  FINF*F(0)')
      WRITE(6,905)(H(I),I=1,IDEM)
  905 FORMAT(10F10.5)
C**** CALCULATION OF THE STARTING VALUES
      DO 400 I=1,NOBS1
      X(I,1)=1.
      X(I,2)=I
      DO 400 II=1,M
      I1=1+2*II
      I2=2+2*II
      AA=ANG*II
      X(I,I1)=DSIN(AA*I)
  400 X(I,I2)=DCOS(AA*I)
      DO 402 I=1,IDEM
      DO 402 J=1,IDEM
      XX(I,J)=0.
      DO 402 K=1,NOBS1
  402 XX(I,J)=XX(I,J)+X(K,I)*X(K,J)
      CALL LINV2F(XX,IDEM,15,XXI,4,WK2,IER)
      WRITE(6,900)IER
```

```
      DO 409 J=1,IDEM
      C(J)=0.
      DO 409 K=1,NOBS1
  409 C(J)=C(J)+X(K,J)*Z(K)
      DO 410 J=1,IDEM
      B(1,J)=0.
      DO 410 K=1,IDEM
  410 B(1,J)=B(1,J)+XXI(J,K)*C(K)
C**** UPDATING OF THE PARAMETER ESTIMATES AND CALCULATION OF THE FORECASTS
      N11=NOBS+1
      DO 500 I=1,N11
      DO 510 J=1,LEAD
      F(1)=1.
      F(2)=J
      DO 511 II=1,M
      AA=ANG*II
      I1=1+2*II
      I2=2+2*II
      F(I1)=DSIN(AA*J)
  511 F(I2)=DCOS(AA*J)
      FO(I,J)=0.
      DO 512 K=1,IDEM
  512 FO(I,J)=FO(I,J)+B(I,K)*F(K)
  510 CONTINUE
      IF(I.EQ.N11) GO TO 500
      DO 560 J=1,IDEM
      B(I+1,J)=0.
      DO 561 K=1,IDEM
  561 B(I+1,J)=B(I+1,J)+T(K,J)*B(I,K)
  560 B(I+1,J)=B(I+1,J)+H(J)*(Z(I)-FO(I,1))
  500 CONTINUE
C**** OUTPUT OF RESULTS
      IND=0
      WRITE(6,910)
  910 FORMAT('    T      Z(T)      Z(T-1,1)   E(T-1,1)   PAR. ESTIMATES')
      WRITE(6,911) IND,(B(1,J),J=1,IDEM)
  911 FORMAT(I5,32X,15F6.3)
      DO 570 I=1,NOBS
      ER(I)=Z(I)-FO(I,1)
  570 WRITE(6,912)I,Z(I),FO(I,1),ER(I),(B(I+1,J),J=1,IDEM)
  912 FORMAT(I5,3F10.3,2X,15F6.3)
      ERR=0.
      DO 580 I=1,NOBS
  580 ERR=ERR+ER(I)*ER(I)
      WRITE(6,915) ERR
  915 FORMAT(' SSE(ALPHA) = ', F10.3)
      WRITE(6,916) NOBS
  916 FORMAT(' L STEP AHEAD FORECASTS FROM TIME ORIGIN',I10)
      DO 590 I=1,LEAD
  590 WRITE(6,917) I,FO(N11,I)
  917 FORMAT(I10,F10.3)
      STOP
      END
```

```
C**** PROGRAM       W I N T E R S 1
C**** WINTERS ADDITIVE SEASONAL EXPONENTIAL SMOOTHING
C**** IS  SEASONAL PERIOD
C**** NOBS  NUMBER OF OBSERVATIONS
C**** NOBS1  NUMBER OF OBSERVATIONS TO DETERMINE STARTING VALUES
C**** NOBS1 IS ASSUMED TO BE INTEGER MULTIPLE OF IS
C**** AL1,AL2,AL3   ARE THE SMOOTHING CONSTANTS
C**** LEAD  IS THE FORECAST LEAD TIME
      DIMENSION Z(200),A(200),B(200),S(200),ER(200),ZA(20),R(20),
     1X(200,20),XX(20,20),XXI(20,20),C(20),D(20),WK(400),FO(200,26)
      READ(5,800) IS,NOBS,NOBS1,LEAD,AL1,AL2,AL3
  800 FORMAT(4I5,3F5.0)
      IDEM=IS+1
      N=NOBS+IS
      READ(5,801)(Z(I),I=IDEM,N)
  801 FORMAT(       )
C**** CALCULATION OF THE STARTING VALUES FOR MU,BETA AND SEAS FACTORS
C**** THE STARTING VALUES ARE STORED IN A(IS),B(IS),S(1),...,S(IS)
C**** K  IS THE NUMBER OF FULL SEASONS USED TO CALCULATE STARTING VALUES
      K=NOBS1/IS
      DO 60 I=1,NOBS1
      DO 60 J=1,IDEM
   60 X(I,J)=0.
      DO 62 I=1,K
      DO 62 J=1,IS
      I1=IS*(I-1)+J
      X(I1,1)=1.
      X(I1,2)=I1
      IF(J.EQ.IS) GO TO 69
      X(I1,J+2)=1.
      GO TO 62
   69 CONTINUE
      DO 63 II=2,IS
   63 X(I1,II+1)=-1.
   62 CONTINUE
      DO 64 I=1,IDEM
      DO 64 J=1,IDEM
      XX(I,J)=0.
      DO 64 I1=1,NOBS1
   64 XX(I,J)=XX(I,J)+X(I1,I)*X(I1,J)
      DO 66 I=1,IDEM
      C(I)=0.
      DO 66 I1=1,NOBS1
   66 C(I)=C(I)+X(I1,I)*Z(IS+I1)
C**** LINV2F IS THE IMSL MATRIX INVERSION ROUTINE
C**** IF NOT AVAILABLE, REPLACE BY ANY OTHER MATRIX INVERSION PROGRAM
      CALL LINV2F(XX,IDEM,20,XXI,4,WK,IER)
      WRITE(6,920) IER
  920 FORMAT(' ERROR DIAGNOSTIC FROM THE IMSL MATRIX INVERSION ',I5)
      DO 68 I=1,IDEM
```

```
      D(I)=0.
      DO 68 J=1,IDEM
 68   D(I)=D(I)+XXI(I,J)*C(J)
      A(IS)=D(1)
      B(IS)=D(2)
      SUM=0.
      DO 70 I=2,IS
 70   SUM=SUM+D(I+1)
      DO 72 J=2,IS
 72   S(J-1)=D(J+1)
      S(IS)=-SUM
C**** UPDATING OF COEFFICIENTS
      DO 100 I=IDEM,N
      A(I)=AL1*(Z(I)-S(I-IS))+(1.-AL1)*(A(I-1)+B(I-1))
      B(I)=AL2*(A(I)-A(I-1))+(1.-AL2)*B(I-1)
 100  S(I)=AL3*(Z(I)-A(I))+(1.-AL3)*S(I-IS)
C**** CALCULATION OF THE L STEP AHEAD FORECAST, STORED IN FO(T,L)
      DO 102 KK=IS,N
      DO 102 L=1,LEAD
      K1=(L-1)/IS
      K2=L-(K1*IS)
 102  FO(KK,L)=A(KK)+B(KK)*L+S(KK+K2-IS)
C**** CALCULATION OF SUM OF SQUARED ONE STEP AHEAD FORCAST ERRORS
      ERR=0.
      DO 110 I=IDEM,N
      ER(I)=Z(I)-FO(I-1,1)
 110  ERR=ERR+ER(I)*ER(I)
C**** OUTPUT OF RESULTS
      WRITE(6,900)
 900  FORMAT(' WINTERS ADDITIVE SEASONAL EXPONENTIAL SMOOTHING')
      WRITE(6,901) AL1,AL2,AL3
 901  FORMAT(' SMOOTHING CONSTANTS',3F5.2)
      WRITE(6,902)
 902  FORMAT('    T   MU(T)  BETA(T)   S(T)    Z(T)  Z(T-1,1) E(T-1,1)')
      DO 120 I=1,N
      IND=I-IS
 120  WRITE(6,905) IND,A(I),B(I),S(I),Z(I),FO(I-1,1),ER(I)
 905  FORMAT(I5,6F8.3)
      WRITE(6,906) ERR
 906  FORMAT(' SSE(AL1,AL2,AL3) = ',F10.3)
      WRITE(6,907) NOBS
 907  FORMAT(' L STEP AHEAD FORCASTS FROM TIME ORIGIN',I10)
      DO 125 I=1,LEAD
 125  WRITE(6,908) I,FO(N,I)
 908  FORMAT(2F10.3)
      STOP
      END
```

```
C**** PROGRAM    W I N T E R S  2
C**** WINTERS MULTIPLICATIVE SEASONAL EXPONENTIAL SMOOTHING
C**** IS  SEASONAL PERIOD
C**** NOBS  NUMBER OF OBSERVATIONS
C**** NOBS1  NUMBER OF OBSERVATIONS TO DETERMINE STARTING VALUES
C**** NOBS1 IS ASSUMED TO BE INTEGER MULTIPLE OF IS
C**** AL1,AL2,AL3   ARE THE SMOOTHING CONSTANTS
C**** LEAD  IS THE FORECAST LEAD TIME
      DIMENSION Z(200),A(200),B(200),S(200),ER(200),ZA(20),R(20),
     1FO(200,26)
      READ(5,800) IS,NOBS,NOBS1,LEAD,AL1,AL2,AL3
  800 FORMAT(4I5,3F5.0)
      IDEM=IS+1
      N=NOBS+IS
      READ(5,801)(Z(I),I=IDEM,N)
  801 FORMAT(6X,F9.3,1X,F9.3,1X,F9.3,1X,F9.3,1X,F9.3,1X,F9.3,1X,F9.3)
C**** CALCULATION OF THE STARTING VALUES FOR MU,BETA AND SEAS FACTORS
C**** THE STARTING VALUES ARE STORED IN A(IS),B(IS),S(1),...,S(IS)
C**** K  IS THE NUMBER OF FULL SEASONS USED TO CALCULATE STARTING VALUES
      K=NOBS1/IS
      DO 60 I=1,K
      ZA(I)=0.
      DO 61 J=1,IS
   61 ZA(I)=ZA(I)+Z(I*IS+J)
   60 ZA(I)=ZA(I)/IS
      B(IS)=(ZA(K)-ZA(1))/((K-1)*IS)
      A(IS)=ZA(1)-(IS/2.)*B(IS)
      DO 70 J=1,IS
      R(J)=0.
      DO 71 I=1,K
   71 R(J)=R(J)+(Z(I*IS+J)/(ZA(I)-(((IS+1.)/2.)-J)*B(IS)))
   70 R(J)=R(J)/K
      SUM=0.
      DO 72 J=1,IS
   72 SUM=SUM+R(J)
      DO 73 J=1,IS
   73 S(J)=IS*R(J)/SUM
C**** UPDATING OF COEFFICIENTS
      DO 100 I=IDEM,N
      A(I)=AL1*(Z(I)/S(I-IS))+(1.-AL1)*(A(I-1)+B(I-1))
      B(I)=AL2*(A(I)-A(I-1))+(1.-AL2)*B(I-1)
  100 S(I)=AL3*(Z(I)/A(I))+(1.-AL3)*S(I-IS)
C**** CALCULATION OF THE L STEP AHEAD FORECAST, STORED IN FO(T,L)
      DO 102 KK=IS,N
      DO 102 L=1,LEAD
      K1=(L-1)/IS
      K2=L-(K1*IS)
  102 FO(KK,L)=(A(KK)+B(KK)*L)*S(KK+K2-IS)
C**** CALCULATION OF SUM OF SQUARED ONE STEP AHEAD FORCAST ERRORS
      ERR=0.
```

```
      DO 110 I=IDEM,N
      ER(I)=Z(I)-FO(I-1,1)
 110  ERR=ERR+ER(I)*ER(I)
C**** OUTPUT OF RESULTS
      WRITE(6,900)
 900  FORMAT(' WINTERS MULTIPLICATIVE SEASONAL EXPONENTIAL SMOOTHING')
      WRITE(6,901) AL1,AL2,AL3
 901  FORMAT(' SMOOTHING CONSTANTS',3F5.2)
      WRITE(6,902)
 902  FORMAT('    T   MU(T)  BETA(T)   S(T)    Z(T)  Z(T-1,1) E(T-1,1)')
      DO 120 I=1,N
      IND=I-IS
 120  WRITE(6,905) IND,A(I),B(I),S(I),Z(I),FO(I-1,1),ER(I)
 905  FORMAT(I5,6F8.3)
      WRITE(6,906) ERR
 906  FORMAT(' SSE(AL1,AL2,AL3) = ',F10.3)
      WRITE(6,907) NOBS
 907  FORMAT(' L STEP AHEAD FORCASTS FROM TIME ORIGIN',I10)
      DO 125 I=1,LEAD
 125  WRITE(6,908) I,FO(N,I)
 908  FORMAT(2F10.3)
      STOP
      END
```

CHAPTER 5

Stochastic Time Series Models

In Chapters 2 through 4 we considered forecast models of the form

$$z_t = f(\mathbf{x}_t; \boldsymbol{\beta}) + \varepsilon_t \tag{5.1}$$

The variables $\mathbf{x}_t$ are either predictor variables or functions of time, in particular polynomials and trigonometric functions. In regression models it is usually assumed that the parameters $\boldsymbol{\beta}$ are constant over time. The smoothing approach relaxes this assumption and discounts the predictive information of past data according to the age of the observations.

In these forecasting models it is usually assumed that the errors ε_t $(t = 1, 2, \ldots, n)$ are uncorrelated, which in turn implies that the observations z_t are uncorrelated. However, this assumption is rarely met in practice. Usually serial correlation can be expected if the data are collected sequentially in time. Thus models that capture the correlation structure have to be considered. In this chapter we discuss a special class of stochastic models called the *autoregressive integrated moving average* (*ARIMA*) *models*, which imply a variety of different correlation structures. Once the correlation structure has been appropriately modeled, it is straightforward to obtain predictions.

5.1. STOCHASTIC PROCESSES

An observed time series $(z_1, z_2, \ldots, z_n)$ can be thought of as a particular realization of a stochastic process. Stochastic processes in general can be described by an n-dimensional probability distribution $p(z_1, z_2, \ldots, z_n)$.

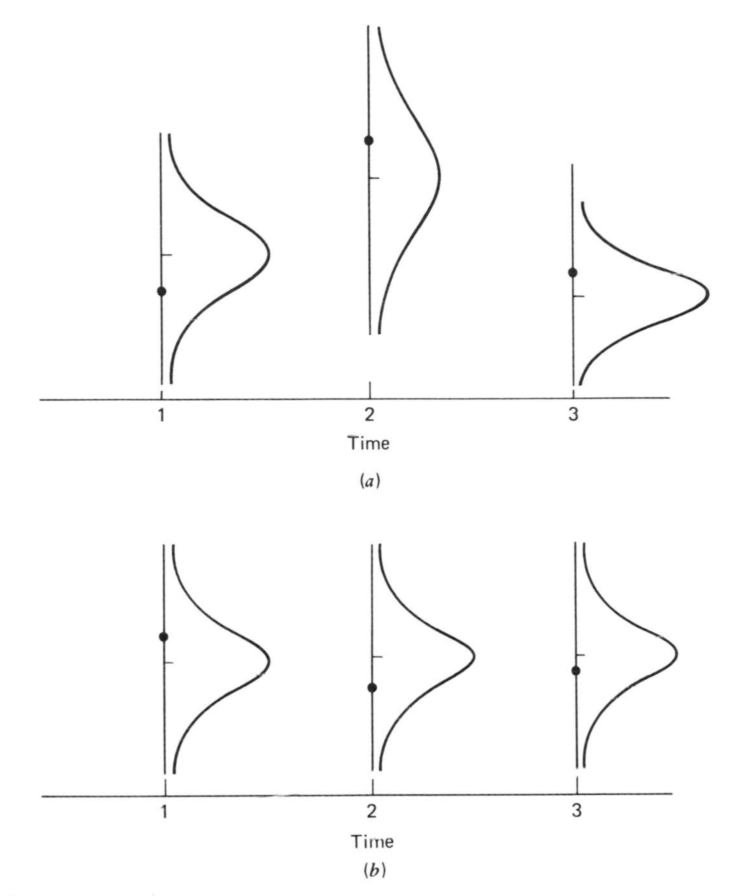

Figure 5.1. (*a*) Probability distributions for a general stochastic process ($n = 3$). The covariances $\text{Cov}(z_1, z_2)$, $\text{Cov}(z_2, z_3)$, $\text{Cov}(z_1, z_3)$ may be different. The dots represent possible realizations. (*b*) Probability distributions for a stationary stochastic process ($n = 3$). Also $\text{Cov}(z_1, z_2) = \text{Cov}(z_2, z_3)$. The dots represent possible realizations.

Assuming joint normality, such a distribution is described by the n means $E(z_1), E(z_2), \ldots, E(z_n)$; n variances $V(z_1), V(z_2), \ldots, V(z_n)$; and $n(n-1)/2$ covariances $\text{Cov}(z_i, z_j)$; $i < j$. The special case $n = 3$ is pictured in Figure 5.1*a*; the distribution is described by nine parameters (three means, three variances, and three covariances).

To infer such a general probability structure from just one realization of the stochastic process will be impossible, since there are only n observations but $n + n(n+1)/2$ unknown parameters. Hence some simplifying assumptions have to be made. One such assumption is *stationarity*.

5.1.1. Stationary Stochastic Processes

This is a special class of stochastic processes that assumes that the processes are in an "equilibrium." For stationarity, the probability distribution at any times $t_1, t_2, \ldots, t_m$ must be the same as the probability distribution at times $t_1 + k, t_2 + k, \ldots, t_m + k$, where k is an arbitrary shift along the time axis. If $m = 1$, this implies that the marginal distribution at time t is the same as the marginal distribution at any other point in time; $p(z_t) = p(z_{t+k})$. Thus the marginal distribution does not depend on time, which in turn implies that the mean $E(z_t) = \mu$ and the variance $V(z_t) = \gamma_0$ are constant [i.e., $E(z_1) = E(z_2) = \cdots = E(z_n)$; $V(z_1) = V(z_2) = \cdots = V(z_n)$]. Furthermore, if $m = 2$, stationarity implies that all bivariate distributions $p(z_t, z_{t-k})$ do not depend on t; thus the covariances $\text{Cov}(z_t, z_{t-k})$ are only functions of the lag k, but not of time t [i.e., $\text{Cov}(z_1, z_{1+k}) = \text{Cov}(z_2 z_{2+k}) = \cdots = \text{Cov}(z_{n-k}, z_n)$ for all k]. In Figure 5.1b we have pictured the marginal distributions of a stationary stochastic process for $n = 3$.

Autocorrelation Function (ACF)

The stationarity condition implies that the mean and the variance of the process are constant and that the *autocovariances*

$$\gamma_k = \text{Cov}(z_t, z_{t-k}) = E(z_t - \mu)(z_{t-k} - \mu) \tag{5.2}$$

and the *autocorrelations*

$$\rho_k = \frac{\text{Cov}(z_t, z_{t-k})}{\left[V(z_t) \cdot V(z_{t-k})\right]^{1/2}} = \frac{\gamma_k}{\gamma_0} \tag{5.3}$$

depend only on the lag (or time difference) k. Since these conditions apply only to the first- and second-order moments of the process, it is also called *second-order or weak stationarity*. If in addition we assume normality, in which case the distribution is characterized by the first two moments, stationarity and weak stationarity are equivalent.

The autocorrelations ρ_k are independent of the scale of the time series. The autocorrelations considered as a function of k are referred to as the *autocorrelation function* (*ACF*) or *correlogram*. Since $\gamma_k = \gamma_{-k}$ [$\gamma_k = \text{Cov}(z_t, z_{t-k}) = \text{Cov}(z_{t-k}, z_t) = \text{Cov}(z_t, z_{t+k}) = \gamma_{-k}$], and $\rho_k = \rho_{-k}$, only the positive half of the ACF is usually given. The ACF plays a major role in modeling the dependencies among observations, since it characterizes, together with the process mean $E(z_t)$ and variance $\gamma_0 = V(z_t)$, the stationary stochastic process.

Sample Autocorrelation Function (SACF)

It is shown in Chapter 2 that an estimate of ρ_k is given by the lag k sample autocorrelation

$$r_k = \frac{\sum_{t=k+1}^{n} (z_t - \bar{z})(z_{t-k} - \bar{z})}{\sum_{t=1}^{n} (z_t - \bar{z})^2} \qquad k = 0, 1, 2, \ldots \tag{5.4}$$

We also discussed the sample variability of this estimate and indicated that for uncorrelated observations the variance of r_k is approximately given by

$$V(r_k) \cong \frac{1}{n} \tag{5.5}$$

For illustration we calculate the autocorrelations for the yield data (series 9 in the Data Appendix). This series consists of monthly differences between the yield on mortgages and the yield on government loans in the Netherlands.

$$\bar{z} = \frac{z_1 + z_2 + \cdots + z_n}{n} = \frac{0.66 + 0.70 + \cdots + 0.91}{159} = 0.993$$

$$\hat{\gamma}_0 = \hat{\sigma}_z^2 = \frac{(0.66 - 0.993)^2 + (0.70 - 0.993)^2 + \cdots + (0.91 - 0.993)^2}{159}$$

$$= 0.085$$

$$r_1 = \frac{(0.66 - 0.993)(0.70 - 0.993) + \cdots + (0.90 - 0.993)(0.91 - 0.993)}{(0.66 - 0.993)^2 + (0.70 - 0.993)^2 + \cdots + (0.91 - 0.993)^2}$$

$$= 0.841$$

$$r_2 = \frac{(0.66 - 0.993)(0.74 - 0.993) + \cdots + (0.74 - 0.993)(0.91 - 0.993)}{(0.66 - 0.993)^2 + (0.70 - 0.993)^2 + \cdots + (0.91 - 0.993)^2}$$

$$= 0.683$$

The sample autocorrelations for higher lags are calculated in a similar way and are shown in Table 5.1. A plot of the SACF is given in Figure 5.2. We notice that the SACF is of a particular form. The autocorrelations decrease

Table 5.1. Sample Autocorrelations for the Yield Data

k	1	2	3	4	5	6	7	8	9	10
r_k	0.841	0.683	0.584	0.515	0.457	0.427	0.405	0.386	0.361	0.321
k	11	12	13	14	15	16	17	18	19	20
r_k	0.329	0.338	0.337	0.294	0.231	0.166	0.126	0.062	0.047	0.042

as the lag k increases, indicating that observations closer together are more correlated than the ones far apart. The shape of the ACF will help us later to specify particular models. The horizontal bands in Figure 5.2 correspond to the $\pm 2\sigma$ limits for r_k ($\pm 2n^{-1/2}$), which are useful in assessing the significance of the sample autocorrelations. Some computer packages use slightly different standard errors for the sample autocorrelations. They are

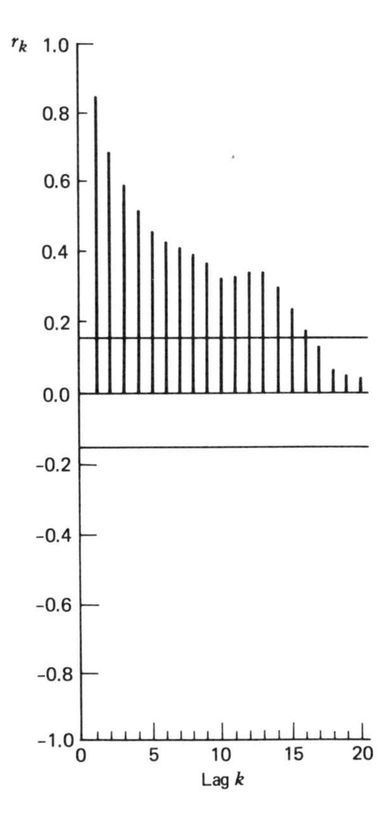

Figure 5.2. SACF for the yield series.

based on a result that was discussed in Section 2.12. There it was shown that for processes in which $\rho_k = 0$ for $k > q$ the variance of r_k (for $k > q$) is given by

$$V(r_k) \cong \frac{1}{n}\left(1 + 2\rho_1^2 + \cdots + 2\rho_q^2\right)$$

Replacing the theoretical autocorrelations by their sample estimates, the variances of $r_1, r_2, \ldots, r_k$ can then be estimated by n^{-1}, $n^{-1}(1 + 2r_1^2)$, $n^{-1}(1 + 2r_1^2 + 2r_2^2), \ldots, n^{-1}(1 + 2r_1^2 + \cdots + 2r_{k-1}^2)$. The corresponding 2σ limits are somewhat larger than the approximation $2n^{-1/2}$ we use.

5.2. STOCHASTIC DIFFERENCE EQUATION MODELS

The models we consider in this chapter are based on an observation by Yule (1921, 1927), who noticed that a time series in which successive values are autocorrelated can be represented as a linear combination of a sequence of uncorrelated random variables. This representation was later confirmed by Wold (1938), who showed that every weakly stationary nondeterministic stochastic process $(z_t - \mu)$ can be written as a linear combination (or linear filter) of a sequence of uncorrelated random variables. In fact, it will be shown later that some of these models can also be expressed in the form (5.1), however, with time-varying coefficients and correlated errors.

The linear filter representation is given by

$$\begin{aligned} z_t - \mu &= a_t + \psi_1 a_{t-1} + \psi_2 a_{t-2} + \cdots \\ &= \sum_{j=0}^{\infty} \psi_j a_{t-j} \qquad \psi_0 = 1 \end{aligned} \tag{5.6}$$

The random variables $\{a_t;\ t = 0, \pm 1, \pm 2, \ldots\}$ are a sequence of uncorrelated random variables from a fixed distribution with mean $E(a_t) = 0$, variance $V(a_t) = E(a_t^2) = \sigma^2$, and $\text{Cov}(a_t, a_{t-k}) = E(a_t a_{t-k}) = 0$ for all $k \neq 0$. Such a sequence is usually referred to as a *white-noise process*. Occasionally we will also call these random variables *random shocks*. The ψ_j weights in (5.6) are the coefficients in this linear combination; their number can be either finite or infinite.

It is easy to see that these models will lead to autocorrelations in the z_t. From (5.6) we find that

$$E(z_t) = \mu$$

$$\begin{aligned}
\gamma_0 &= V(z_t) = E(z_t - \mu)^2 \\
&= E(a_t + \psi_1 a_{t-1} + \psi_2 a_{t-2} + \cdots)^2 \\
&= E(a_t^2) + \psi_1^2 E(a_{t-1}^2) + \psi_2^2 E(a_{t-2}^2) + \cdots \\
&= \sigma^2 \sum_{j=0}^{\infty} \psi_j^2
\end{aligned}$$

Here we used the fact that $E(a_{t-i}a_{t-j}) = 0$ for $i \neq j$.

$$\begin{aligned}
\gamma_k &= E(z_t - \mu)(z_{t-k} - \mu) \\
&= E(a_t + \psi_1 a_{t-1} + \cdots + \psi_k a_{t-k} + \psi_{k+1} a_{t-k-1} + \cdots) \\
&\quad \times (a_{t-k} + \psi_1 a_{t-k-1} + \cdots) \\
&= \sigma^2 (1.\psi_k + \psi_1 \psi_{k+1} + \psi_2 \psi_{k+2} + \cdots) \\
&= \sigma^2 \sum_{j=0}^{\infty} \psi_j \psi_{j+k}
\end{aligned}$$

This implies that

$$\rho_k = \frac{\sum_{j=0}^{\infty} \psi_j \psi_{j+k}}{\sum_{j=0}^{\infty} \psi_j^2} \tag{5.7}$$

If the number of ψ weights in (5.6) is infinite, then some assumptions concerning the convergence of these coefficients are needed. In fact, we have to assume that the weights converge absolutely ($\sum_{j=0}^{\infty} |\psi_j| < \infty$). This condition, which is equivalent to the stationarity assumption, guarantees that all moments exist and are independent of time t.

At first sight, expression (5.6) appears complicated and nonparsimonious. However, many realistic models will result from proper choices of these ψ weights.

For example, if we choose $\psi_1 = -\theta$ and $\psi_j = 0, j \geqslant 2$, (5.6) will lead to the model

$$z_t - \mu = a_t - \theta a_{t-1} \tag{5.8}$$

This is usually referred to as the *first-order moving average process*.

Similarly, the choice of $\psi_j = \phi^j$ will lead to a process of the form

$$\begin{aligned} z_t - \mu &= a_t + \phi a_{t-1} + \phi^2 a_{t-2} + \cdots \\ &= a_t + \phi\left(a_{t-1} + \phi a_{t-2} + \phi^2 a_{t-3} + \cdots\right) \\ &= \phi(z_{t-1} - \mu) + a_t \end{aligned} \tag{5.9}$$

This is like the regression model in which the deviation from the mean at time t is regressed on itself, but with a lag of one time period. Thus it is usually referred to as the *first-order autoregressive process*. To satisfy the stationarity condition, we have to restrict the autoregressive parameter such that $|\phi| < 1$; otherwise, $\Sigma_{j=0}^{\infty}|\psi_j|$ would not converge. In the following sections the autoregressive and moving average processes are considered in more detail.

5.2.1. Autoregressive Processes

First-Order Autoregressive Process [AR(1)]

Consider again the model (5.9), which can be written as

$$(z_t - \mu) - \phi(z_{t-1} - \mu) = a_t \tag{5.10}$$

We now introduce a *backward-shift* (or *backshift*) *operator* B, which shifts time one step back, such that $Bz_t = z_{t-1}$; or in general $B^m z_t = z_{t-m}$. Using this notation we can write the AR(1) model as

$$(1 - \phi B)(z_t - \mu) = a_t \tag{5.11}$$

Note that the application of the backward-shift operator on a constant (which is the same for all t) results in the constant itself ($B^m\mu = \mu$).

We can successively substitute for $z_{t-j} - \mu$ $(j = 1, 2, \dots)$ in (5.10) and obtain

$$\begin{aligned} z_t - \mu &= \phi(z_{t-1} - \mu) + a_t \\ &= \phi[\phi(z_{t-2} - \mu) + a_{t-1}] + a_t = a_t + \phi a_{t-1} + \phi^2(z_{t-2} - \mu) \\ &= \cdots = a_t + \phi a_{t-1} + \phi^2 a_{t-2} + \cdots \end{aligned}$$

In fact, this is the linear filter representation of the AR(1) process seen earlier. An alternative, and in general simpler, way to obtain this representation is to consider the operator $(1 - \phi B)^{-1}$ as an expression in B and write

$$\begin{aligned} z_t - \mu &= (1 - \phi B)^{-1} a_t = (1 + \phi B + \phi^2 B^2 + \cdots) a_t \\ &= a_t + \phi a_{t-1} + \phi^2 a_{t-2} + \cdots \end{aligned}$$

In this expansion it is important that $|\phi| < 1$ (stationarity condition), since otherwise the ψ weights would not converge.

Application of the general results in (5.7) leads to

$$\gamma_0 = \sigma^2 \sum_{j=0}^{\infty} \psi_j^2 = \sigma^2 \sum_{j=0}^{\infty} \phi^{2j} = \frac{\sigma^2}{1 - \phi^2}$$

$$\gamma_k = \sigma^2 \sum_{j=0}^{\infty} \psi_j \psi_{j+k} = \sigma^2 \sum_{j=0}^{\infty} \phi^j \phi^{j+k} = \frac{\sigma^2 \phi^k}{1 - \phi^2} = \phi^k \gamma_0 \qquad k = 0, 1, 2, \dots$$

and

$$\rho_k = \phi^k \qquad k = 0, 1, 2, \dots \tag{5.12}$$

Plots of the ACF for $\phi = .5$ and $\phi = -.5$ are given in Figure 5.3. For $\phi > 0$ the autocorrelations decay geometrically (or exponentially) to zero, and for $\phi < 0$ the autocorrelations decay in an oscillatory pattern. The decay is slow if ϕ is close to the nonstationarity boundary ($+1$ or -1).

It is instructive to derive the autocorrelations from a different approach, since it can be extended more readily to other processes. Furthermore, since autocovariances are independent of the level, we can set $\mu = 0$. Multiplying both sides of model (5.9) by z_{t-k} $(k \geqslant 0)$ and taking expectations, we obtain

$$E(z_t z_{t-k}) = \phi E(z_{t-1} z_{t-k}) + E(a_t z_{t-k})$$

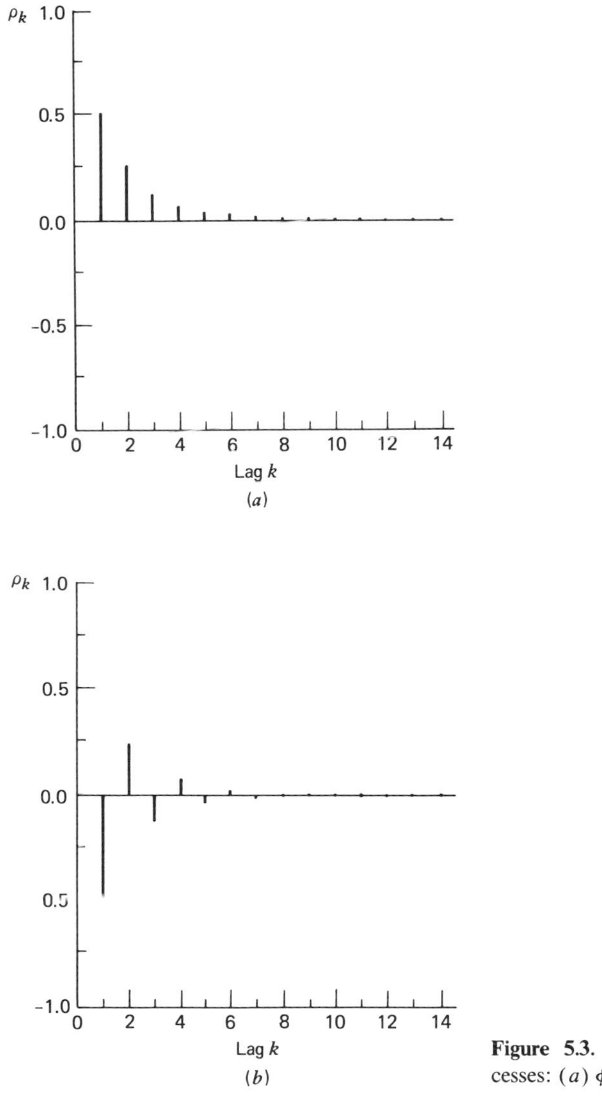

Figure 5.3. ACF for two AR(1) processes: (*a*) $\phi = .5$; (*b*) $\phi = -.5$.

or

$$\gamma_k = \phi\gamma_{k-1} + E(a_t z_{t-k}) \tag{5.13}$$

Furthermore,

$$\text{for } k = 0,\ E(a_t z_t) = E[a_t(a_t + \phi a_{t-1} + \phi^2 a_{t-2} + \cdots)] = \sigma^2$$

$$\text{for } k > 0,\ E(a_t z_{t-k}) = E[a_t(a_{t-k} + \phi a_{t-k-1} + \phi^2 a_{t-k-2} + \cdots)] = 0$$

In other words, this implies that a_t is uncorrelated with the past z_t's. Hence the autocovariances are given by

$$\gamma_0 = \phi\gamma_{-1} + \sigma^2 = \phi\gamma_1 + \sigma^2$$

$$\gamma_k = \phi\gamma_{k-1} \qquad k = 1, 2, \ldots \tag{5.14}$$

Substitution of $\gamma_1 = \phi\gamma_0$ into the first equation of (5.14) leads to

$$\gamma_0 = \frac{\sigma^2}{1 - \phi^2}$$

Dividing the second equation in (5.14) by γ_0 yields the difference equation

$$\rho_k = \phi\rho_{k-1} \qquad k = 1, 2, \ldots$$

Solving this equation gives

$$\rho_k = \phi\rho_{k-1} = \phi\phi\rho_{k-2} = \cdots$$

$$= \phi^{k-1}\rho_1 = \phi^k\rho_0 = \phi^k \qquad k = 0, 1, 2, \ldots$$

Second-Order Autoregressive Process [AR(2)]

The deviations from the process mean (without loss of generality, we take $\mu = 0$) can be written as

$$z_t = \phi_1 z_{t-1} + \phi_2 z_{t-2} + a_t \tag{5.15}$$

or using the backward-shift operator notation as

$$(1 - \phi_1 B - \phi_2 B^2) z_t = a_t$$

This process can also be written in terms of a linear filter (or moving average) representation

$$z_t = a_t + \psi_1 a_{t-1} + \psi_2 a_{t-2} + \cdots$$

$$= (1 + \psi_1 B + \psi_2 B^2 + \cdots) a_t = \psi(B) a_t \tag{5.16}$$

For the model (5.15), $\psi(B) = (1 - \phi_1 B - \phi_2 B^2)^{-1}$. Although the expansion of $(1 - \phi_1 B - \phi_2 B^2)^{-1}$ in terms of B is cumbersome, the ψ weights can be calculated easily by equating coefficients in $(1 - \phi_1 B - \phi_2 B^2)(1 + \psi_1 B$

$+ \psi_2 B^2 + \cdots) = 1$. For this equality to hold, the coefficients of B^j ($j \geq 0$) on each side of the equation have to be the same:

$$B^1: \quad \psi_1 - \phi_1 = 0 \qquad\qquad \psi_1 = \phi_1$$

$$B^2: \quad \psi_2 - \phi_1\psi_1 - \phi_2 = 0 \qquad\qquad \psi_2 = \phi_1^2 + \phi_2$$

$$B^3: \quad \psi_3 - \phi_1\psi_2 - \phi_2\psi_1 = 0 \qquad\qquad \psi_3 = \phi_1^3 + 2\phi_1\phi_2$$

For $j \geq 2$, the ψ weights can be derived recursively from

$$\psi_j = \phi_1\psi_{j-1} + \phi_2\psi_{j-2}$$

For stationarity we require that these ψ_j weights converge, which in turn implies that conditions on ϕ_1 and ϕ_2 have to be imposed. For the AR(1) process the stationarity condition requires that $|\phi| < 1$, or equivalently that the solution of $1 - \phi B = 0$ (which is $1/\phi$) has to be bigger than 1 in absolute value. For the AR(2) process, we have to look at the solutions G_1^{-1} and G_2^{-1} of $1 - \phi_1 B - \phi_2 B^2 = 0$. These solutions are frequently called the roots of the characteristic equation

$$1 - \phi_1 B - \phi_2 B^2 = (1 - G_1 B)(1 - G_2 B) = 0$$

They can both be real, or they can be a pair of complex numbers. For stationarity, we require that the roots are such that $|G_1^{-1}| > 1$ and $|G_2^{-1}| > 1$. The notation $|x|$ means the absolute value of x if x is real, and the modulus $(a^2 + b^2)^{1/2}$ if $x = a + bi$ is a complex number. The stationarity condition requires the roots of the characteristic equation to lie outside the unit circle.

Example 1

$$\phi_1 = 0.8 \qquad \phi_2 = -0.15$$

The solution of $(1 - 0.8B + 0.15B^2) = (1 - 0.5B)(1 - 0.3B) = 0$ is given by $G_1^{-1} = 1/0.5 = 2$ and $G_2^{-1} = 1/0.3 = 10/3$, which are both larger than 1 in absolute value. Hence the process is stationary.

Example 2

$$\phi_1 = 1.5 \qquad \phi_2 = -0.5$$

The characteristic equation $(1 - 1.5B + 0.5B^2) = (1 - B)(1 - 0.5B) = 0$ has one root at 1; the process is not stationary.

Example 3

$$\phi_1 = 1 \qquad \phi_2 = -0.5$$

The solutions of the characteristic equation $1 - B + 0.5B^2 = 0$ are complex and given by $G_1^{-1} = 1 + i$ and $G_2^{-1} = 1 - i$. In this case, $|G_1^{-1}| = |G_2^{-1}| = \sqrt{2}$. The process is stationary.

The stationarity conditions on the roots of the characteristic equation can also be restated in terms of ϕ_1 and ϕ_2. They are equivalent to checking whether

$$\phi_1 + \phi_2 < 1 \qquad \phi_2 - \phi_1 < 1 \qquad -1 < \phi_2 < 1 \tag{5.17}$$

These conditions restrict the parameters (ϕ_1, ϕ_2) to be within a triangular region determined by the endpoints $(-2, -1)$, $(0, 1)$, and $(2, -1)$ [see Box and Jenkins (1976), p. 59].

Autocorrelations. An approach similar to the one used in (5.13) for the AR(1) process can be adopted. Multiplying Equation (5.15) by z_{t-k} and taking expectations leads to

$$E(z_t z_{t-k}) = \phi_1 E(z_{t-1} z_{t-k}) + \phi_2 E(z_{t-2} z_{t-k}) + E(a_t z_{t-k})$$

or

$$\gamma_k = \phi_1 \gamma_{k-1} + \phi_2 \gamma_{k-2} + E(a_t z_{t-k}) \tag{5.18}$$

Since z_{t-k} depends only on $a_{t-k}, a_{t-k-1}, \ldots,$

$$E(a_t z_{t-k}) = \begin{cases} \sigma^2 & \text{for } k = 0 \\ 0 & \text{for } k = 1, 2, \ldots \end{cases}$$

Hence we find that

$$\gamma_0 = \phi_1 \gamma_{-1} + \phi_2 \gamma_{-2} + \sigma^2 = \phi_1 \gamma_1 + \phi_2 \gamma_2 + \sigma^2 \qquad k = 0$$

$$\gamma_k = \phi_1 \gamma_{k-1} + \phi_2 \gamma_{k-2} \qquad k > 0 \tag{5.19}$$

The autocovariances γ_k and the autocorrelations

$$\rho_k = \phi_1 \rho_{k-1} + \phi_2 \rho_{k-2} \qquad k = 1, 2, \ldots \tag{5.20}$$

or

$$\left(1 - \phi_1 B - \phi_2 B^2\right)\rho_k = 0$$

follow a difference equation. In this equation the backward-shift operator operates on k.

Now for $k = 1$,

$$\rho_1 = \phi_1\rho_0 + \phi_2\rho_{-1} = \phi_1 + \phi_2\rho_1 = \phi_1/(1 - \phi_2)$$

For $k = 2$,

$$\rho_2 = \phi_1\rho_1 + \phi_2\rho_0 = \frac{\phi_1^2}{1 - \phi_2} + \phi_2$$

For all other lags k, we can use Equation (5.20) to calculate ρ_k recursively.

The first equation in (5.19) can be written as $\gamma_0(1 - \phi_1\rho_1 - \phi_2\rho_2) = \sigma^2$. Substituting for ρ_1 and ρ_2, we obtain the variance of z_t as

$$\gamma_0 = \frac{1 - \phi_2}{1 + \phi_2}\,\frac{\sigma^2}{\left(1 - \phi_2\right)^2 - \phi_1^2} \tag{5.21}$$

It is instructive to learn more about the nature of the ACF by studying the solution of the difference equation in (5.20). It can be shown that this solution is given by

$$\rho_k = A_1 G_1^k + A_2 G_2^k \qquad k = 0, 1, 2, \ldots \tag{5.22}$$

where A_1 and A_2 are constants that are determined by the initial condition $\rho_0 = 1$, $\rho_{-1} = \rho_1$, and where G_1^{-1} and G_2^{-1} are the roots of the characteristic equation $1 - \phi_1 B - \phi_2 B^2 = (1 - G_1 B)(1 - G_2 B) = 0$. Here we have assumed that they are distinct. Furthermore, since the difference equation in (5.20) holds for $k = 1, 2, \ldots$, and the two starting values ρ_0 and $\rho_{-1} = \rho_1$ have to lie on this solution, the result (5.22) holds for $k = 0, 1, 2, \ldots$.

If the roots of the characteristic equation are equal ($G_1^{-1} = G_2^{-1} = G^{-1}$), the solution of the difference equation becomes

$$\rho_k = \left(A_1 + A_2 k\right)G^k \qquad k = 0, 1, 2, \ldots \tag{5.23}$$

The behavior of the ACF for several combinations of (ϕ_1, ϕ_2) is shown in Figure 5.4. In the case of two real G_1 and G_2, which for stationarity have to be between -1 and $+1$, the ACF is a mixture of two damped exponentials. Depending on the sign of G_i, the autocorrelations can also damp out in an oscillatory manner. If the roots are complex, the ACF follows a damped sine wave.

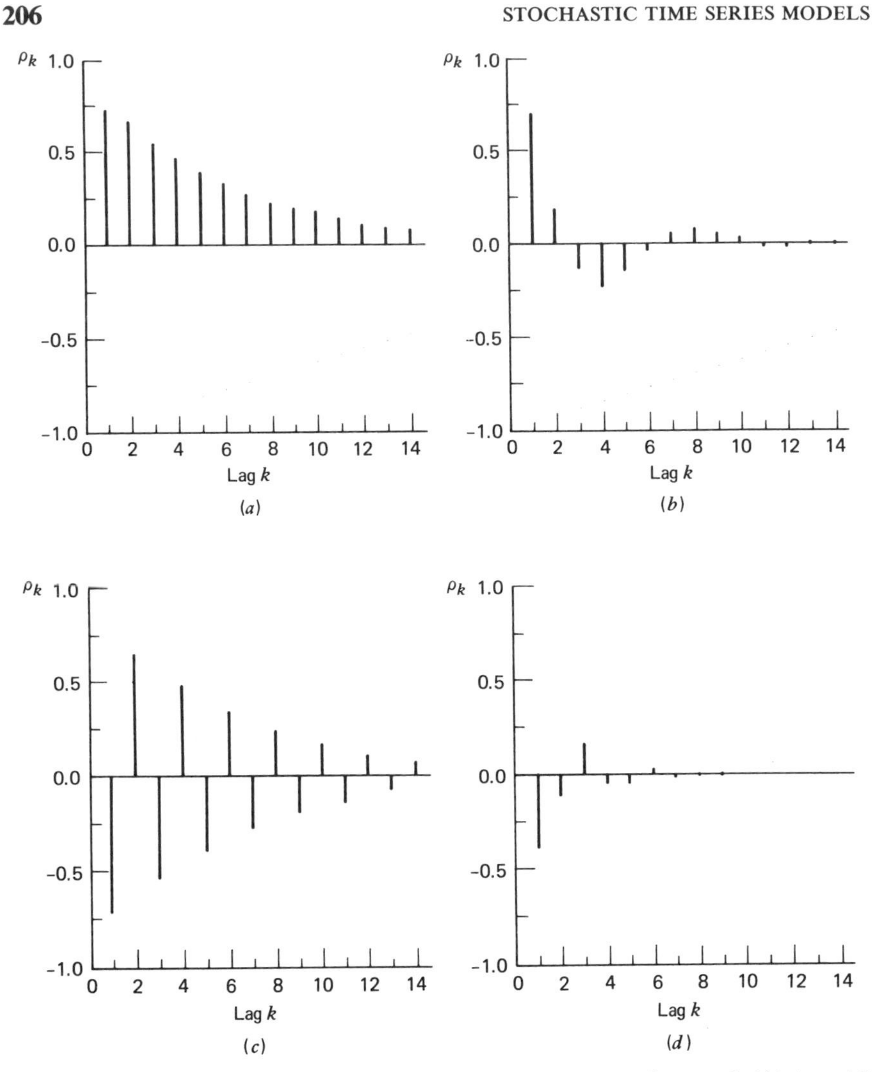

Figure 5.4. ACF for various stationary AR(2) processes: (*a*) $\phi_1 = .5$, $\phi_2 = .3$; (*b*) $\phi_1 = 1.0$, $\phi_2 = -.5$; (*c*) $\phi_1 = -.5$, $\phi_2 = .3$; (*d*) $\phi_1 = -.5$, $\phi_2 = -.3$.

Autoregressive Process of Order p [AR(p)]

In this model we treat z_t, the deviation from the mean at time t, as being regressed on the p previous deviations $z_{t-1}, \ldots, z_{t-p}$:

$$z_t = \phi_1 z_{t-1} + \cdots + \phi_p z_{t-p} + a_t \tag{5.24}$$

In backshift operator notation, this model can be written as

$$(1 - \phi_1 B - \cdots - \phi_p B^p) z_t = a_t \qquad \text{or} \qquad \phi(B) z_t = a_t \tag{5.25}$$

It can also be expressed in the linear filter (moving average) representation $z_t = \psi(B) a_t$. The ψ weights are determined from $\psi(B) = (1 - \phi_1 B - \cdots - \phi_p B^p)^{-1}$; they can be calculated by equating coefficients in $(1 - \phi_1 B - \cdots - \phi_p B^p)\psi(B) = 1$. The ψ weights will converge only if we impose certain stationarity conditions on the autoregressive parameters. As in the AR(1) and AR(2) models, these conditions put restrictions on the roots of the characteristic equation $\phi(B) = (1 - G_1 B)(1 - G_2 B) \cdots (1 - G_p B) = 0$. For stationarity, we require that all roots G_i^{-1} lie strictly outside the unit circle.

Autocorrelations. Proceeding as in the simpler AR models, we can show that

$$\gamma_0 = \phi_1 \gamma_1 + \cdots + \phi_p \gamma_p + \sigma^2$$

which implies that

$$\gamma_0 = \frac{\sigma^2}{1 - \phi_1 \rho_1 - \cdots - \phi_p \rho_p} \tag{5.26}$$

Furthermore, we can show that the autocovariances and autocorrelations follow a pth-order difference equation

$$\gamma_k = \phi_1 \gamma_{k-1} + \cdots + \phi_p \gamma_{k-p} \qquad k > 0$$

$$\rho_k = \phi_1 \rho_{k-1} + \cdots + \phi_p \rho_{k-p} \qquad k > 0 \tag{5.27}$$

The first p equations ($k = 1, 2, \ldots, p$) in (5.27) are called the *Yule-Walker equations* and are shown below:

$$k = 1: \quad \rho_1 = \phi_1 + \rho_1 \phi_2 + \cdots + \rho_{p-1} \phi_p$$

$$k = 2: \quad \rho_2 = \rho_1 \phi_1 + \phi_2 + \cdots + \rho_{p-2} \phi_p$$

$$\vdots$$

$$k = p: \quad \rho_p = \rho_{p-1} \phi_1 + \rho_{p-2} \phi_2 + \cdots + \phi_p$$

In matrix notation we can write these equations as

$$\boldsymbol{\rho} = \mathbf{P}\boldsymbol{\phi} \tag{5.28}$$

where $\boldsymbol{\rho} = (\rho_1, \rho_2, \ldots, \rho_p)'$, $\boldsymbol{\phi} = (\phi_1, \phi_2, \ldots, \phi_p)'$, and

$$\mathbf{P} = \begin{bmatrix} 1 & \rho_1 & \rho_2 & \cdots & \rho_{p-1} \\ \rho_1 & 1 & \rho_1 & \cdots & \rho_{p-2} \\ \vdots & \vdots & \vdots & & \vdots \\ \rho_{p-1} & \rho_{p-2} & \rho_{p-3} & \cdots & 1 \end{bmatrix}$$

The autoregressive parameters $\boldsymbol{\phi}$ can be expressed as a function of the first p autocorrelations by solving the equation system (5.28)

$$\boldsymbol{\phi} = \mathbf{P}^{-1}\boldsymbol{\rho} \tag{5.29}$$

Frequently, initial estimates for $\boldsymbol{\phi}$ are obtained by replacing the theoretical autocorrelations ρ_j in (5.29) by their sample estimates r_j. The resulting estimates are called *moment estimates*, since they involve the estimated moments r_j.

For the AR(1) model, the Yule-Walker equation is given by $\rho_1 = \phi$. For the AR(2) model, the Yule-Walker equations are

$$\rho_1 = \phi_1 + \rho_1\phi_2$$

$$\rho_2 = \rho_1\phi_1 + \phi_2$$

which leads to

$$\phi_1 = \frac{\rho_1(1 - \rho_2)}{1 - \rho_1^2}$$

$$\phi_2 = \frac{\rho_2 - \rho_1^2}{1 - \rho_1^2}$$

The difference equation in (5.27), $\phi(B)\rho_k = 0$, $k = 1, 2, \ldots$, determines the behavior of the autocorrelation function. It can be shown that its solution is

$$\rho_k = A_1G_1^k + \cdots + A_pG_p^k \qquad k = 0, 1, 2, \ldots \tag{5.30}$$

where G_i^{-1} $(i = 1, 2, \ldots, p)$ are the distinct roots of $\phi(B) = (1 - G_1B)(1 -$

$G_2 B) \cdots (1 - G_p B) = 0$, and the A_i's are constants. The stationarity conditions imply that the G_i's lie inside the unit circle. Hence the ACF is described by a mixture of damped exponentials (for real roots) and damped sine waves (for complex roots).

Since all autoregressive processes imply ACF's that damp out, it is sometimes difficult to differentiate among models of different orders. In particular, it is often difficult to distinguish between a single exponential decay and a sum of several.

We now introduce a particular function of the autocorrelations to help discriminate among different models. This function, which is referred to as the partial autocorrelation function (PACF), is described below.

5.2.2. Partial Autocorrelations

The correlation between two random variables is often due only to the fact that both variables are correlated with the same third variable. In the time series context, a large portion of the correlation between z_t and z_{t-k} can be due to the correlation these variables have with $z_{t-1}, z_{t-2}, \ldots, z_{t-k+1}$. To adjust for this correlation, one can calculate *partial autocorrelations*. The partial autocorrelation of lag k can be thought of as the partial regression coefficient ϕ_{kk} in the representation

$$z_t = \phi_{k1} z_{t-1} + \cdots + \phi_{kk} z_{t-k} + a_t \tag{5.31}$$

It measures the additional correlation between z_t and z_{t-k} after adjustments have been made for the intermediate variables $z_{t-1}, \ldots, z_{t-k+1}$.

The lag 1 partial autocorrelation is given by the partial regression coefficient ϕ_{11} in the representation

$$z_t = \phi_{11} z_{t-1} + a_t$$

From the Yule-Walker equation for the AR(1) model, it follows that $\phi_{11} = \rho_1$. This was to be expected, since there are no intermediate variables between z_{t-1} and z_t. The lag 2 partial autocorrelation is the partial regression coefficient ϕ_{22} in the representation

$$z_t = \phi_{21} z_{t-1} + \phi_{22} z_{t-2} + a_t$$

From the Yule-Walker equations for the AR(2) process,

$$\rho_1 = \phi_{21} + \rho_1 \phi_{22}$$

$$\rho_2 = \rho_1 \phi_{21} + \phi_{22}$$

we can express this coefficient as

$$\phi_{22} = \frac{\rho_2 - \rho_1^2}{1 - \rho_1^2}$$

In general, we can obtain the lag k partial autocorrelation ϕ_{kk} from the Yule-Walker equations that correspond to the representation in (5.31). These equations are given by

$$\begin{bmatrix} \rho_1 \\ \rho_2 \\ \vdots \\ \rho_k \end{bmatrix} = \begin{bmatrix} 1 & \rho_1 & \cdots & \rho_{k-1} \\ \rho_1 & 1 & \cdots & \rho_{k-2} \\ \vdots & \vdots & & \vdots \\ \rho_{k-1} & \rho_{k-2} & \cdots & 1 \end{bmatrix} \begin{bmatrix} \phi_{k1} \\ \phi_{k2} \\ \vdots \\ \phi_{kk} \end{bmatrix}$$

Solving these equations for the last coefficient ϕ_{kk} leads to

$$\phi_{kk} = \frac{\begin{vmatrix} 1 & \rho_1 & \cdots & \rho_{k-2} & \rho_1 \\ \rho_1 & 1 & \cdots & \rho_{k-3} & \rho_2 \\ \vdots & \vdots & & \vdots & \vdots \\ \rho_{k-1} & \rho_{k-2} & \cdots & \rho_1 & \rho_k \end{vmatrix}}{\begin{vmatrix} 1 & \rho_1 & \cdots & \rho_{k-2} & \rho_{k-1} \\ \rho_1 & 1 & \cdots & \rho_{k-3} & \rho_{k-2} \\ \vdots & \vdots & & \vdots & \vdots \\ \rho_{k-1} & \rho_{k-2} & \cdots & \rho_1 & 1 \end{vmatrix}} \qquad (5.32)$$

where $|\mathbf{A}|$ denotes the determinant of the matrix $\mathbf{A}$. The partial autocorrelations ϕ_{kk} as a function of k are called the *partial autocorrelation function* (PACF).

It follows from the definition of ϕ_{kk} that the PACF's of autoregressive processes are of a particular form:

$$\text{AR}(1): \quad \phi_{11} = \rho_1 = \phi \qquad \phi_{kk} = 0 \qquad \text{for } k > 1$$

$$\text{AR}(2): \quad \phi_{11} = \rho_1 \qquad \phi_{22} = \frac{\rho_2 - \rho_1^2}{1 - \rho_1^2} \qquad \phi_{kk} = 0 \qquad \text{for } k > 2$$

$$\text{AR}(p): \quad \phi_{11} \neq 0, \ldots, \phi_{pp} \neq 0 \qquad \phi_{kk} = 0 \qquad \text{for } k > p$$

The partial autocorrelations for lags that are larger than the order of the process are zero. This fact, together with the structure of the autocorrelations, makes it easy to recognize autoregressive processes.

In summary, the AR(p) process is described by:

1. An autocorrelation function that is infinite in extent and that is a combination of damped exponentials and damped sine waves.
2. Partial autocorrelations that are zero for lags larger than the autoregressive order p.

Alternative Interpretation of Partial Autocorrelations

The partial correlation coefficient between two random variables X and Y, conditional on a third variable W, is the ordinary correlation coefficient calculated from the conditional distribution $p(x, y|w)$. It can be thought of as the correlation between $X - E(X|W)$ and $Y - E(Y|W)$, and under the assumption of joint normality of (X, Y, W) it is given by

$$\rho_{XY\cdot W} = \frac{E[X - E(X|W)][Y - E(Y|W)]}{\{E[X - E(X|W)]^2 E[Y - E(Y|W)]^2\}^{1/2}}$$

$$= \frac{\rho_{XY} - \rho_{XW}\rho_{YW}}{[(1 - \rho_{XW}^2)(1 - \rho_{YW}^2)]^{1/2}} \tag{5.33}$$

In the context of a lag 2 partial autocorrelation, the variables are $X = z_t$, $Y = z_{t-2}$, $W = z_{t-1}$, and $\rho_{XY} = \rho_2$, $\rho_{XW} = \rho_{YW} = \rho_1$. Hence,

$$\phi_{22} = \rho_{z_t z_{t-2}\cdot z_{t-1}}$$

$$= \text{Corr}[z_t - E(z_t|z_{t-1}), z_{t-2} - E(z_{t-2}|z_{t-1})]$$

$$= \frac{\rho_2 - \rho_1^2}{1 - \rho_1^2}$$

which is the expression obtained earlier. Similarly, we can interpret the lag k partial autocorrelation as the correlation between $z_t - E(z_t|z_{t-1}, \ldots, z_{t-k+1})$ and $z_{t-k} - E(z_{t-k}|z_{t-1}, \ldots z_{t-k+1})$, which can be shown to equal the expression in (5.32).

Sample Partial Autocorrelation Function (SPACF)

Two different approaches can be used to estimate ϕ_{kk}.

1. We can replace the theoretical autocorrelations ρ_j in (5.32) by their sample estimates r_j and obtain the sample partial autocorrelations $\hat{\phi}_{kk}$. When k is large, this approach requires the evaluation of many determinants of high dimensions.
2. Alternatively, we can fit autoregressive models of increasing orders; then the estimate of the last coefficient in each model is the sample partial autocorrelation. Levinson (1946) and later Durbin (1960) have shown that at step k the estimates $\hat{\phi}_{kj}$ $(1 \leqslant j \leqslant k)$ in $z_t = \phi_{k1}z_{t-1} + \cdots + \phi_{kk}z_{t-k} + a_t$ can be updated recursively from the estimates $\hat{\phi}_{k-1,j}$ at step $k-1$, which are obtained from $z_t = \phi_{k-1,1}z_{t-1} + \cdots + \phi_{k-1,k-1}z_{t-k+1} + a_t$. The updating equations are given by

$$\hat{\phi}_{kk} = \frac{r_k - \sum_{j=1}^{k-1} \hat{\phi}_{k-1,j} r_{k-j}}{1 - \sum_{j=1}^{k-1} \hat{\phi}_{k-1,j} r_j}$$

$$\hat{\phi}_{k,j} = \hat{\phi}_{k-1,j} - \hat{\phi}_{kk}\hat{\phi}_{k-1,k-j} \qquad j = 1,2,\ldots,k-1 \quad (5.34)$$

These recursive equations avoid the inversion of $k \times k$ matrices. However, they are sensitive to rounding errors, and most computer packages use double precision arithmetic when computing these estimates.

Example. We now illustrate this algorithm on the yield data. The sample autocorrelations r_k were calculated previously (see Table 5.1). Starting from $\hat{\phi}_{11} = r_1 = .841$, we can calculate

$$\hat{\phi}_{22} = \frac{r_2 - r_1^2}{1 - r_1^2} = \frac{.683 - .841^2}{1 - .841^2} = -.083$$

$$\hat{\phi}_{21} = \hat{\phi}_{11} - \hat{\phi}_{22}\hat{\phi}_{11} = .841 - (-.083)(.841) = .911$$

Table 5.2. Sample Partial Autocorrelations for the Yield Data

k	1	2	3	4	5	6	7	8	9	10
$\hat{\phi}_{kk}$	0.841	−0.083	0.111	0.036	0.018	0.091	0.025	0.035	0.003	−0.044
k	11	12	13	14	15	16	17	18	19	20
$\hat{\phi}_{kk}$	0.168	0.001	0.027	−0.110	−0.080	−0.057	0.007	−0.152	0.122	−0.071

and

$$\hat{\phi}_{33} = \frac{r_3 - \hat{\phi}_{21}r_2 - \hat{\phi}_{22}r_1}{1 - \hat{\phi}_{21}r_1 - \hat{\phi}_{22}r_2}$$

$$= \frac{.584 - (.911)(.683) + (.083)(.841)}{1 - (.911)(.841) + (.083)(.683)} = .111$$

$$\hat{\phi}_{31} = \hat{\phi}_{21} - \hat{\phi}_{33}\hat{\phi}_{22} = .913$$

$$\hat{\phi}_{32} = \hat{\phi}_{22} - \hat{\phi}_{33}\hat{\phi}_{21} = -.112$$

Similarly, these calculations can be carried further. The first 20 sample partial autocorrelations are given in Table 5.2. A plot of the SPACF is shown in Figure 5.5.

Standard Errors of $\hat{\phi}_{kk}$

If the data follow an AR(p) process, then for lags greater than p the variance of $\hat{\phi}_{kk}$ can be approximated by

$$V(\hat{\phi}_{kk}) \cong n^{-1} \qquad \text{for } k > p \tag{5.35}$$

Approximate standard errors for $\hat{\phi}_{kk}$ are then given by $n^{-1/2}$.

From Table 5.2 we find that $\hat{\phi}_{11}$ is significant. However, for $k > 1$ all estimates are within the $\pm 2\sigma$ limits ($\pm 2n^{-1/2}$). The pattern of the partial autocorrelations is thus consistent with that of a first-order autoregressive process.

5.2.3. Moving Average Processes

Another class of stochastic models arises from the general linear filter model (5.6) by specifying only a finite number of nonzero ψ weights ($\psi_1 = -\theta_1$,

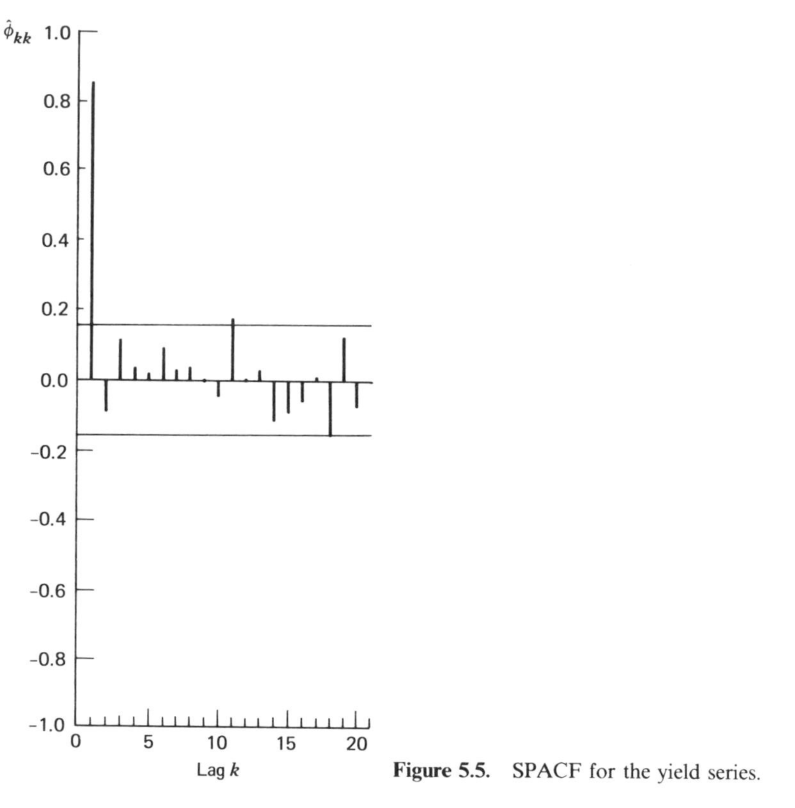

Figure 5.5. SPACF for the yield series.

$\psi_2 = -\theta_2, \ldots, \psi_q = -\theta_q, \psi_j = 0;\ j > q$). The resulting process is said to follow a moving average model of order q [MA(q)] and is given by

$$z_t - \mu = a_t - \theta_1 a_{t-1} - \cdots - \theta_q a_{t-q}$$

or

$$z_t - \mu = (1 - \theta_1 B - \cdots - \theta_q B^q) a_t = \theta(B) a_t \qquad (5.36)$$

Since there are only a finite number of ψ weights, these processes are always stationary. Special cases are considered below.

First-Order Moving Average Process [MA(1)]

The first-order moving average model is given by

$$z_t - \mu = a_t - \theta a_{t-1} \qquad \text{or} \qquad z_t - \mu = (1 - \theta B) a_t \qquad (5.37)$$

Using the general results in (5.7) and setting $\psi_0 = 1$, $\psi_1 = -\theta$, and $\psi_j = 0$ for $j > 1$, it follows that

$$\gamma_0 = (1 + \theta^2)\sigma^2 \qquad \gamma_1 = -\theta\sigma^2 \qquad \gamma_k = 0 \qquad \text{for } k > 1$$

The autocorrelation function is thus described by

$$\rho_1 = \frac{-\theta}{1 + \theta^2} \qquad \rho_k = 0 \qquad \text{for } k > 1 \tag{5.38}$$

It implies that observations one step apart are correlated. However, observations more than one step apart are uncorrelated.

We notice that ρ_1 is always between $-.5$ and $.5$. Furthermore, θ and $1/\theta$ both satisfy the quadratic equation $\theta^2\rho_1 + \theta + \rho_1 = 0$. In other words, we can always find two MA(1) processes that correspond to the same ACF. To establish a one-to-one correspondence between the ACF and the model and to obtain a converging autoregressive representation, we restrict the moving average parameter such that $|\theta| < 1$. This restriction is known as the *invertibility condition* and is similar to the stationarity condition in autoregressive models. Invertibility implies that the process can be written in terms of an autoregressive representation

$$z_t - \mu = \pi_1(z_{t-1} - \mu) + \pi_2(z_{t-2} - \mu) + \cdots + a_t$$

in which $\sum_{j=1}^{\infty}|\pi_j|$ converges.

The MA(1) model, for example, can be written as $(1 - \theta B)^{-1}(z_t - \mu) = a_t$. Expanding $(1 - \theta B)^{-1} = 1 + \theta B + \theta^2 B^2 + \cdots$, we find that z_t can be expressed as

$$z_t - \mu = -\theta(z_{t-1} - \mu) - \theta^2(z_{t-2} - \mu) - \cdots + a_t$$

The weights $\pi_j = -\theta^j$ converge if the model is invertible ($|\theta| < 1$). This implies that the effect of the past observations decreases with their age. This is an intuitively reasonable assumption.

Second-Order Moving Average Process [MA(2)]

The second-order moving average process is described by

$$z_t - \mu = a_t - \theta_1 a_{t-1} - \theta_2 a_{t-2} \tag{5.39}$$

or

$$z_t - \mu = (1 - \theta_1 B - \theta_2 B^2) a_t$$

The autocovariances can be obtained by substituting $\psi_0 = 1$, $\psi_1 = -\theta_1$, $\psi_2 = -\theta_2$, $\psi_j = 0; j > 2$ in (5.7). They are given by

$$\gamma_0 = \left(1 + \theta_1^2 + \theta_2^2\right)\sigma^2 \qquad \gamma_1 = \left(-\theta_1 + \theta_1\theta_2\right)\sigma^2 \qquad \gamma_2 = -\theta_2\sigma^2$$

$$\gamma_k = 0 \qquad \text{for } k > 2$$

The autocorrelations are

$$\rho_1 = \frac{-\theta_1 + \theta_1\theta_2}{1 + \theta_1^2 + \theta_2^2} \qquad \rho_2 = \frac{-\theta_2}{1 + \theta_1^2 + \theta_2^2}$$

$$\rho_k = 0 \qquad \text{for } k > 2 \tag{5.40}$$

This model implies that observations more than two steps apart are uncorrelated. As in the MA(1) model, we can write the MA(2) process in terms of an infinite autoregressive representation,

$$z_t - \mu = \pi_1(z_{t-1} - \mu) + \pi_2(z_{t-2} - \mu) + \cdots + a_t$$

The π weights can be obtained from

$$\pi(B) = 1 - \pi_1 B - \pi_2 B^2 - \cdots = \left(1 - \theta_1 B - \theta_2 B^2\right)^{-1}$$

They can be calculated by equating coefficients of B^j in

$$\left(1 - \pi_1 B - \pi_2 B^2 - \cdots\right)\left(1 - \theta_1 B - \theta_2 B^2\right) = 1$$

and are given by

$$B^1\!: \quad -\pi_1 - \theta_1 = 0 \qquad\qquad \pi_1 = -\theta_1$$

$$B^2\!: \quad -\pi_2 + \theta_1\pi_1 - \theta_2 = 0 \qquad\qquad \pi_2 = \theta_1\pi_1 - \theta_2 = -\theta_1^2 - \theta_2$$

$$B^j\!: \quad -\pi_j + \theta_1\pi_{j-1} + \theta_2\pi_{j-2} = 0 \qquad \pi_j = \theta_1\pi_{j-1} + \theta_2\pi_{j-2} \qquad j > 2$$

For invertibility we require that the π weights converge, which in turn implies conditions on the parameters θ_1 and θ_2. Similar to the stationarity conditions for an AR(2) process, invertibility for an MA(2) process requires that the roots of $1 - \theta_1 B - \theta_2 B^2 = (1 - H_1 B)(1 - H_2 B) = 0$ lie outside the unit circle (i.e., $|H_1^{-1}| > 1$ and $|H_2^{-1}| > 1$). In terms of the parameters, the invertibility conditions become

$$\theta_1 + \theta_2 < 1 \qquad \theta_2 - \theta_1 < 1 \qquad -1 < \theta_2 < 1 \tag{5.41}$$

The π weights converge if the parameters are within the triangular region described by the inequalities in (5.41).

Moving Average Process of Order q [MA(q)]

This process is described by

$$z_t - \mu = a_t - \theta_1 a_{t-1} - \cdots - \theta_q a_{t-q}$$

or

$$z_t - \mu = (1 - \theta_1 B - \cdots - \theta_q B^q) a_t = \theta(B) a_t \tag{5.42}$$

Its autocovariances and autocorrelations can be derived from (5.7) by setting $\psi_0 = 1$, $\psi_1 = -\theta_1, \ldots, \psi_q = -\theta_q$, $\psi_j = 0$; $j > q$, or by considering $E(z_t - \mu)(z_{t-k} - \mu)$ directly. The autocovariances are

$$\gamma_0 = (1 + \theta_1^2 + \cdots + \theta_q^2)\sigma^2$$

$$\gamma_k = (-\theta_k + \theta_1\theta_{k+1} + \cdots + \theta_{q-k}\theta_q)\sigma^2 \qquad k = 1, 2, \ldots, q$$

$$\gamma_k = 0 \qquad k > q$$

and the autocorrelations are

$$\rho_k = \frac{-\theta_k + \theta_1\theta_{k+1} + \cdots + \theta_{q-k}\theta_q}{1 + \theta_1^2 + \cdots + \theta_q^2} \qquad k = 1, 2, \ldots, q$$

$$\rho_k = 0 \qquad k > q \tag{5.43}$$

The ACF of the MA(q) process cuts off after lag q. The memory of such a process extends only q steps; observations more than q steps apart are uncorrelated.

In the infinite autoregressive representation $(1 - \pi_1 B - \pi_2 B^2 - \cdots)(z_t - \mu) = a_t$, the π weights are given by $\pi(B) = \theta^{-1}(B)$, where $\theta(B) = 1 - \theta_1 B - \cdots - \theta_q B^q$. These weights can be obtained by equating the coefficients of B^j in $\pi(B)\theta(B) = 1$. Again, the invertibility conditions impose some restrictions on the moving average parameters. For invertibility we require that the roots H_i^{-1} $(i = 1, 2, \ldots, q)$ of $1 - \theta_1 B - \cdots - \theta_q B^q = (1 - H_1 B) \cdots (1 - H_q B) = 0$ lie outside the unit circle.

Partial Autocorrelation Function for MA Processes

The ACF of a moving average process of order q cuts off after lag q ($\rho_k = 0$; $k > q$). However, its PACF is infinite in extent (it tails off). In general, the PACF is dominated by combinations of damped exponentials and/or damped sine waves. For example, consider the partial autocorrelations of the MA(1) process for which $\rho_1 = -\theta/(1+\theta^2)$ and $\rho_k = 0$, $k > 1$. It follows from the definition of partial autocorrelations in (5.32) that

$$\phi_{11} = \rho_1 = \frac{-\theta}{1+\theta^2} = \frac{-\theta(1-\theta^2)}{1-\theta^4}$$

$$\phi_{22} = \frac{\rho_2 - \rho_1^2}{1-\rho_1^2} = \frac{-\rho_1^2}{1-\rho_1^2} = \frac{-\theta^2}{1+\theta^2+\theta^4} = \frac{-\theta^2(1-\theta^2)}{1-\theta^6}$$

$$\phi_{33} = \frac{\begin{vmatrix} 1 & \rho_1 & \rho_1 \\ \rho_1 & 1 & \rho_2 \\ \rho_2 & \rho_1 & \rho_3 \end{vmatrix}}{\begin{vmatrix} 1 & \rho_1 & \rho_2 \\ \rho_1 & 1 & \rho_1 \\ \rho_2 & \rho_1 & 1 \end{vmatrix}} = \frac{\begin{vmatrix} 1 & \rho_1 & \rho_1 \\ \rho_1 & 1 & 0 \\ 0 & \rho_1 & 0 \end{vmatrix}}{\begin{vmatrix} 1 & \rho_1 & 0 \\ \rho_1 & 1 & \rho_1 \\ 0 & \rho_1 & 1 \end{vmatrix}} = \frac{\rho_1^3}{1-2\rho_1^2} = \frac{-\theta^3(1-\theta^2)}{1-\theta^8}$$

and, in general,

$$\phi_{kk} = \frac{-\theta^k(1-\theta^2)}{1-\theta^{2(k+1)}} \qquad k > 0$$

With increasing lag k, the partial autocorrelations ϕ_{kk} are dominated by an exponential decay. The ACF and PACF for an MA(1) process with $\theta = .7$ and $\theta = -.7$ are shown in Figure 5.6.

Expressions for the partial autocorrelations of higher order moving average processes become complicated. In general, they are dominated by combinations of exponential decays [for the real roots of $\theta(B)$] and/or damped sine waves (for the complex roots). Their patterns are thus very similar to the autocorrelation patterns of autoregressive processes.

We notice a duality between AR and MA processes. While the ACF of an AR(p) process is infinite in extent, the PACF cuts off after lag p. The ACF of an MA(q) process cuts off after lag q, while the PACF is infinite in extent and dominated by damped exponentials and sine waves.

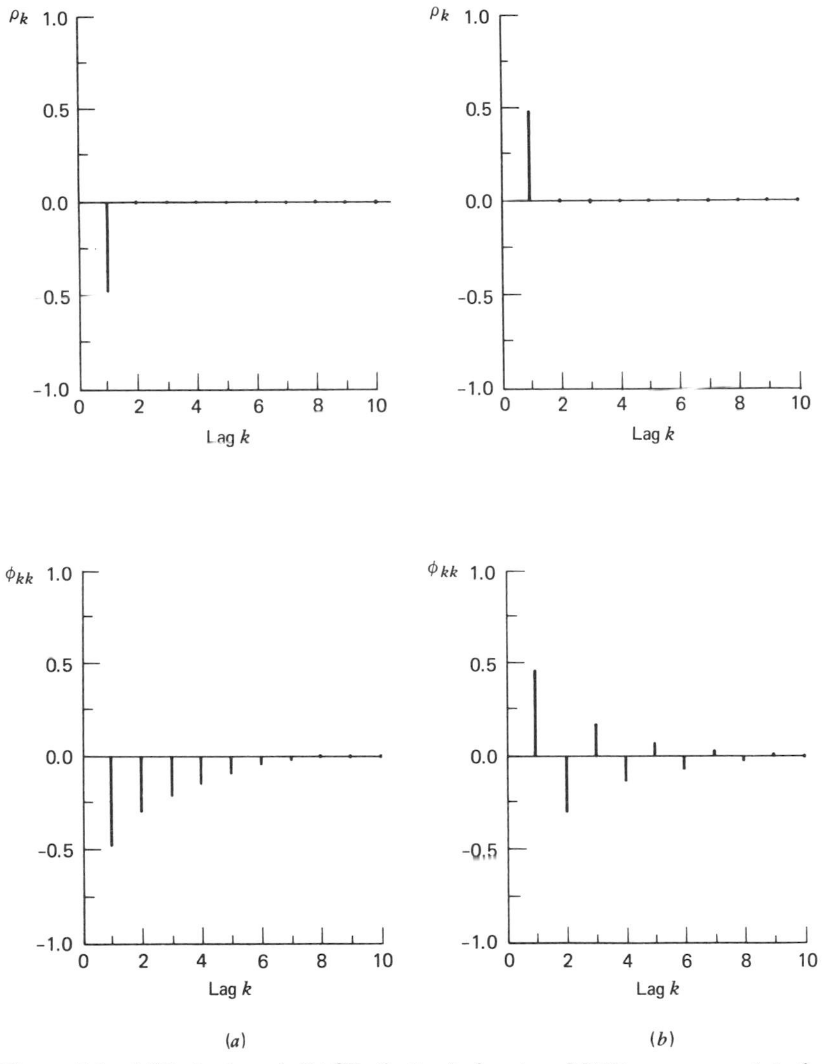

Figure 5.6. ACF (top) and PACF (bottom) for two MA(1) processes: (a) $\theta = .7$; (b) $\theta = -.7$.

5.2.4. Autoregressive Moving Average (ARMA) Processes

In time series modeling we look for parsimonious representations of the ψ weights (or equivalently the π weights) in the linear filter representation,

$$z_t - \mu = \psi(B)a_t = a_t + \psi_1 a_{t-1} + \psi_2 a_{t-2} + \cdots$$

In the pure MA model we truncate the ψ weights and consider $\psi(B) = 1 - \theta_1 B - \cdots - \theta_q B^q$. In the pure autoregressive model we truncate the π weights in $\pi(B)(z_t - \mu) = a_t$ and consider $\psi(B) = (1 - \phi_1 B - \cdots - \phi_p B^p)^{-1}$.

A more general approach to representing the ψ weights is to consider $\psi(B)$ as a ratio of two finite polynomials in B:

$$\psi(B) = \frac{\theta(B)}{\phi(B)} = \frac{1 - \theta_1 B - \cdots - \theta_q B^q}{1 - \phi_1 B - \cdots - \phi_p B^p}$$

This leads to the *autoregressive moving average, or ARMA(p, q) model*

$$(1 - \phi_1 B - \cdots - \phi_p B^p)(z_t - \mu) = (1 - \theta_1 B - \cdots - \theta_q B^q) a_t \tag{5.44}$$

or

$$\begin{aligned} z_t - \mu = {} & \phi_1(z_{t-1} - \mu) + \cdots + \phi_p(z_{t-p} - \mu) \\ & + a_t - \theta_1 a_{t-1} - \cdots - \theta_q a_{t-q} \end{aligned}$$

In cases where one would need a large number of parameters in either "pure MA" or "pure AR" models, a ratio of low-order MA and AR operators frequently leads to parsimonious representations of the ψ and π weights.

ARMA(1, 1) Process

The simplest example of an autoregressive moving average process is the ARMA(1, 1) process

$$(1 - \phi B)(z_t - \mu) = (1 - \theta B) a_t \tag{5.45}$$

This process can be written in the linear filter representation

$$z_t - \mu = a_t + \psi_1 a_{t-1} + \psi_2 a_{t-2} + \cdots$$

where the ψ weights are given by $\psi(B) = (1 - \theta B)/(1 - \phi B)$. Equating coefficients of B^j in $(1 - \phi B)(1 + \psi_1 B + \psi_2 B^2 + \cdots) = 1 - \theta B$ leads to

$$\begin{aligned}
&B^1: \quad \psi_1 - \phi = -\theta && \psi_1 = \phi - \theta \\
&B^2: \quad \psi_2 - \phi\psi_1 = 0 && \psi_2 = \phi\psi_1 = (\phi - \theta)\phi \\
&\quad \vdots && \quad \vdots \\
&B^j: \quad \psi_j - \phi\psi_{j-1} = 0 && \psi_j = \phi\psi_{j-1} = (\phi - \theta)\phi^{j-1} \qquad j > 0
\end{aligned}$$

Equivalently we can represent the ARMA(1, 1) process in terms of an infinite autoregression

$$z_t - \mu = \pi_1(z_{t-1} - \mu) + \pi_2(z_{t-2} - \mu) + \cdots + a_t$$

where the π weights are given by $\pi(B) = (1 - \phi B)/(1 - \theta B)$. Equating coefficients of B^j in $(1 - \theta B)(1 - \pi_1 B - \pi_2 B^2 - \cdots) = 1 - \phi B$ yields

$$B^1: \quad -\pi_1 - \theta = -\phi \qquad \pi_1 = \phi - \theta$$

$$B^2: \quad -\pi_2 + \theta\pi_1 = 0 \qquad \pi_2 = \theta\pi_1 = (\phi - \theta)\theta$$

$$\vdots \qquad\qquad \vdots$$

$$B^j: \quad -\pi_j + \theta\pi_{j-1} = 0 \qquad \pi_j = \theta\pi_{j-1} = (\phi - \theta)\theta^{j-1} \qquad j > 0$$

The simple ARMA(1, 1) model leads to both moving average and autoregressive representations with an infinite number of weights. The ψ weights converge for $|\phi| < 1$ (stationarity condition), and the π weights converge for $|\theta| < 1$ (invertibility condition). The stationarity condition for the ARMA(1, 1) model is the same as that of an AR(1) model; the invertibility condition is the same as that of an MA(1) model.

Autocorrelation function. The autocovariances of an ARMA(1, 1) process can be derived from (5.7), using the ψ weights $\psi_j = (\phi - \theta)\phi^{j-1}, j > 0$. An alternative approach, which is somewhat easier to extend to higher order models, starts from the difference equation

$$z_t = \phi z_{t-1} + a_t - \theta a_{t-1}$$

where $E(z_t) = \mu$ is assumed to be zero without loss of generality. Multiplying this equation by z_{t-k} and taking expectations, we obtain

$$\gamma_k = \phi\gamma_{k-1} + E(a_t z_{t-k}) - \theta E(a_{t-1} z_{t-k}) \tag{5.46}$$

If $k > 1$, $E(a_t z_{t-k}) = E(a_{t-1} z_{t-k}) = 0$, since a_t and a_{t-1} are uncorrelated with z_{t-k}. Therefore,

$$\gamma_k = \phi\gamma_{k-1} \qquad \text{for } k > 1$$

Furthermore, since

$$E(a_t z_t) = E[a_t(a_t + \psi_1 a_{t-1} + \psi_2 a_{t-2} + \cdots)] = \sigma^2$$

and

$$E(a_{t-1}z_t) = E[a_{t-1}(a_t + \psi_1 a_{t-1} + \psi_2 a_{t-2} + \cdots)]$$
$$= \psi_1 \sigma^2 = (\phi - \theta)\sigma^2$$

we obtain for

$$k = 0: \quad \gamma_0 = \phi\gamma_1 + \sigma^2 - \theta(\phi - \theta)\sigma^2$$

and

$$k = 1: \quad \gamma_1 = \phi\gamma_0 - \theta\sigma^2$$

Solving these two equations leads to

$$\gamma_0 = \frac{1 + \theta^2 - 2\phi\theta}{1 - \phi^2}\sigma^2 \qquad \gamma_1 = \frac{(1 - \phi\theta)(\phi - \theta)}{1 - \phi^2}\sigma^2$$

Hence the ACF is given by

$$\rho_1 = \frac{(1 - \phi\theta)(\phi - \theta)}{1 + \theta^2 - 2\phi\theta}$$

$$\rho_k = \phi\rho_{k-1} \qquad \text{for } k > 1 \tag{5.47}$$

The ACF of an ARMA(1, 1) process is similar to that of an AR(1) process; it is characterized by an exponential decay ($\rho_k = \phi\rho_{k-1} = \phi^2\rho_{k-2} = \cdots = \phi^{k-1}\rho_1$). However, this decay starts from ρ_1, and not from $\rho_0 = 1$ as in the AR(1) case. The ACF of an ARMA(1, 1) process is thus characterized by an exponential decay from lag 1 onwards and by one initial autocorrelation $\rho_0 = 1$ that does not lie on this exponential decay. The decay is smooth for $\phi > 0$ and oscillating if $\phi < 0$. Figure 5.7 shows the ACF for ($\phi = .5$, $\theta = .3$) and for ($\phi = -.5$, $\theta = .3$).

Partial autocorrelation function. The PACF consists of a single initial value $\phi_{11} = \rho_1$. From there it behaves like the PACF of an MA(1) process that is dominated by an exponential decay.

ARMA(p,q) Process

This process is written as

$$(1 - \phi_1 B - \cdots - \phi_p B^p)(z_t - \mu) = (1 - \theta_1 B - \cdots - \theta_q B^q)a_t$$

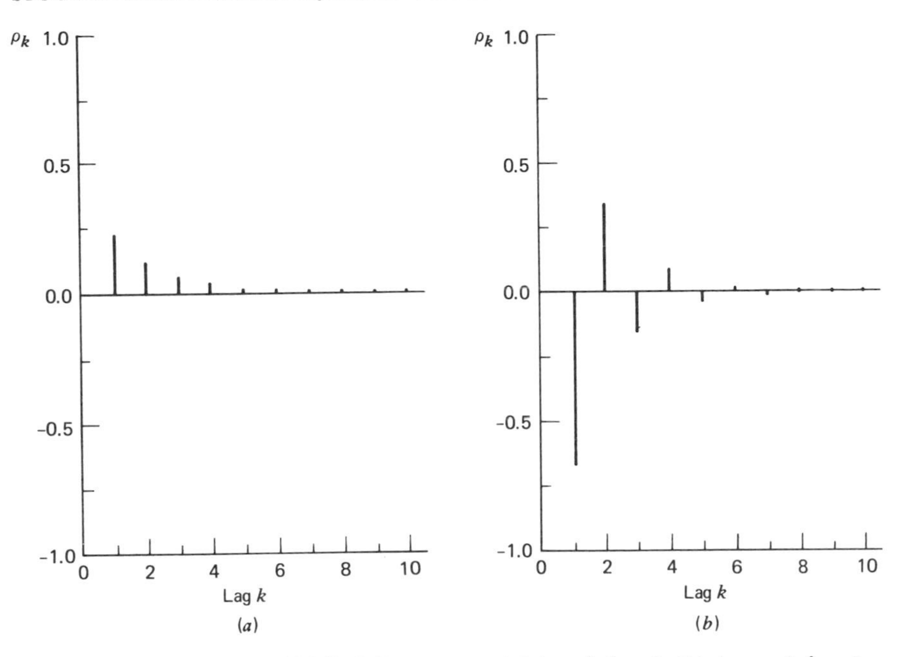

Figure 5.7. ACF for two ARMA(1, 1) processes: (a) $\phi = .5, \theta = .3$; (b) $\phi = -.5, \theta = .3$.

or

$$\phi(B)(z_t - \mu) = \theta(B)a_t \tag{5.48}$$

where $\phi(B) = 1 - \phi_1 B - \cdots - \phi_p B^p$ and $\theta(B) = 1 - \theta_1 B - \cdots - \theta_q B^q$ are the AR and MA operators, respectively. It can also be expressed in a pure MA representation:

$$z_t - \mu = \psi(B)a_t \qquad \psi(B) = \frac{\theta(B)}{\phi(B)}$$

and in a pure AR representation:

$$\pi(B)(z_t - \mu) = a_t \qquad \pi(B) = \frac{\phi(B)}{\theta(B)}$$

The ψ and π weights can be derived by equating coefficients as described before. For stationarity we have to assume that the roots of $\phi(B) = 0$ are outside the unit circle. For invertibility we have to assume that the roots of $\theta(B) = 0$ are outside the unit circle.

Autocorrelations. Multiplying the difference equation in (5.48) by $z_{t-k} - \mu$ and taking expectations, we find that

$$\gamma_k = \phi_1 \gamma_{k-1} + \cdots + \phi_p \gamma_{k-p} \qquad k > q$$

$$\rho_k = \phi_1 \rho_{k-1} + \cdots + \phi_p \rho_{k-p} \qquad k > q \tag{5.49}$$

This implies that eventually the autocorrelations of an ARMA(p, q) process follow the same pattern as that of an AR(p) process. They are described by combinations of damped exponentials and/or damped sine waves. The difference equation (5.49) holds for $k > q$. Since the p starting values $\rho_q, \rho_{q-1}, \ldots, \rho_{q-p+1}$ have to lie on the solution of this equation, the solution itself holds for $k > q - p$. Therefore, there will be $q - p + 1$ initial values $\rho_0, \rho_1, \ldots, \rho_{q-p}$ that do not follow this general pattern.

Some special cases

ARMA(1, 1): $q - p + 1 = 1$; one initial value ρ_0; exponential decay for $k > 0$

ARMA(1, 2): $q - p + 1 = 2$; two initial values ρ_0, ρ_1; exponential decay for $k > 1$

ARMA(2, 2): $q - p + 1 = 1$; one initial value ρ_0; exponential decay or damped sine wave for $k > 0$

ARMA(2, 1): $q - p + 1 = 0$; no initial values; all autocorrelations (for $k \geqslant 0$) are described by either a sum of two damped exponentials or a damped sine wave.

Partial autocorrelations. Eventually (for $k > p - q$), the partial autocorrelations of an ARMA(p, q) process behave like those of an MA(q) process. However, for $k \leqslant p - q$ the partial autocorrelations do not follow this general pattern.

Summary

The ACF and PACF of a mixed ARMA(p, q) process are both infinite in extent and tail off (die down) as the lag k increases. Eventually (for $k > q - p$), the ACF is determined from the autoregressive part of the model. The PACF is eventually (for $k > p - q$) determined from the moving average part of the model.

5.3. NONSTATIONARY PROCESSES

So far we have assumed that the underlying process is stationary. This implies that the mean, the variance, and the autocovariances of the process are invariant under time translations. Thus the mean and the variance are constant, and the autocovariances depend only on the time lag.

Many observed time series, however, are not stationary. In particular, most economic and business series exhibit time-changing levels and/or variances. For example, plots of the growth rates of Iowa nonfarm income (Fig. 5.8*a*) and the demand for repair parts (Fig. 5.9*a*) show that the level in each series changes with time. In addition, Figure 5.9*a* indicates that the variance also increases with the level of the series.

5.3.1. Nonstationarity, Differencing, and Transformations

A changing mean can often be described by low-order polynomials in time. However, frequently the coefficients in these polynomials are not constant but vary randomly with time. Such nonstationarity, in which the observations are described by random (or stochastic) trends, is usually referred to as *homogeneous nonstationarity*. It is characterized by a behavior in which, apart from local level and/or local trend, one part of the series behaves like the others.

Tintner (1940), Yaglom (1955), and Box and Jenkins (1976) argue that homogeneous nonstationary sequences can be transformed into stationary sequences by taking successive differences of the series; that is, by considering the series $\nabla z_t, \nabla^2 z_t, \ldots$, where ∇ is the difference operator $\nabla = 1 - B$. The use of these differences will be illustrated later.

Deterministic polynomial trends are those in which the coefficients of the polynomial stay constant for all times. In stochastic trends the coefficients are subject to random variation; thus the trend changes stochastically according to random shocks that enter the system. The distinction can be explained more fully by considering the following three linear trend models.

The first model (M1) is the deterministic linear trend model

$$z_t = \beta_0 + \beta_1 t + a_t \tag{5.50}$$

where a_t is a white-noise sequence or, more generally, a zero-mean stationary stochastic process. In this model the level $\mu_t = E(z_t) = \beta_0 + \beta_1 t$ grows according to a deterministic linear function of time.

The second model (M2) involves the first differences of the series and a constant trend parameter β_1. It is given by

$$(1 - B)z_t = \beta_1 + a_t \tag{5.51}$$

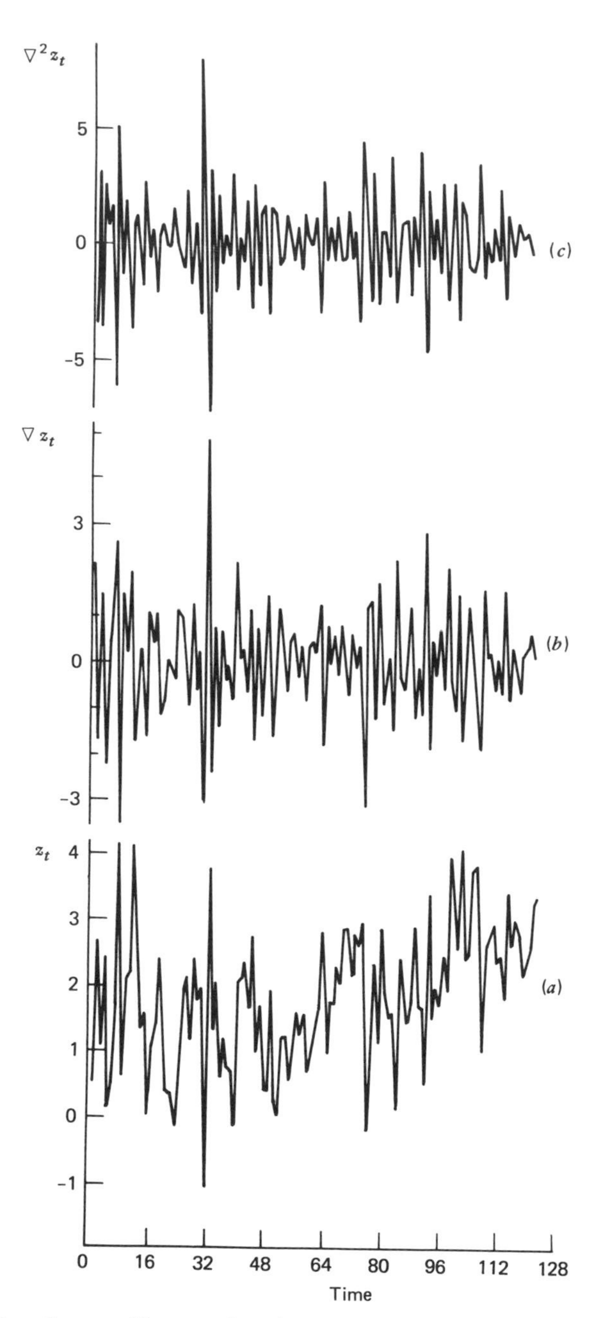

Figure 5.8. Growth rates of Iowa nonfarm income: (a) original series z_t; (b) first differences ∇z_t; (c) second differences $\nabla^2 z_t$.

or

$$z_t = z_{t-1} + \beta_1 + a_t$$

This model also leads to a linear trend. The level of the series at time t (or, more formally, the conditional expectation of z_t given $z_{t-1}, z_{t-2}, \ldots$) is $\mu_t = z_{t-1} + \beta_1$. Apart from the constant slope parameter β_1, the level depends on the previous observation. Since z_{t-1} is subject to random shocks, the trend changes stochastically.

The third model (M3) involves the second differences of the data

$$(1 - B)^2 z_t = a_t$$

$$z_t = 2z_{t-1} - z_{t-2} + a_t \tag{5.52}$$

Since z_{t-1} and z_{t-2} in the level $\mu_t = 2z_{t-1} - z_{t-2}$ are influenced by random shocks, both the intercept and the slope of the linear trend, which passes through z_{t-1} and z_{t-2}, change stochastically.

In Figure 5.10 we have generated sample paths for each of the three series. M1 is characterized by a deterministic trend, M2 and M3 by

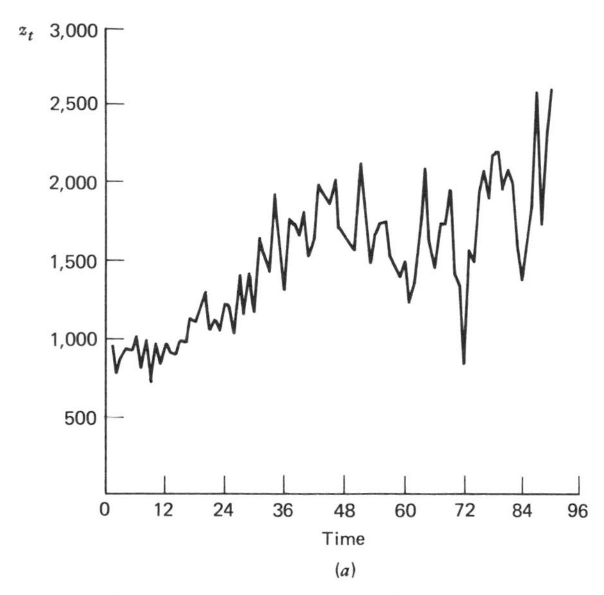

Figure 5.9. Demand for repair parts: (a) original series z_t; (b) $\ln z_t$; (c) first differences $\nabla \ln z_t$; (d) second differences $\nabla^2 \ln z_t$.

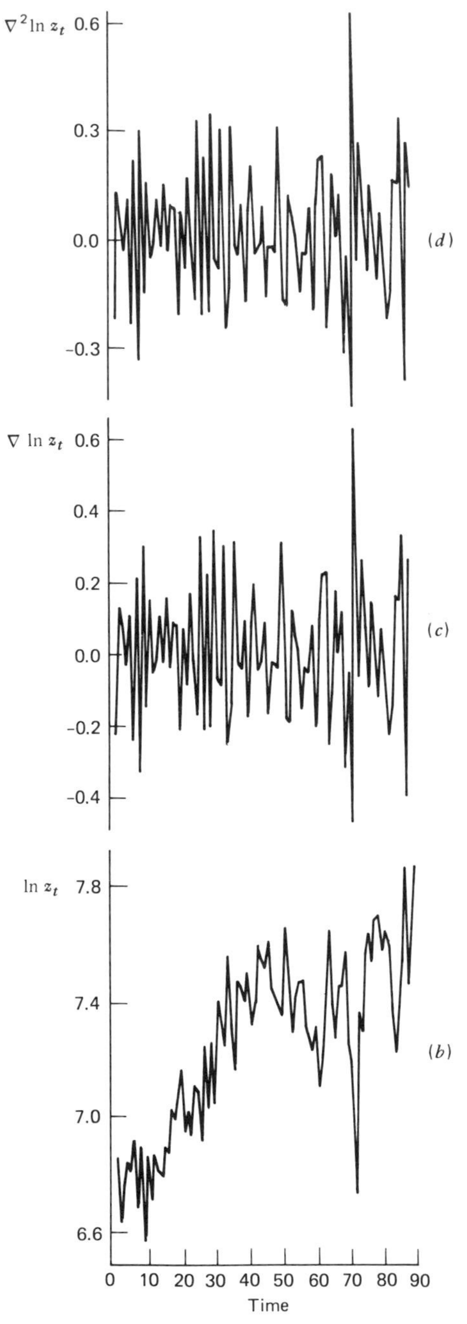

Figure 5.9. (*Continued*).

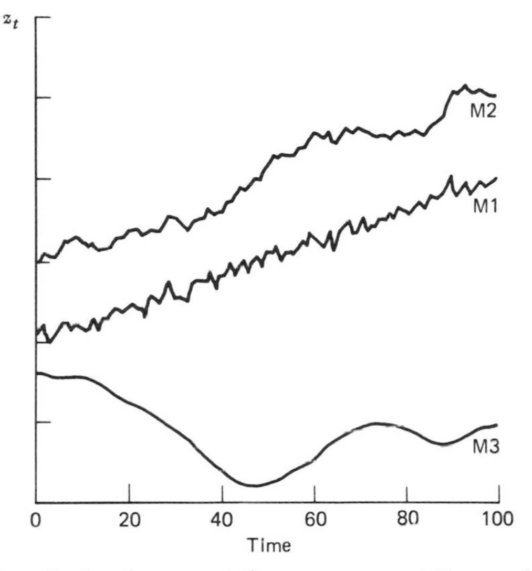

Figure 5.10. Sample paths for three nonstationary processes. M1: $z_t = \beta_0 + \beta_1 t + a_t$. M2: $(1 - B)z_t = \beta_1 + a_t$. M3: $(1 - B)^2 z_t = a_t$.

stochastic trends. Whereas in model M2 only the intercept in the trend model is subject to variation, in model M3 the slope parameter also changes randomly.

Nonstationary Variance

Apart from a nonstationary mean level, many economic and business data exhibit nonstationary variance. Frequently the variation in the data increases with the level. It was shown in Chapter 2 that in cases where the standard deviation of the data is proportional to the level a logarithmic transformation will stabilize the variance. In other cases, a more general class of transformations, such as the power transformations described in Section 2.11, may be needed.

We illustrate the use of differences and transformations on two examples.

Example 1: Growth rates. The plot of Iowa nonfarm income growth rates in Figure 5.8*a* shows nonstationarity in the level. However, a plot of $(1 - B)z_t$ in Figure 5.8*b* indicates that the first differences $w_2 = z_2 - z_1$, $w_3 = z_3 - z_2, \ldots, w_n = z_n - z_{n-1}$ can be considered stationary. The second diferences $\nabla^2 z_t$ ($w_t = z_t - 2z_{t-1} + z_{t-2}$; $w_3 = z_3 - 2z_2 + z_1$, $w_4 = z_4 - 2z_3 + z_2, \ldots, w_n = z_n - 2z_{n-1} + z_{n-2}$) in Figure 5.8*c* are also stationary.

In fact, it can be shown that any linear combination of a stationary sequence is again stationary. Thus, additional differences of a stationary sequence will always be stationary.

Example 2: Demand for repair parts. Figure 5.9*b* gives a plot of the logarithm of the demand for repair parts. A logarithmic transformation is necessary, since Figure 5.9*a* has shown that the variation in the original observations is roughly proportional to the level of the series. The variance of $\ln z_t$ is approximately constant. However, the level of this series changes with time. Hence the first difference, $\nabla \ln z_t$, is considered (Fig. 5.9*c*). The mean and variance of this series are constant. Thus, stationarity has been achieved, and any further differences are also stationary.

An Alternative Argument for Differencing

Let us assume that the data z_t are described by a nonstationary level $\mu_t = E(z_t) = \beta_0 + \beta_1 t$ and that we wish to entertain an ARMA model for a function of z_t that has a level independent of time. Suppose we consider the model

$$\phi_{p+1}(B) z_t = \theta_0 + \theta(B) a_t$$

where $\phi_{p+1}(B) = 1 - \phi_1 B - \phi_2 B^2 - \cdots - \phi_{p+1} B^{p+1}$, $\theta(B) = 1 - \theta_1 B - \cdots - \theta_q B^q$, and θ_0 represents the level of $\phi_{p+1}(B) z_t$. To achieve a θ_0 that is constant, independent of time, we have to choose $\phi_{p+1}(B)$ such that it includes $(1 - B)$ as a factor. If it does, then let

$$\phi_{p+1}(B) = (1 - B)\phi(B) = (1 - B)(1 - \phi_1 B - \cdots - \phi_p B^p)$$

and

$$\begin{aligned} \theta_0 &= \phi_{p+1}(B)(\beta_0 + \beta_1 t) = \phi(B)(1 - B)(\beta_0 + \beta_1 t) \\ &= \phi(B)\beta_1 = (1 - \phi_1 - \cdots - \phi_p)\beta_1 \end{aligned}$$

which is a constant. In such a case, the first differences of the original observations should be analyzed.

This argument can be extended to the general case, where the mean is a polynomial in time of degree d, $\mu_t = \beta_0 + \beta_1 t + \cdots + \beta_d t^d$. If we now look for a function of z_t that has a level independent of time, we have to choose the autoregressive operator in $\phi_{p+d}(B) z_t = \theta_0 + \theta(B) a_t$ such that $\phi_{p+d}(B)$ includes a factor $(1 - B)^d$. Suppose that

$$\phi_{p+d}(B) = (1 - B)^d \phi(B) = (1 - B)^d (1 - \phi_1 B - \cdots - \phi_p B^p)$$

Then

$$\begin{aligned}\theta_0 &= \phi_{p+d}(B)(\beta_0 + \beta_1 t + \cdots + \beta_d t^d) \\ &= \phi(B)(1-B)^d(\beta_0 + \beta_1 t + \cdots + \beta_d t^d) \\ &= \phi(B)\beta_d\, d! = (1 - \phi_1 - \cdots - \phi_p)\beta_d d!,\end{aligned}$$

which is again a constant. In such a case the dth difference of the original observations should be analyzed. A similar argument can be made if the coefficients in μ_t change randomly.

5.3.2. Autoregressive Integrated Moving Average (ARIMA) Models

The discussion in the last section indicates that a nonstationary series z_t should be first transformed into a stationary one by considering relevant differences $w_t = \nabla^d z_t = (1 - B)^d z_t$. Since w_t is then a stationary process, we can use the ARMA models to describe w_t. The corresponding model can be written as

$$\phi(B)w_t = \theta_0 + \theta(B)a_t$$

or

$$\phi(B)(1 - B)^d z_t = \theta_0 + \theta(B)a_t \tag{5.53}$$

where $\phi(B)$, $\theta(B)$, and $\{a_t\}$ were defined before and θ_0 is a constant usually referred to as a *trend parameter*. The model in (5.53) is called the *autoregressive integrated moving average* (*ARIMA*) *model* of order (p, d, q) and is denoted by ARIMA(p, d, q). The orders p, d, q are usually small. The model is called "integrated," since z_t can be thought of as the summation (integration) of the stationary series w_t. For example, if $d = 1$,

$$z_t = (1 - B)^{-1}w_t = \sum_{k=-\infty}^{t} w_k$$

Choice of the Proper Degree of Differencing

Homogeneous nonstationarity can be recognized in several ways.

Plot of the series. The discussion in the previous section indicates that, in general, nonstationary series exhibit time-varying levels. Such behavior is demonstrated in the plot of the growth rates (Fig. 5.8*a*) and the logarithms of the demand data (Fig. 5.9*b*).

Examination of the SACF. A stationary AR(p) process requires that all G_i in

$$(1 - \phi_1 B - \cdots - \phi_p B^p) = (1 - G_1 B) \cdots (1 - G_p B)$$

are such that $|G_i| < 1$ ($i = 1, 2, \ldots, p$). Now suppose that one of them, say G_1, approaches 1, that is, $G_1 = 1 - \delta$, where δ is a small positive number. Then the autocorrelations

$$\rho_k = A_1 G_1^k + \cdots + A_p G_p^k \cong A_1 G_1^k$$

will be dominated by $A_1 G_1^k$, since all other terms will go to zero more rapidly. Furthermore, since G_1 is close to 1, the exponential decay will be slow and almost linear [i.e., $A_1 G_1^k = A_1(1 - \delta)^k \cong A_1(1 - \delta k)$]. Failure of the ACF to die down quickly can then be taken as an indication of nonstationarity.

As a further illustration, consider the ARMA(1, 1) model

$$(1 - \phi B) z_t = (1 - \theta B) a_t$$

The ACF of this process is shown in (5.47) and is given by

$$\rho_1 = \frac{(1 - \phi\theta)(\phi - \theta)}{1 + \theta^2 - 2\phi\theta}$$

$$\rho_k = \phi^{k-1} \rho_1 \qquad k > 0$$

If we let $\phi \to 1$, then $\rho_k \to 1$ for any finite $k > 0$. Again the autocorrelations decay very slowly when ϕ is near 1. Hence a slow and almost linear decay in the sample ACF may be taken as an indication of nonstationarity and hence of the need for differencing. It should be pointed out that the slow decay in the SACF can start at values of r_1 considerably smaller than 1. This is illustrated by Wichern (1973), who showed that due to sampling variability the sample autocorrelations of some nonstationary processes can be expected to decay from a value $r_1 = .5$ or even less.

If the original series is found to be nonstationary, we analyze the first difference. If the first difference is stationary (constant level, rapidly decaying autocorrelations), we entertain ARMA(p, q) models for this stationary difference. In cases where the first difference is still not stationary, we consider second differences and repeat this strategy until a stationary difference is found.

Figure 5.11*a* shows the SACF of the growth rate series. The fact that the sample autocorrelations fail to die out quickly leads us to examine the first difference of the series. Similar observations can be made from the SACF of the logarithms of the demand series (Fig. 5.12*a*).

The SACF of the first and second differences (Figs. 5.11*b*, *c* and 5.12*b*, *c*) die out rapidly. This conveys the same information as the data plots in Figures 5.8 and 5.9, which showed that first and second differences are stationary. Thus we consider the first differences for further modeling.

Dangers of Overdifferencing

Although further differences of a stationary series will again be stationary, overdifferencing can lead to serious difficulties. In particular, it can unnecessarily complicate the autocorrelation structure and increase the variance of the overdifferenced series. These difficulties are illustrated in the following examples.

Effect of overdifferencing on the autocorrelation structure. Consider the stationary MA(1) process $z_t = (1 - \theta B)a_t$. The autocorrelation function of this process is given by [see (5.38)]

$$\rho_k = \begin{cases} \dfrac{-\theta}{1+\theta^2} & k = 1 \\ 0 & k > 1 \end{cases}$$

The first difference of the process is

$$(1 - B)z_t = \left[1 - (1+\theta)B + \theta B^2\right]a_t$$

and the autocorrelations of this difference are given by [see (5.40)]

$$\rho_k = \begin{cases} \dfrac{-(1+\theta)^2}{2(1+\theta+\theta^2)} & k = 1 \\ \dfrac{\theta}{2(1+\theta+\theta^2)} & k = 2 \\ 0 & k > 2 \end{cases}$$

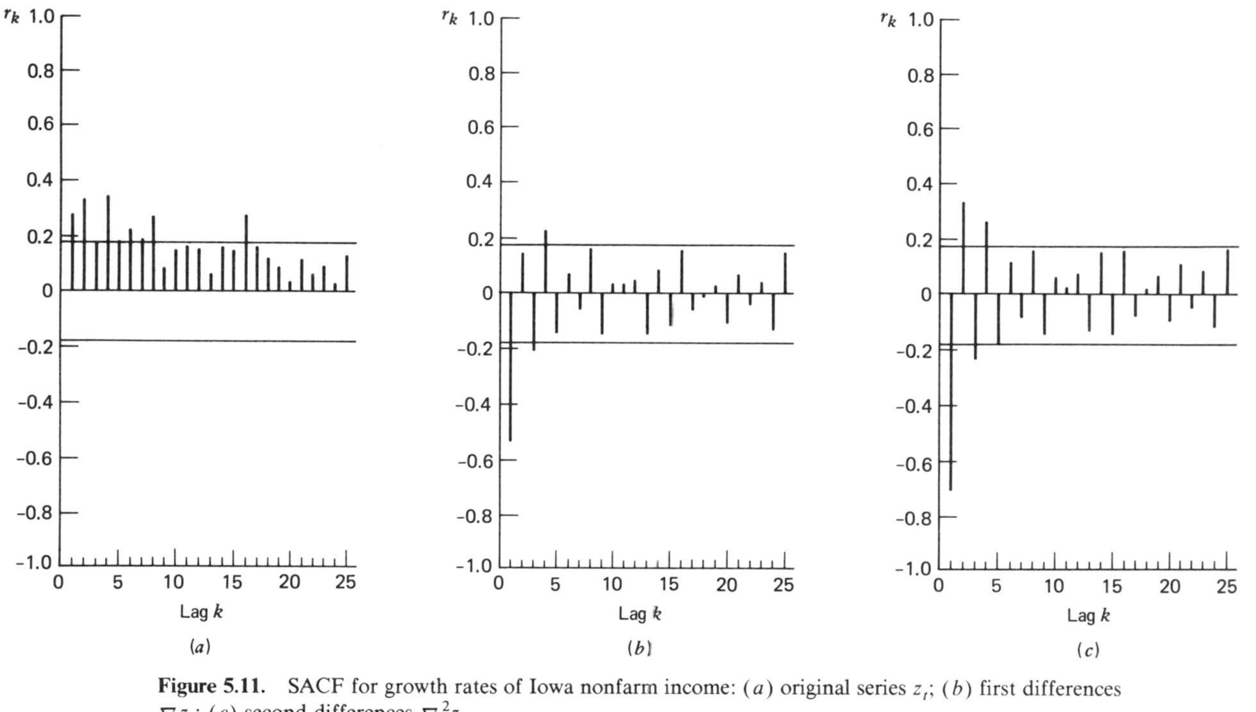

Figure 5.11. SACF for growth rates of Iowa nonfarm income: (a) original series z_t; (b) first differences ∇z_t; (c) second differences $\nabla^2 z_t$.

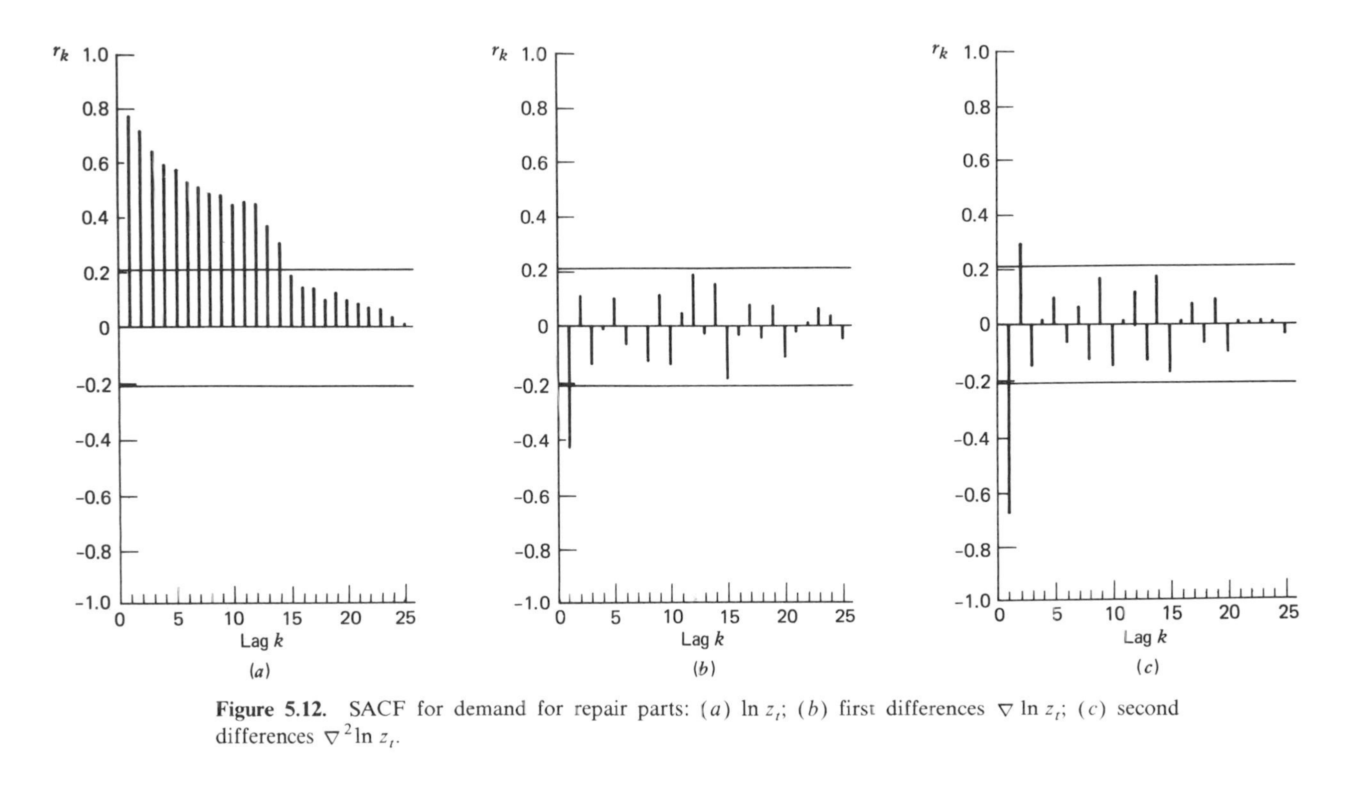

Figure 5.12. SACF for demand for repair parts: (*a*) $\ln z_t$; (*b*) first differences $\nabla \ln z_t$; (*c*) second differences $\nabla^2 \ln z_t$.

It can be seen that the structure of the ACF of the overdifferenced series is more complicated than that of the original process. This leads to a non-parsimonious representation, since it requires the estimation of two parameters as compared to one in the original MA(1) model. Furthermore, overdifferencing will force the moving average operator to be of noninvertible form. Problems associated with the estimation of such models will be discussed in Section 5.6.

Effect of overdifferencing on the variance. Suppose we consider the same MA(1) process. Its variance is given by $\gamma_0(z) = (1 + \theta^2)\sigma^2$. The variance of the overdifferenced series $w_t = (1 - B)z_t$, which follows an MA(2) process, is given by

$$\gamma_0(w) = 2(1 + \theta + \theta^2)\sigma^2$$

Hence,

$$\gamma_0(w) - \gamma_0(z) = (1 + \theta)^2\sigma^2 > 0$$

which shows that the variance of the overdifferenced process will always be larger than that of the original process.

As another example, consider the stationary AR(1) process $(1 - \phi B)z_t = a_t$ with variance $\gamma_0(z) = \sigma^2/(1 - \phi^2)$. The first difference w_t follows the ARMA(1, 1) process $(1 - \phi B)w_t = (1 - B)a_t$. The variance of this process is given by (see Sec. 5.2.4)

$$\gamma_0(w) = \frac{2(1 - \phi)\sigma^2}{1 - \phi^2}.$$

From

$$\gamma_0(w) - \gamma_0(z) = \frac{(1 - 2\phi)\sigma^2}{1 - \phi^2}$$

we find that for $\phi < \frac{1}{2}$ overdifferencing will increase the variance.

Tintner (1940), in fact, has used the changes in the sample variances of successive differences to determine the proper degree of differencing. For nonstationary sequences the sample variances will be large, since the squared deviations are taken from a constant mean. Considerable bias will be introduced, since nonstationary sequences do not have a fixed level. The variances will continue to decrease until a stationary sequence has been found. Overdifferencing, however, can introduce additional variation, and the sample variances will begin to increase.

5.3.3. Regression and ARIMA Models

In this section we explore the similarities between regression models in which $E(z_t)$ is modeled as a deterministic function of time and ARIMA models, which use differences to model stochastic trends in nonstationary sequences. Two examples will be considered here, the ARIMA(0, 1, 1) and ARIMA(0, 2, 2) models.

ARIMA(0, 1, 1) model. This model can be written as

$$(1 - B)z_t = (1 - \theta B)a_t \tag{5.54}$$

Using the ψ weights of this model [$\psi_j = 1 - \theta$ for all j, since $\psi(B) = (1 - \theta B)/(1 - B) = 1 + (1 - \theta)B + (1 - \theta)B^2 + \cdots$], we can relate the variable z_{t+l} at time $t + l$ to a fixed origin t and express it as

$$z_{t+l} = \beta_0^{(t)} + e_t(l) \tag{5.55}$$

In this equation, $e_t(l) = a_{t+l} + (1 - \theta)[a_{t+l-1} + \cdots + a_{t+1}]$ is a function of the random shocks that enter the system after time t, and $\beta_0^{(t)} = (1 - \theta)\Sigma_{j=-\infty}^{t} a_j$ depends on all the shocks (observations) up to and including time t. The level $\beta_0^{(t)}$ can be updated according to

$$\beta_0^{(t+1)} = \beta_0^{(t)} + (1 - \theta)a_{t+1}$$

Thus this model can be described by a mean level that changes stochastically and error sequences $e_t(l)$ that are correlated. Equation (5.55) allows the mean level to adapt as time evolves and furthermore allows correlations in the errors. In the special case when $\theta = 1$, we find that $e_t(l) = a_{t+l}$ and $\beta_0^{(t+1)} = \beta_0^{(t)} = \beta_0$. In this case, the model is equivalent to the constant mean model $z_{t+l} = \beta_0 + a_{t+l}$.

ARIMA(0, 2, 2) model. Using an argument similar to that followed for the ARIMA(0, 1, 1) model [see also Box and Jenkins (1976) and Abraham and Box (1978)], we can write the model

$$(1 - B)^2 z_{t+l} = \left(1 - \theta_1 B - \theta_2 B^2\right)a_{t+l} \tag{5.56}$$

as

$$z_{t+l} = \beta_0^{(t)} + \beta_1^{(t)} l + e_t(l) \tag{5.57}$$

In this representation the errors $e_t(l)$ are correlated and are given by

$$e_t(l) = a_{t+l} + \psi_1 a_{t+l-1} + \cdots + \psi_{l-1} a_{t+1}$$

where the ψ weights may be obtained from

$$\left(1 + \psi_1 B + \psi_2 B^2 + \cdots\right)(1 - B)^2 = 1 - \theta_1 B - \theta_2 B^2$$

These ψ weights can be given as $\psi_j = (1 + \theta_2) + j(1 - \theta_1 - \theta_2)$, $j \geqslant 1$. Furthermore, the trend parameters $\beta_0^{(t)}$ and $\beta_1^{(t)}$ adapt with time according to

$$\beta_0^{(t+1)} = \beta_0^{(t)} + \beta_1^{(t)} + (1 + \theta_2) a_{t+1}$$

$$\beta_1^{(t+1)} = \beta_1^{(t)} + (1 - \theta_1 - \theta_2) a_{t+1}$$

Again model (5.57) can be thought of as a generalization of the deterministic trend regression model

$$z_{t+l} = \beta_0 + \beta_1 l + a_{t+l} \tag{5.58}$$

since model (5.57) allows for time-varying coefficients $\beta_0^{(t)}$ and $\beta_1^{(t)}$ and also for correlated errors $e_t(l)$. In the special case when $\theta_1 \to 2$ and $\theta_2 \to -1$, model (5.56) is equivalent to model (5.58) with constant coefficients.

In these two examples we find that the stochastic difference equation models in fact include the deterministic trend models as special cases. These results are generalized by Abraham and Box (1978).

5.4. FORECASTING

Suppose we have obtained an ARIMA model $\phi(B)(1 - B)^d z_t = \theta(B) a_t$ for a given time series and wish to use it to predict future observations. Let us write the model in its autoregressive representation:

$$z_t = \pi_1 z_{t-1} + \pi_2 z_{t-2} + \cdots + a_t$$

where the π weights are given by

$$1 - \pi_1 B - \pi_2 B^2 - \cdots = \frac{\phi(B)(1 - B)^d}{\theta(B)}$$

We are interested in predicting z_{n+l} $(l \geqslant 1)$ as a linear combination of past observations $z_n, z_{n-1}, \ldots$. Hence we consider forecasts of the form

$$z_n(l) = \eta_0 z_n + \eta_1 z_{n-1} + \eta_2 z_{n-2} + \cdots \tag{5.59}$$

Equivalently we could express the forecast as a linear combination of past shocks:

$$z_n(l) = \xi_0 a_n + \xi_1 a_{n-1} + \xi_2 a_{n-2} + \cdots \tag{5.60}$$

The coefficients η_j and ξ_j $(j \geqslant 0)$ are constants that have to be determined. They are usually chosen such that the mean square error $E[z_{n+l} - z_n(l)]^2$ is minimized. This leads to *minimum mean square error* (*MMSE*) forecasts.

Using the ψ weights in $\psi(B) = \theta(B)\phi^{-1}(B)(1 - B)^{-d}$ to write the model in its linear filter representation,

$$z_{n+l} = a_{n+l} + \psi_1 a_{n+l-1} + \cdots + \psi_{l-1} a_{n+1} + \psi_l a_n + \psi_{l+1} a_{n-1} + \cdots$$

we can express the mean square error as

$$\begin{aligned} E\left[z_{n+l} - z_n(l)\right]^2 &= E\left[a_{n+l} + \psi_1 a_{n+l-1} + \cdots + \psi_{l-1} a_{n+1}\right. \\ &\qquad \left. + (\psi_l - \xi_0) a_n + (\psi_{l+1} - \xi_1) a_{n-1} + \cdots\right]^2 \\ &= \left(1 + \psi_1^2 + \cdots + \psi_{l-1}^2\right)\sigma^2 + \sum_{j=0}^{\infty} \left(\psi_{l+j} - \xi_j\right)^2 \sigma^2 \end{aligned} \tag{5.61}$$

As a function of ξ_j, this expression is minimized if $\xi_j = \psi_{l+j}$ $(j = 0, 1, 2, \ldots)$. This leads to the minimum mean square error (MMSE) forecast

$$z_n(l) = \psi_l a_n + \psi_{l+1} a_{n-1} + \cdots \tag{5.62}$$

The forecast error is given by

$$e_n(l) = z_{n+l} - z_n(l) = a_{n+l} + \psi_1 a_{n+l-1} + \cdots + \psi_{l-1} a_{n+1} \tag{5.63}$$

and its variance by

$$V\left[e_n(l)\right] = \sigma^2\left(1 + \psi_1^2 + \psi_2^2 + \cdots + \psi_{l-1}^2\right) \tag{5.64}$$

The forecast error is a linear combination of the unobservable future shocks entering the system after time n. In particular, if $l = 1$, the one-step-ahead forecast error is $e_n(1) = z_{n+1} - z_n(1) = a_{n+1}$. For different time origins n, the one-step-ahead forecast errors are uncorrelated. In general, the forecast errors for higher lead times are correlated. For example,

$$\operatorname{Cov}[e_n(2), e_{n+1}(2)] = E(a_{n+2} + \psi_1 a_{n+1})(a_{n+3} + \psi_1 a_{n+2}) = \psi_1 \sigma^2$$

Equation (5.62) expresses the MMSE forecast as a linear combination of present and past random shocks $a_n, a_{n-1}, \dots$. We could, in principle, substitute $a_t = \pi(B) z_t$ into Equation (5.62) and express the forecast as a linear combination of the present and past observations $z_n, z_{n-1}, \dots$. However, this is tedious and unnecessary, since another, simpler, method can be used to compute the forecasts.

From (5.62) we find that

$$z_n(l) = E(z_{n+l} | z_n, z_{n-1}, \dots) \tag{5.65}$$

where the expression on the right-hand side is the conditional expectation of z_{n+l} given $z_n, z_{n-1}, \dots$ (or equivalently $a_n, a_{n-1}, \dots$). This follows from the fact that the conditional expectation of a_{n+j} given $z_n, z_{n-1}, \dots$ is given by

$$E(a_{n+j} | z_n, z_{n-1}, \dots) = \begin{cases} a_{n+j} & j \leqslant 0 \\ 0 & j > 0 \end{cases}$$

and hence

$$\begin{aligned} E(z_{n+l} | z_n, z_{n-1}, \dots) &= E[(a_{n+l} + \psi_1 a_{n+l-1} + \cdots + \psi_{l-1} a_{n+1} \\ &\quad + \psi_l a_n + \psi_{l+1} a_{n-1} + \cdots) | z_n, z_{n-1}, \dots] \\ &= \psi_l a_n + \psi_{l+1} a_{n-1} + \cdots \end{aligned}$$

which is exactly the expression in (5.62). The conditional expectation representation simplifies the actual computation of the forecasts.

5.4.1. Examples

AR(1) process. As the first example we consider the process

$$z_t - \mu = \phi(z_{t-1} - \mu) + a_t$$

Suppose we are given past observations $z_n, z_{n-1}, \ldots$ and wish to predict z_{n+l}. For $l = 1$ (one step ahead),

$$z_n(1) = E(z_{n+1}|z_n, z_{n-1}, \ldots)$$
$$= E\{[\mu + \phi(z_n - \mu) + a_{n+1}]|z_n, z_{n-1}, \ldots\} = \mu + \phi(z_n - \mu)$$

since $E(z_n|z_n, z_{n-1}, \ldots) = z_n$ and $E(a_{n+1}|z_n, z_{n-1}, \ldots) = 0$.

For $l = 2$,

$$z_n(2) = E(z_{n+2}|z_n, z_{n-1}, \ldots)$$
$$= E\{[\mu + \phi(z_{n+1} - \mu) + a_{n+2}]|z_n, z_{n-1}, \ldots\}$$
$$= \mu + \phi[z_n(1) - \mu] = \mu + \phi^2(z_n - \mu)$$

The l-step-ahead prediction can be written as

$$z_n(l) = E(z_{n+l}|z_n, z_{n-1}, \ldots)$$
$$= E\{[\mu + \phi(z_{n+l-1} - \mu) + a_{n+l}]|z_n, z_{n-1}, \ldots\}$$
$$= \mu + \phi[z_n(l-1) - \mu] = \cdots = \mu + \phi^l(z_n - \mu) \quad (5.66)$$

This indicates that for stationary processes ($|\phi| < 1$) and large lead time l the best forecast of a future observation is eventually the mean of the process.

The forecast errors corresponding to the above forecasts are

$$e_n(1) = z_{n+1} - z_n(1) = \mu + \phi(z_n - \mu) + a_{n+1} - [\mu + \phi(z_n - \mu)] = a_{n+1}$$
$$e_n(2) = z_{n+2} - z_n(2) = \mu + \phi(z_{n+1} - \mu) + a_{n+2} - [\mu + \phi^2(z_n - \mu)]$$
$$= a_{n+2} + \phi[(z_{n+1} - \mu) - \phi(z_n - \mu)] = a_{n+2} + \phi a_{n+1}$$

Similarly, it can be shown that

$$e_n(l) = a_{n+l} + \phi a_{n+l-1} + \cdots + \phi^{l-1}a_{n+1} \quad (5.67)$$

and

$$V[e_n(l)] = \sigma^2(1 + \phi^2 + \cdots + \phi^{2(l-1)})$$
$$= \sigma^2 \frac{1 - \phi^{2l}}{1 - \phi^2} \quad (5.68)$$

For the AR(1) model, the forecast error $e_n(l)$ and its variance could have been derived directly from (5.63) and (5.64). The ψ weights for the AR(1) process are given by the coefficients in $\psi(B) = (1 - \phi B)^{-1} = 1 + \phi B + \phi^2 B^2 + \cdots$. Substituting the ψ weights $\psi_j = \phi^j$ into Equations (5.63) and (5.64) leads directly to (5.67) and (5.68).

The forecast in (5.66) and the forecast error variance in (5.68) depend on the parameters μ, ϕ, and σ^2. Usually these parameters are estimated from past observations $z_n, z_{n-1}, \ldots$ and are replaced by their estimates $\hat{\mu}$, $\hat{\phi}$, and $\hat{\sigma}^2$. (A detailed discussion of estimation is given in Sec. 5.6.) Then the estimated forecast becomes

$$\hat{z}_n(l) = \hat{\mu} + \hat{\phi}^l(z_n - \hat{\mu}) \tag{5.69}$$

The use of estimates in place of the parameters will affect the variance of the forecast error given in (5.68). In long time series, however, this effect is usually negligible [see Box and Jenkins (1976), p. 267]. Hence the estimated variance may be taken as

$$\hat{V}[e_n(l)] \cong \hat{\sigma}^2 \frac{1 - \hat{\phi}^{2l}}{1 - \hat{\phi}^2} \tag{5.70}$$

For a numerical example, consider the first 156 observations of the yield series. It is shown in Section 5.8 that this series can be described by an AR(1) model with $\hat{\mu} = 0.97$, $\hat{\phi} = 0.85$, and $\hat{\sigma}^2 = 0.024$. Since the last observation is $z_{156} = 0.49$, the forecasts are

$$\begin{aligned}
\hat{z}_{156}(1) &= 0.97 + 0.85(0.49 - 0.97) = 0.56 \\
\hat{z}_{156}(2) &= 0.97 + 0.85^2(0.49 - 0.97) = 0.62 \\
\hat{z}_{156}(3) &= 0.97 + 0.85^3(0.49 - 0.97) = 0.68
\end{aligned} \tag{5.71}$$

and their variances are

$$\hat{V}[e_{156}(1)] = 0.024$$

$$\hat{V}[e_{156}(2)] = 0.024\frac{1 - .85^4}{1 - .85^2} = 0.041$$

$$\hat{V}[e_{156}(3)] = 0.024\frac{1 - .85^6}{1 - .85^2} = 0.054$$

AR(2) process. As a second example we consider the forecast calculation for the AR(2) process, in which the deviations from the mean follow the model $z_t = \phi_1 z_{t-1} + \phi_2 z_{t-2} + a_t$. The one-step-ahead ($l = 1$) forecast, given the observations $z_n, z_{n-1}, \ldots$, can be expressed as

$$\begin{aligned} z_n(1) &= E(z_{n+1}|z_n, z_{n-1}, \ldots) \\ &= E[(\phi_1 z_n + \phi_2 z_{n-1} + a_{n+1})|z_n, z_{n-1}, \ldots] = \phi_1 z_n + \phi_2 z_{n-1} \end{aligned}$$

For $l = 2$,

$$\begin{aligned} z_n(2) &= E(z_{n+2}|z_n, z_{n-1}, \ldots) \\ &= E[(\phi_1 z_{n+1} + \phi_2 z_n + a_{n+2})|z_n, z_{n-1}, \ldots] = \phi_1 z_n(1) + \phi_2 z_n \end{aligned}$$

and, in general,

$$z_n(l) = \phi_1 z_n(l-1) + \phi_2 z_n(l-2) \tag{5.72}$$

This difference equation can also be written as

$$(1 - \phi_1 B - \phi_2 B^2) z_n(l) = 0 \qquad l > 0$$

where B now operates on l, and $z_n(l) = z_{n+l}$ for $l \leqslant 0$. This difference equation form makes the recursive forecast calculation very convenient. For the AR(2) model with nonzero mean, the forecasts follow the difference equation

$$z_n(l) = \mu + \phi_1[z_n(l-1) - \mu] + \phi_2[z_n(l-2) - \mu]$$

The forecast error and its variance can be calculated from (5.63) and (5.64) by substituting the ψ weights of the AR(2) model. It is easily seen that the ψ weights are

$$\psi_1 = \phi_1 \qquad \psi_2 = \phi_1^2 + \phi_2 \qquad \psi_j = \phi_1 \psi_{j-1} + \phi_2 \psi_{j-2} \qquad j \geqslant 2$$

ARIMA(0, 1, 1) process. Given the observations $z_n, z_{n-1}, \ldots$, the predictions from the model $z_t = z_{t-1} + a_t - \theta a_{t-1}$ can be obtained from the conditional expectation form:

$$\begin{aligned} z_n(1) &= E(z_{n+1}|z_n, z_{n-1}, \ldots) \\ &= E[(z_n + a_{n+1} - \theta a_n)|z_n, z_{n-1}, \ldots] = z_n - \theta a_n \\ z_n(2) &= E(z_{n+2}|z_n, z_{n-1}, \ldots) = z_n(1) = z_n - \theta a_n \end{aligned}$$

and, in general,

$$z_n(l) = z_n(l-1) \qquad \text{or} \qquad (1-B)z_n(l) = 0 \qquad l > 1 \qquad (5.73)$$

This model implies that all future forecasts are the same.

The ψ weights can be obtained from $\psi(B) = (1 - \theta B)/(1 - B)$, and, as we saw before, they are given by $\psi_j = 1 - \theta$ for all $j > 0$. Hence the forecast error is given by

$$e_n(l) = a_{n+l} + (1-\theta)(a_{n+l-1} + \cdots + a_{n+1}) \qquad (5.74)$$

and its variance by

$$V[e_n(l)] = \sigma^2\left[1 + (l-1)(1-\theta)^2\right] \qquad (5.75)$$

Alternatively, the forecasts can be expressed as a linear combination of the past observations. Such a representation is instructive, since one can see how the forecasts discount past information. Writing the model in its autoregressive representation

$$z_t = \sum_{j=1}^{\infty} \pi_j z_{t-j} + a_t$$

where $\pi_j = (1-\theta)\theta^{j-1}$ $(j \geqslant 1)$ are the coefficients in $\pi(B) = (1 - B)/(1 - \theta B)$, leads to

$$z_t = (1-\theta)(z_{t-1} + \theta z_{t-2} + \theta^2 z_{t-3} + \cdots) + a_t$$

Taking the conditional expectation of z_{n+1} given $z_n, z_{n-1}, \ldots,$ we find that

$$z_n(1) = (1-\theta)(z_n + \theta z_{n-1} + \theta^2 z_{n-2} + \cdots) \qquad (5.76)$$

This forecast is an exponentially weighted average of present and past observations and is the same as that obtained from single exponential smoothing with a smoothing constant $\alpha = 1 - \theta$ (see Sec. 3.3).

The forecasts in (5.73) involve the unknown quantities θ and a_n. Once the model is estimated and the parameter estimate $\hat{\theta}$ is obtained, we can calculate the residuals $\hat{a}_t$ that can be used in place of a_t. Hence the estimated forecast becomes

$$\hat{z}_n(l) = \hat{z}_n(1) = z_n - \hat{\theta}\hat{a}_n \qquad (5.77)$$

As mentioned before, in moderately long series the effect of the estimation on the forecast error variance is negligible.

ARIMA(1, 1, 1) process. We now consider the forecasts from the model $(1 - \phi B)(1 - B)z_t = \theta_0 + (1 - \theta B)a_t$. We have included a deterministic trend parameter θ_0 to illustrate its effect on the forecast. The model can be rewritten as

$$z_t = \theta_0 + (1 + \phi)z_{t-1} - \phi z_{t-2} + a_t - \theta a_{t-1}$$

Taking conditional expectations, we can calculate the forecasts according to

$$z_n(1) = E(z_{n+1}|z_n, z_{n-1}, \dots) = \theta_0 + (1 + \phi)z_n - \phi z_{n-1} - \theta a_n$$

$$z_n(2) = E(z_{n+2}|z_n, z_{n-1}, \dots) = \theta_0 + (1 + \phi)z_n(1) - \phi z_n$$

$$\vdots \qquad \vdots \qquad \vdots$$

$$z_n(l) = E(z_{n+l}|z_n, z_{n-1}, \dots) = \theta_0 + (1 + \phi)z_n(l - 1) - \phi z_n(l - 2) \quad (5.78)$$

For $l \geqslant 2$, we can write (5.78) as

$$\left[1 - (1 + \phi)B + \phi B^2\right]z_n(l) = \theta_0$$

or

$$(1 - \phi B)(1 - B)z_n(l) = \theta_0 \quad (5.79)$$

Equations (5.78) or (5.79) provide a convenient recurrence relation for calculating the forecasts.

ARIMA(0, 2, 2) process. This model can be written as

$$(1 - B)^2 z_t = \left(1 - \theta_1 B - \theta_2 B^2\right)a_t$$

or

$$z_t = 2z_{t-1} - z_{t-2} + a_t - \theta_1 a_{t-1} - \theta_2 a_{t-2}$$

Given the observations $z_n, z_{n-1}, z_{n-2}, \dots$, the forecasts are

$$z_n(1) = E(z_{n+1}|z_n, z_{n-1}, \dots) = 2z_n - z_{n-1} - \theta_1 a_n - \theta_2 a_{n-1}$$

$$z_n(2) = E(z_{n+2}|z_n, z_{n-1}, \dots) = 2z_n(1) - z_n - \theta_2 a_n$$

$$z_n(3) = E(z_{n+3}|z_n, z_{n-1}, \dots) = 2z_n(2) - z_n(1)$$

and

$$z_n(l) = E(z_{n+l}|z_n, z_{n-1},\dots) = 2z_n(l-1) - z_n(l-2) \qquad l \geqslant 3 \tag{5.80}$$

Forecasts can be calculated from the recurrence relation (5.80). For $l \geqslant 3$, we can write the forecasts in terms of the difference equation

$$(1 - 2B + B^2)z_n(l) = 0 \qquad \text{or} \qquad (1-B)^2 z_n(l) = 0 \tag{5.81}$$

This implies that eventually the forecasts are determined by a straight line that passes through the points $z_n(1)$ and $z_n(2)$.

Previous examples illustrate that the eventual behavior of the forecasts is dictated by the autoregressive polynomial [see Eqs. (5.72), (5.73), (5.79), and (5.81)]. For $l > q$, the forecasts from ARIMA(p, d, q) models follow difference equations of the form

$$\phi(B)(1-B)^d z_n(l) = 0 \tag{5.82}$$

where B operates on l. The solution of this difference equation is usually referred to as the *eventual forecast function*. The moving average polynomial dictates the starting values of the difference equation that generates the forecasts. For a detailed discussion, see Chapter 7.

5.4.2. Prediction Limits

It is usually not sufficient to calculate point forecasts alone. The uncertainty associated with a forecast should be expressed, and the prediction limits should always be computed. It is possible, but not very easy, to calculate a simultaneous "prediction region" for a set of future observations. Therefore, we give probability limits only for individual forecasts. Assuming that the a_t's (and hence the z_t's) are normally distributed, we can calculate $100(1-\lambda)$ percent prediction limits from

$$z_n(l) \pm u_{\lambda/2}\{V[e_n(l)]\}^{1/2} \tag{5.83}$$

where $u_{\lambda/2}$ is the $100(1-\lambda/2)$ percentage point of the standard normal distribution. Here it is assumed that $z_n(l)$ and $V[e_n(l)]$ are known exactly. This is not true in general, as we need to estimate the parameters of the model. Hence the estimated prediction limits are given by

$$\hat{z}_n(l) \pm u_{\lambda/2}\{\hat{V}[e_n(l)]\}^{1/2} \tag{5.84}$$

In a short time series ($n \leqslant 50$), the estimation may have a nonnegligible effect on the prediction limits. However, in long series this effect is usually negligible. For additional discussion, see Ansley and Newbold (1980), Yamamoto (1981), and Ledolter and Abraham (1981).

To illustrate the computation of the prediction limits, we consider the AR(1) process of Section 5.4.1. There we found $\hat{z}_{156}(1) = 0.56$, $\hat{z}_{156}(2) = 0.62$, $\hat{z}_{156}(3) = 0.68$, $\hat{V}[e_{156}(1)] = .024$, $\hat{V}[e_{156}(2)] = 0.041$, and $\hat{V}[e_{156}(3)] = .054$. Thus the 95 percent prediction limits for z_{156+l}, $l = 1, 2, 3$, are given by

$$\begin{aligned}
z_{157}:\quad \hat{z}_{156}(1) \pm 1.96\{\hat{V}[e_{156}(1)]\}^{1/2} &= .56 \pm 1.96(.15)\\
&= .56 \pm .29\\
z_{158}:\quad \hat{z}_{156}(2) \pm 1.96\{\hat{V}[e_{156}(2)]\}^{1/2} &= .62 \pm 1.96(.20)\\
&= .62 \pm .39\\
z_{159}:\quad \hat{z}_{156}(3) \pm 1.96\{\hat{V}[e_{156}(3)]\}^{1/2} &= .68 \pm 1.96(.23)\\
&= .68 \pm .45
\end{aligned}$$

5.4.3. Forecast Updating

Forecasting is usually an ongoing process, and forecasts have to be updated as new observations become available. Suppose we are at time n and we are predicting $l + 1$ steps ahead (i.e., predicting z_{n+l+1}). From the results in (5.62), the forecast is given by

$$z_n(l + 1) = \psi_{l+1}a_n + \psi_{l+2}a_{n-1} + \cdots$$

After z_{n+1} has become available, we need to update our prediction of z_{n+l+1}. Since the updated forecast $z_{n+1}(l)$ can be written as

$$z_{n+1}(l) = \psi_l a_{n+1} + \psi_{l+1}a_n + \psi_{l+2}a_{n-1} + \cdots$$

we find that

$$z_{n+1}(l) = z_n(l + 1) + \psi_l a_{n+1} = z_n(l + 1) + \psi_l[z_{n+1} - z_n(1)] \tag{5.85}$$

The updated forecast is a linear combination of the previous forecast of z_{n+l+1} made at time n and the most recent one-step-ahead forecast error $e_n(1) = z_{n+1} - z_n(1) = a_{n+1}$.

We use the example of Section 5.4.2 to illustrate forecast updating. We already obtained the forecasts for $z_{157}, z_{158}, z_{159}$ from time origin $n = 156$. Suppose now that $z_{157} = 0.74$ is observed and we need to update the forecasts for z_{158} and z_{159}. Since the model under consideration is AR(1) with the parameter estimate $\hat{\phi} = .85$, the estimated ψ weights are given by $\hat{\psi}_j = \hat{\phi}^j = .85^j$. Then the updated forecasts are given by

$$z_{158}: \quad \hat{z}_{157}(1) = \hat{z}_{156}(2) + \hat{\psi}_1[z_{157} - \hat{z}_{156}(1)]$$

$$= 0.62 + .85(0.74 - 0.56) = 0.77$$

$$z_{159}: \quad \hat{z}_{157}(2) = \hat{z}_{156}(3) + \hat{\psi}_2[z_{157} - \hat{z}_{156}(1)]$$

$$= 0.68 + .85^2(0.74 - 0.56) = 0.81$$

The updated 95 percent prediction limits for z_{158} and z_{159} become

$$z_{158}: \quad \hat{z}_{157}(1) \pm 1.96\{\hat{V}[e_{157}(1)]\}^{1/2} = .77 \pm .29$$

$$z_{159}: \quad \hat{z}_{157}(2) \pm 1.96\{\hat{V}[e_{157}(2)]\}^{1/2} = .81 \pm .39$$

5.5. MODEL SPECIFICATION

It was emphasized in Chapter 1 that the first step in designing a forecast system is the construction of a model. Statistical model building is iterative and consists of model specification, estimation, and model checking.

At the model specification stage we use past data and any information on how the series was generated to suggest a subclass of parsimonious models that are worthy of consideration.

Step 1. Variance-stabilizing transformations. If the variance of the series changes with the level, then a logarithmic transformation will often stabilize the variance. If the logarithmic transformation still does not stabilize the variance, we can adopt a more general approach and employ the class of power transformations (see Sec. 2.11).

Step 2. Degree of differencing. If the series or its appropriate transformation is not mean stationary, then we have to determine the proper degree of differencing. For the series and its differences, we examine

1. Plots of the time series
2. Plots of the SACF
3. Sample variances of the successive differences

Nonstationary processes exhibit changing levels and slowly decaying autocorrelations. For a detailed discussion of the choice of the appropriate differences, refer to Section 5.3. There the dangers of overdifferencing are illustrated and it is emphasized that the first available stationary difference should be used for further modeling. Usually d, the degree of differencing, is either 0, 1, or 2.

Step 3. Specification of p and q. Once we obtain a stationary difference we must specify the orders of the autoregressive (p) and moving average (q) polynomials. The orders can be specified by matching the patterns in the sample autocorrelations and partial autocorrelations with the theoretical patterns of known models. The orders p and q will usually be small. In Table 5.3 we summarize the theoretical properties of several commonly used models.

The sample variability and the correlation among the estimates r_k can disguise the theoretical ACF patterns. For example, in the AR(1) process with $\phi = .5$, the correlation among adjacent sample autocorrelations r_k and r_{k+1} (for large k) is about 0.8 [see Kendall (1976), p. 89].

Step 4. Inclusion of a trend parameter. If the series requires differencing, we should check whether it is necessary to include a deterministic trend θ_0 in the model. This can be done by comparing the sample mean $\bar{w}$ of the stationary difference with its standard error $s(\bar{w})$. This standard error can be approximated by

$$s(\bar{w}) \cong \left[\frac{c_0}{n}(1 + 2r_1 + 2r_2 + \cdots + 2r_K)\right]^{1/2} \tag{5.86}$$

where c_0 is the sample variance and $r_1, \ldots, r_K$ are the first K significant sample autocorrelations of the stationary difference. This approximation is appropriate only if $s(\bar{w}) > 0$.

Table 5.3. Properties of the ACF and the PACF for Various ARMA Models

Model	ACF	PACF
$(1, d, 0)$ AR(1)	Exponential or oscillatory decay	$\phi_{kk} = 0$ for $k > 1$
$(2, d, 0)$ AR(2)	Exponential or sine wave decay	$\phi_{kk} = 0$ for $k > 2$
$(p, d, 0)$ AR(p)	Exponential and/or sine wave decay	$\phi_{kk} = 0$ for $k > p$
$(0, d, 1)$ MA(1)	$\rho_k = 0$ for $k > 1$	Dominated by damped exponential
$(0, d, 2)$ MA(2)	$\rho_k = 0$ for $k > 2$	Dominated by damped exponential or sine wave
$(0, d, q)$ MA(q)	$\rho_k = 0$ for $k > q$	Dominated by linear combination of damped exponentials and/or sine waves
$(1, d, 1)$ ARMA(1, 1)	Tails off. Exponential decay from lag 1	Tails off. Dominated by exponential decay from lag 1
(p, d, q) ARMA(p, q)	Tails off after $q - p$ lags. Exponential and/or sine wave decay after $q - p$ lags	Tails off after $p - q$ lags. Dominated by damped exponentials and/or sine waves after $p - q$ lags

5.6. MODEL ESTIMATION

After specifying the form of the model, it is necessary to estimate its parameters. Let $(z_1, \ldots, z_N)$ represent the vector of the original observations and $\mathbf{w} = (w_1, \ldots, w_n)'$ the vector of the $n = N - d$ stationary differences.

5.6.1. Maximum Likelihood Estimates

The ARIMA model can be written as

$$a_t = \theta_1 a_{t-1} + \cdots + \theta_q a_{t-q} + w_t - \phi_1 w_{t-1} - \cdots - \phi_p w_{t-p} \tag{5.87}$$

Since we have assumed that the a_t are independent and come from a normal distribution with mean zero and variance σ^2, the joint probability density function of $\mathbf{a} = (a_1, a_2, \ldots, a_n)'$ is given by

$$p(\mathbf{a} | \boldsymbol{\phi}, \boldsymbol{\theta}, \sigma^2) = (2\pi\sigma^2)^{-n/2} \exp\left[-\frac{1}{2\sigma^2} \sum_{t=1}^{n} a_t^2\right] \tag{5.88}$$

where $\boldsymbol{\phi} = (\phi_1, \ldots, \phi_p)'$ and $\boldsymbol{\theta} = (\theta_1, \ldots, \theta_q)'$. The joint probability density function of $\mathbf{w}$ [or, equivalently, the likelihood function of the parameters $(\boldsymbol{\phi}, \boldsymbol{\theta}, \sigma^2)$] can be written down, at least in principle. It was shown by Newbold (1974) that it is of the form

$$L(\boldsymbol{\phi}, \boldsymbol{\theta}, \sigma^2 | \mathbf{w}) = g_1(\boldsymbol{\phi}, \boldsymbol{\theta}, \sigma^2) \exp\left[-\frac{1}{2\sigma^2} S(\boldsymbol{\phi}, \boldsymbol{\theta})\right] \tag{5.89}$$

where g_1 is a function of the parameters $(\boldsymbol{\phi}, \boldsymbol{\theta}, \sigma^2)$ and

$$S(\boldsymbol{\phi}, \boldsymbol{\theta}) = \sum_{t=1-p-q}^{n} E^2(u_t | \mathbf{w}) \tag{5.90}$$

In (5.90), $E(u_t|\mathbf{w}) = E(u_t|\mathbf{w}, \boldsymbol{\phi}, \boldsymbol{\theta}, \sigma^2)$ is the conditional expectation of u_t given $\mathbf{w}$, $\boldsymbol{\phi}$, $\boldsymbol{\theta}$ and σ^2, and

$$u_t = \begin{cases} a_t & t = 1, 2, \ldots, n \\ g_2(\mathbf{a}_*, \mathbf{w}_*) & t \leqslant 0 \end{cases}$$

where g_2 is a linear function of the initial unobservable values $\mathbf{a}_* = (a_{1-q}, \ldots, a_{-1}, a_0)'$ and $\mathbf{w}_* = (w_{1-p}, \ldots, w_{-1}, w_0)'$, which are needed for the evaluation of $a_1, a_2, \ldots, a_n$ in Equation (5.87).

The functions g_1 and g_2 depend on the particular model. As illustration we have derived the exact likelihood functions for three important models [AR(1), MA(1), and ARMA(1, 1)]. Details are given in Appendix 5.

1. AR(1): $a_t = w_t - \phi w_{t-1}$

$$L(\phi, \sigma^2 | \mathbf{w}) = (2\pi\sigma^2)^{-n/2}(1 - \phi^2)^{1/2} \times \exp\left\{-\frac{1}{2\sigma^2}\left[(1 - \phi^2)w_1^2 + \sum_{t=2}^{n}(w_t - \phi w_{t-1})^2\right]\right\} \tag{5.91}$$

2. MA(1): $a_t = \theta a_{t-1} + w_t$

$$L(\theta, \sigma^2 | \mathbf{w}) = (2\pi\sigma^2)^{-n/2}\left[\frac{1 - \theta^2}{1 - \theta^{2(n+1)}}\right]^{1/2} \times \exp\left[-\frac{1}{2\sigma^2}\sum_{t=0}^{n} E^2(a_t | \mathbf{w})\right] \tag{5.92}$$

3. ARMA(1, 1): $a_t = \theta a_{t-1} + w_t - \phi w_{t-1}$

$$L(\phi, \theta, \sigma^2|\mathbf{w}) = (2\pi\sigma^2)^{-n/2}|\mathbf{Z}'\mathbf{Z}|^{-1/2}\exp\left[-\frac{1}{2\sigma^2}S(\phi, \theta)\right] \quad (5.93)$$

where

$$|\mathbf{Z}'\mathbf{Z}| = \frac{(1-\theta^2)(1-\phi^2) + (1-\theta^{2n})(\theta-\phi)^2}{(1-\theta^2)(1-\phi^2)} \quad (5.94)$$

and

$$S(\phi, \theta) = E^2(a_0|\mathbf{w}) + \frac{1-\phi^2}{(\theta-\phi)^2}\left[E(w_0|\mathbf{w}) - E(a_0|\mathbf{w})\right]^2 + \sum_{t=1}^{n} E^2(a_t|\mathbf{w}) \quad (5.95)$$

Maximum likelihood estimates (MLE) of the parameters (ϕ, θ, σ^2) for the model in (5.87) can be obtained by maximizing the function (5.89). In general, closed-form solutions cannot be found. However, various algorithms [see, for example, Ansley (1979), Dent (1977), and Ljung and Box (1979)] are available to compute the MLE's, or close approximations, numerically. Generally the maximization of (5.89) is difficult; some of these difficulties can be appreciated through the study of several special cases.

AR(1). After setting the first derivative of (5.91) equal to zero, we observe that the MLE is a solution of a cubic equation. Closed-form solutions of cubic equations are cumbersome, and iterative methods have to be employed. This is quite different from the linear regression model situation, where the derivatives of the log-likelihood function are linear in the parameters.

MA(1). In the calculation of the likelihood function of the MA(1) process, we need $E(a_t|\mathbf{w})$, $t = 0, 1, \ldots, n$. For given θ and $\mathbf{w}$, these can be calculated using the recurrence relation

$$E(a_t|\mathbf{w}) = \theta E(a_{t-1}|\mathbf{w}) + w_t \qquad t = 1, 2, \ldots, n \quad (5.96)$$

provided $E(a_0|\mathbf{w})$ is known. The difficulties of obtaining the MLE are thus twofold: (1) we have to calculate $E(a_0|\mathbf{w})$ and (2) the likelihood function is a complicated function of θ.

ARMA(1, 1). For the calculation of the likelihood function (5.93), we need $E(a_0|\mathbf{w})$ and $E(w_0|\mathbf{w})$. From these we can calculate

$$E(a_t|\mathbf{w}) = \theta E(a_{t-1}|\mathbf{w}) + w_t - \phi E(w_{t-1}|\mathbf{w}) \qquad t = 1, 2, \ldots, n \quad (5.97)$$

Apart from the difficulties of obtaining these initial expectations, we face a complicated maximization problem.

In summary, these special cases illustrate two major difficulties with the maximum likelihood estimation approach.

1. The presence of the function $g_1(\boldsymbol{\phi}, \boldsymbol{\theta}, \sigma^2)$ in the likelihood function (5.89) makes the maximization difficult.
2. This approach requires the calculation of the conditional expectations $E(\mathbf{u}_*|\mathbf{w})$, where $\mathbf{u}_* = (u_{1-p-q}, \ldots, u_{-1}, u_0)'$. This in turn requires the calculation of

$$E(a_t|\mathbf{w}) \qquad t = 1 - q, \ldots, -1, 0$$

and

$$E(w_t|\mathbf{w}) \qquad t = 1 - p, \ldots, -1, 0 \qquad (5.98)$$

5.6.2. Unconditional Least Squares Estimates

Many approximations have been considered in the literature to ease these difficulties. A common approximation is to ignore the function $g_1(\boldsymbol{\phi}, \boldsymbol{\theta}, \sigma^2)$ and maximize $\exp[-(1/2\sigma^2)S(\boldsymbol{\phi}, \boldsymbol{\theta})]$ or, equivalently, minimize $S(\boldsymbol{\phi}, \boldsymbol{\theta})$. The resulting estimates are called *least squares estimates* (LSE). This approximation is satisfactory if the parameters are not close to the invertibility boundaries. In this case it can be shown that the likelihood function is dominated by the exponential part in (5.89) and that the removal of g_1 has only negligible effect.

Even in the minimization of $S(\boldsymbol{\phi}, \boldsymbol{\theta})$ there are two major difficulties. Initially we face the same difficulty outlined in (5.98): the evaluation of the initial expectations. Furthermore, $S(\boldsymbol{\phi}, \boldsymbol{\theta})$ can be a complicated (not necessarily quadratic) function of the parameters.

Evaluation of the Initial Expectations

There are two methods available for the evaluation of $E(u_t|\mathbf{w})$ ($t \leqslant 0$), the method of least squares and backforecasting.

Method of least squares. As shown in Appendix 5 for the ARMA(1, 1) model, it is true in general that $E(\mathbf{u}_*|\mathbf{w}) = \hat{\mathbf{u}}_*$, where $\hat{\mathbf{u}}_*$ is the least squares estimator of $\mathbf{u}_*$ in

$$\mathbf{Lw} = -\mathbf{Zu}_* + \mathbf{u} \tag{5.99}$$

In Equation (5.99), **L** and **Z** are $(n+p+q) \times n$ and $(n+p+q) \times (p+q)$ matrices involving the parameters $\boldsymbol{\phi}$ and $\boldsymbol{\theta}$. These matrices are given in Appendix 5 for the ARMA(1, 1) process. For given values of $(\boldsymbol{\phi}, \boldsymbol{\theta})$, we can calculate $\hat{\mathbf{u}}_*$ and hence $S(\boldsymbol{\phi}, \boldsymbol{\theta})$. Then we can search for a minimum value of $S(\boldsymbol{\phi}, \boldsymbol{\theta})$ in the space of $(\boldsymbol{\phi}, \boldsymbol{\theta})$.

Method of backforecasting. In the forecasting section we found that the MMSE forecast of z_{n+l} made at time n is given by the conditional expectation $E(z_{n+l}|z_n, z_{n-1}, \dots)$. This result may be used to compute the unknown quantities $E(a_t|\mathbf{w})$ $(t = 1-q, \dots, -1, 0)$ and $E(w_t|\mathbf{w})$ $(t = 1-p, \dots, -1, 0)$. The MMSE forecasts of a_t and w_t $(t \leqslant 0)$ can be found by considering the "backward" model

$$e_t = \theta_1 e_{t+1} + \theta_2 e_{t+2} + \cdots + \theta_q e_{t+q} + w_t - \phi_1 w_{t+1} - \cdots - \phi_p w_{t+p} \tag{5.100}$$

The ACF of this backward model is the same as the one of model (5.87). Furthermore, e_t is a white-noise sequence with $V(e_t) = \sigma^2$. This method of evaluating $E(a_t|\mathbf{w})$ and $E(w_t|\mathbf{w})$ $(t \leqslant 0)$ is called *backforecasting* (or *backcasting*).

We illustrate the method of backforecasting with a simple MA(1) process that can be written as

$$a_t = \theta a_{t-1} + w_t \tag{5.101}$$

This process can be expressed in the backward form as

$$e_t = \theta e_{t+1} + w_t \tag{5.102}$$

When we consider the process as going backwards, then $e_0, e_{-1}, \dots$ are random shocks that are "future" to the observations $w_n, w_{n-1}, \dots, w_1$. Hence,

$$E(e_t|\mathbf{w}) = 0 \qquad t \leqslant 0 \tag{5.103}$$

Furthermore, since the MA(1) process has a memory of only one period, the correlation between a_{-1} and $w_1, w_2, \dots, w_n$ is zero, and

$$E(a_t|\mathbf{w}) = 0 \qquad t \leqslant -1 \tag{5.104}$$

For illustration, consider the last 10 observations z_t (or the nine differenced observations w_t) of the growth rates series. These ten observations are shown in Table 5.4. Suppose we take $\theta = .8$. Since $E(w_t|\mathbf{w}) = w_t$ $(t = 1, 2, \dots, 9)$, it is easily seen that

$$E(w_1|\mathbf{w}) = w_1 = 1.58$$

$$E(w_2|\mathbf{w}) = w_2 = -.80$$

$$\vdots$$

$$E(w_9|\mathbf{w}) = w_9 - .14$$

However, the computation of $E(w_0|\mathbf{w})$ is different. For this we use Equation (5.102) with $t = 0$:

$$E(e_0|\mathbf{w}) = \theta E(e_1|\mathbf{w}) + E(w_0|\mathbf{w}) \tag{5.105}$$

This leads to

$$E(w_0|\mathbf{w}) = -\theta E(e_1|\mathbf{w}) \tag{5.106}$$

since $E(e_0|\mathbf{w}) = 0$ [see Eq. (5.103)]. Hence we need $E(e_1|\mathbf{w})$, which in turn

Table 5.4. Computation of $S(\theta)$ Through Backforecasting

t	z_t	$w_t = \nabla z_t$	$E(w_t\|\mathbf{w})$	$E(e_t\|\mathbf{w})$	$E(a_t\|\mathbf{w})$
0	1.84		−0.88	0	−0.88
1	3.42	1.58	1.58	1.10	0.88
2	2.62	−0.80	−0.80	−0.60	−0.10
3	3.02	0.40	0.40	0.25	0.32
4	2.76	−0.26	−0.26	−0.19	0.00
5	2.16	−0.60	−0.60	0.09	−0.60
6	2.32	0.16	0.16	0.86	0.32
7	2.59	0.27	0.27	0.88	0.01
8	3.24	0.65	0.65	0.76	0.66
9	3.38	0.14	0.14	0.14	0.67
10				0	

requires $E(e_2|\mathbf{w})$, and so on. Therefore, we start at the end of the series and assume that $E(e_{10}|\mathbf{w}) = 0$. This approximation introduces a transient into the system. However, for an invertible series its effect will be negligible by the time the computation reaches $E(e_1|\mathbf{w})$. Under this assumption,

$$E(e_9|\mathbf{w}) = \theta E(e_{10}|\mathbf{w}) + w_9 = 0 + w_9 = .14$$

$$E(e_8|\mathbf{w}) = \theta E(e_9|\mathbf{w}) + w_8 = .8(.14) + .65 = .76$$

$$\vdots$$

$$E(e_1|\mathbf{w}) = \theta E(e_2|\mathbf{w}) + w_1 = .8(-.60) + 1.58 = 1.10$$

Hence from (5.106) we obtain

$$E(w_0|\mathbf{w}) = -.8(1.10) = -.88$$

Now using (5.101) and (5.104) we can compute $E(a_t|\mathbf{w})$, $(t = 0, 1, \ldots, 9)$.

$$E(a_0|\mathbf{w}) = \theta E(a_{-1}|\mathbf{w}) + E(w_0|\mathbf{w}) = .8(0) - .88 = -.88$$

$$E(a_1|\mathbf{w}) = \theta E(a_0|\mathbf{w}) + w_1 = .8(-.88) + 1.58 = .88$$

$$E(a_2|\mathbf{w}) = \theta E(a_1|\mathbf{w}) + w_2 = .8(.88) + (-.80) = -.10$$

$$\vdots \qquad \vdots \qquad \vdots$$

$$E(a_9|\mathbf{w}) = \theta E(a_8|\mathbf{w}) + w_9 = .8(.66) + .14 = .67$$

All computed quantities, $E(w_t|\mathbf{w})$, $E(e_t|\mathbf{w})$, and $E(a_t|\mathbf{w})$ $(t = 0, 1, \ldots, 9)$ are shown in Table 5.4. The sum of squares $S(\theta)$ can then be obtained for $\theta = .8$ as

$$S(.8) = \sum_{t=0}^{9} E^2(a_t|\mathbf{w}) = (-.88)^2 + (.88)^2 + \cdots + (.67)^2 = 3.01$$

Similarly, $S(\theta)$ can be calculated for other values of θ, and its minimum can be found.

This method can be used to backforecast $E(a_t|\mathbf{w})$ and $E(w_t|\mathbf{w})$ $(t \leqslant 0)$ and to evaluate the function $S(\boldsymbol{\phi}, \boldsymbol{\theta})$ for any given ARMA model. This process can be repeated for several $(\boldsymbol{\phi}, \boldsymbol{\theta})$ values, and the minimum can be located. The parameter estimates obtained by the minimization of $S(\boldsymbol{\phi}, \boldsymbol{\theta})$, using either least squares or backforecasting to calculate $\hat{\mathbf{u}}_*$, are usually called *unconditional least squares estimates* or simply *least squares estimates* (LSE).

5.6.3. Conditional Least Squares Estimates

Computationally simpler estimates can be obtained by minimizing the conditional sum of squares $S_c(\boldsymbol{\phi}, \boldsymbol{\theta}) = \Sigma_{t=p+1}^{n} a_t^2$, where the starting values $a_p, a_{p-1}, \ldots, a_{p+1-q}$ in the calculation of the appropriate a_t's are set equal to zero. The resulting estimates are called *conditional least squares estimates* (CLSE).

If the model is autoregressive, then CLSE can be obtained from linear least squares. In the AR(p) case, the CLSE is obtained by minimizing

$$S_c(\boldsymbol{\phi}) = \sum_{t=p+1}^{n} a_t^2 = \sum_{t=p+1}^{n} \left(w_t - \phi_1 w_{t-1} - \cdots - \phi_p w_{t-p}\right)^2 \qquad (5.107)$$

which leads to the normal equations

$$\mathbf{X}'\mathbf{X}\hat{\boldsymbol{\phi}}_c = \mathbf{X}'\mathbf{y} \qquad (5.108)$$

where

$$\mathbf{X} = \begin{bmatrix} w_p & w_{p-1} & \cdots & w_1 \\ w_{p+1} & w_p & \cdots & w_2 \\ \vdots & \vdots & & \vdots \\ w_{n-1} & w_{n-2} & \cdots & w_{n-p} \end{bmatrix}$$

and

$$\mathbf{y} = \left(w_{p+1}, w_{p+2}, \ldots, w_n\right)'$$

Hence $\hat{\boldsymbol{\phi}}_c = (\mathbf{X}'\mathbf{X})^{-1}\mathbf{X}'\mathbf{y}$ is the CLSE of $\boldsymbol{\phi}$.

In the special case of an AR(1) process, this becomes

$$\hat{\phi}_c = \frac{\sum_{t=2}^{n} w_t w_{t-1}}{\sum_{t=2}^{n} w_{t-1}^2} \qquad (5.109)$$

For comparison, the unconditional least squares estimator of ϕ in the AR(1) process is obtained by minimizing

$$S(\phi) = (1 - \phi^2)w_1^2 + \sum_{t=2}^{n} (w_t - \phi w_{t-1})^2$$

which yields

$$\hat{\phi} = \frac{\sum_{t=2}^{n} w_t w_{t-1}}{\sum_{t=3}^{n} w_{t-1}^2} \tag{5.110}$$

5.6.4. Nonlinear Estimation

If the model contains moving average terms, then the conditional and unconditional sums of squares are not quadratic functions of the parameters $(\boldsymbol{\phi}, \boldsymbol{\theta})$. This is because $E(u_t|\mathbf{w})$ in $S(\boldsymbol{\phi}, \boldsymbol{\theta})$ is nonlinear in the parameters. Hence nonlinear least squares procedures must be used to minimize $S(\boldsymbol{\phi}, \boldsymbol{\theta})$ or $S_c(\boldsymbol{\phi}, \boldsymbol{\theta})$.

We now illustrate the nonlinear estimation procedure and derive the unconditional least squares estimates of the parameters in the ARMA(1, 1) model

$$a_t = \theta a_{t-1} + w_t - \phi w_{t-1} \tag{5.111}$$

for which $S(\phi, \theta)$ is given in (5.95).

Step 1. Computation at the starting point. For given $\mathbf{w}$ and a starting point $(\phi, \theta) = (\phi_0, \theta_0)$, we compute $E(a_0|\mathbf{w}, \phi_0, \theta_0)$ and $E(w_0|\mathbf{w}, \phi_0, \theta_0)$ using backforecasting or the method of least squares. Then we can compute

$$\hat{u}_{-10} = E(a_0|\mathbf{w}, \phi_0, \theta_0)$$

and

$$\hat{u}_{00} = \frac{(1 - \phi_0^2)^{1/2}}{\theta_0 - \phi_0}\left[E(w_0|\mathbf{w}, \phi_0, \theta_0) - E(a_0|\mathbf{w}, \phi_0, \theta_0)\right]$$

The remaining elements in $\hat{\mathbf{u}}_0 = (\hat{u}_{-10}, \hat{u}_{00}, \hat{u}_{10}, \ldots, \hat{u}_{n0})'$ can be obtained from $\hat{u}_{t0} = E(a_t|\mathbf{w}, \phi_0, \theta_0)$ for $t = 1, 2, \ldots, n$.

Step 2. Linearization. We consider $\hat{u}_t = E(a_t|\mathbf{w}, \phi, \theta)$ $(t = -1, 0, 1, \ldots, n)$ as a function of $\boldsymbol{\beta} = (\phi, \theta)'$ and linearize this function at the starting point $\boldsymbol{\beta}_0 = (\phi_0, \theta_0)'$ in a first-order Taylor series expansion:

$$\hat{u}_t \cong \hat{u}_{t0} + (\phi - \phi_0)\frac{\partial \hat{u}_t}{\partial \phi}\bigg|_{\boldsymbol{\beta}=\boldsymbol{\beta}_0} + (\theta - \theta_0)\frac{\partial \hat{u}_t}{\partial \theta}\bigg|_{\boldsymbol{\beta}=\boldsymbol{\beta}_0}$$

$$= \hat{u}_{t0} - x_{1,t}(\phi - \phi_0) - x_{2,t}(\theta - \theta_0) \tag{5.112}$$

where

$$x_{1,t} = -\frac{\partial \hat{u}_t}{\partial \phi}\bigg|_{\boldsymbol{\beta}=\boldsymbol{\beta}_0} \quad \text{and} \quad x_{2,t} = -\frac{\partial \hat{u}_t}{\partial \theta}\bigg|_{\boldsymbol{\beta}=\boldsymbol{\beta}_0}$$

Equation (5.112) may be rewritten as

$$\hat{u}_{t0} = (x_{1,t}, x_{2,t})\begin{bmatrix} \phi - \phi_0 \\ \theta - \theta_0 \end{bmatrix} + \hat{u}_t$$

Considering this equation for $t = -1, 0, \ldots, n$, we obtain

$$\hat{\mathbf{u}}_0 = \mathbf{X}(\boldsymbol{\beta} - \boldsymbol{\beta}_0) + \hat{\mathbf{u}} \tag{5.113}$$

where

$$\mathbf{X} = \begin{bmatrix} x_{1,-1} & x_{2,-1} \\ x_{1,0} & x_{2,0} \\ x_{1,1} & x_{2,1} \\ \vdots & \vdots \\ x_{1,n} & x_{2,n} \end{bmatrix}$$

and where the errors $\hat{\mathbf{u}} = (\hat{u}_{-1}, \hat{u}_0, \ldots, \hat{u}_n)'$ satisfy the usual least squares assumptions. The least squares estimate of $\boldsymbol{\delta} = \boldsymbol{\beta} - \boldsymbol{\beta}_0$ may be obtained from $\hat{\boldsymbol{\delta}} = (\mathbf{X}'\mathbf{X})^{-1}\mathbf{X}'\hat{\mathbf{u}}_0$. Hence we obtain a modified estimator of $\boldsymbol{\beta}$ as

$$\hat{\boldsymbol{\beta}}_1 = \boldsymbol{\beta}_0 + \hat{\boldsymbol{\delta}}$$

Step 3. We now repeat steps 1 and 2 with $\hat{\boldsymbol{\beta}}_1$ as the new point of expansion in the Taylor series and continue the process until convergence occurs. Convergence is usually determined by one of the following criteria:

1. The relative change in the sum of squares $\sum_{t=-1}^{n} \hat{u}_t^2$ from two consecutive iterations is less than a prespecified small value.
2. The largest relative change in the parameter estimate of two consecutive iterations is less than a prespecified small value.

When convergence occurs, we have an estimate $\hat{\boldsymbol{\beta}}$ based on local linear least squares, and this will be taken as the unconditional least squares estimate of $\hat{\boldsymbol{\beta}}$. The conditional least squares estimate can be obtained in a similar fashion by replacing the vector $\hat{\mathbf{u}}$ by $\mathbf{a} = (a_2, a_3, \ldots, a_n)'$ and setting $a_1 = 0$.

The method is easily generalized to ARMA(p, q) processes. In this case, $\boldsymbol{\beta} = (\phi_1, \phi_2, \ldots, \phi_p, \theta_1, \theta_2, \ldots, \theta_q)'$ and $x_{i,t} = -(\partial \hat{u}_t / \partial \beta_i)|_{\boldsymbol{\beta}=\boldsymbol{\beta}_0}$ ($i = 1, 2, \ldots, p + q$; $t = 1 - p - q, \ldots, n$), which leads to an $\mathbf{X}$ matrix of dimension $(n + p + q) \times (p + q)$. All other operations remain basically the same.

For machine computation, various nonlinear least squares algorithms are available [e.g., see Marquardt (1963)]. In most algorithms the derivatives are computed numerically and many convenient modifications are implemented to assure faster convergence. The details of the algorithms, computations, and programs are not given, since many easy-to-use programs are now commercially available.

Standard Errors of the Estimates

Since we are using successive linear least squares operations, we can obtain the covariance matrix of the estimates $\hat{\boldsymbol{\beta}}$ from the $\mathbf{X}$ matrix in the final iteration of the minimization (maximization) procedure. Let this matrix be denoted by $\mathbf{X}_{\hat{\boldsymbol{\beta}}}$. Then the covariance matrix of $\hat{\boldsymbol{\beta}}$ can be estimated as

$$\hat{\mathbf{V}}(\hat{\boldsymbol{\beta}}) = \hat{\sigma}^2 (\mathbf{X}'_{\hat{\boldsymbol{\beta}}} \mathbf{X}_{\hat{\boldsymbol{\beta}}})^{-1} \tag{5.114}$$

where the estimate of σ^2 is given by

$$\hat{\sigma}^2 = \frac{S(\hat{\boldsymbol{\phi}}, \hat{\boldsymbol{\theta}})}{n}$$

Some computer programs, such as MINITAB, divide by $n - p - q$ rather than n. The square roots of the diagonal elements in $\hat{\mathbf{V}}(\hat{\boldsymbol{\beta}})$ are the estimated standard errors. The estimated covariance between the parameter estimates $\hat{\beta}_i$ and $\hat{\beta}_j$ is given by the (i, j)th element of $\hat{\mathbf{V}}(\hat{\boldsymbol{\beta}})$; thus the correlations between the parameter estimates are easily obtained.

Possible Difficulties in Nonlinear Estimation

In some computer packages, starting values $\boldsymbol{\beta}_0$ have to be provided. In some cases the convergence to the final estimates can be rather slow. This can be due to poorly chosen starting values or to misspecified models with too many parameters. For illustration, consider fitting an ARMA(1, 1) model $(1 - \phi B)w_t = (1 - \theta B)a_t$ to an independent sequence $w_t = a_t$. In this case any constant $\phi = \theta$ will lead to the model $w_t = a_t$, and the parameters ϕ and θ cannot be estimated separately. This implies a likelihood function that is constant for all points on the line $\phi - \theta = 0$. Such a situation is indicated by slow convergence and high correlation among parameter estimates.

In some rare cases the likelihood function can have more than one maximum, and the estimates can converge to a local maximum rather than a global one. In such a case, the forecaster should experiment with several starting points.

5.7. MODEL CHECKING

After the parameters in a model have been estimated it is necessary to check whether the model assumptions are satisfied. If the assumptions are not met, the model must be respecified. This phase in the model building, usually referred to as *diagnostic checking*, relies heavily on the analysis of the residuals $\hat{a}_t$.

The basic assumption in ARIMA models is that the a_t's are uncorrelated random variables with mean zero and constant variance. In long series we expect the behavior of the residuals $\hat{a}_t$ to be similar to that of the errors a_t. This implies that (1) the mean of the residuals should be close to zero, (2) the variance of the residuals should be approximately constant, and (3) the autocorrelations of the residuals should be negligible. To check whether the mean of the residuals is zero, we can compare the sample mean $\bar{\hat{a}}$ with its standard error. To check whether the variance is constant, we examine the plot of the residuals. Such a plot is also useful in detecting possible outliers and other systematic patterns. To check whether the residuals are uncorrelated, we calculate their sample autocorrelations

$$r_{\hat{a}}(k) = \frac{\sum_{t=k+1}^{n} (\hat{a}_t - \bar{\hat{a}})(\hat{a}_{t-k} - \bar{\hat{a}})}{\sum_{t=1}^{n} (\hat{a}_t - \bar{\hat{a}})^2} \tag{5.115}$$

and compare them with their standard errors. The standard error of $r_{\hat{a}}(k)$ is usually approximated as

$$s[r_{\hat{a}}(k)] \cong n^{-1/2} \tag{5.116}$$

However, for small k the true standard error can be much smaller. For example, for the residuals from an estimated AR(1) process with parameter ϕ, the standard error of $r_{\hat{a}}(1)$ is given by

$$s\left[r_{\hat{a}}(1)\right] \cong |\phi| n^{-1/2} \tag{5.117}$$

which can be much smaller than $n^{-1/2}$ [see Box and Pierce (1970)]. In general, the true standard errors depend on (1) the form of the fitted model, (2) the true parameter values, and (3) the value of k.

5.7.1. An Improved Approximation of the Standard Error

Let $\mathbf{r}_{\hat{a}} = [r_{\hat{a}}(1), \ldots, r_{\hat{a}}(K)]'$ denote the vector of the first K sample autocorrelations of the residuals, and let

$$\mathbf{T} = \left[\begin{array}{cccc|cccc} \phi_0^* & \phi_{-1}^* & \cdots & \phi_{1-p}^* & \theta_0^* & \theta_{-1}^* & \cdots & \theta_{1-q}^* \\ \phi_1^* & \phi_0^* & \cdots & \phi_{2-p}^* & \theta_1^* & \theta_0^* & \cdots & \theta_{2-q}^* \\ \vdots & \vdots & & \vdots & \vdots & \vdots & & \vdots \\ \phi_{K-1}^* & \phi_{K-2}^* & \cdots & \phi_{K-p}^* & \theta_{K-1}^* & \theta_{K-2}^* & \cdots & \theta_{K-q}^* \end{array}\right]$$

denote a $K \times (p + q)$ matrix, where ϕ^* and θ^* are defined by

$$\phi_s^* = \theta_s^* = 0 \qquad s < 0$$

and

$$-\phi^{-1}(B) = \sum_{s=0}^{\infty} \phi_s^* B^s \qquad \theta^{-1}(B) = \sum_{s=0}^{\infty} \theta_s^* B^s \qquad s \geqslant 0 \tag{5.118}$$

Then it can be shown [see, e.g., McLeod (1977a, 1978)] that the estimated covariance matrix of $\mathbf{r}_{\hat{a}}$ is given by

$$\mathbf{V}(\mathbf{r}_{\hat{a}}) = \frac{1}{n}\mathbf{I}_K - \hat{\mathbf{T}}\hat{\mathbf{V}}\hat{\mathbf{T}}' \tag{5.119}$$

where $\mathbf{I}_K$ is the $K \times K$ identity matrix, $\hat{\mathbf{V}}$ is the estimated large-sample covariance matrix of the estimated parameters $(\hat{\boldsymbol{\phi}}, \hat{\boldsymbol{\theta}})$, and $\hat{\mathbf{T}}$ is the matrix obtained by replacing the parameters in $\mathbf{T}$ by their estimates. Square roots of the diagonal elements of the covariance matrix $\mathbf{V}(\mathbf{r}_{\hat{a}})$ provide estimated standard errors for $r_{\hat{a}}(k)$. These can be used to construct approximate probability limits for the autocorrelations of the residuals. These limits, which are given by

$$r_{\hat{a}}(k) \pm 2s\left[r_{\hat{a}}(k)\right] \tag{5.120}$$

are used in the examples in Section 5.8.

If these standard errors are not available, one may use the standard errors given in (5.116). However, one should be aware that for small k these may overestimate the true standard errors.

5.7.2. Portmanteau Test

Box and Pierce (1970) show that, under the null hypothesis of model adequacy, the large-sample distribution of $\mathbf{r}_{\hat{a}}$ is multivariate normal and that

$$Q^* = n \sum_{k=1}^{K} r_{\hat{a}}^2(k) \tag{5.121}$$

has a large-sample chi-square distribution with $K - p - q$ degrees of freedom. In (5.121), n represents the number of observations after differencing, and $p + q$ represents the total number of parameters in the model. If a mean is estimated, the degrees of freedom are $K - p - q - 1$. Davies et al. (1977) found that in small samples this test is quite conservative; that is, the chance of incorrectly rejecting the null hypothesis of model adequacy is smaller than the chosen significance level. Ljung and Box (1978) and Ansley and Newbold (1979) modify Q^* and propose the statistic

$$Q = n(n + 2) \sum_{k=1}^{K} \frac{r_{\hat{a}}^2(k)}{n - k} \tag{5.122}$$

and show that this modification results in a much better approximation to the $\chi^2(K - p - q)$ distribution. The statistic Q is used in model checking; if it exceeds the tabulated value $\chi_\alpha^2(K - p - q)$, usually $\alpha = .05$, one would question the adequacy of the model.

5.8. EXAMPLES

In this section we consider some examples to illustrate the methods discussed in this chapter. Several computer programs such as IDA, MINITAB, the PACK-SYSTEM, TS [see McLeod (1977a)], and WMTS [see Tiao et al. (1979)] are available for the implementation of these methods. These packages use slightly different estimation methods. For example, IDA and MINITAB calculate the unconditional least squares estimates, TS computes a certain approximation to the maximum likelihood estimates [see McLeod (1977b)], and WMTS obtains the exact maximum likelihood estimates. It should be noted that, depending on the estimation method, slightly different estimates may be obtained, especially if the series is short. Here we compute the exact maximum likelihood estimates.

5.8.1. Yield Data

This series consists of monthly differences between the yield on mortgages and the yield on government loans in the Netherlands from January 1961 to December 1973 (series 9 in the Data Appendix). The observations for January to March 1974 are also available, but they are held back to illustrate the forecast computations and updating. A time series plot is given in Figure 5.13. An initial examination of the data does not call for any variance-stabilizing transformations. The next step is to see whether the series is stationary. The plot of the data and the SACF in Figure 5.2 indicate that the original series can be taken as stationary. The SACF decays exponentially, and the sample partial autocorrelations in Figure 5.5 are negligible after lag 1. This suggests an AR(1) model:

$$z_t - \mu = \phi(z_{t-1} - \mu) + a_t$$

The maximum likelihood estimates are

$$\hat{\mu} = 0.97\ (0.08) \qquad \hat{\phi} = 0.85\ (0.04) \qquad \hat{\sigma}^2 = 0.024$$

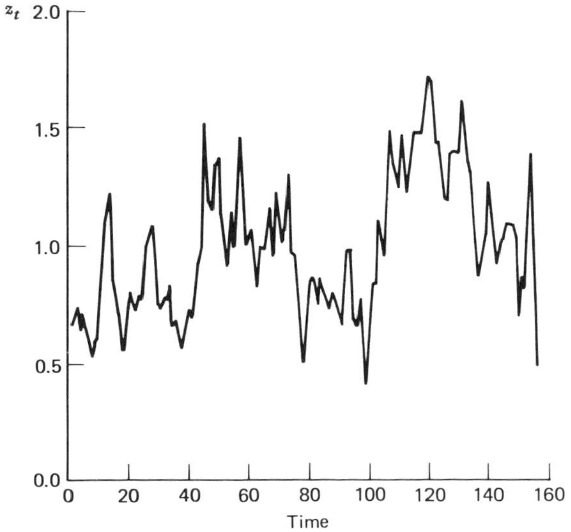

Figure 5.13. Yield series.

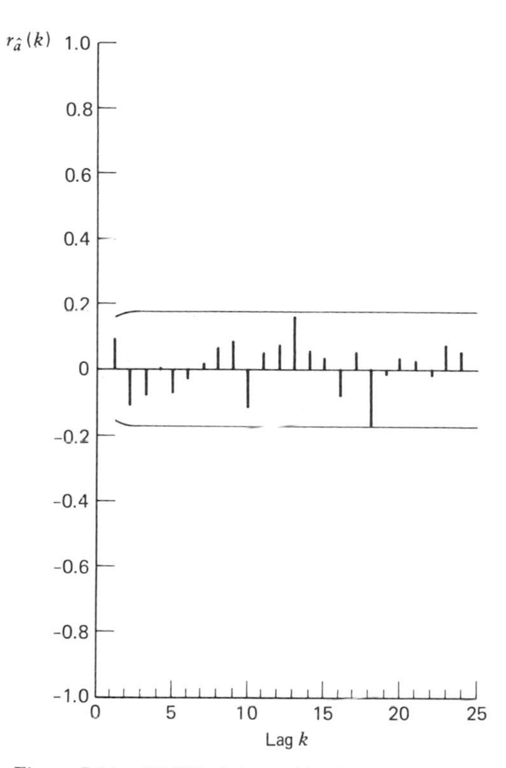

Figure 5.14. SACF of the residuals—yield series.

The quantities in parentheses are the estimated standard errors. All estimated residual autocorrelations in Figure 5.14 are within the probability limits $r_{\hat{a}}(k) \pm 2s[r_{\hat{a}}(k)]$ and therefore there is no reason to reject the model. This is confirmed by the Q statistic. Suppose we take $K = 24$. Then,

$$Q = 156 \times 158\left(\frac{(.09)^2}{155} + \frac{(-.10)^2}{154} + \cdots + \frac{(.06)^2}{132} \right) = 22.4$$

We find from Table C in the Table Appendix that $\chi^2_{.05}(22) = 33.9$. Since $Q = 22.4$ is less than 33.9, our conclusion regarding model adequacy is confirmed. Thus we take

$$z_t = 0.97 + 0.85(z_{t-1} - 0.97) + a_t$$

as the forecasting model. It should be emphasized that there are other models that can describe the series. In fact, the estimate of ϕ is quite close to the nonstationary region, and a random walk model with $\phi = 1$, $z_t = z_{t-1} + a_t$ would also lead to an adequate representation.

Forecasting

Using this model we can generate the forecasts $\hat{z}_{156}(l)$. The first three ($l = 1, 2, 3$) forecasts and the 95% prediction limits were calculated in Section 5.4.

1. $\hat{z}_{156}(1) = \hat{\mu} + \hat{\phi}(z_{156} - \hat{\mu}) = .97 + .85(.49 - .97) = 0.56$

Prediction limits:

$$\hat{z}_{156}(1) \pm 1.96\hat{\sigma} = 0.56 \pm 1.96(.024)^{1/2} = 0.56 \pm 0.29$$

2. $\hat{z}_{156}(2) = \hat{\mu} + \hat{\phi}^2(z_{156} - \hat{\mu}) = .97 + .85^2(.49 - .97) = 0.62$

Prediction limits:

$$\hat{z}_{156}(2) \pm 1.96\hat{\sigma}(1 + \hat{\psi}_1^2)^{1/2} = .62 \pm 1.96(.024)^{1/2}(1 + .85^2)^{1/2}$$
$$= 0.62 \pm 0.39$$

3. $\hat{z}_{156}(3) = \hat{\mu} + \hat{\phi}^3(z_{156} - \hat{\mu}) = .97 + .85^3(.49 - .97) = 0.68$

Prediction limits:

$$\hat{z}_{156}(3) \pm 1.96\hat{\sigma}(1 + \hat{\psi}_1^2 + \hat{\psi}_2^2)^{1/2} = 0.68 \pm 1.96(0.024)^{1/2}$$
$$\times (1 + .85^2 + .85^4)^{1/2}$$
$$= 0.68 \pm 0.45$$

We notice that the width of the interval increases with the lead time. This is intuitively reasonable, since the uncertainty grows with the forecast horizon.

Suppose $z_{157} = 0.74$ has now become available. Forecasts for z_{158} and z_{159} are updated according to:

$$z_{158}: \quad \hat{z}_{157}(1) = \hat{z}_{156}(2) + \hat{\psi}_1[z_{157} - \hat{z}_{156}(1)]$$
$$= 0.62 + .85(.74 - .56) = 0.77$$

$$z_{159}: \quad \hat{z}_{157}(2) = \hat{z}_{156}(3) + \hat{\psi}_2[z_{157} - \hat{z}_{156}(1)]$$
$$= 0.68 + .85^2(.74 - .56) = 0.81$$

The prediction interval for z_{158} is 0.77 ± 0.29, and that for z_{159} is 0.81 ± 0.39. These intervals are narrower than the ones obtained at $t = 156$. This is because of the new information z_{157}.

5.8.2. Growth Rates

Quarterly growth rates of Iowa nonfarm income from the second quarter of 1948 to the fourth quarter of 1978 ($N = 123$) are shown in Figure 5.8*a*. The same data were used in Chapter 3 to illustrate simple exponential smoothing. The plot of the series indicates that no variance-stabilizing transformation is necessary. In Section 5.3 it was found that z_t is nonstationary and that the first difference ∇z_t should be analyzed. The second difference is also stationary but would lead to overdifferencing. This is also indicated by the increase in the sample variance [$s^2(\nabla z_t) = 1.58$, $s^2(\nabla^2 z_t) = 4.82$]. The SACF of ∇z_t in Figure 5.11*b* indicates a cutoff after lag 1. The SPACF in Figure 5.15 shows a rough exponential decay. These patterns lead us to

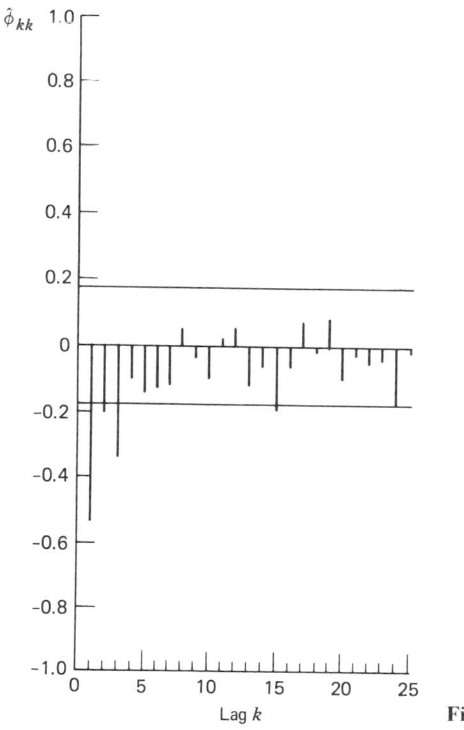

Figure 5.15. SPACF of ∇z_t— growth rates.

consider the model

$$\nabla z_t = \theta_0 + (1 - \theta B) a_t$$

The ML estimates and their standard errors are

$$\hat{\theta}_0 = 0.012(.008) \qquad \hat{\theta} = 0.92(.04) \qquad \hat{\sigma}^2 = 0.93$$

Dropping the insignificant θ_0 from the model and reestimating the parameters leads to

$$\hat{\theta} = 0.88(.04) \qquad \hat{\sigma}^2 = 0.95$$

The estimated residual autocorrelations together with their probability limits are presented in Figure 5.16. All autocorrelations are within these

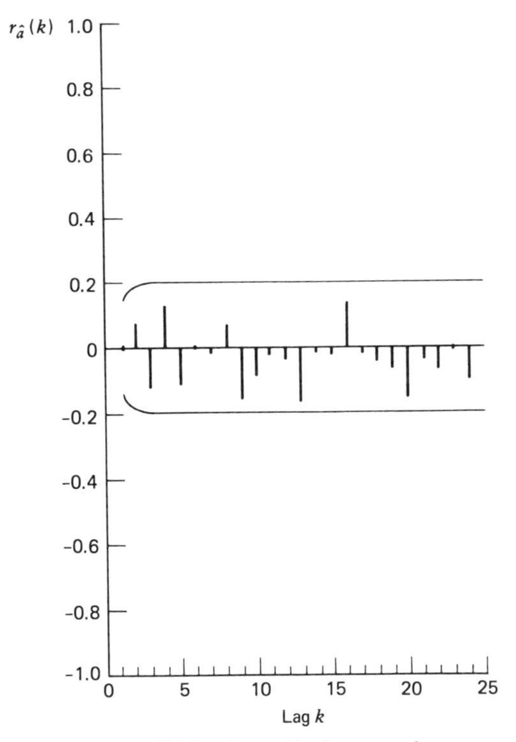

Figure 5.16. SACF for the residuals—growth rates.

limits, which indicates that there is not sufficient evidence to reject the model. This conclusion is confirmed by the Q statistic based on the first $K = 24$ autocorrelations ($n = 123 - 1 = 122$)

$$Q = 122 \times 124\left(\frac{(-.01)^2}{121} + \frac{(.08)^2}{120} + \cdots + \frac{(-.09)^2}{98}\right) = 20.5$$

which is insignificant compared with $\chi^2_{.05}(23) = 35.2$. Hencc we take

$$z_t = z_{t-1} + a_t - 0.88a_{t-1}$$

as the forecasting model.

Forecasting

The predictions $\hat{z}_{123}(l)$ are described below.

$$\hat{z}_{123}(l) = \hat{z}_{123}(1) = z_{123} - \hat{\theta}\hat{a}_{123}$$

$$= 3.38 - .88(0.718) = 2.75$$

Note that $\hat{a}_{123} = 0.718$ is the last residual. For this model the forecasts $\hat{z}_{123}(l)$ are the same for all l. They are also the same as those derived by simple exponential smoothing with a smoothing constant $\alpha = 1 - \hat{\theta} = 1 - .88 = .12$. This is very close to the smoothing constant $\alpha = .11$ we used in Section 3.3.

The estimated variance of the l-step-ahead forecast error is

$$\hat{V}[e_{123}(l)] = \left[1 + (l-1)(1-\hat{\theta})^2\right]\hat{\sigma}^2$$

$$= \left[1 + (l-1)(1-.88)^2\right](.95) = [1 + (l-1)(.01)](.95)$$

The forecasts, forecast error variances, and 95 percent prediction limits for z_{123+l} ($l = 1, 2, 3, 4$) are given in Table 5.5. The forecast error variances and the width of the prediction intervals increase gradually with l.

After a new observation $z_{124} = 1.55$ has become available, we can update the forecasts:

$$\hat{z}_{124}(l) = \hat{z}_{123}(l+1) + \hat{\psi}_l\left[z_{124} - \hat{z}_{123}(1)\right]$$

$$= 2.75 + (1 - .88)(1.55 - 2.75) = 2.61$$

Table 5.5. Forecasts and Prediction Limits—Growth Rates

l	$\hat{z}_{123}(l)$	$\hat{V}[e_{123}(l)]$	95% Prediction Limits
1	2.75	0.95	2.75 ± 1.91
2	2.75	0.96	2.75 ± 1.92
3	2.75	0.97	2.75 ± 1.93
4	2.75	0.98	2.75 ± 1.94

Furthermore, after $z_{125} = 2.93$ has become available,

$$\hat{z}_{125}(l) = \hat{z}_{124}(l+1) + \hat{\psi}_l[z_{125} - \hat{z}_{124}(1)]$$

$$= 2.61 + (1 - .88)(2.93 - 2.61) = 2.65$$

In addition, the prediction limits are easily updated.

5.8.3. Demand for Repair Parts

This series consists of the monthly demand for repair parts for a large Midwestern production company from January 1972 to June 1979 ($N = 90$). The series is shown in Figure 5.9*a*. In Section 5.3 we found that a logarithmic transformation stabilizes the variance and that the first differences $\nabla \ln z_t$ are stationary. Further differencing is not necessary since it results in an increase in the sample variance [$s^2(\nabla \ln z_t) = .034$, $s^2(\nabla^2 \ln z_t) = .099$]. The sample autocorrelations and partial autocorrelations are shown in Figures 5.12*b* and 5.17. They indicate a cutoff in the SACF after lag 1 and a rough exponential decay in the SPACF. Thus we consider the ARIMA(0, 1, 1) model

$$(1 - B)y_t = (1 - \theta B)a_t$$

where $y_t = \ln z_t$. The ML estimates are

$$\hat{\theta} = 0.57(.09) \qquad \hat{\sigma}^2 = .027$$

A constant θ_0 was originally included in the model, but it was found to be insignificant. The estimated residual autocorrelations, together with their probability limits, are shown in Figure 5.18. All autocorrelations except $r_{\hat{a}}(12)$ are small. The large autocorrelation $r_{\hat{a}}(12)$ could be due to "seasonal variation" in the data.

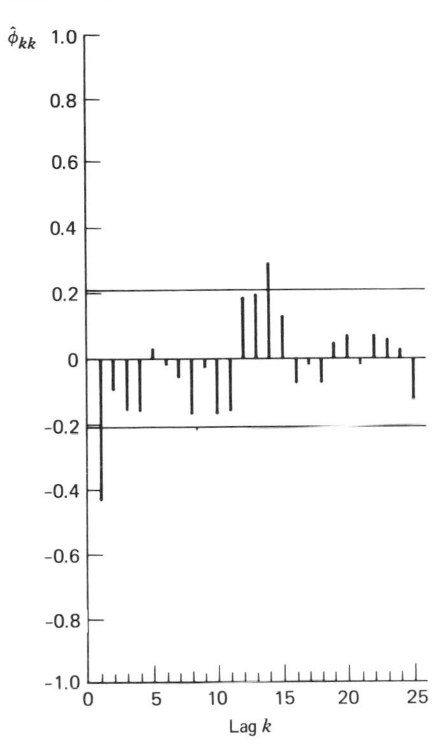

Figure 5.17. SPACF of ∇ ln z_t—demand for repair parts.

The Q statistic based on the first $K = 24$ autocorrelations is given by $Q = 37.0$ and is slightly bigger than $\chi^2_{.05}(23) = 35.2$. This is due mainly to one large autocorrelation at lag 12. For the time being, we ignore this correlation and attribute it to chance. However, we will return to this series in Chapter 6 and consider a modification that takes account of the seasonal component.

Forecasting

The forecasting model is given in terms of the transformed variable $y_t = \ln z_t$. The MMSE forecasts $\hat{y}_n(l)$ and the associated prediction intervals (c_1, c_2) for y_{n+l} can be derived as usual. However, to derive the forecasts of the original observations $z_{n+l} = \exp(y_{n+l})$, we have to backtransform the forecasts and the prediction intervals. They are given by

$$\hat{z}_n(l) = \exp[\hat{y}_n(l)] \qquad \text{and} \qquad [\exp(c_1), \exp(c_2)]$$

The forecasts of the transformed series y_{n+l} are MMSE forecasts. However,

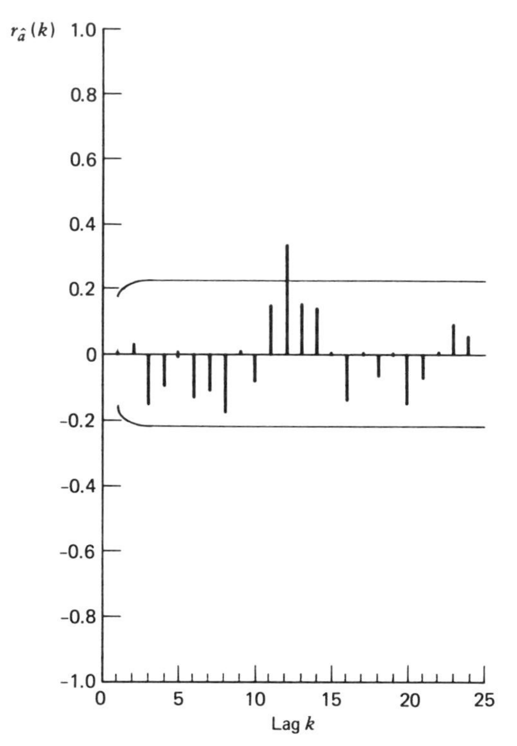

Figure 5.18. SACF of the residuals—demand for repair parts.

if the transformation is nonlinear, as in the case of the logarithmic transformation, the MMSE property will not be preserved for $\hat{z}_n(l) = \exp[\hat{y}_n(l)]$ [see Granger and Newbold (1976)]. Nevertheless, the forecast $\hat{z}_n(l)$ will be the median of the predictive distribution of z_{n+l}.

For the present example, the prediction $\hat{y}_{90}(l)$ and the corresponding 95 percent prediction limits are given by

$$\hat{y}_{90}(l) = \hat{y}_{90}(1) = y_{90} - \hat{\theta}\hat{a}_{90}$$
$$= 7.864 - .57(0.234) = 7.731$$

and

$$\hat{y}_{90}(l) \pm 1.96\left[1 + (l-1)(1-\hat{\theta})^2\right]^{1/2}\hat{\sigma}$$
$$= 7.731 \pm 1.96\left[1 + (l-1)(1-.57)^2\right]^{1/2}(.027)^{1/2}$$

Table 5.6. Forecasts and Prediction Limits—Demand for Repair Parts[a]

l	95% Prediction Limits for $y_{90+l} = \ln z_{90+l}$	Approximate 95% Prediction Limits for z_{90+l}
1	7.731 ± 0.322	[exp(7.409), exp(8.053)]
	$= (7.409, 8.053)$	$= (1651, 3143)$
2	7.731 ± 0.351	[exp(7.380), exp(8.082)]
	$= (7.380, 8.082)$	$= (1604, 3236)$

[a] $\hat{y}_{90}(l) = 7.731$; $\hat{z}_{90}(l) = \exp(7.731) = 2278$.

Forecasts and 95 percent prediction intervals for y_{91}, y_{92} and z_{91}, z_{92} are given in Table 5.6.

The prediction intervals are quite large. This is due partially to the large uncertainty ($\hat{\sigma}^2$) in the past data and partly to the unexplained correlation at lag 12.

APPENDIX 5. EXACT LIKELIHOOD FUNCTIONS FOR THREE SPECIAL MODELS

I. Exact Likelihood Function for an ARMA(1, 1) Process

The model is given by

$$w_t = \phi w_{t-1} + a_t - \theta a_{t-1} \tag{1}$$

where the a_t's are assumed to be independent normal random variables with mean zero and variance σ^2. The data $\mathbf{w} = (w_1, w_2, \ldots, w_n)'$ are available; however, the starting values (a_0, w_0) are unknown.

Suppose

$$b_{-1} = a_0 \qquad b_0 = w_0 \qquad b_t = a_t \qquad t = 1, 2, \ldots, n$$

$$\mathbf{b}_* = (b_{-1}, b_0)' \qquad \mathbf{b}_n = (b_1, \ldots, b_n)'$$

and

$$\mathbf{b} = (b_{-1}, b_0, \ldots, b_n)' = (\mathbf{b}'_*, \mathbf{b}'_n)' \tag{2}$$

Then it follows that

$$b_1 = a_1 = w_1 - \phi b_0 + \theta b_{-1}$$

$$b_2 = a_2 = w_2 - \phi w_1 + \theta b_1$$

$$= w_2 - \phi w_1 + \theta(w_1 - \phi b_0 + \theta b_{-1}) = w_2 + (\theta - \phi)w_1 - \phi\theta b_0 + \theta^2 b_{-1}$$

$$\vdots$$

$$b_n = a_n = w_n - \phi w_{n-1} + \theta b_{n-1} = \text{a linear function of } \mathbf{b}_* \text{ and } \mathbf{w}$$

In general, each b_t can be expressed as a linear function of $\mathbf{b}_*$ and $\mathbf{w}$. Therefore, we obtain

$$\mathbf{b} = \mathbf{L}\mathbf{w} + \mathbf{X}\mathbf{b}_* \tag{3}$$

where the $(n+2) \times n$ matrix $\mathbf{L}$ and the $(n+2) \times 2$ matrix $\mathbf{X}$ involve only the parameters ϕ and θ. These are given by

$$\mathbf{L} = \begin{bmatrix} 0 & 0 & 0 & \cdots & 0 & 0 \\ 0 & 0 & 0 & \cdots & 0 & 0 \\ 1 & 0 & 0 & \cdots & 0 & 0 \\ (\theta - \phi) & 1 & 0 & \cdots & 0 & 0 \\ \theta(\theta - \phi) & (\theta - \phi) & 1 & \cdots & 0 & 0 \\ \vdots & \vdots & \vdots & \vdots & \vdots & \vdots \\ \theta^{n-2}(\theta - \phi) & \theta^{n-3}(\theta - \phi) & \theta^{n-4}(\theta - \phi) & \cdots & (\theta - \phi) & 1 \end{bmatrix}$$

$$\mathbf{X}' = \begin{bmatrix} 1 & 0 & \theta & \theta^2 & \cdots & \theta^n \\ 0 & 1 & -\phi & -\theta\phi & \cdots & -\theta^{n-1}\phi \end{bmatrix} \tag{4}$$

Now Equation (3) can also be written as

$$\begin{bmatrix} \mathbf{b}_* \\ \hline \mathbf{b}_n \end{bmatrix} = \begin{bmatrix} \mathbf{0} \\ \hline \mathbf{L}_n \end{bmatrix} \mathbf{w} + \begin{bmatrix} \mathbf{I} \\ \hline \mathbf{X}_n \end{bmatrix} \mathbf{b}_* \tag{5}$$

where $\mathbf{L}_n$ is $n \times n$, $\mathbf{X}_n$ is $n \times 2$, $\mathbf{0}$ is a $2 \times n$ matrix of zeros, and $\mathbf{I}$ is the 2×2 identity matrix.

The covariance matrix of $\mathbf{b}_*$ is given by

$$E(\mathbf{b}_*\mathbf{b}_*') = \begin{bmatrix} E(b_{-1}^2) & E(b_{-1}b_0) \\ E(b_{-1}b_0) & E(b_0^2) \end{bmatrix} = \begin{bmatrix} 1 & 1 \\ 1 & (1+\theta^2-2\phi\theta)\delta^2 \end{bmatrix}\sigma^2 = \sigma^2\boldsymbol{\Sigma} \tag{6}$$

where $\delta = (1-\phi^2)^{-1/2}$. Then one can find a matrix $\mathbf{P}$ such that $\mathbf{P\Sigma P}' = \mathbf{I}$. It can be shown that such a $\mathbf{P}$ is given by

$$\mathbf{P} = \begin{bmatrix} 1 & 0 \\ \dfrac{-1}{(\theta-\phi)\delta} & \dfrac{1}{(\theta-\phi)\delta} \end{bmatrix} \tag{7}$$

Now, premultiplying Equation (5) by the $(n+2)\times(n+2)$ matrix

$$\left[\begin{array}{c|c} \mathbf{P} & \mathbf{0} \\ \hline \mathbf{0} & \mathbf{I}_n \end{array}\right]$$

where $\mathbf{I}_n$ is the $n \times n$ identity matrix, we obtain

$$\mathbf{u} = \mathbf{Lw} + \mathbf{Zu}_* \tag{8}$$

where

$$\mathbf{u} = \left[\begin{array}{c} \mathbf{u}_* \\ \hline \mathbf{u}_n \end{array}\right] \qquad \mathbf{L} = \left[\begin{array}{c} \mathbf{0} \\ \hline \mathbf{L}_n \end{array}\right] \qquad \mathbf{Z} = \left[\begin{array}{c} \mathbf{I} \\ \hline \mathbf{X}_n\mathbf{P}^{-1} \end{array}\right]$$

$$\mathbf{u}_* = \begin{bmatrix} u_{-1} \\ u_0 \end{bmatrix} = \mathbf{P}\begin{bmatrix} b_{-1} \\ b_0 \end{bmatrix} = \mathbf{Pb}_*$$

and $\mathbf{u}_n = (u_1, u_2, \dots u_n)' = \mathbf{b}_n$. The exact expression for $\mathbf{Z}$ in Equation (8) can be given as

$$\mathbf{Z}' = \begin{bmatrix} 1 & 0 & \theta-\phi & \theta(\theta-\phi) & \cdots & \theta^{n-1}(\theta-\phi) \\ 0 & 1 & -\phi(\theta-\phi)\delta & -\theta\phi(\theta-\phi)\delta & \cdots & -\theta^{n-1}\phi(\theta-\phi)\delta \end{bmatrix}$$

It is now important to note that

1. $E(\mathbf{u}_*\mathbf{u}_*') = E(\mathbf{Pb}_*\mathbf{b}_*'\mathbf{P}') = \mathbf{P}E(\mathbf{b}_*\mathbf{b}_*')\mathbf{P}' = \mathbf{P\Sigma P}'\sigma^2 = \mathbf{I}\sigma^2$
2. Since $\mathbf{u}_*$ involves only a_0 and w_0, it is independent of $\mathbf{u}_n$.

Thus the elements of $\mathbf{u} = (u_{-1}, u_0, u_1, \ldots, u_n)'$ are independent normal random variables with mean zero and variance σ^2. Then the joint probability density function of $\mathbf{u}$ is given by

$$p(\mathbf{u}|\phi, \theta, \sigma^2) = (2\pi\sigma^2)^{-(n+2)/2} \exp\left[-\frac{1}{2\sigma^2}\mathbf{u}'\mathbf{u}\right] \tag{9}$$

Since the Jacobian of the transformation in (8) is unity, we find that

$$p(\mathbf{w}, \mathbf{u}_*|\phi, \theta, \sigma^2) = (2\pi\sigma^2)^{-(n+2)/2} \exp\left[-\frac{1}{2\sigma^2}S(\phi, \theta, \mathbf{u}_*)\right] \tag{10}$$

where

$$S(\phi, \theta, \mathbf{u}_*) = (\mathbf{Lw} + \mathbf{Zu}_*)'(\mathbf{Lw} + \mathbf{Zu}_*)$$

is a sum of squares that depends on, among others, $\mathbf{u}_*$, which in turn is a function of the unknown initial values a_0 and w_0. Now it can be shown that

$$S(\phi, \theta, \mathbf{u}_*) = S(\phi, \theta) + (\mathbf{u}_* - \hat{\mathbf{u}}_*)'\mathbf{Z}'\mathbf{Z}(\mathbf{u}_* - \hat{\mathbf{u}}_*)$$

where

$$\hat{\mathbf{u}}_* = -(\mathbf{Z}'\mathbf{Z})^{-1}\mathbf{Z}'\mathbf{Lw} \tag{11}$$

and

$$S(\phi, \theta) = S(\phi, \theta, \hat{\mathbf{u}}_*) = (\mathbf{Lw} + \mathbf{Z}\hat{\mathbf{u}}_*)'(\mathbf{Lw} + \mathbf{Z}\hat{\mathbf{u}}_*) \tag{12}$$

Equation (10) then becomes

$$p(\mathbf{w}, \mathbf{u}_*|\phi, \theta, \sigma^2) = (2\pi\sigma^2)^{-(n+2)/2} \exp\left[-\frac{1}{2\sigma^2}\{S(\phi, \theta) + (\mathbf{u}_* - \hat{\mathbf{u}}_*)'\mathbf{Z}'\mathbf{Z}(\mathbf{u}_* - \hat{\mathbf{u}}_*)\}\right] \tag{13}$$

and the left-hand side of this equation can be written as

$$p(\mathbf{w}, \mathbf{u}_*|\phi, \theta, \sigma^2) = p(\mathbf{w}|\phi, \theta, \sigma^2)\, p(\mathbf{u}_*|\mathbf{w}, \phi, \theta, \sigma^2) \tag{14}$$

Integrating Equation (13) over $\mathbf{u}_*$, we obtain

$$p(\mathbf{w}|\phi, \theta, \sigma^2) = (2\pi\sigma^2)^{-n/2}\, |\mathbf{Z}'\mathbf{Z}|^{-1/2} \exp\left[-\frac{1}{2\sigma^2}S(\phi, \theta)\right] \tag{15}$$

where $|\mathbf{Z}'\mathbf{Z}|$ is the determinant of $\mathbf{Z}'\mathbf{Z}$ and is given by

$$|\mathbf{Z}'\mathbf{Z}| = \frac{(1-\theta^2)(1-\phi^2) + (1-\theta^{2n})(\theta-\phi)^2}{(1-\theta^2)(1-\phi^2)} \tag{16}$$

Equation (15) is the exact likelihood function of (ϕ, θ, σ^2).

Interpretation of $\hat{\mathbf{u}}_*$

For given values of ϕ and θ, Equation (8) can be written as

$$\mathbf{Lw} = -\mathbf{Z}\mathbf{u}_* + \mathbf{u}$$

which is in the form of the usual linear model. Given (ϕ, θ), (1) $\mathbf{Lw}$ is known and takes the place of the response vector, (2) $-\mathbf{Z}$ is the known design matrix, (3) $\mathbf{u}_*$ represents the parameter vector, and (4) elements of $\mathbf{u}$ are the random errors satisfying the least squares assumptions. Hence the least squares estimate of $\mathbf{u}_*$ conditional on (ϕ, θ) is given by $-(\mathbf{Z}'\mathbf{Z})^{-1}\mathbf{Z}'\mathbf{Lw}$, which is $\hat{\mathbf{u}}_*$ given in Equation (11).

From Equations (13) and (14) it can also be seen that

$$p(\mathbf{u}_*|\mathbf{w}, \phi, \theta, \sigma^2) = (2\pi\sigma^2)^{-1}|\mathbf{Z}'\mathbf{Z}|^{1/2}\exp\left[-\frac{1}{2\sigma^2}(\mathbf{u}_* - \hat{\mathbf{u}}_*)'\mathbf{Z}'\mathbf{Z}(\mathbf{u}_* - \hat{\mathbf{u}}_*)\right] \tag{17}$$

This implies that

$$\hat{\mathbf{u}}_* = E(\mathbf{u}_*|\mathbf{w}, \phi, \theta, \sigma^2) = E(\mathbf{u}_*|\mathbf{w}) \tag{18}$$

is the conditional expectation of $\mathbf{u}_*$ given the data and the parameters.

Two Other Representations of the Likelihood Function

1. From Equations (8) and (18) it follows that

$$E(\mathbf{u}|\mathbf{w}, \phi, \theta, \sigma^2) = \mathbf{Lw} + \mathbf{Z}E(\mathbf{u}_*|\mathbf{w}) = \mathbf{Lw} + \mathbf{Z}\hat{\mathbf{u}}_*$$

Then the sum of squares in Equation (12) can be written as

$$S(\phi, \theta) = \sum_{t=-1}^{n} E^2(u_t|\mathbf{w}) \tag{19}$$

Hence the likelihood function in (15) can be expressed as

$$L(\phi, \theta, \sigma^2|\mathbf{w}) = (2\pi\sigma^2)^{-n/2} |\mathbf{Z}'\mathbf{Z}|^{-1/2} \exp\left[-\frac{1}{2\sigma^2} \sum_{t=-1}^{n} E^2(u_t|\mathbf{w})\right] \tag{20}$$

2. From Equations (2), (5), (7), and (8), it follows that $u_t = a_t$ $(t = 1, 2, \ldots, n)$ and

$$\begin{bmatrix} u_{-1} \\ u_0 \end{bmatrix} = \mathbf{P}\begin{bmatrix} b_{-1} \\ b_0 \end{bmatrix} = \mathbf{P}\begin{bmatrix} a_0 \\ w_0 \end{bmatrix} = \begin{bmatrix} a_0 \\ (w_0 - a_0)/\delta(\theta - \phi) \end{bmatrix} \tag{21}$$

Therefore,

$$E(u_{-1}|\mathbf{w}) = E(a_0|\mathbf{w})$$

and

$$E(u_0|\mathbf{w}) = \frac{(1 - \phi^2)^{1/2}}{\theta - \phi}[E(w_0|\mathbf{w}) - E(a_0|\mathbf{w})]$$

Hence the likelihood function in (20) may be written as

$$L(\phi, \theta, \sigma^2|\mathbf{w}) = (2\pi\sigma^2)^{-n/2} |\mathbf{Z}'\mathbf{Z}|^{-1/2} \exp\left[-\frac{1}{2\sigma^2} S(\phi, \theta)\right]$$

where

$$S(\phi, \theta) = E^2(a_0|\mathbf{w}) + \frac{1 - \phi^2}{(\theta - \phi)^2}[E(w_0|\mathbf{w}) - E(a_0|\mathbf{w})]^2 + \sum_{t=1}^{n} E^2(a_t|\mathbf{w}) \tag{22}$$

The calculation of $E(a_0|\mathbf{w})$ and $E(w_0|\mathbf{w})$ is discussed in Section 5.6.

II. Exact Likelihood Function for an AR(1) Process

In this case there is only one unknown starting value, $b_0 = w_0$ with variance $\sigma^2/(1 - \phi^2)$. Hence $\mathbf{P}$ in Equation (7) is a scalar given by $P = (1 - \phi^2)^{1/2}$, and $u_0 = Pw_0 = (1 - \phi^2)^{1/2} w_0$. Therefore, the sum of squares in Equation

(19) becomes

$$S(\phi) = \sum_{t=0}^{n} E^2(u_t|\mathbf{w}) = E^2(u_0|\mathbf{w}) + E^2(a_1|\mathbf{w}) + \sum_{t=2}^{n} E^2(a_t|\mathbf{w})$$

$$= (1-\phi^2)E^2(w_0|\mathbf{w}) + [w_1 - \phi E(w_0|\mathbf{w})]^2 + \sum_{t=2}^{n} (w_t - \phi w_{t-1})^2 \tag{23}$$

since $u_t = a_t = w_t - \phi w_{t-1}$ for $t = 1, 2, \ldots, n$. $E(w_0|\mathbf{w})$ can be found by considering the backward representation of the process w_t, which can be written as $w_t = \phi w_{t+1} + e_t$, $V(e_t) = V(a_t) = \sigma^2$. This implies that

$$w_0 = \phi w_1 + e_0$$

and

$$E(w_0|\mathbf{w}) = \phi w_1 + E(e_0|\mathbf{w}) = \phi w_1 \tag{24}$$

since the shock e_0 is "future" to the data $\mathbf{w}$, and thus $E(e_0|\mathbf{w}) = 0$. Substituting for $E(w_0|\mathbf{w})$ in (23), we obtain

$$S(\phi) = (1-\phi^2)w_1^2 + \sum_{t=2}^{n} (w_t - \phi w_{t-1})^2 \tag{25}$$

Now $\mathbf{Z}$ in Equation (8) will be an $n + 1$ vector given by

$$\mathbf{Z}' = \begin{bmatrix} 1 & \dfrac{-\phi}{(1-\phi^2)^{1/2}} & 0 & \cdots & 0 \end{bmatrix}$$

which implies that $|\mathbf{Z}'\mathbf{Z}| = (1-\phi^2)^{-1}$. Therefore, the likelihood function can be written as

$$L(\phi, \sigma^2|\mathbf{w}) = (2\pi\sigma^2)^{-n/2}(1-\phi^2)^{1/2}$$

$$\times \exp\left[-\frac{1}{2\sigma^2}\left\{ (1-\phi^2)w_1^2 + \sum_{t=2}^{n} (w_t - \phi w_{t-1})^2 \right\} \right] \tag{26}$$

III. Exact Likelihood Function for an MA(1) Process

In this case there is one unobserved starting value a_0; $\mathbf{P}$ in Equation (7) is a scalar and is 1, implying that $u_0 = Pa_0 = a_0$. Hence the sum of squares in

Equation (19) can be written as

$$S(\theta) = \sum_{t=0}^{n} E^2(a_t|\mathbf{w}) \tag{27}$$

The calculation of $E(a_0|\mathbf{w})$ is discussed in Section 5.6. $E(a_t|\mathbf{w})$ ($t = 1, 2, \ldots n$) can be calculated using

$$E(a_t|\mathbf{w}) = \theta E(a_{t-1}|\mathbf{w}) + w_t \qquad t = 1, 2, \ldots, n$$

Here the $n + 1$ vector $\mathbf{Z}$ is given by

$$\mathbf{Z}' = \begin{bmatrix} 1 & \theta & \theta^2 & \cdots & \theta^n \end{bmatrix}$$

which implies that

$$|\mathbf{Z}'\mathbf{Z}| = \frac{1 - \theta^{2(n+1)}}{1 - \theta^2}$$

Therefore the likelihood function can be expressed as

$$L(\theta, \sigma^2|\mathbf{w}) = (2\pi\sigma^2)^{-n/2}\left[\frac{1 - \theta^2}{1 - \theta^{2(n+1)}}\right]^{1/2} \exp\left[-\frac{1}{2\sigma^2}\sum_{t=0}^{n} E^2(a_t|\mathbf{w})\right] \tag{28}$$

CHAPTER 6

Seasonal Autoregressive Integrated Moving Average Models

In Chapter 4 we considered regression and smoothing methods for predicting seasonal time series. In this chapter we describe a seasonal extension of the parametric time series models of Chapter 5.

High correlations among observations from the same season (week, month, quarter) are the main features of seasonal time series. Usually, for monthly data, the January observations are alike, the February observations are alike, and so on. This implies that the autocorrelations r_k at lags $k = 12, 24, \ldots$ are large. For quarterly data the observations corresponding to each quarter are alike, which leads to large autocorrelations at lags $4, 8, \ldots$. In addition to autocorrelations at seasonal lags, one can also expect correlations at adjacent (nonseasonal) lags. For example, in monthly data the December observation is usually correlated with the one from November, October, and so on.

In Figure 6.1 we show the SACF of the housing start series in Figure 4.3. We notice large autocorrelations at seasonal lags $12, 24, \ldots$ and at nonseasonal lags $1, 2, \ldots$, resulting in a cyclical pattern. Such ACF's are characteristic of many seasonal series. In this chapter we construct seasonal forecast models that capture the correlations at the seasonal and nonseasonal lags, and describe the seasonal pattern exhibited by the historical series.

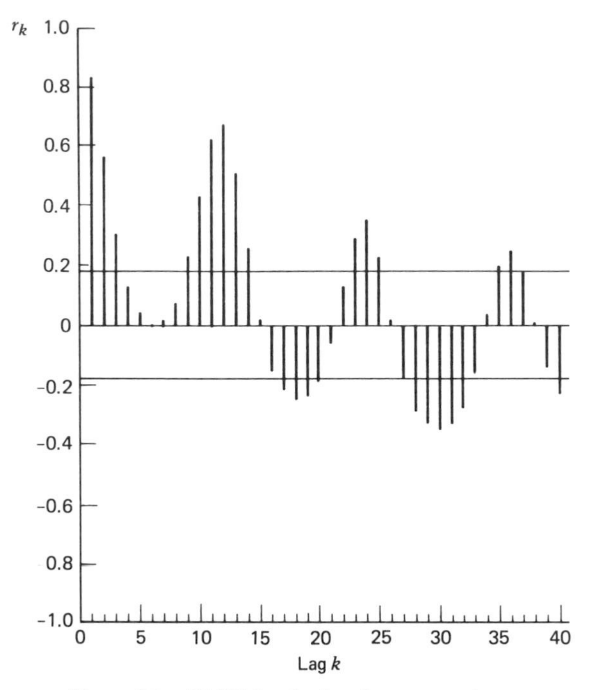

Figure 6.1. SACF for the housing start series.

A traditional method for modeling monthly seasonal time series data is to employ regression models of the form

$$z_{t+l} = \beta_1 \sin\frac{2\pi l}{12} + \beta_2 \cos\frac{2\pi l}{12} + a_{t+l} \tag{6.1a}$$

$$\begin{aligned} z_{t+l} &= \beta_0 + \sum_{j=1}^{6}\left(\beta_{1j}\sin\frac{2\pi jl}{12} + \beta_{2j}\cos\frac{2\pi jl}{12}\right) + a_{t+l} \\ &= \beta_0 + \beta_{26}(-1)^l + \sum_{j=1}^{5}\left(\beta_{1j}\sin\frac{2\pi jl}{12} + \beta_{2j}\cos\frac{2\pi jl}{12}\right) + a_{t+l} \end{aligned} \tag{6.1b}$$

Since $\sin \pi l = 0$ ($l = 1, 2, \ldots$), the model in (6.1b) contains only 12 coefficients.

Models of this form have been considered in Chapter 4. There it was assumed that the β coefficients are fixed constants and the errors $\{a_t\}$ are uncorrelated. These models are capable of reproducing seasonal patterns with fixed amplitudes and phases and are quite appropriate to describe the seasonality in stable series, such as monthly temperatures or other meteorological variables. However, for economic and business time series it may be more realistic to employ models that allow for adaptive amplitudes and phases. In other words, the β coefficients should adapt as time evolves. Also, it may be more realistic to adopt models that allow for correlated errors.

It was shown that for nonseasonal time series both these features are incorporated in certain difference equation models. This is also true in the seasonal case. For example, the difference equations

$$(1 - \sqrt{3}B + B^2)z_{t+l} = (1 - \theta_1 B - \theta_2 B^2)a_{t+l} \tag{6.2a}$$

$$(1 - B^{12})z_{t+l} = (1 - \Theta B^{12})a_{t+l} \tag{6.2b}$$

have, with respect to the time origin t, the following solutions:

$$z_{t+l} = \beta_1^{(t)}\sin\frac{2\pi l}{12} + \beta_2^{(t)}\cos\frac{2\pi l}{12} + e_t(l) \tag{6.3a}$$

$$z_{t+l} = \beta_0^{(t)} + \beta_{26}^{(t)}(-1)^l + \sum_{j=1}^{5}\left(\beta_{1j}^{(t)}\sin\frac{2\pi jl}{12} + \beta_{2j}^{(t)}\cos\frac{2\pi jl}{12}\right) + e_t(l) \tag{6.3b}$$

The errors $e_t(l) = a_{t+l} + \psi_1 a_{t+l-1} + \cdots + \psi_{l-1}a_{t+1}$, where the ψ weights are defined in Chapter 5, are correlated over time. Furthermore, the β coefficients are stochastic and change with the time origin t for which the solution is obtained.

Operators such as $(1 - \sqrt{3}B + B^2)$ or $(1 - B^{12})$, which are usually referred to as *simplifying operators*, introduce adaptive seasonal patterns into the forecasting equation. In the nonseasonal case, series with nonstationary means can be made stationary by means of simplifying operators of the form $(1 - B)^d$. Likewise, series with periodic means (either stochastic or deterministic) can be made stationary by considering differences of the form $(1 - B^s)^D$ ($D = 1, 2, \ldots$). Sometimes specific factors of $(1 - B^s)$, such as $(1 - \sqrt{3}B + B^2)$, are sufficient.

6.1. MULTIPLICATIVE SEASONAL MODELS

We use monthly housing starts (series 6 in the Data Appendix) to explain the principles behind seasonal models. For the time being, consider only the

January observations and think of them as a time series by themselves. For simplicity, suppose that the ARIMA model

$$(1 - L)z_t = (1 - \Theta L)\alpha_t \tag{6.4}$$

describes the January observations. The seasonal backshift operator $L = B^{12}$, $Lz_t = z_{t-12}$, serves the same purpose as the backshift operator B but shifts time by 12 units. The errors $\{\alpha_t\}$ correspond to the January observations. Since it is assumed that the model in (6.4) describes these observations, the errors $\alpha_t, \alpha_{t-12}, \alpha_{t-24}, \ldots$ are uncorrelated.

Now suppose that the same model holds for the data from the other months: February, March,..., December. This implies that all errors that correspond to a fixed month in different years are uncorrelated. However, the errors corresponding to adjacent months need not be uncorrelated. In other words, the "residual" series $\alpha_t, \alpha_{t-1}, \alpha_{t-2}, \ldots$ may be autocorrelated. These autocorrelations can be represented by nonseasonal ARIMA models. For simplicity, let us suppose that the model is of the form

$$(1 - B)\alpha_t = (1 - \theta B)a_t \tag{6.5}$$

where a_t is a white-noise sequence. Then combining the models (6.4) and (6.5), we obtain

$$(1 - B)(1 - L)z_t = (1 - \theta B)(1 - \Theta L)a_t \tag{6.6}$$

which, as it is shown later, is an appropriate model for the housing start series. Such models are referred to as *multiplicative* models.

The general multiplicative model can be introduced similarly. The observations corresponding to any particular season may be thought of as a time series, and ARIMA models can be used to model its autocorrelation structure. Suppose that the model is of the form

$$\Phi(L)(1 - L)^D z_t = \Theta(L)\alpha_t \tag{6.7}$$

where $L = B^s$, $Lz_t = z_{t-s}$, s is the period of seasonality, $\Phi(L) = 1 - \Phi_1 L - \cdots - \Phi_P L^P$ is a stationary seasonal autoregressive polynomial, $\Theta(L) = 1 - \Theta_1 L - \cdots - \Theta_Q L^Q$ is an invertible seasonal moving average polynomial, and α_t is a noise series such that $\alpha_t, \alpha_{t-s}, \alpha_{t-2s}, \ldots$ are uncorrelated. The polynomials of the ARIMA model in (6.7) are in terms of $L = B^s$.

One can think of α_t as a "deseasonalized" series. In general, the errors $\alpha_t, \alpha_{t-1}, \ldots$ are correlated, and ordinary ARIMA models of the form

$$\phi(B)(1 - B)^d \alpha_t = \theta(B)a_t \tag{6.8}$$

may be used to capture these correlations (see Chap. 5). The errors a_t are uncorrelated.

Combining Equations (6.7) and (6.8), we obtain the *multiplicative seasonal autoregressive integrated moving average model* of order $(p, d, q)(P, D, Q)_s$:

$$\phi(B)\Phi(L)(1-B)^d(1-L)^D z_t = \theta(B)\Theta(L)a_t \tag{6.9}$$

Models of this form have been introduced by Box and Jenkins (1976) and are widely used in practice. A constant θ_0 can also be included in (6.9); this introduces deterministic trend components into the model.

Our previous discussion focused on the conceptual construction of seasonal models. However, in actual model building we rely on the autocorrelation and partial autocorrelation patterns of these processes. Therefore, it is important to study the nature of these functions.

6.2. AUTOCORRELATION AND PARTIAL AUTOCORRELATION FUNCTIONS OF MULTIPLICATIVE SEASONAL MODELS

It is difficult to give explicit expressions for the ACF's and PACF's that are implied by the general multiplicative model. For this reason, we study only the most commonly used models:

Model 1: $w_t = (1-\Theta B^s)a_t$; $(0, d, 0)(0, D, 1)_s$ (6.10a)

Model 2: $(1-\Phi B^s)w_t = a_t$; $(0, d, 0)(1, D, 0)_s$ (6.10b)

Model 3: $w_t = (1-\theta B)(1-\Theta B^s)a_t$; $(0, d, 1)(0, D, 1)_s$ (6.10c)

Model 4: $(1-\Phi B^s)w_t = (1-\Theta B^s)a_t$; $(0, d, 0)(1, D, 1)_s$ (6.10d)

Model 5: $(1-\Phi B^s)w_t = (1-\theta B)a_t$; $(0, d, 1)(1, D, 0)_s$ (6.10e)

Model 6: $w_t = (1-\theta_1 B-\theta_2 B^2)(1-\Theta B^s)a_t$; $(0, d, 2)(0, D, 1)_s$ (6.10f)

Model 7: $(1-\Phi B^s)w_t = (1-\theta B)(1-\Theta B^s)a_t$; $(0, d, 1)(1, D, 1)_s$ (6.10g)

Note that $w_t = (1 - B)^d(1 - B^s)^D z_t$ is a stationary difference. To simplify the discussion, we use $s = 12$. Modifications for other values are straightforward.

6.2.1. Autocorrelation Function

The ACF for model 7 in (6.10g) is derived in Appendix 6. Models 1 to 5 are special cases. Model 6 is a generalization of model 3. The autocorrelations that are implied by these processes are summarized below. They are illustrated in Figures 6.2*a–f*.

Model 1. $w_t = (1 - \Theta B^{12})a_t$

$$\rho_k = \begin{cases} 1 & k = 0 \\ -\dfrac{\Theta}{1 + \Theta^2} & k = 12 \\ 0 & \text{otherwise} \end{cases}$$

Model 2. $(1 - \Phi B^{12})w_t = a_t$

$$\rho_k = \begin{cases} 1 & k = 0 \\ \Phi^{k/12} & k = 12, 24, \ldots \\ 0 & \text{otherwise} \end{cases}$$

Model 3. $w_t = (1 - \theta B)(1 - \Theta B^{12})a_t$

$$\rho_k = \begin{cases} 1 & k = 0 \\ -\dfrac{\theta}{1 + \theta^2} & k = 1 \\ \dfrac{\theta\Theta}{(1 + \theta^2)(1 + \Theta^2)} & k = 11 \\ -\dfrac{\Theta}{1 + \Theta^2} & k = 12 \\ \rho_{11} & k = 13 \\ 0 & \text{otherwise} \end{cases}$$

Model 4. $(1 - \Phi B^{12})w_t = (1 - \Theta B^{12})a_t$

$$\rho_k = \begin{cases} 1 & k = 1 \\ -\dfrac{(\Theta - \Phi)(1 - \Phi\Theta)}{1 + \Theta^2 - 2\Phi\Theta}\Phi^{k/12-1} & k = 12, 24, 36, \ldots \\ 0 & \text{otherwise} \end{cases}$$

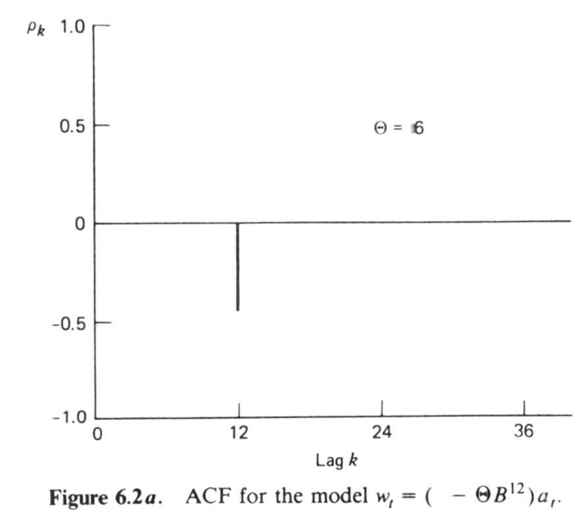

Figure 6.2a. ACF for the model $w_t = (\ - \Theta B^{12})a_t$.

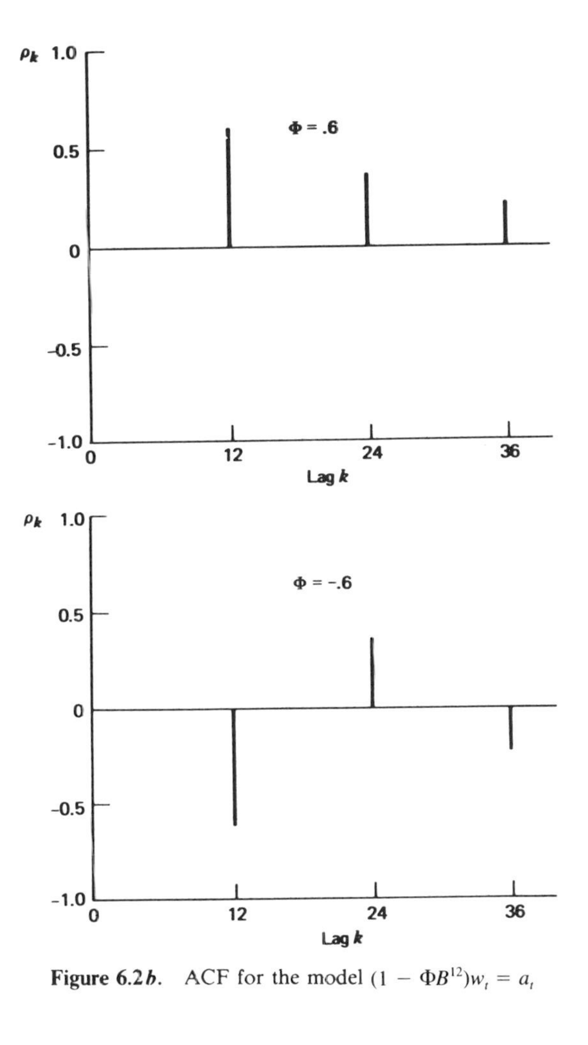

Figure 6.2b. ACF for the model $(1 - \Phi B^{12})w_t = a_t$

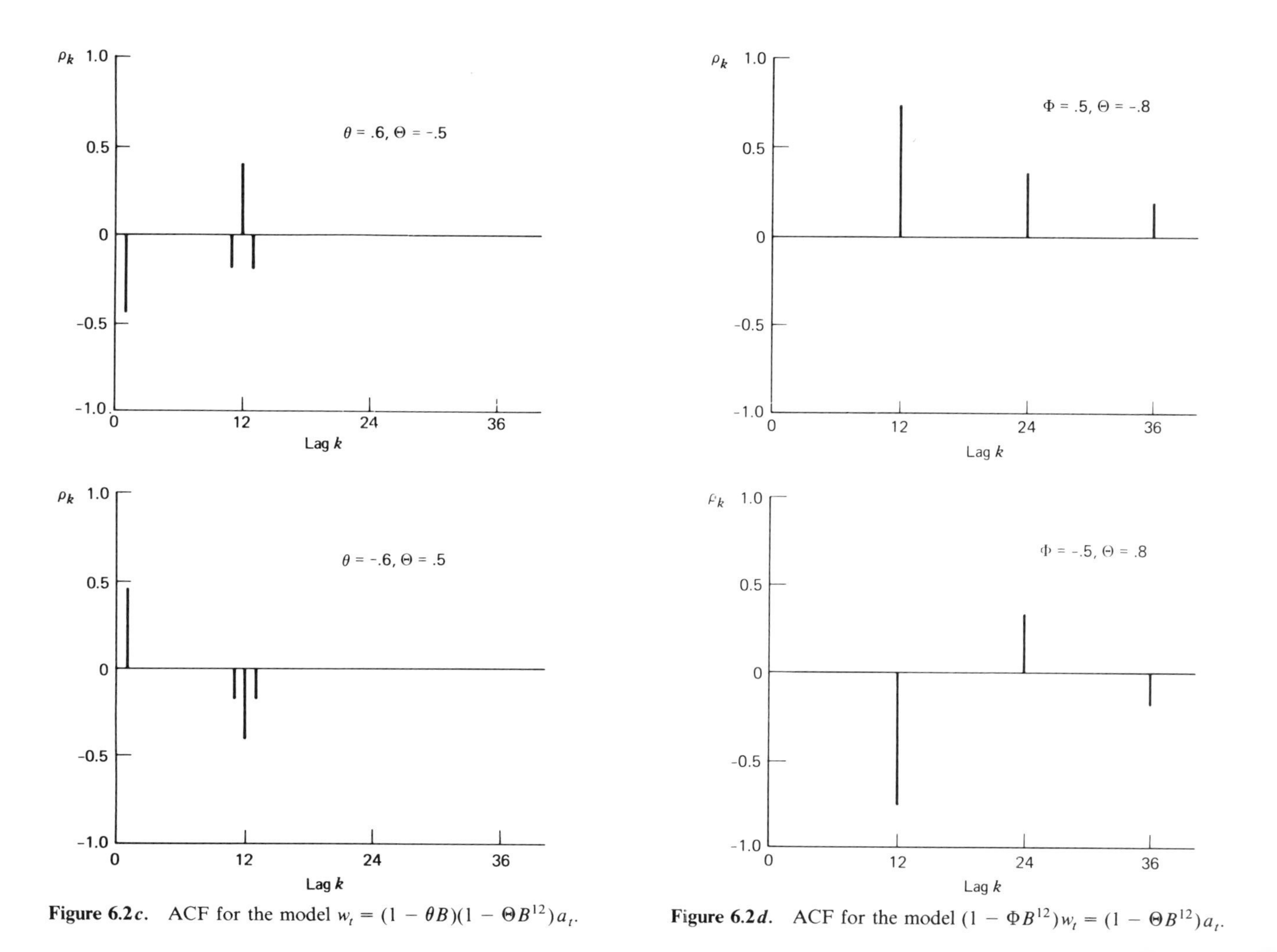

Figure 6.2*c*. ACF for the model $w_t = (1 - \theta B)(1 - \Theta B^{12})a_t$.

Figure 6.2*d*. ACF for the model $(1 - \Phi B^{12})w_t = (1 - \Theta B^{12})a_t$.

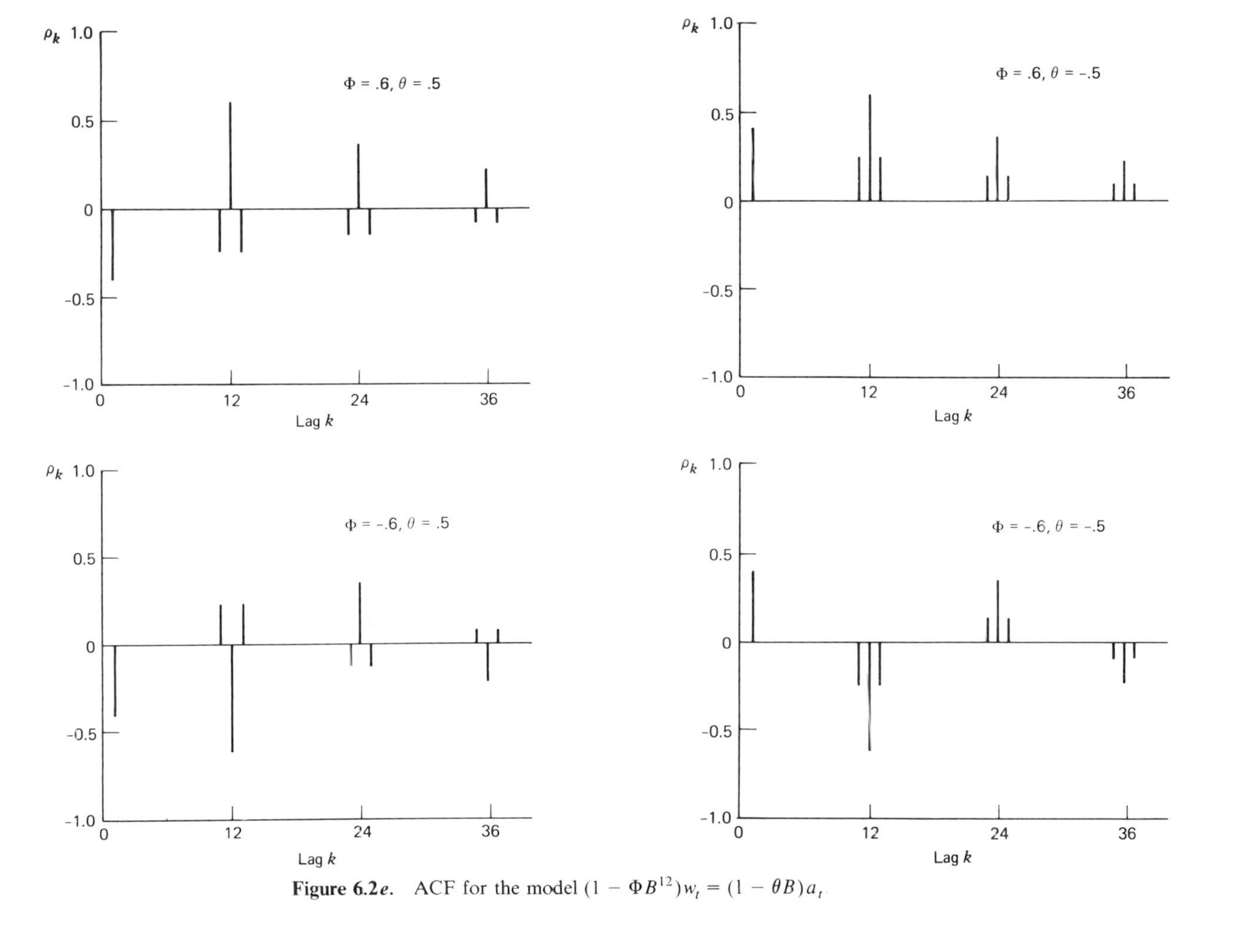

Figure 6.2e. ACF for the model $(1 - \Phi B^{12})w_t = (1 - \theta B)a_t$

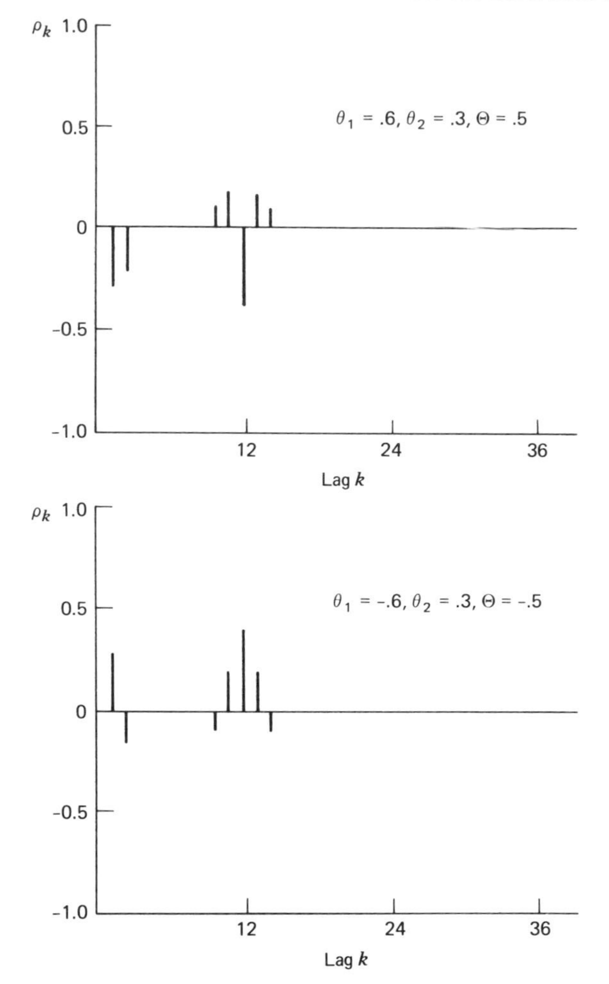

Figure 6.2 *f*. ACF for the model $w_t = (1 - \theta_1 B - \theta_2 B^2)(1 - \Theta B^{12})a_t$.

Model 5. $(1 - \Phi B^{12})w_t = (1 - \theta B)a_t$

$$\rho_k = \begin{cases} 1 & k = 0 \\ -\dfrac{\theta}{1 + \theta^2} & k = 1 \\ 0 & k = 2, \dots, 10 \\ -\dfrac{\theta\Phi}{(1 + \theta^2)} & k = 11 \\ \Phi & k = 12 \\ \rho_{11} & k = 13 \\ \Phi\rho_{k-12} & k > 13 \end{cases}$$

Model 6. $w_t = (1 - \theta_1 B - \theta_2 B^2)(1 - \Theta B^{12})a_t$

$$\rho_k = \begin{cases} 1 & k = 0 \\ -\dfrac{\theta_1(1-\theta_2)}{1+\theta_1^2+\theta_2^2} & k = 1 \\ -\dfrac{\theta_2}{1+\theta_1^2+\theta_2^2} & k = 2 \\ \dfrac{\theta_2\Theta}{(1+\theta_1^2+\theta_2^2)(1+\Theta^2)} & k = 10 \\ \dfrac{\theta_1\Theta(1-\theta_2)}{(1+\theta_1^2+\theta_2^2)(1+\Theta^2)} & k = 11 \\ -\dfrac{\Theta}{(1+\Theta^2)} & k = 12 \\ \rho_{11} & k = 13 \\ \rho_{10} & k = 14 \\ 0 & \text{otherwise} \end{cases}$$

6.2.2. Partial Autocorrelation Function

The patterns in the partial autocorrelations are more difficult to explain. However, in general, the seasonal and nonseasonal moving average components introduce exponential decays and damped sine waves at the seasonal and nonseasonal lags. In autoregressive processes the partial autocorrelations cut off. Two special cases are illustrated in Figure 6.3.

6.3. NONMULTIPLICATIVE MODELS

Our previous discussion does not imply that the multiplicative models in Equations (6.9) and (6.10) are appropriate for all seasonal time series. One useful modification is to make the autoregressive and/or moving average operators nonmultiplicative. For example, corresponding to the multiplicative model 3 of Equation (6.10c), we could consider the nonmultiplicative model

$$w_t = (1 - \theta_1 B - \theta_{12} B^{12})a_t \tag{6.11}$$

Whereas the multiplicative model with its two parameters can accommodate correlations at lags 1, 11, 12, and 13 with the restriction $\rho_{11} = \rho_{13}$, model

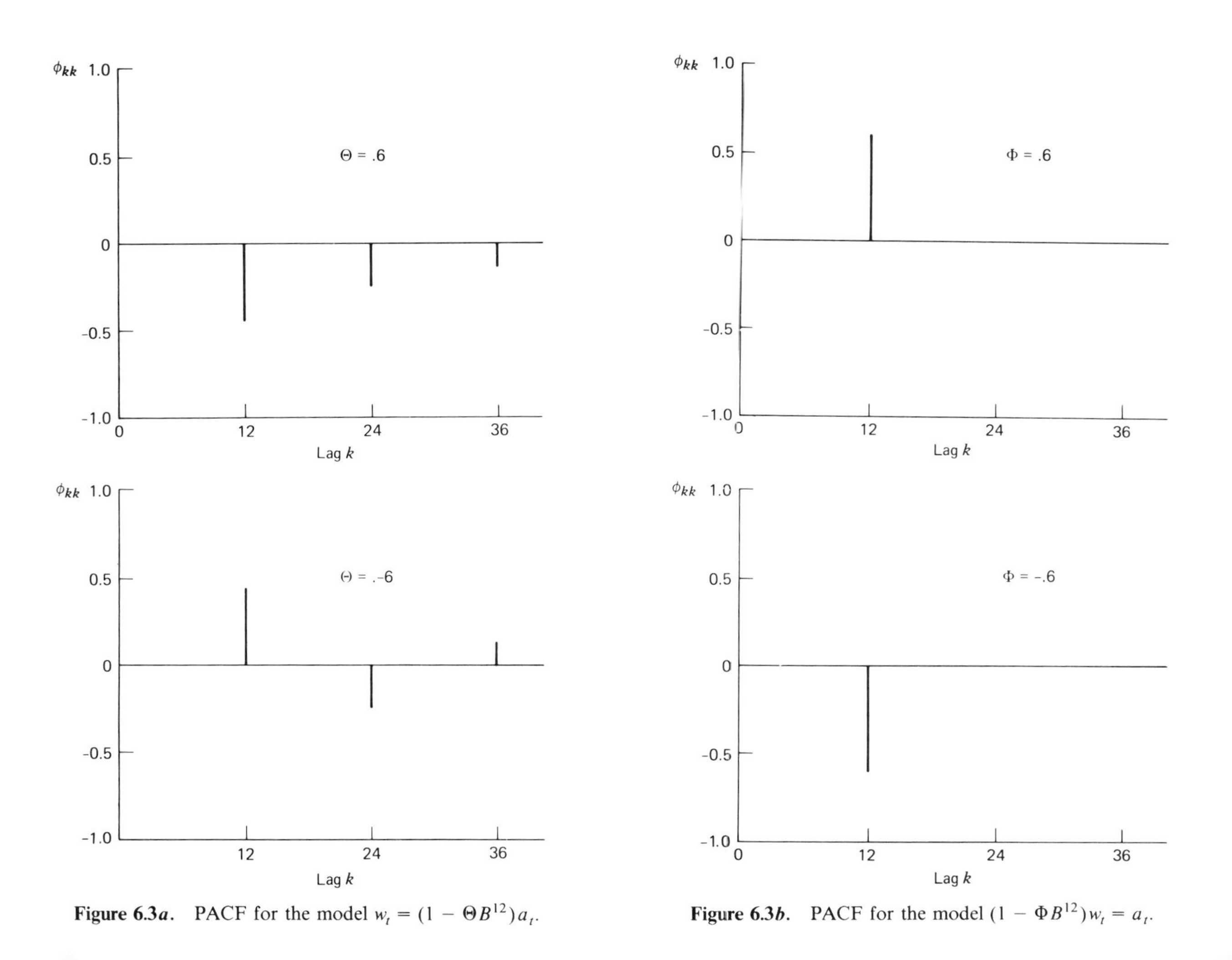

Figure 6.3a. PACF for the model $w_t = (1 - \Theta B^{12})a_t$.

Figure 6.3b. PACF for the model $(1 - \Phi B^{12})w_t = a_t$.

(6.11) acommodates only the ones at lags 1, 11, and 12. In practice it may be difficult to recognize whether a multiplicative or nonmultiplicative model should be used. In cases where nonmultiplicative models are useful, it has been found that usually the best-fitting multiplicative model provides a good starting point from which better nonmultiplicative models can be built.

6.4. MODEL BUILDING

A forecast model for seasonal data can be obtained by the iterative strategy of specification, estimation, and diagnostic checking described in Chapter 5.

6.4.1. Model Specification

The data in Chapter 4 (Figs. 4.1 to 4.5), and the series in Figures 6.4 (monthly gas usage in Iowa City), 6.5 (monthly gasoline demand in Ontario), and 6.6 (monthly percentage change in Canadian wages and salaries) illustrate strong seasonal patterns. This is also confirmed by the selected SACF's given in Figure 6.7. In all these cases the autocorrelations at lags 12, 24, 36,... are large and illustrate various periodic patterns. The nonde-

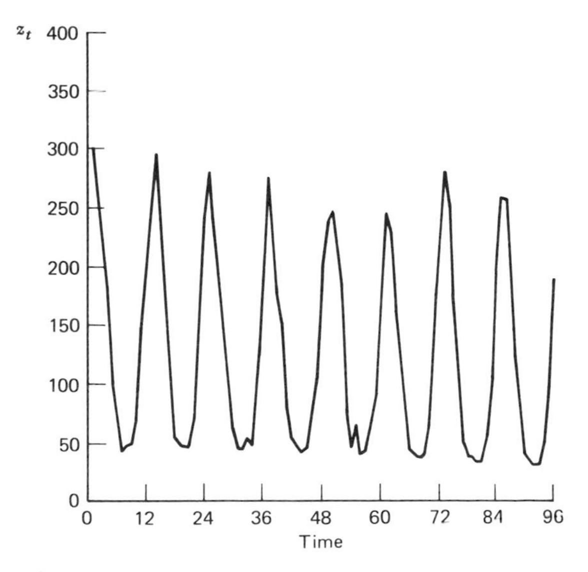

Figure 6.4. Monthly average residential gas usage in Iowa City (in hundreds of cubic feet), January 1971 to December 1978.

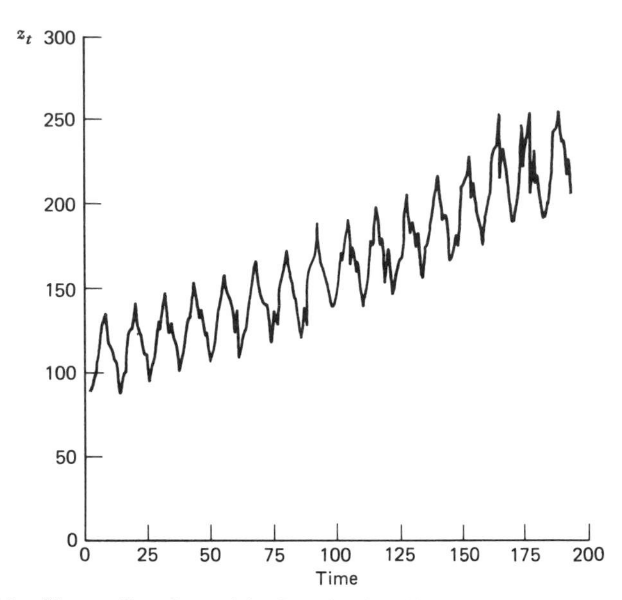

Figure 6.5. Monthly gasoline demand in Ontario (in millions of gallons), January 1960 to December 1975.

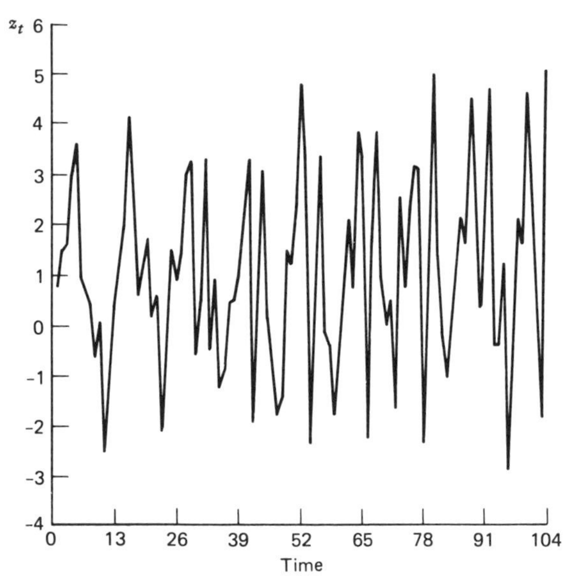

Figure 6.6. Monthly percentage changes in Canadian wages and salaries, February 1967 to September 1975.

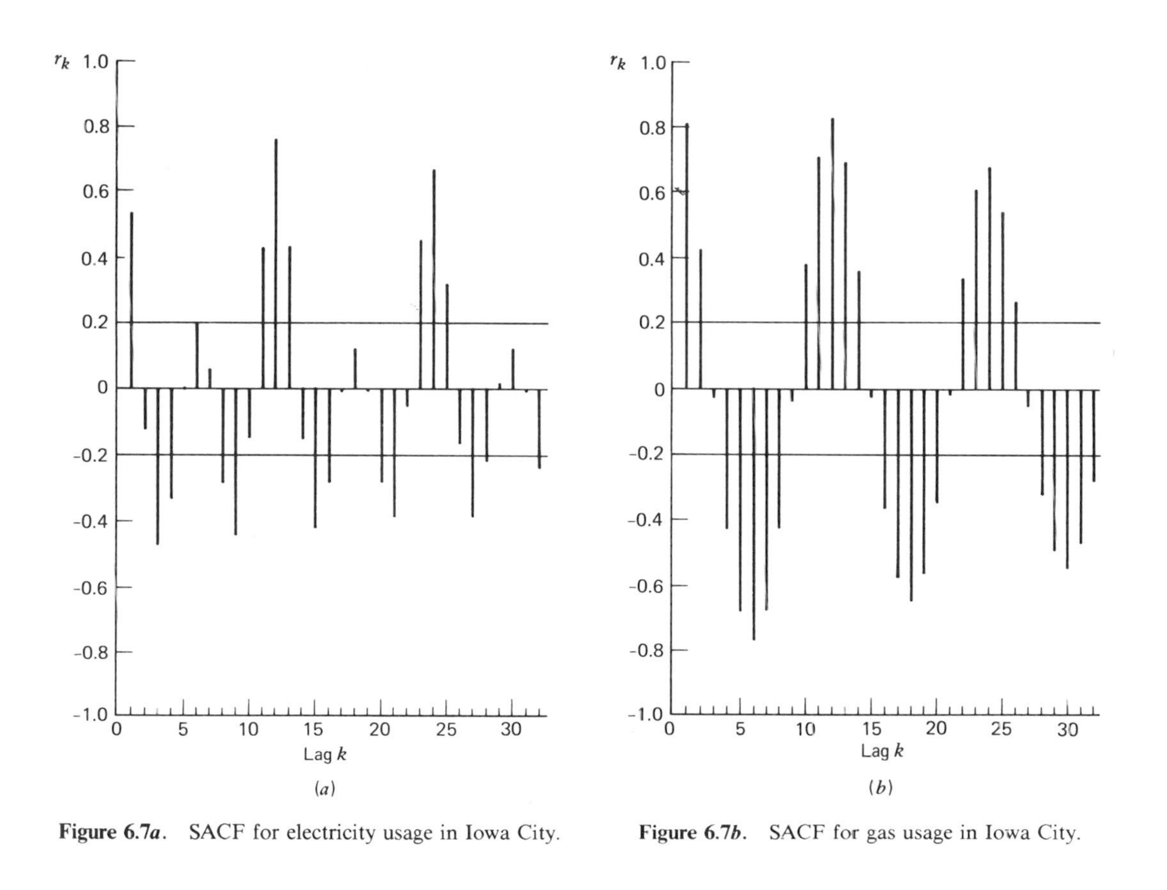

Figure 6.7*a*. SACF for electricity usage in Iowa City.

Figure 6.7*b*. SACF for gas usage in Iowa City.

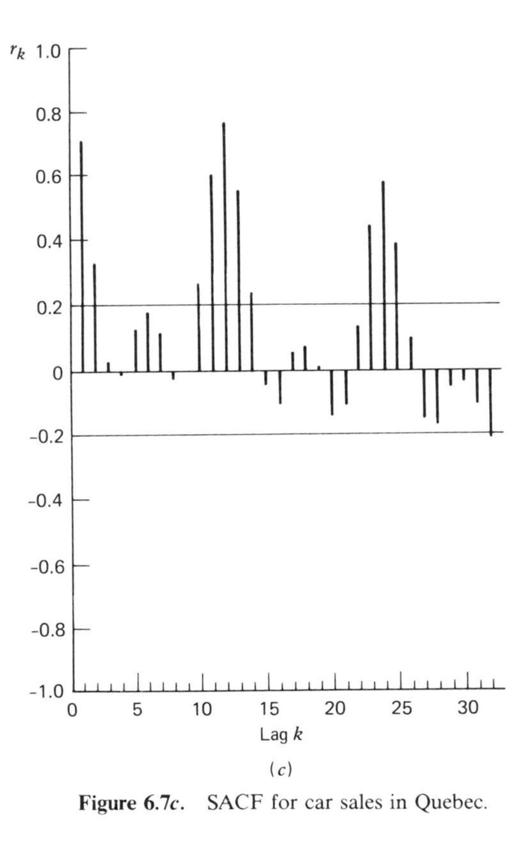

(c)

Figure 6.7c. SACF for car sales in Quebec.

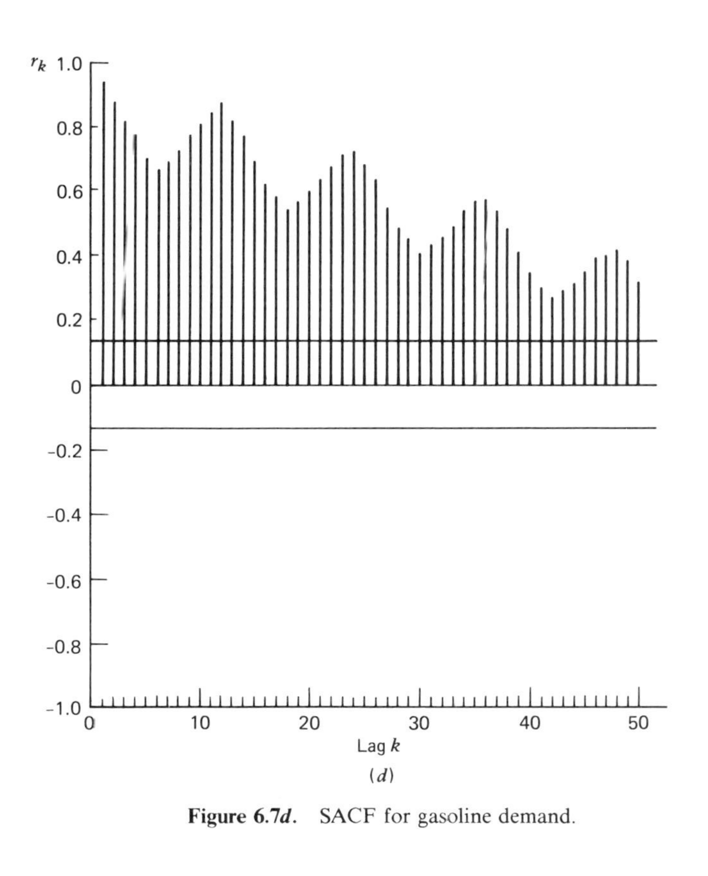

(d)

Figure 6.7d. SACF for gasoline demand.

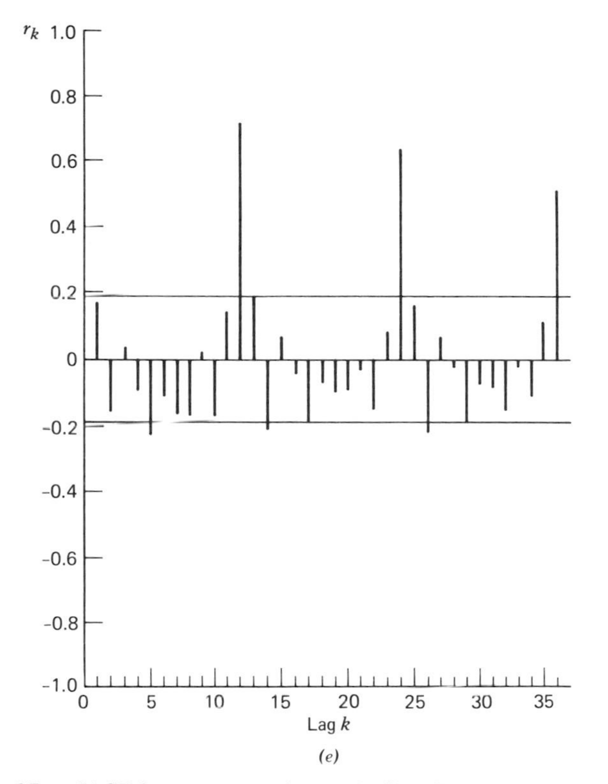

Figure 6.7e. SACF for percentage changes in Canadian wages and salaries.

caying nature of $r_{12}, r_{24}, r_{36}, \ldots$ calls for seasonal differences of the form $(1 - B^{12})^D$ $(D = 1, 2, \ldots)$. In addition, for time series with a nonseasonal trend, differences of the form $(1 - B)^d$ $(d = 1, 2, \ldots)$ may be necessary. Generally, one considers the series z_t, $(1 - B^s)z_t$, $(1 - B)(1 - B^s)z_t$, and so on, and studies their sample autocorrelations. If the SACF exhibits a slow decay (either at seasonal or nonseasonal lags), one continues to take appropriate differences until stationarity is achieved. Once a stationary difference $w_t = (1 - B)^d(1 - B^s)^D$ $(d, D = 0, 1, 2$, usually) has been found, the orders of the autoregressive and moving average polynomials can be specified by matching the patterns in the SACF of w_t with those of the models in Equations (6.10). In some cases it may also help to identify the patterns of the seasonal and nonseasonal parts separately.

Simplifying Operators

The model in (6.9) includes differencing operators of the form $(1 - B)^d$, $(1 - B^s)^D$, or their products. These are special cases of what are called

Table 6.1. Factors of $1 - B^{12}$

Factor	Root[a]	Period	Frequency in Cycles per Year
1. $1 - \sqrt{3}B + B^2$	$(\sqrt{3} \pm i)/2$	12	1
2. $1 - B + B^2$	$(1 \pm i\sqrt{3})/2$	6	2
3. $1 + B^2$	$\pm i$	4	3
4. $1 + B + B^2$	$(-1 \pm i\sqrt{3})/2$	3	4
5. $1 + \sqrt{3}B + B^2$	$(-\sqrt{3} \pm i)/2$	$\frac{12}{5}$	5
6. $1 + B$	-1	2	6
7. $1 - B$	1		Constant

[a]*Note*: $i = \sqrt{-1}$.

simplifying operators [Abraham and Box (1978)]. Their roots are spaced equally on the unit circle. For illustration, let us take the simplifying operator $1 - B^{12}$, which can be factored as

$$1 - B^{12} = (1 - \sqrt{3}B + B^2)(1 - B + B^2)(1 + B^2)(1 + B + B^2)$$
$$\times (1 + \sqrt{3}B + B^2)(1 + B)(1 - B)$$

The roots, periods, and frequencies associated with these factors are presented in Table 6.1. The unit circle with the corresponding roots of $1 - B^{12} = 0$ is shown in Figure 6.8. Usually, seasonal differences are taken to transform seasonal nonstationarity. However, Abraham and Box (1978) illustrate that sometimes simplifying operators of the form $(1 - \sqrt{3}B + B^2)$, $(1 - \sqrt{3}B + B^2)(1 - B + B^2)$, etc. are sufficient.

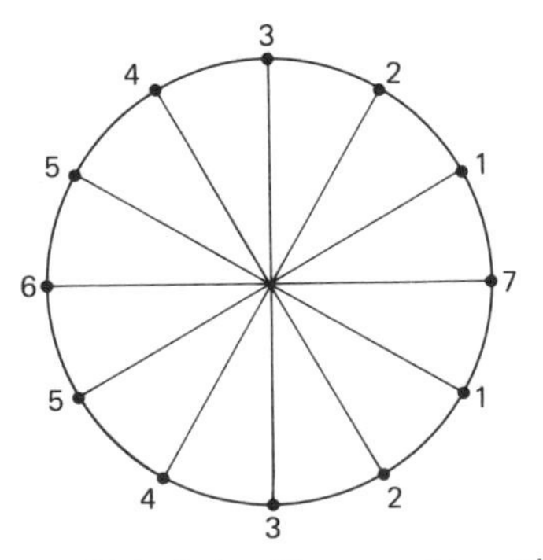

Figure 6.8. Unit circle with roots of $1 - B^{12} = 0$.

6.4.2. Model Estimation

Maximum likelihood (ML) estimation of the parameters in seasonal models is similar to the estimation for nonseasonal models in Chapter 5; the computer programs mentioned in Chapter 5 can be used to calculate the estimates. The computation of the likelihood function is slightly more complicated, since one has to evaluate more starting values $E(a_t|\mathbf{w})$, $E(w_t|\mathbf{w})$ ($t \leqslant 0$). For example, for the model $w_t = a_t - \Theta a_{t-12}$ one has to calculate 12 starting values $E(a_t|\mathbf{w})$, $t = 0, -1, \ldots, -11$. Similarly, for a model of the form $w_t = \Phi w_{t-12} + a_t - \Theta a_{t-12}$ one has to evaluate $E(w_t|\mathbf{w})$ and $E(a_t|\mathbf{w})$ for $t = 0, 1, \ldots, -11$.

In our examples we have calculated the exact maximum likelihood estimates. Readers who use conditional or unconditional least squares estimation procedures (i.e., the ones in MINITAB or IDA) will find that these programs give somewhat different estimates, especially for the seasonal parameters. In seasonal series of length N, there are only $N/12$ observations for each month. Thus the weights in the autoregressive representation $a_t = \pi(B)z_t$ may extend far back into the series, and the way in which the starting values are treated may influence the estimates.

6.4.3. Diagnostic Checking

We adopt the same diagnostic procedures as in the nonseasonal case and examine whether the residuals are uncorrelated. We compare the residual autocorrelations with their corresponding standard errors given in Chapter 5. Note that the standard error of the autocorrelation $r_{\hat{a}}(k)$ is somewhat smaller than $1/\sqrt{n}$ if the model includes a parameter at lag k.

Also the portmanteau test in (5.122) is used here. The statistic

$$Q_1 = n(n+2) \sum_{k=1}^{K} \frac{r_{\hat{a}}^2(k)}{n-k}$$

is approximately χ^2 distributed with $K - p - q - P - Q$ degrees of freedom; $p + q + P + Q$ is the number of estimated parameters, and n is the number of observations after differencing.

6.5. REGRESSION AND SEASONAL ARIMA MODELS

There are similarities between regression models in which $E(z_t)$ is taken as a deterministic function of time and the seasonal time series models in (6.9).

To discuss these similarities we consider several special cases.

Case 1

$$(1 - \sqrt{3}B + B^2)z_{t+l} = (1 - \theta_1 B - \theta_2 B^2)a_{t+l}$$

In (6.3a) we relate the variable z_{t+l} to a fixed time origin t and express it as

$$z_{t+l} = \beta_1^{(t)} \sin \frac{2\pi l}{12} + \beta_2^{(t)} \cos \frac{2\pi l}{12} + e_t(l) \tag{6.12}$$

The error $e_t(l) = a_{t+l} + \psi_1 a_{t+l-1} + \cdots + \psi_{l-1} a_{t+1}$ is a function of the ψ weights $[\psi_1 = \sqrt{3} - \theta_1, \psi_2 = 2 - \sqrt{3}\theta_1 - \theta_2, \psi_j = \sqrt{3}\psi_{j-1} - \psi_{j-2} (j > 2)]$, and of the random shocks that enter the system after time t. The coefficients $\beta_1^{(t)}$ and $\beta_2^{(t)}$ depend on all shocks (observations) up to and including time t. Following Abraham and Box (1978) or Box and Jenkins (1976, App. A5.3), these coefficients can be updated according to

$$\begin{bmatrix} \beta_1^{(t+1)} \\ \beta_2^{(t+1)} \end{bmatrix} = \begin{bmatrix} \frac{\sqrt{3}}{2} & \frac{-1}{2} \\ \frac{1}{2} & \frac{\sqrt{3}}{2} \end{bmatrix} \begin{bmatrix} \beta_1^{(t)} \\ \beta_2^{(t)} \end{bmatrix} + \begin{bmatrix} \sqrt{3}(1 - \theta_2) - 2\theta_1 \\ 1 + \theta_2 \end{bmatrix} a_{t+1}$$

or

$$\boldsymbol{\beta}^{(t+1)} = \mathbf{T}\boldsymbol{\beta}^{(t)} + \mathbf{g}a_{t+1} \tag{6.13}$$

Here $\mathbf{T}$ is a "transition" matrix that takes into account the change of the time origin from t to $t + 1$, and $\mathbf{g}$ is a vector that adjusts the coefficients proportional to the most recent error.

Equation (6.12) describes a regression model with correlated errors and stochastic coefficients. In the special case when $\theta_1 \to \sqrt{3}$ and $\theta_2 \to -1$, it can be seen from (6.13) that $\mathbf{g} \to \mathbf{0}$ and $\boldsymbol{\beta}^{(t+1)} \to \mathbf{T}\boldsymbol{\beta}^{(t)}$. Furthermore, $e_t(l)$ in (6.12) tends to a_{t+l}. Then the coefficients do not change stochastically, but only so as to express the fitting functions in terms of the new origin. Thus (6.12) reduces to the usual regression model in (6.1a).

Case 2

$$(1 - B^{12})z_{t+l} = (1 - \Theta B^{12})a_{t+l} \tag{6.14}$$

With respect to the time origin t, this model can be written as

$$z_{t+l} = \beta_0^{(t)} + \beta_{26}^{(t)}(-1)^l + \sum_{j=1}^{5}\left(\beta_{1j}^{(t)}\sin\frac{2\pi jl}{12} + \beta_{2j}^{(t)}\cos\frac{2\pi jl}{12}\right) + e_t(l) \tag{6.15}$$

The ψ weights in $e_t(l)$ are given by $\psi_j = 1 - \Theta$ ($j = 12, 24, \ldots$), and zero otherwise. The $\boldsymbol{\beta}^{(t)}$ coefficients are stochastic and depend on the shocks a_{t-j} ($j \geqslant 0$). They can be updated as

$$\beta_0^{(t+1)} = \beta_0^{(t)} + \frac{1-\Theta}{12}a_{t+1}$$

$$\beta_{26}^{(t+1)} = -\beta_{26}^{(t)} + \frac{1-\Theta}{12}a_{t+1}$$

$$\beta_{1j}^{(t+1)} = \left(\cos\frac{2\pi j}{12}\right)\beta_{1j}^{(t)} - \left(\sin\frac{2\pi j}{12}\right)\beta_{2j}^{(t)}$$

$$\beta_{2j}^{(t+1)} = \left(\sin\frac{2\pi j}{12}\right)\beta_{1j}^{(t)} + \left(\cos\frac{2\pi j}{12}\right)\beta_{2j}^{(t)} + \frac{1-\Theta}{6}a_{t+1} \qquad j = 1, 2, \ldots, 5 \tag{6.16}$$

We can obtain these updating equations by comparing the representation of z_{t+l} in (6.15) with the one for time origin $t+1$ and lead time $l-1$ (see Exercises 6.4 and 6.5).

Model (6.14) can be expressed in yet another form. Suppose $l = (r, m)$ represents a lead time of r years ($r = 0, 1, 2, \ldots$) and m months ($m = 1, 2, \ldots, 12$); then the model can be written as

$$z_{t+l} = \beta_{0m}^{(t)} + e_t(l) \tag{6.17}$$

where $\beta_{0m}^{(t)}$ is a stochastic level corresponding to month m.

Representations (6.15) and (6.17) are expressed in terms of stochastically changing coefficients and correlated errors. However, as Θ tends to 1, $e_t(l)$ tends to a_{t+l}, and the parameters $\boldsymbol{\beta}^{(t)}$ [or equivalently the levels $\beta_{0m}^{(t)}$ ($m = 1, 2, \ldots, 12$)] become nonstochastic. Thus models (6.15) and (6.17) simplify to the ordinary regression model given in (6.1b).

Case 3

$$(1-B)(1-B^{12})z_{t+l} = (1-\theta B)(1-\Theta B^{12})a_{t+l} \tag{6.18}$$

Proceeding as before, this model can be written as

$$z_{t+l} = \beta_0^{(t)} + \beta_1^{(t)}l + \beta_{26}^{(t)}(-1)^l + \sum_{j=1}^{5}\left(\beta_{1j}^{(t)}\sin\frac{2\pi jl}{12} + \beta_{2j}^{(t)}\cos\frac{2\pi jl}{12}\right) + e_t(l) \tag{6.19}$$

Equivalently, it can be expressed as

$$z_{t+l} = \beta_{0m}^{(t)} + \beta_{1*}^{(t)}r + e_t(l) \tag{6.20}$$

where $l = (r, m)$ and $m = 1, 2, \ldots, 12$; $r = 0, 1, 2, \ldots$. The coefficient $\beta_{0m}^{(t)}$ is a stochastic level corresponding to month m, and $\beta_{1*}^{(t)}$ is a stochastic yearly trend term that is related to the one in (6.19). These models are similar to the traditional regression models for data containing a linear trend and additive seasonal components. However, they are more general, since they allow for stochastic coefficients and correlated errors. The coefficients in (6.20) can be updated as

$$\begin{aligned}
\beta_{0m}^{(t+1)} &= \beta_{0m+1}^{(t)} + (1-\theta)a_{t+1} \qquad m = 1, 2, \ldots, 11 \\
\beta_{0,12}^{(t+1)} &= \beta_{01}^{(t)} + \beta_{1*}^{(t)} + (2 - \theta - \Theta)a_{t+1} \\
\beta_{1*}^{(t+1)} &= \beta_{1*}^{(t)} + (1-\theta)(1-\Theta)a_{t+1}
\end{aligned} \tag{6.21}$$

For the special case $\theta = \Theta = 1$, the model in (6.18) simplifies to the usual regression model with nonstochastic coefficients and uncorrelated errors.

6.6. FORECASTING

Once a time series model has been obtained, the minimum mean square error forecasts are easily calculated from the difference equation of the model. The MMSE forecast of z_{n+l} from time origin n is given by the conditional expectation

$$z_n(l) = E(z_{n+l}|z_n, z_{n-1}, \ldots)$$

The calculation of such forecasts is explained in Chapter 5. Here we consider two special cases, the $(0, 0, 0)(0, 1, 1)_{12}$ and the $(0, 1, 1)(0, 1, 1)_{12}$ models.

Case 1. The $(0, 0, 0)(0, 1, 1)_{12}$ model is given by

$$(1 - B^{12})z_t = (1 - \Theta B^{12})a_t$$

From the difference equation

$$z_{n+l} = z_{n+l-12} + a_{n+l} - \Theta a_{n+l-12} \tag{6.22}$$

the forecasts can be obtained as follows:

$$\begin{aligned} z_n(1) &= z_{n-11} - \Theta a_{n-11} \\ z_n(2) &= z_{n-10} - \Theta a_{n-10} \\ &\vdots \\ z_n(12) &= z_n - \Theta a_n \\ z_n(l) &= z_n(l-12) \qquad \text{or} \qquad (1 - B^{12})z_n(l) = 0 \qquad l > 12 \end{aligned} \tag{6.23}$$

where the backshift B operates on l. This implies that

$$\begin{aligned} z_n(1) &= z_n(13) = z_n(25) = \cdots \\ z_n(2) &= z_n(14) = z_n(26) = \cdots \\ &\vdots \\ z_n(12) &= z_n(24) = z_n(36) = \cdots \end{aligned}$$

For example, if n refers to December, then the forecasts for all future Januarys are the same, those for future Februarys are the same, and so on. The forecast function consists of 12 levels, one corresponding to each month of the year.

After writing the model (6.22) in its autoregressive form

$$(1 - B^{12})(1 - \Theta B^{12})^{-1} z_n = a_n$$

$$z_n = (1 - \Theta)\left(z_{n-12} + \Theta z_{n-24} + \Theta^2 z_{n-36} + \cdots\right) + a_n$$

we can express the forecasts as

$$z_n(1) = (1 - \Theta) \sum_{j \geq 0} \Theta^j z_{n-11-12j}$$

$$z_n(2) = (1 - \Theta) \sum_{j \geq 0} \Theta^j z_{n-10-12j}$$

$$\vdots$$

$$z_n(12) = (1 - \Theta) \sum_{j \geq 0} \Theta^j z_{n-12j}$$

and

$$z_n(l) = z_n(l - 12) \qquad l > 12 \tag{6.24}$$

The forecast for any future January is an exponentially weighted average of all past Januarys, that for February an exponentially weighted average of all the previous Februarys, and so on.

According to Equation (5.64) in Chapter 5, the forecast error variance is given by

$$V[e_n(l)] = (1 + \psi_1^2 + \cdots + \psi_{l-1}^2)\sigma^2$$

Since for this model, $\psi_j = 1 - \Theta$ for $j = 12, 24, \ldots$ and zero otherwise, this variance is given by

$$V[e_n(l)] = \sigma^2\left\{1 + \left[\frac{l-1}{12}\right](1 - \Theta)^2\right\} \tag{6.25}$$

where $[x]$ denotes the integer part of x.

The forecasts in (6.23) can be interpreted in yet another form. The difference equation

$$(1 - B^{12})z_n(l) = 0 \qquad l > 12$$

has the solution

$$z_n(l) = z_n(r, m) = \beta_{0m}^{(n)} \qquad l = (r, m) > 0 \tag{6.26}$$

were the coefficients $\beta_{0m}^{(n)}$ are described in (6.17); $\beta_{0m}^{(n)}$ is an exponentially weighted average of previous observations in month m. The solution (6.26) is referred to as the *eventual forecast function*.

So far we have assumed that the parameter Θ and a_{n-j} $(j \geqslant 0)$ are known. In practice, Θ has to be replaced by its estimate $\hat{\Theta}$ and a_{n-j} by the residual $\hat{a}_{n-j}$.

Case 2. The $(0, 1, 1)(0, 1, 1)_{12}$ model is given by

$$(1 - B)(1 - B^{12})z_t = (1 - \theta B)(1 - \Theta B^{12})a_t$$

From the difference equation,

$$\begin{aligned} z_{n+l} = z_{n+l-1} &+ z_{n+l-12} - z_{n+l-13} + a_{n+l} - \theta a_{n+l-1} \\ &- \Theta a_{n+l-12} + \theta\Theta a_{n+l-13} \end{aligned} \tag{6.27}$$

the forecasts can be obtained as follows:

$$\begin{aligned} z_n(1) &= z_n + z_{n-11} - z_{n-12} - \theta a_n - \Theta a_{n-11} + \theta\Theta a_{n-12} \\ z_n(2) &= z_n(1) + z_{n-10} - z_{n-11} - \Theta a_{n-10} + \theta\Theta a_{n-11} \\ &\vdots \\ z_n(12) &= z_n(11) + z_n - z_{n-1} - \Theta a_n + \theta\Theta a_{n-1} \\ z_n(13) &= z_n(12) + z_n(1) - z_n + \theta\Theta a_n \\ z_n(l) &= z_n(l-1) + z_n(l-12) - z_n(l-13) \end{aligned}$$

or

$$(1 - B)(1 - B^{12})z_n(l) = 0 \qquad l > 13 \tag{6.28}$$

Although the forecasts are easily calculated from (6.28), their nature can be better understood by studying the eventual forecast function

$$z_n(l) = z_n(r, m) = \beta_{0m}^{(n)} + \beta_{1*}^{(n)} r \qquad l = (r, m) > 0 \tag{6.29}$$

[See Eqs. (6.19) and (6.20).] The forecast function is described by 12 monthly levels $\beta_{0m}^{(n)}$, and a coefficient $\beta_{1*}^{(n)}$ for the yearly trend change.

Using the autoregressive representation of this model, the forecasts can also be interpreted in terms of exponentially weighted averages [see Box and Jenkins (1976, p. 312) and Exercise 6.6].

6.7. EXAMPLES

6.7.1. Electricity Usage

Figure 4.1 shows the average monthly residential electricity usage in Iowa City from January 1971 to December 1978 ($N = 96$). The sample autocorrelation function is given in Figure 6.7*a*. Since the autocorrelations at lags 12, 24,... are large and decay slowly, we consider seasonal differences. The SACF of $(1 - B^{12})z_t$ is given in Figure 6.9*a*; only the autocorrelations at lags 1, 11, 12 are outside the 2σ limits ($2/\sqrt{n} = .22$, where $n = N - 12 = 84$); furthermore, $r_{11} \cong r_{13}$. Hence we entertain a model of the form

$$(1 - B^{12})z_t = \theta_0 + (1 - \theta B)(1 - \Theta B^{12})a_t \tag{6.30}$$

The maximum likelihood estimates and their standard errors are given by

$$\hat{\theta}_0 = 6.86(2.73) \qquad \hat{\theta} = -.30(.10) \qquad \hat{\Theta} = .69(.07) \qquad \hat{\sigma}^2 = 1654$$

The estimate $\hat{\theta}_0$, which measures the yearly increase in electricity usage, is significant. The SACF of the residuals in Figure 6.9*b* reveals no serious model inadequacies. Also, the portmanteau statistic $Q_1 = 21.7$ is insignificant compared with $\chi^2_{.05}(21) = 32.7$.

Forecasts from time origin $N = 96$ are generated from the difference equation

$$\begin{aligned}\hat{z}_{96}(l) = \hat{\theta}_0 + \hat{z}_{96}(l-12) &+ E(a_{96+l}|z_{96}, z_{95},\ldots)\\ &-\hat{\theta}E(a_{96+l-1}|z_{96}, z_{95},\ldots) - \Theta E(a_{96+l-12}|z_{96}, z_{95},\ldots)\\ &+\hat{\theta}\hat{\Theta}E(a_{96+l-13}|z_{96}, z_{95},\ldots)\end{aligned}$$

where

$$\hat{z}_{96}(j) = z_{96+j} \qquad j \leqslant 0$$

$$E(a_{96+j}|z_{96}, z_{95},\ldots) = \begin{cases}\hat{a}_{96+j} & j \leqslant 0\\ 0 & j > 0\end{cases}$$

The first 12 forecasts and their standard errors are shown in Table 6.2. They can be compared with the actual observations $z_{97},\ldots, z_{106}$. Note that the forecasts are made from the same forecast origin; thus the corresponding forecast errors are correlated (see Exercise 5.18). One consequence of this is that there is a tendency for the forecast function to lie either completely above or completely below the actual observations.

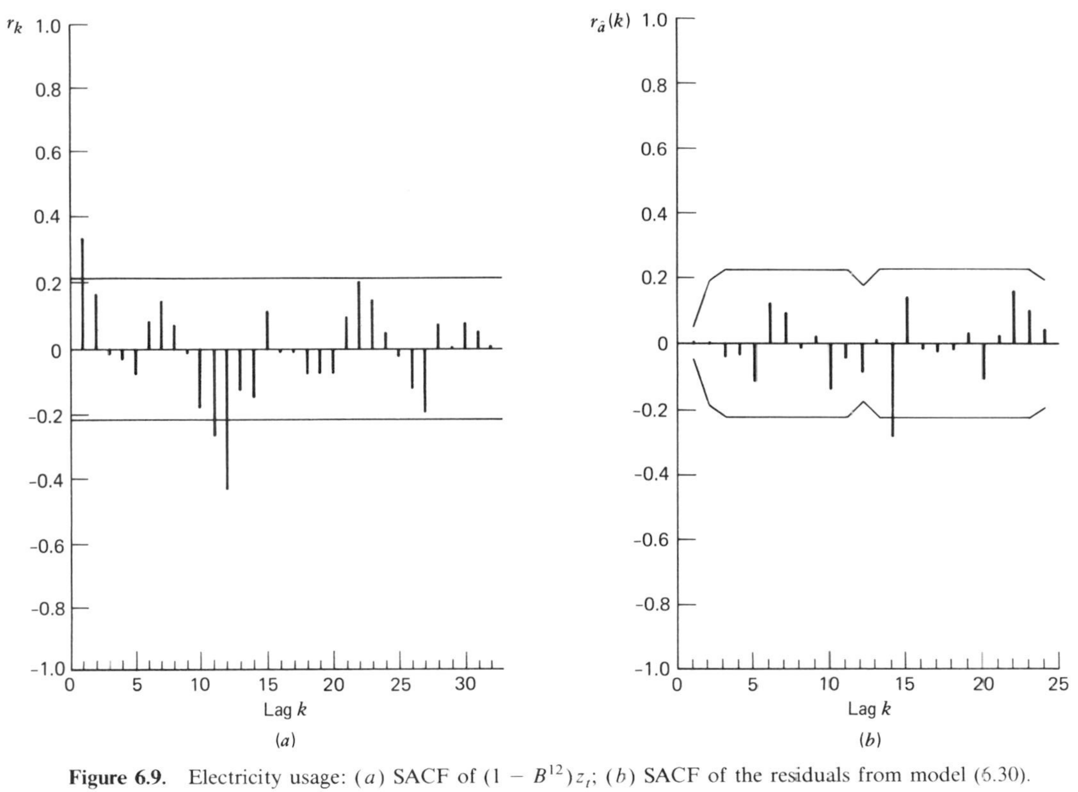

Figure 6.9. Electricity usage: (*a*) SACF of $(1 - B^{12})z_t$; (*b*) SACF of the residuals from model (6.30).

Table 6.2. Forecasts and Standard Errors for Electricity Usage[a]

Lead Time l	Forecasts $\hat{z}_{96}(l)$	Standard Errors	Actuals z_{96+l}
1	541	40.7	536
2	509	42.5	506
3	463		477
4	433		422
5	408	·	385
6	498	·	438
7	692	·	546
8	667		642
9	606		566
10	451		399
11	443		
12	514	42.5	

[a] Forecast origin is December 1978.

6.7.2. Gas Usage

Figure 6.4 shows the average monthly residential gas usage in Iowa City from January 1971 to December 1978 ($N = 96$). This plot, as well as the SACF in Figure 6.7*b*, reveals the seasonal nature of the series. The seasonal pattern in the series and the slowly decaying autocorrelations at lags $12, 24, \ldots$ indicate a need for seasonal differencing. The pattern in the SACF of $(1 - B^{12})z_t$ in Figure 6.10*a* is similar to the one in the SACF of the differenced electricity usage series; there are large autocorrelations at lags 1, 11, and 12. Thus we consider the model

$$(1 - B^{12})z_t = \theta_0 + (1 - \theta B)(1 - \Theta B^{12})a_t \tag{6.31}$$

Maximum likelihood estimation yields

$$\hat{\theta}_0 = -4.08(1.04) \qquad \hat{\theta} = -.41(.10) \qquad \hat{\Theta} = .89(.08) \qquad \hat{\sigma}^2 = 276.5$$

The fitted model passes all diagnostic checks; the residual autocorrelations in Figure 6.10*b* are negligible, and the portmanteau statistic $Q_1 = 22.7$ is insignificant compared with the percentage points from a χ^2 distribution with 21 degrees of freedom.

Forecasts and standard errors are generated from the time origin $N = 96$ and are shown in Table 6.3.

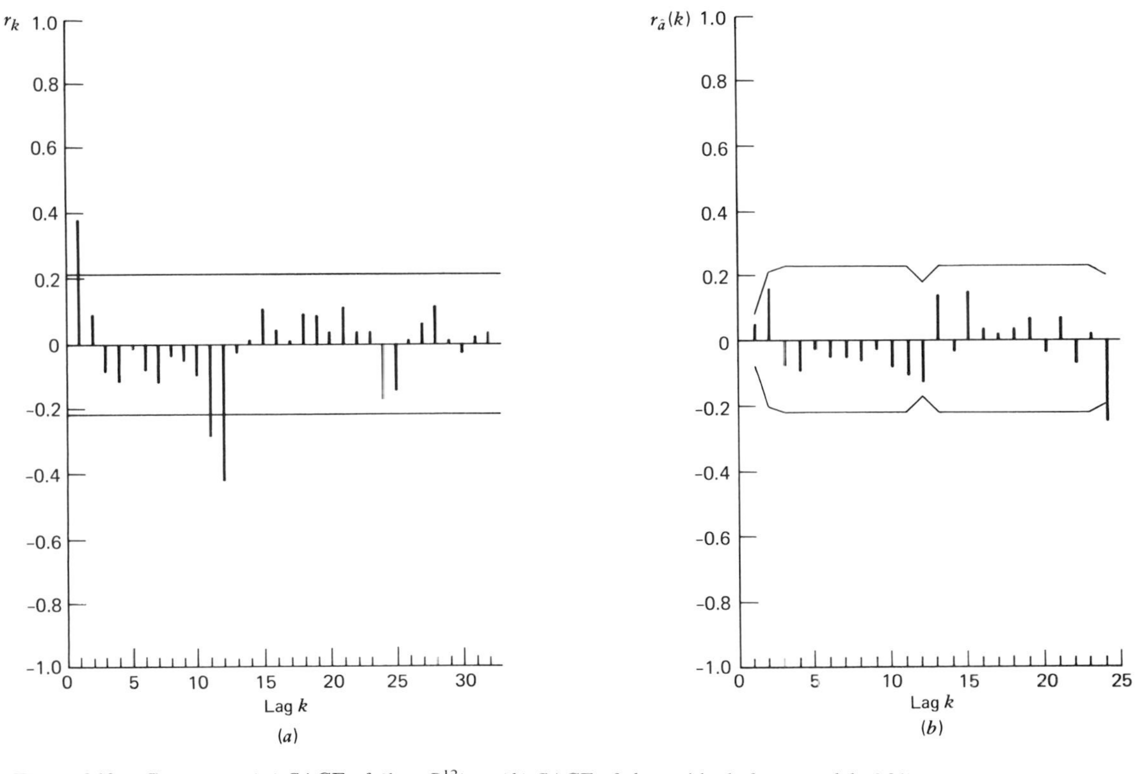

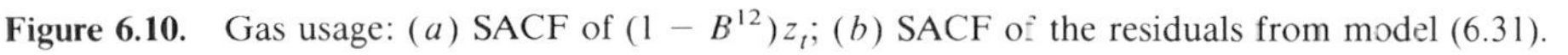

Figure 6.10. Gas usage: (a) SACF of $(1 - B^{12})z_t$; (b) SACF of the residuals from model (6.31).

Table 6.3. Forecasts and Standard Errors for Gas Usage[a]

Lead Time l	Forecasts $\hat{z}_{96}(l)$	Standard Errors	Actuals z_{96+l}
1	249	16.6	256
2	232	18.0	250
3	179		198
4	128		136
5	66	.	73
6	34	.	39
7	28	.	32
8	23		30
9	26		31
10	45		45
11	99		
12	190	18.0	

[a] Forecast origin is December 1978.

6.7.3. Housing Starts

The monthly U.S. housing starts of single-family structures for the period January 1965 to December 1974 are shown in Figure 4.3 ($N = 120$). The data exhibit seasonality as well as a changing mean level. The SACF's of z_t and $(1 - B^{12})z_t$ (see Figs. 6.1 and 6.11a) decay very slowly. However, the SACF of $(1 - B)(1 - B^{12})z_t$ in Figure 6.11b has large values only at lags 1 and 12. Hence we entertain the $(0, 1, 1)(0, 1, 1)_{12}$ model

$$(1 - B)(1 - B^{12})z_t = (1 - \theta B)(1 - \Theta B^{12})a_t \qquad (6.32)$$

The ML estimates are

$$\hat{\theta} = .28(.09) \qquad \hat{\Theta} = .88(.07) \qquad \hat{\sigma}^2 = 42.35$$

A constant θ_0 was originally included but was found to be insignificant. The residual autocorrelations are within the probability limits, and the portmanteau statistic $Q_1 = 31.4$ is insignificant compared with $\chi^2_{.05}(22)$. The forecasts from time origin $N = 120$, their standard errors, and the actual realizations are shown in Table 6.4.

6.7.4. Car Sales

The new car sales in Quebec from January 1960 to December 1968 are shown in Figure 4.4. Here we analyze the first $N = 96$ observations; the last

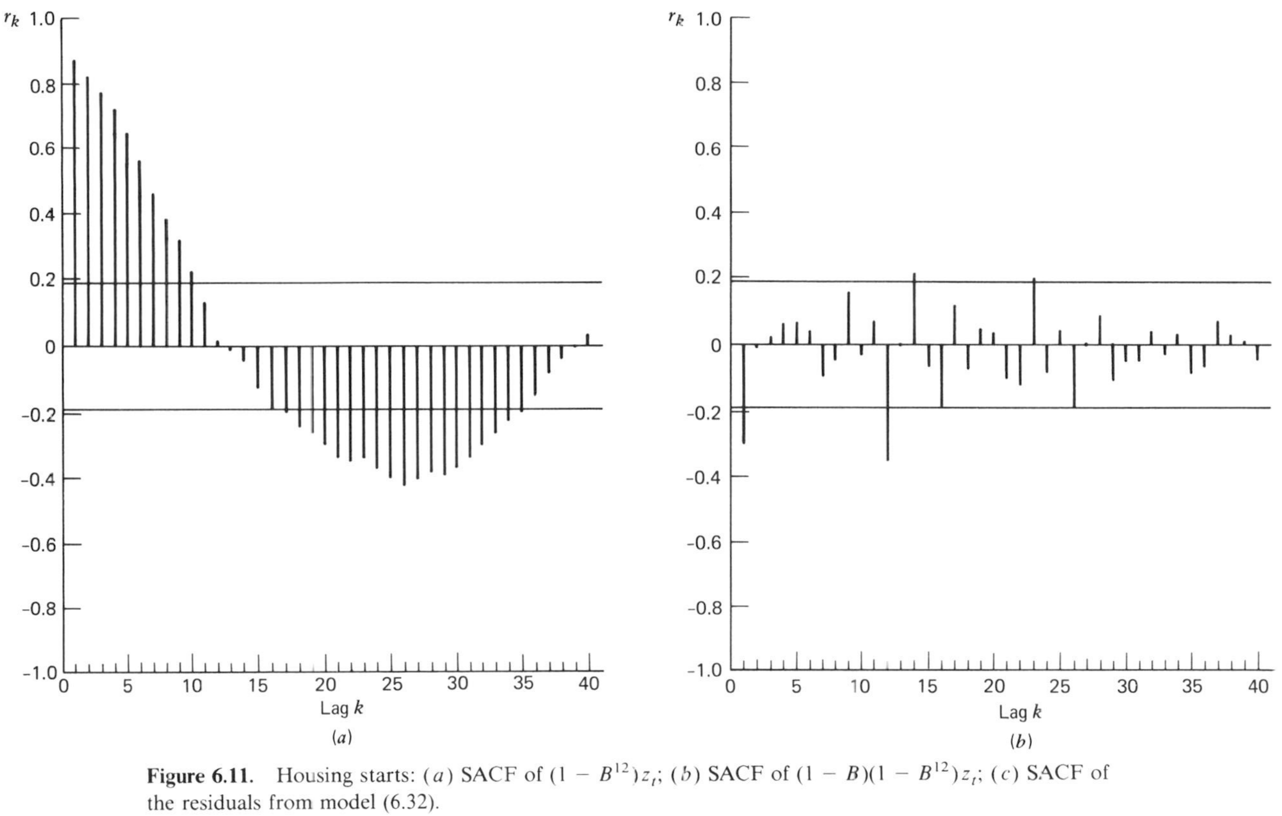

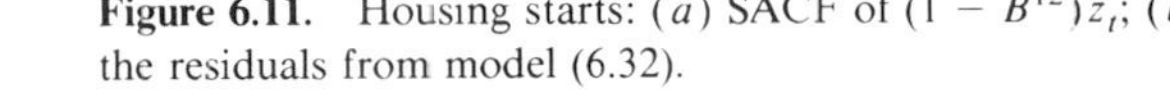

Figure 6.11. Housing starts: (*a*) SACF of $(1 - B^{12})z_t$; (*b*) SACF of $(1 - B)(1 - B^{12})z_t$; (*c*) SACF of the residuals from model (6.32).

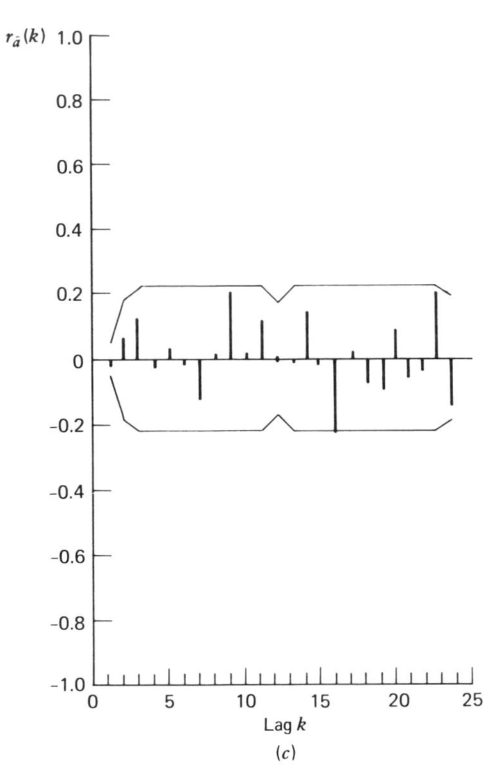

Figure 6.11. (*Continued*).

Table 6.4. Forecasts and Standard Errors for Housing Starts[a]

Lead Time l	Forecasts $\hat{z}_{120}(l)$	Standard Errors	Actuals z_{120+l}
1	37.06	6.51	39.791
2	40.25	8.04	39.959
3	68.21	9.32	62.498
4	84.61	10.44	77.777
5	86.29	11.46	92.782
6	83.44	12.39	90.284
7	77.29	13.26	92.782
8	75.60	14.07	90.655
9	67.63	14.84	84.517
10	67.64	15.57	93.826
11	54.30	16.27	71.646
12	39.29	16.94	55.650

[a] Forecast origin is December 1974.

12 observations are kept as a holdout period. The plot displays seasonality and an upward trend. The SACF of z_t in Figure 6.7c shows slowly decaying autocorrelations. Thus we consider the seasonal difference $w_t = (1 - B^{12})z_t$. The SACF of w_t in Figure 6.12a exhibits large autocorrelations at lags, 1, 2, 12, and 17. Ignoring r_{17}, we could consider models of the form $(0, 0, 2)(0, 1, 1)_{12}$ or $(2, 0, 0)(0, 1, 1)_{12}$. These models differ only in the nonseasonal component; however, we know that MA(2) models can be approximated by AR(2) models and vice versa, provided that the parameters in these models are small. These two models are also confirmed by the SPACF in Figure 6.12b, which has large values at lags 1, 2, and 12. The observations in Figure 4.4 increase with time; this implies that the mean of the seasonal differences is nonzero and that a trend parameter θ_0 should be included in the model. We have fitted both models to the data and have found that

$$(1 - \phi_1 B - \phi_2 B^2)(1 - B^{12})z_t = \theta_0 + (1 - \Theta B^{12})a_t \qquad (6.33)$$

leads to a better fit. The ML estimates of the parameters are

$$\hat{\theta}_0 = .51(.17) \qquad \hat{\phi}_1 = .27(.11) \qquad \hat{\phi}_2 = .21(.11)$$

$$\hat{\Theta} = .49(.10) \qquad \hat{\sigma}^2 = 2.27$$

Apart from a rather large value at lag 17, the residual autocorrelations in Figure 6.12c do not indicate serious model inadequacies. Also the portmanteau test statistic $Q_1 = 28.8$ is insignificant compared to $\chi^2_{.05}(20) = 31.4$.

The forecasts from time origin $N = 96$, their standard errors, and the actual realizations are given in Table 6.5.

6.7.5. Demand for Repair Parts

This data set has been analyzed in Chapter 5. There we fitted the model

$$(1 - B)\ln z_t = (1 - \theta B)b_t \qquad (6.34)$$

and found that the lag 12 residual autocorrelation was large and outside the probability limits. This indicates that the residuals 12 steps apart are still correlated. To capture this correlation among the residuals, we consider the model

$$b_t = (1 - \Theta B^{12})a_t \qquad (6.35)$$

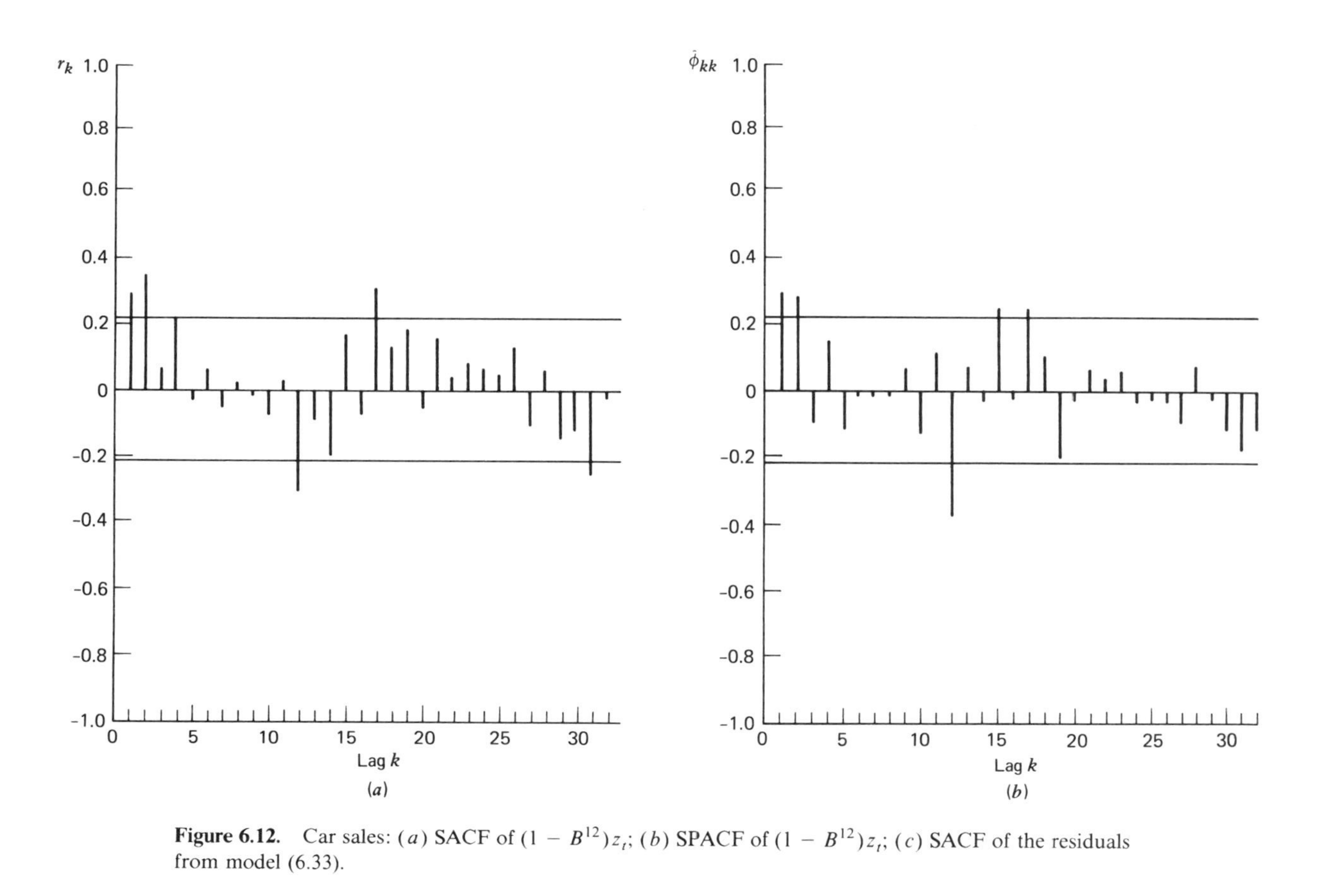

Figure 6.12. Car sales: (*a*) SACF of $(1 - B^{12})z_t$; (*b*) SPACF of $(1 - B^{12})z_t$; (*c*) SACF of the residuals from model (6.33).

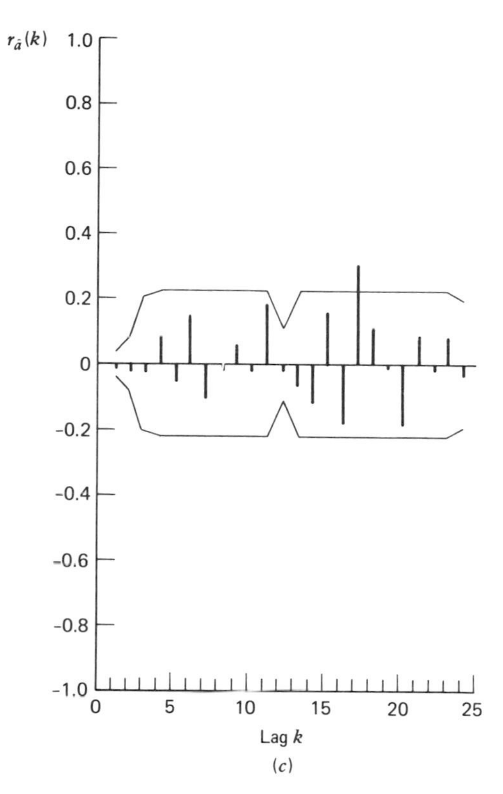

(c)

Figure 6.12. (*Continued*).

Table 6.5. Forecasts and Standard Errors for Car Sales[a]

Lead Time l	Forecasts $\hat{z}_{96}(l)$	Standard Errors	Actuals z_{96+l}
1	12.91	1.51	13.210
2	13.11	1.56	14.251
3	21.59	1.62	20.139
4	22.34	1.63	21.725
5	24.50	1.63	26.099
6	22.73	1.64	21.084
7	16.41		18.024
8	15.43	·	16.722
9	14.58	·	14.385
10	18.40	·	21.342
11	18.32		17.180
12	15.98	1.64	14.577

[a]Forecast origin is December 1967.

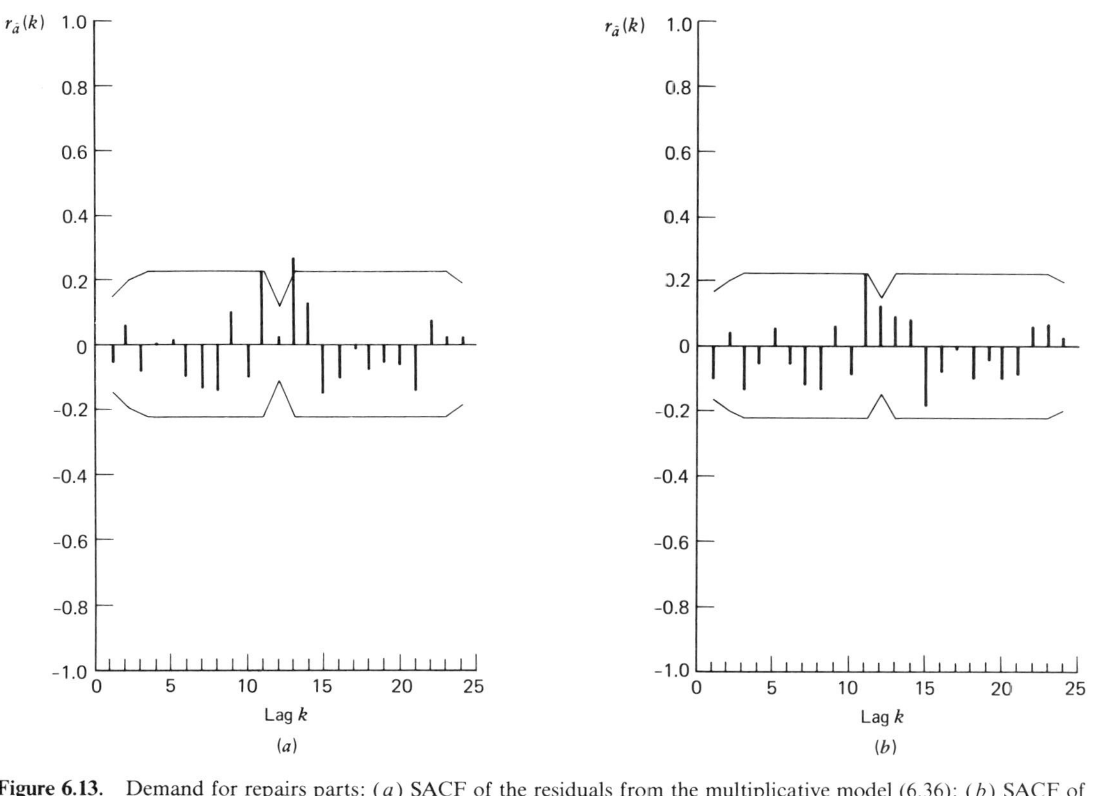

Figure 6.13. Demand for repairs parts: (*a*) SACF of the residuals from the multiplicative model (6.36); (*b*) SACF of the residuals from the nonmultiplicative model (6.37).

Combining (6.34) and (6.35) leads to a revised model

$$(1 - B)\ln z_t = (1 - \theta B)(1 - \Theta B^{12})a_t \tag{6.36}$$

This model is fitted and the new residual autocorrelations are given in Figure 6.13*a*. The lag 12 residual autocorrelation is negligible now. However, the autocorrelations at lags 11 and 13 are slightly outside the probability limits. This could be an indication that the model may not be multiplicative. Fitting the nonmultiplicative model

$$(1 - B)\ln z_t = \left(1 - \theta_1 B - \theta_{12} B^{12}\right)a_t \tag{6.37}$$

yields the estimates

$$\hat{\theta}_1 = .66(.08) \qquad \hat{\theta}_{12} = -.37(.07) \qquad \hat{\sigma}^2 = .021$$

The residual autocorrelations are given in Figure 6.13*b*; this model appears somewhat better than the one in (6.36).

6.8. SEASONAL ADJUSTMENT USING SEASONAL ARIMA MODELS

6.8.1. X-11-ARIMA

In Section 4.5 we discussed seasonal adjustment procedures, in particular the X-11 program. One of the drawbacks of X-11 is its use of asymmetric MA's to smooth the initial and final parts of the series. This implies that sometimes large revisions of the most recently adjusted data are necessary when new observations become available. X-11-ARIMA attempts to correct this drawback by adopting the following three-step strategy:

1. Obtain a seasonal ARIMA model for the series. The automatic option of this program selects one of the following three models: $(0,1,1)(0,1,1)_s$, $(0,2,2)(0,1,1)_s$, or $(2,1,2)(0,1,1)_s$; $s = 12$ for monthly series and $s = 4$ for quarterly series. If the decomposition is multiplicative, these models are applied to logarithmically transformed data.
2. Generate forecasts and backcasts to extend the observed series at both ends by one year.
3. Use the X-11 method to seasonally adjust the extended series.

6.8.2. Signal Extraction or Model-Based Seasonal Adjustment Methods

In this approach it is assumed that the decomposition is additive, $z_t = S_t + N_t$, or additive for some suitable transformation of z_t. The objective of model-based seasonal adjustment methods is to decompose the given series z_t into two mutually independent seasonal and nonseasonal components S_t and N_t. Details of this approach can be found in Box, Hillmer, and Tiao (1978); Burman (1979); Cleveland and Tiao (1976); Hillmer and Tiao (1982); Hillmer, Bell, and Tiao (1982); Pierce (1978); Tiao and Hillmer (1978); and Whittle (1963). We summarize the approach given in Hillmer, Bell, and Tiao (1982).

Suppose the components S_t and N_t can be represented by ARIMA models

$$\phi_S(B)S_t = \theta_S(B)b_t \qquad \text{and} \qquad \phi_N(B)N_t = \theta_N(B)c_t \tag{6.38}$$

where the pairs $\{\phi_S(B), \theta_S(B)\}$, $\{\phi_N(B), \theta_N(B)\}$, and $\{\phi_S(B), \phi_N(B)\}$ have no common roots, and b_t and c_t are independent white-noise sequences with variances σ_b^2 and σ_c^2, respectively. Cleveland (1972) has shown that the model for z_t is then given by

$$\phi(B)z_t = \theta(B)a_t \tag{6.39}$$

where a_t is a white-noise sequence with variance σ^2, where

$$\phi(B) = \phi_S(B)\phi_N(B) \tag{6.40}$$

and where $\theta(B)$ and σ^2 are to be determined from the equation

$$\sigma^2\frac{\theta(B)\theta(F)}{\phi(B)\phi(F)} = \sigma_b^2\frac{\theta_S(B)\theta_S(F)}{\phi_S(B)\phi_S(F)} + \sigma_c^2\frac{\theta_N(B)\theta_N(F)}{\phi_N(B)\phi_N(F)} \tag{6.41}$$

and $F = B^{-1}$. Furthermore, if all roots of $\phi_S(B) = 0$ and $\phi_N(B) = 0$ are on or outside the unit circle, the estimated components are given by

$$\hat{N}_t = \pi_N(B)z_t \qquad \text{and} \qquad \hat{S}_t = \pi_S(B)z_t \tag{6.42}$$

where

$$\pi_N(B) = \frac{\sigma_c^2\theta_N(B)\theta_N(F)}{\sigma^2\theta(B)\theta(F)}\phi_S(B)\phi_S(F)$$

and

$$\pi_S(B) = \frac{\sigma_b^2 \theta_S(B)\theta_S(F)}{\sigma^2 \theta(B)\theta(F)} \phi_N(B)\phi_N(F)$$

In practice, the S_t and N_t series are unobservable. Thus, without additional assumptions, the weight functions $\pi_S(B)$ and $\pi_N(B)$ and hence $\hat{S}_t$ and $\hat{N}_t$ cannot be computed. However, an estimate of model (6.39) can be obtained from the observable z_t series. Consequently, it is of interest to investigate to what extent the component models can be obtained from the observable z_t series. It turns out that if we allow some prior restrictions (see below), the components S_t and N_t can be estimated uniquely.

1. For monthly data, Hillmer, Bell, and Tiao (1982) take $\phi_S(B) = (1 + B + B^2 + \cdots + B^{11}) = (1 - B^{12})/(1 - B)$ and $\theta_S(B)$ of degree at most 11. This implies that the sum of S_t over any 12 consecutive months varies around zero. Since seasonal adjustment procedures restrict this sum to zero, this is a very reasonable assumption. This particular choice of $\phi_S(B)$ enables us to obtain $\phi_N(B)$ from

$$\phi(B) = \phi_S(B)\phi_N(B) \tag{6.43}$$

2. It remains to determine σ_b^2, σ_c^2, and the MA polynomials $\theta_S(B)$ and $\theta_N(B)$. Hillmer, Bell, and Tiao (1982) adopt a canonical decomposition and choose these components such that they satisfy (6.41) and (6.43) and furthermore minimize σ_b^2, the innovation variance for the seasonal component. This is a plausible assumption, since minimizing σ_b^2 selects a model for the seasonal component that is as deterministic as possible while remaining consistent with the data. The resulting polynomials, together with $\phi_S(B)$, $\phi(B)$, and σ^2, can be used to estimate $\pi_S(B)$. Application of this weight function to the original data yields a seasonally adjusted series.

It is also suggested that adjustments for outliers and trading-day variation be done at the modeling stage prior to seasonal adjustment.

APPENDIX 6. AUTOCORRELATIONS OF THE MULTIPLICATIVE $(0, d, 1)(1, D, 1)_{12}$ MODEL

The difference equation for this model is given by

$$w_t = \Phi w_{t-12} + a_t - \theta a_{t-1} - \Theta a_{t-12} + \theta\Theta a_{t-13} \tag{1}$$

Multiplying both sides by w_t and taking expectations leads to

$$\begin{aligned}\gamma_0 &= \Phi\gamma_{12} + \sigma^2 + \theta^2\sigma^2 - \Theta(\Phi - \Theta)\sigma^2 + \theta\Theta(-\Phi\theta + \theta\Theta)\sigma^2 \\ &= \Phi\gamma_{12} + \sigma^2\big[(1 + \theta^2) + \Theta(\Theta - \Phi)(1 + \theta^2)\big] \\ &= \Phi\gamma_{12} + \sigma^2(1 + \theta^2)[1 + \Theta(\Theta - \Phi)] \end{aligned} \tag{2}$$

Multiplying both sides of Equation (1) by w_{t-12} and taking expectations, we obtain

$$\gamma_{12} = \Phi\gamma_0 - \Theta\sigma^2 + \theta\Theta(-\theta)\sigma^2 = \Phi\gamma_0 - \Theta\sigma^2(1 + \theta^2) \tag{3}$$

Solving Equations (2) and (3), we get

$$\begin{aligned}\gamma_0 &= \sigma^2(1 + \theta^2)\frac{1 + \Theta^2 - 2\Phi\Theta}{1 - \Phi^2} \\ \gamma_{12} &= \sigma^2(1 + \theta^2)\left[\Phi - \Theta + \frac{\Phi(\Theta - \Phi)^2}{1 - \Phi^2}\right]\end{aligned} \tag{4}$$

Furthermore,

$$\begin{aligned}\gamma_1 &= E(w_t w_{t-1}) \\ &= \Phi\gamma_{11} - \theta\sigma^2 - \Theta E(a_{t-12}w_{t-1}) + \theta\Theta E(a_{t-13}w_{t-1}) \\ &= \Phi\gamma_{11} - \theta\sigma^2 + \theta\Theta(\Phi - \Theta)\sigma^2\end{aligned} \tag{5}$$

and

$$\gamma_{11} = E(w_t w_{t-11}) = \Phi\gamma_1 + \Theta\theta\sigma^2 \tag{6}$$

Solving Equations (5) and (6) leads to

$$\begin{aligned}\gamma_1 &= -\theta\sigma^2\left[1 + \frac{(\Theta - \Phi)^2}{1 - \Phi^2}\right] \\ \gamma_{11} &= \theta\sigma^2\left[\Theta - \Phi - \frac{\Phi(\Theta - \Phi)^2}{1 - \Phi^2}\right]\end{aligned} \tag{7}$$

It can also be seen that $\gamma_2 = \gamma_3 = \cdots = \gamma_{10} = 0$, $\gamma_{13} = \gamma_{11}$, and

$$\gamma_k = \Phi\gamma_{k-12} \qquad k > 13$$

Hence, the autocorrelation function is given by

$$\rho_k = \frac{\gamma_k}{\gamma_0} = \begin{cases} 1 & k = 0 \\ -\dfrac{\theta}{1 + \theta^2} & k = 1 \\ 0 & k = 2, \ldots, 10 \\ \dfrac{\theta}{1 + \theta^2} \dfrac{(\Theta - \Phi)(1 - \Phi\Theta)}{1 + \Theta^2 - 2\Phi\Theta} & k = 11 \\ \dfrac{-(\Theta - \Phi)(1 - \Phi\Theta)}{1 + \Theta^2 - 2\Phi\Theta} & k = 12 \\ \rho_{11} & k = 13 \\ \Phi\rho_{k-12} & k > 13 \end{cases}$$

CHAPTER 7

Relationships Between Forecasts from General Exponential Smoothing and Forecasts from ARIMA Time Series Models

General exponential smoothing forecast procedures for nonseasonal and seasonal series are discussed in Chapters 3 and 4. Forecasts from parametric time series models are considered in Chapters 5 and 6. In this chapter we explore the relationships between these two approaches.

7.1. PRELIMINARIES

7.1.1. General Exponential Smoothing

Regression models of the form

$$z_{n+j} = \sum_{i=1}^{m} \beta_i f_i(j) + \varepsilon_{n+j} = \mathbf{f}'(j)\boldsymbol{\beta} + \varepsilon_{n+j} \tag{7.1}$$

are considered in Chapter 3. There we assume that the forecast or fitting functions $\mathbf{f}(j) = [f_1(j), \dots, f_m(j)]'$ satisfy the difference equation $\mathbf{f}(j) = \mathbf{L}\mathbf{f}(j-1)$. Exponential, polynomial, and trigonometric functions follow such difference equations. In Section 3.5 we estimate the coefficients $\boldsymbol{\beta} = (\beta_1, \dots, \beta_m)'$ by discounted least squares, and choose the estimates $\hat{\boldsymbol{\beta}}_n$ such that

$$\sum_{j=0}^{n-1} \omega^j \left[z_{n-j} - \mathbf{f}'(-j)\boldsymbol{\beta}\right]^2$$

is minimized; $\omega = 1 - \alpha$ is a chosen discount coefficient. It is shown in (3.25) and (3.28) that in the steady state these estimates are given by

$$\hat{\boldsymbol{\beta}}_n = \mathbf{F}^{-1} \sum_{j=0}^{n-1} \omega^j \mathbf{f}(-j) z_{n-j} \tag{7.2}$$

where $\mathbf{F} = \sum_{j \geqslant 0} \omega^j \mathbf{f}(-j)\mathbf{f}'(-j)$. The forecasts are then obtained from

$$\hat{z}_n(l) = \mathbf{f}'(l)\hat{\boldsymbol{\beta}}_n \tag{7.3}$$

Whenever a new observation z_{n+1} becomes available, these forecasts can be updated by revising the parameter estimates according to

$$\hat{\boldsymbol{\beta}}_{n+1} = \mathbf{L}'\hat{\boldsymbol{\beta}}_n + \mathbf{F}^{-1}\mathbf{f}(0)\left[z_{n+1} - \hat{z}_n(1)\right] \tag{7.4}$$

This relationship reduces the computer storage requirements, since only the current estimates and the most recent forecast error have to be stored. It also simplifies the calculation in (7.2).

7.1.2. ARIMA Time Series Models

Forecasts from the autoregressive integrated moving average, ARIMA (p, d, q), model

$$\varphi(B) z_t = \theta(B) a_t \tag{7.5}$$

where $\varphi(B) = \phi(B)(1 - B)^d = 1 - \varphi_1 B - \cdots - \varphi_{p+d} B^{p+d}$ is the generalized autoregressive operator, are discussed in Chapters 5 and 6. In seasonal models this operator also includes seasonal differences and seasonal autoregressive polynomials. Since the expected value of a future error a_{n+j} is zero, the forecasts $z_n(l)$ from model (7.5) follow the difference equation

$$\varphi(B) z_n(l) = 0 \qquad l > q \tag{7.6}$$

For illustration, refer to the results and special examples in Sections 5.4 and

6.6. Here the backshift B operates on the lead time l. Now it can be shown that the general solution of the difference equation (7.6) is of the form

$$z_n(l) = \beta_1^{(n)} f_1^*(l) + \cdots + \beta_{p+d}^{(n)} f_{p+d}^*(l) \qquad l > q - p - d \tag{7.7}$$

The forecast functions $f_i^*(l)$ $(1 \leqslant i \leqslant p + d)$ satisfy the difference equation $\varphi(B) f_i^*(l) = 0$. Thus the forecast functions $\mathbf{f}^*(l) = [f_1^*(l), \ldots, f_{p+d}^*(l)]'$ are functions of l that depend only on the form of the autoregressive operator in model (7.5). For example, if the roots G_i^{-1} $(|G_i^{-1}| > 1)$ of $\phi(B) = (1 - G_1 B)(1 - G_2 B) \cdots (1 - G_p B) = 0$ are distinct, the solutions of $\phi(B) f_i^*(l) = 0$ are given by exponentials $f_i^*(l) = G_i^l$ for $1 < i < p$. This follows since $(1 - G_i B) f_i^*(l) = (1 - G_i B) G_i^l = G_i^l - G_i^l = 0$. If some of the roots are equal (let us assume that there are k roots at G^{-1}), then the solutions corresponding to these roots are $G^l, lG^l, \ldots, l^{k-1}G^l$. If the autoregressive operator has d unit roots, the solutions of $(1 - B)^d f_i^*(l) = 0$ are given by the polynomials $f_1^*(l) = 1$, $f_2^*(l) = l, \ldots, f_d^*(l) = l^{d-1}$.

The solution in (7.7) holds for $l > q - p - d$. For this reason it is called the *eventual forecast function*. If the order of the moving average operator on the right-hand side of the model in (7.5) is the same as the order of the generalized autoregressive operator (i.e., $q = p + d$), the eventual forecast function holds for all lead times $l > 0$.

Examples of Eventual Forecast Functions

ARIMA (1, 0, 0) model. The forecasts satisfy the difference equation $(1 - \phi B) z_n(l) = 0$ for $l > 0$. The solution is given by

$$z_n(l) = \beta_1^{(n)} \phi^l \qquad l > -1 \tag{7.8}$$

In fact $\beta_1^{(n)} = z_n$, since the solution (7.8) holds also for $l = 0$. Thus Equation (7.8) passes through the last observation z_n.

ARIMA (0, 1, 1) model. The solution of the difference equation $(1 - B) z_n(l) = 0$ for $l > 1$ is given by

$$z_n(l) = \beta_1^{(n)} \qquad l > 0 \tag{7.9}$$

Here $\beta_1^{(n)} = z_n(1)$ is an exponentially weighted average of past observations; see Equation (5.76).

ARIMA (0, 2, 2) model. The solution of $(1 - B)^2 z_n(l) = 0$ for $l > 2$ is

$$z_n(l) = \beta_1^{(n)} + \beta_2^{(n)} l \qquad l > 0 \tag{7.10}$$

All forecasts lie on a straight line which is determined by the first two forecasts $z_n(1)$ and $z_n(2)$.

ARIMA (1, 1, 1) model. The solution of $(1 - \phi B)(1 - B) z_n(l) = 0$ for $l > 1$ is given by

$$z_n(l) = \beta_1^{(n)} + \beta_2^{(n)} \phi^l \qquad l > -1 \tag{7.11}$$

The eventual forecast function passes through the last observation z_n and the first forecast $z_n(1)$.

Seasonal $(0, 0, 0)(0, 1, 1)_{s=12}$ model. In this case the forecasts satisfy the difference equation $(1 - B^{12}) z_n(l) = 0$ for $l > 12$. Thus the eventual forecast function [see Eq. (6.26)] can be written as

$$z_n(l) = z_n(r, m) = \beta_{0m}^{(n)} \qquad l > 0 \tag{7.12}$$

Here we have expressed l as $l = 12r + m$ $(1 \leqslant m \leqslant 12)$, where r is the year and m denotes the month. The coefficients $\beta_{0m}^{(n)} = z_n(m)$ are 12 seasonal constants, which are seasonal exponentially weighted averages of past observations; see Equation (6.24). Alternatively, we can express the eventual forecast function (7.12) in terms of trigonometric functions:

$$z_n(l) = \beta_0^{(n)} + \sum_{j=1}^{5} \left(\beta_{1j}^{(n)} \sin \frac{2\pi j}{12} l + \beta_{2j}^{(n)} \cos \frac{2\pi j}{12} l \right) + \beta_{26}(-1)^l \tag{7.13}$$

Seasonal multiplicative $(0, 1, 1)(0, 1, 1)_{s=12}$ model. The solution of the difference equation $(1 - B)(1 - B^{12}) z_n(l) = 0$ for $l > 13$ is given in (6.29):

$$z_n(l) = z_n(r, m) = \beta_{0m}^{(n)} + \beta_{1*}^{(n)} r \qquad l > 0 \tag{7.14}$$

Equivalently, the eventual forecast function can be written as

$$z_n(l) = \beta_0^{(n)} + \beta_1^{(n)} l + \sum_{j=1}^{5} \left(\beta_{1j}^{(n)} \sin \frac{2\pi j}{12} l + \beta_{2j}^{(n)} \cos \frac{2\pi j}{12} l \right) + \beta_{26}(-1)^l \tag{7.15}$$

The coefficients in (7.14) and (7.15) are functions of $z_1, z_2, \ldots, z_n$, and can be determined from the first 13 forecasts.

Updating the Coefficients in the Eventual Forecast Function

The coefficients $\boldsymbol{\beta}^{(n)} = (\beta_1^{(n)}, \ldots, \beta_{p+d}^{(n)})'$ in the eventual forecast function (7.7) can be obtained from the initial conditions of the difference equation in (7.6). They depend on the past history up to time n, are fixed for all forecast lead times l, and are easily updated as each new observation becomes available. A convenient updating relationship for the forecasts is given in (5.85). There it is shown that

$$z_{n+1}(l) = z_n(l+1) + \psi_l[z_{n+1} - z_n(1)] \tag{7.16}$$

where ψ_l is the coefficient of B^l in $\psi(B) = \theta(B)/\varphi(B)$. Substitution of the eventual forecast function (7.7) into (7.16) leads to the equation

$$\mathbf{f}^*(l)'\boldsymbol{\beta}^{(n+1)} = \mathbf{f}^*(l+1)'\boldsymbol{\beta}^{(n)} + \psi_l[z_{n+1} - z_n(1)] \qquad l > q - p - d \tag{7.17}$$

By expressing this equation for $l, l+1, \ldots, l+p+d-1$, where any $l > q-p-d$ can be chosen, we can rewrite it as a difference equation for the coefficients:

$$\boldsymbol{\beta}^{(n+1)} = \mathbf{F}_l^{*-1}\mathbf{F}_{l+1}^*\boldsymbol{\beta}^{(n)} + \mathbf{g}^*[z_{n+1} - z_n(1)] \tag{7.18}$$

where

$$\mathbf{F}_l^* = \begin{bmatrix} f_1^*(l) & \cdots & f_{p+d}^*(l) \\ f_1^*(l+1) & \cdots & f_{p+d}^*(l+1) \\ \vdots & & \vdots \\ f_1^*(l+p+d-1) & \cdots & f_{p+d}^*(l+p+d-1) \end{bmatrix}$$

and

$$\mathbf{g}^* = \mathbf{F}_l^{*-1}\begin{bmatrix} \psi_l \\ \psi_{l+1} \\ \vdots \\ \psi_{l+p+d-1} \end{bmatrix}$$

Special cases of these updating equations are discussed in Chapters 5 and 6.

7.2. RELATIONSHIPS AND EQUIVALENCE RESULTS

Comparing (1) the forecast function for general exponential smoothing in (7.3) with the eventual forecast function of ARIMA models in (7.7) and (2) the respective updating equations for their coefficients in Equations (7.4) and (7.18), we notice certain similarities.

1. It is possible to find an autoregressive operator $\varphi(B)$ such that the forecast functions $f_i^*(l)$ coincide with the m fitting functions $f_i(l)$ in general exponential smoothing. For this, the autoregressive order has to be m, and the autoregressive coefficients φ_j in $\varphi_m(B) = 1 - \varphi_1 B - \cdots - \varphi_m B^m$ have to be chosen such that

$$\varphi_m(B) f_i(l) = \prod_{j=1}^{m} (1 - G_j B) f_i(l) = 0 \qquad 1 \leqslant i \leqslant m \qquad (7.19)$$

   For example, the autoregressive operator that leads to polynomial fitting functions is given by $(1 - B)^m$. Additional examples relating autoregressive operators and specific forecast functions are given in Equations (7.8) to (7.15).
2. For the eventual forecast function in (7.7) to hold for all $l > 0$ [as in general exponential smoothing; see Eq. (7.3)], we must choose the moving average operator in the ARIMA model (7.5) of order m [i.e., the same order as the autoregressive operator $\varphi_m(B)$].
3. So far we have established the equivalence of the forecast functions in exponential smoothing and certain ARIMA models by determining the form of the autoregressive operator and by requiring that the moving average operator be of the same order. It remains to choose the m moving average coefficients in $\theta_m(B)$, such that the coefficients $\hat{\boldsymbol{\beta}}_n$ in exponential smoothing [see Eq. (7.4)] and $\boldsymbol{\beta}^{(n)}$ in the eventual forecast function [see Eq. (7.18)] are the same. In exponential smoothing, the coefficients depend, apart from the fitting functions and past observations, only on the discount coefficient ω. However, the coefficients in the eventual forecast function depend on all autoregressive and moving average coefficients. It can be shown that in order to get the same coefficients the moving average operator must be of the form

$$\theta_m(B) = \prod_{j=1}^{m} \left(1 - \frac{\omega}{G_j} B\right) \qquad (7.20)$$

   where G_j^{-1} are the roots of $\varphi_m(B) = 0$.

This leads to the equivalence result expressed in the following theorem.

Theorem. Forecasts from general exponential smoothing with m fitting functions $f_i(l)$ and discount coefficient ω are the same as those from the equivalent ARIMA model

$$\prod_{j=1}^{m}(1 - G_j B)z_t = \prod_{j=1}^{m}\left(1 - \frac{\omega}{G_j}B\right)a_t \tag{7.21}$$

where $\varphi(B) = \prod_{j=1}^{m}(1 - G_j B)$ is chosen such that

$$\varphi(B)f_i(l) = 0 \qquad 1 \leqslant i \leqslant m$$

Proof. See Appendix 7.

In general, the roots G_j^{-1} of the autoregressive operator can be either real or complex numbers. Since the autoregressive coefficients φ_j are real, the complex roots have to occur in conjugate pairs.

Also note that for models with stationary autoregressive components with $|G_j| < 1$, the discount coefficient ω has to be restricted. As shown in Section 3.5, we have to restrict the discount coefficient such that $|\omega| < G_j^2$. This assures the convergence of the elements in

$$\mathbf{F} = \sum_{j \geqslant 0} \omega^j \mathbf{f}(-j)\mathbf{f}'(-j)$$

and implies the invertibility of the equivalent ARIMA model in (7.21).

Furthermore, the complex roots $c^{-1}e^{ix}$ and $c^{-1}e^{-ix}$, where $|c| \leqslant 1$ is a real number, lead to forecast functions $f_1(l) = c^l e^{-ixl}$ and $f_2(l) = c^l e^{ixl}$. Alternatively, since

$$\begin{bmatrix} c^l e^{-ixl} \\ c^l e^{ixl} \end{bmatrix} = \begin{bmatrix} -i & 1 \\ i & 1 \end{bmatrix} \begin{bmatrix} c^l \sin xl \\ c^l \cos xl \end{bmatrix}$$

these forecast functions can be expressed in terms of discounted trigonometric functions $f_1^*(l) = c^l \sin xl$ and $f_2^*(l) = c^l \cos xl$. Also, since this is a linear transformation, $\varphi(B)f_i^*(l) = 0$.

The case when all roots of the autoregressive operator lie on the unit circle is interesting, since simplifying operators such as $(1 - B)^d$ and $(1 - B^s)^D$ lead to the polynomial and trigonometric forecast functions usually considered in general exponential smoothing.

Corollary. Forecasts from general exponential smoothing with a discount coefficient ω and m fitting functions $f_i(l)$, which satisfy $\varphi(B)f_i(l) = 0$ for $1 \leqslant i \leqslant m$, and where all roots of $\varphi(B) = 0$ lie on the unit circle, are the same as those from the equivalent ARIMA model,

$$\varphi(B)z_t = \varphi(\omega B)a_t \tag{7.22}$$

Proof. The roots of $\varphi(B) = 0$ are either real, $+1$ or -1, or complex in conjugate pairs $e^{\pm ix}$. Therefore, $1 - (\omega/G)B = 1 - \omega GB$ for real roots at ± 1, and

$$\left(1 - \frac{\omega}{G_1}B\right)\left(1 - \frac{\omega}{G_2}B\right) = (1 - \omega e^{-ix}B)(1 - \omega e^{ix}B)$$

$$= (1 - \omega G_2 B)(1 - \omega G_1 B)$$

for complex pairs $G_1 = e^{ix}$ and $G_2 = e^{-ix}$. Hence the result follows from the above theorem.

7.2.1. Illustrative Examples

Example 1. For kth *order exponential smoothing* [$m = k$ and $f_i(l) = l^{i-1}/(i-1)!$ for $1 \leqslant i \leqslant k$], the equivalent ARIMA $(0, k, k)$ model is given by

$$(1 - B)^k z_t = (1 - \omega B)^k a_t \tag{7.23}$$

For *simple exponential smoothing* ($k = 1$), the equivalent model is

$$(1 - B) z_t = (1 - \omega B) a_t \tag{7.24}$$

For an early proof of this result, see Muth (1960).

For *double exponential smoothing* ($k = 2$), the equivalent model is

$$(1 - B)^2 z_t = (1 - \omega B)^2 a_t \tag{7.25}$$

For *triple exponential smoothing* ($k = 3$), the equivalent model is

$$(1 - B)^3 z_t = (1 - \omega B)^3 a_t \tag{7.26}$$

Example 2. For the *12-point sinusoidal model* [$m = 3$; $f_1(l) = 1$, $f_2(l) = \sin(2\pi l/12)$, $f_3(l) = \cos(2\pi l/12)$], the equivalent ARIMA model is given by

$$(1 - B)(1 - \sqrt{3}B + B^2) z_t = (1 - \omega B)(1 - \omega\sqrt{3}B + \omega^2 B^2) a_t \tag{7.27}$$

Example 3. For the exponential smoothing model with $m = 6$ fitting functions $f_1(l) = 1$, $f_2(l) = l$, $f_3(l) = \sin(2\pi l/12)$, $f_4(l) = \cos(2\pi l/12)$, $f_5(l) = l\sin(2\pi l/12)$, $f_6(l) = l\cos(2\pi l/12)$, the equivalent ARIMA model is given by

$$(1 - B)^2(1 - \sqrt{3}\,B + B^2)^2 z_t = (1 - \omega B)^2(1 - \omega\sqrt{3}\,B + \omega^2 B^2)^2 a_t \tag{7.28}$$

Example 4. For the additive seasonal trend model in Section 4.3,

$$z_{n+l} = \beta_0 + \beta_1 l + \sum_{i=1}^{s-1} \delta_i \mathrm{IND}_{li} + \varepsilon_{n+l}$$

where IND_{li} ($i = 1, \ldots, s - 1$) are $s - 1$ seasonal indicators, the equivalent ARIMA model is given by

$$(1 - B)(1 - B^s) z_t = (1 - \omega B)(1 - \omega^s B^s) a_t \tag{7.29}$$

The fitting functions we have discussed so far arise from autoregressive operators with roots on the unit circle. In the following examples the fitting functions correspond to stationary autoregressive operators.

Example 5. For exponential smoothing with $m = 1$ fitting function $f(l) = \phi^l$, where $|\phi| < 1$, the equivalent time series model is given by

$$(1 - \phi B) z_t = \left(1 - \frac{\omega}{\phi} B\right) a_t \tag{7.30}$$

Since $\mathbf{F} = \Sigma \omega^j \mathbf{f}(-j)\mathbf{f}'(-j) = \Sigma \omega^j \phi^{-2j} = \Sigma(\omega\phi^{-2})^j$ must converge, we have to restrict the discount coefficient ω such that $|\omega| < \phi^2$. This restriction implies that the model in (7.30) is invertible.

Example 6. For exponential smoothing with $m = 2$ fitting functions, $f_1(l) = 1$ and $f_2(l) = \phi^l$, with $|\phi| < 1$, the equivalent time series model is of the form

$$(1 - \phi B)(1 - B) z_t = \left(1 - \frac{\omega}{\phi} B\right)(1 - \omega B) a_t \tag{7.31}$$

7.3. INTERPRETATION OF THE RESULTS

The equivalence results imply that the shortcomings of the exponential smoothing forecast procedures are essentially threefold:

1. In exponential smoothing, the fitting functions are chosen by visual inspection of the data. These fitting functions then determine the

autoregressive operator in the corresponding ARIMA model. However, as discussed by Box and Jenkins (1976, p. 166), visual inspection of the time series alone can lead to incorrectly specified models. Thus more reliable identification tools such as the sample autocorrelation and sample partial autocorrelation functions must be considered.

2. Exponentially discounted least squares forces the moving average operator to be of a very restricted form. It is determined by the autoregressive operator and the discount coefficient ω. For example, if $\varphi(B)$ has all its roots on the unit circle, the moving average operator is simply given by $\varphi(\omega B)$.
3. Furthermore, in exponential smoothing it is usually assumed that the discount coefficient ω, or the smoothing constant $\alpha = 1 - \omega$, is known. In practice, however, this constant is rarely given. Thus, Brown (1962) has suggested that this coefficient be chosen arbitrarily, such that ω^m lies somewhere between .70 and .95. This implies a very special moving average operator.

Exponential smoothing forecasts thus arise as forecasts from certain restricted ARIMA models. They will lead to minimum mean square error forecasts only if the underlying process that has generated the data is within this subclass of restricted ARIMA processes. Users of exponential smoothing thus tacitly assume that the stochastic processes that occur in the real world are from this limited class.

Consider, for example, triple exponential smoothing, which according to (7.26) implies a very restricted ARIMA (0, 3, 3) model. It is difficult to find a data series for which a third successive difference is necessary. Even if such a series can be found, it is very unlikely that the moving average operator is of this particular form.

Study of actual time series gives little empirical support for these restricted models. In fact, many time series that have been modeled by the three-stage iterative procedure discussed in Chapters 5 and 6 lead to quite different models. We believe that the data themselves should determine the form of the model and the values of its parameters.

APPENDIX 7. PROOF OF THE EQUIVALENCE THEOREM

The autoregressive operator and the order of the moving average operator are chosen such that the forecast functions in (7.7) and the fitting functions in exponential smoothing in (7.3) coincide. It remains to show that the coefficients in these equations are also the same.

Since $\mathbf{f}^*(l) = \mathbf{f}(l)$, and thus $\mathbf{f}^*(l) = \mathbf{L}\mathbf{f}^*(l-1)$, it follows that

$$\mathbf{F}_{l+1}^* = \begin{bmatrix} \mathbf{f}^*(l+1)' \\ \vdots \\ \mathbf{f}^*(l+m)' \end{bmatrix} = \begin{bmatrix} \mathbf{f}^*(l)'\mathbf{L}' \\ \vdots \\ \mathbf{f}^*(l+m-1)'\mathbf{L}' \end{bmatrix} = \mathbf{F}_l^*\mathbf{L}'$$

This implies that $\mathbf{F}_l^{*-1}\mathbf{F}_{l+1}^* = \mathbf{L}'$, and thus the equality of the transition matrices in the updating equations (7.4) and (7.18).

It remains to prove that the adjustment vectors $\mathbf{g} = \mathbf{F}^{-1}\mathbf{f}(0)$ and $\mathbf{g}^* = \mathbf{F}_1^{*-1}\boldsymbol{\psi}$, where $\boldsymbol{\psi} = (\psi_1, \ldots, \psi_m)'$ are the ψ weights in the equivalent ARIMA model, are the same. To show this result, we distinguish between distinct and multiple roots of the autoregressive operator. Here only the proof for distinct roots is given. We follow an approach by Ledolter and Box (1978). The proof for the multiple-roots case can be found in papers by Cogger (1974), McKenzie (1974), and Godolphin and Harrison (1975), who show the equivalence for models in which the autoregressive operator has multiple roots at 1. Additional discussion is given by McKenzie (1976).

Dobbie (1963) has shown that for distinct exponential fitting functions $\mathbf{f}(l) = (G_1^l, \ldots, G_m^l)'$ the elements in the adjustment vector $\mathbf{g} = \mathbf{F}^{-1}\mathbf{f}(0)$ are given by

$$g_i = \left(1 - \omega G_i^{-2}\right) \prod_{\substack{k=1 \\ k \neq i}}^{m} \frac{1 - \omega G_i^{-1} G_k^{-1}}{1 - G_k G_i^{-1}} \tag{1}$$

For the ARIMA model in (7.21), the vector $\mathbf{g}^*$ in (7.18) is given by

$$\mathbf{g}^* = \begin{bmatrix} G_1 & G_2 & \cdots & G_m \\ G_1^2 & G_2^2 & \cdots & G_m^2 \\ \vdots & \vdots & & \vdots \\ G_1^m & G_2^m & \cdots & G_m^m \end{bmatrix}^{-1} \begin{bmatrix} \psi_1 \\ \psi_2 \\ \vdots \\ \psi_m \end{bmatrix} \tag{2}$$

where the ψ weights are the coefficients in the expansion

$$\psi(B) = \prod_{k=1}^{m} \frac{1 - \omega G_k^{-1} B}{1 - G_k B} \tag{3}$$

To prove the equality of **g** and **g***, we have to show that

$$\mathbf{d} = \begin{bmatrix} 1 & 1 & \cdots & 1 \\ G_1 & G_2 & \cdots & G_m \\ \vdots & \vdots & & \vdots \\ G_1^{m-1} & G_2^{m-1} & \cdots & G_m^{m-1} \end{bmatrix}^{-1} \begin{bmatrix} \psi_1 \\ \psi_2 \\ \vdots \\ \psi_m \end{bmatrix} = \begin{bmatrix} G_1 g_1 \\ G_2 g_2 \\ \vdots \\ G_m g_m \end{bmatrix} \tag{4}$$

where the g_i $(1 \leqslant i \leqslant m)$ are given in Equation (1). The matrix on the left is a Vandermonde matrix; the elements a_{ij} of its inverse **A** are the coefficients in the expansion

$$P_i(B) = \prod_{\substack{k=1 \\ k \neq i}}^{m} \frac{B - G_k}{G_i - G_k} = a_{i1}B^0 + a_{i2}B^1 + \cdots + a_{im}B^{m-1} \tag{5}$$

This result can be shown by multiplying **A** with its inverse. For details, see Appendix 2.1 of Ledolter (1975).

Thus the elements d_i in $\mathbf{d} = (d_1, \ldots, d_m)'$ in Equation (4) are the coefficients of B^0 in

$$B^{-1}P_i(B^{-1})\psi(B) = \frac{1}{B^m(1 - G_iB)} \frac{\prod_{k=1}^{m}(1 - \omega G_k^{-1}B)}{\prod_{k \neq i}(G_i - G_k)} \tag{6}$$

To establish the equality in (4), we have to show that

$$G_i g_i \prod_{k \neq i}(G_i - G_k) = G_i^m(1 - \omega G_i^{-2}) \prod_{k \neq i}(1 - \omega G_i^{-1}G_k^{-1}) \tag{7}$$

is the coefficient of B^0 in

$$\begin{aligned} V_i(B) &= \frac{\prod_{k=1}^{m}(1 - \omega G_k^{-1}B)}{B^m(1 - G_iB)} \\ &= (1 - G_iB)^{-1} \prod_{k=1}^{m}(B^{-1} - \omega G_k^{-1}) \\ &= (1 + G_iB + G_i^2B^2 + \cdots) \\ &\quad \times (B^{-m} + c_1B^{-m+1} + \cdots + c_{m-1}B^{-1} + c_m) \end{aligned} \tag{8}$$

where the coefficients c_j are given by

$$c_1 = (-\omega)\sum_{k=1}^{m} G_k^{-1}$$

$$c_2 = (-\omega)^2 \sum_{k<l} G_k^{-1}G_l^{-1}$$

$$c_3 = (-\omega)^3 \sum_{k<l<r} G_k^{-1}G_l^{-1}G_r^{-1}$$

$$\vdots$$

$$c_m = (-\omega)^m \prod_{k=1}^{m} G_k^{-1} \tag{9}$$

Thus the coefficient of B^0 in $V_i(B)$ is given by

$$G_i^m + G_i^{m-1}c_1 + \cdots + G_i c_{m-1} + c_m \tag{10}$$

On the other hand, the expression in (7) can be written as

$$G_i^m\left(1 - \omega G_i^{-2}\right)\left(1 + e_1^{(i)} + \cdots + e_{m-1}^{(i)}\right) \tag{11}$$

where the coefficients $e_j^{(i)}$ $(1 \leqslant j \leqslant m-1)$ are given by

$$e_1^{(i)} = (-\omega)G_i^{-1}\sum_{k\neq i} G_k^{-1}$$

$$e_2^{(i)} = (-\omega)^2 G_i^{-2} \sum_{\substack{k<l \\ k,l\neq i}} G_k^{-1}G_l^{-1}$$

$$e_3^{(i)} = (-\omega)^3 G_i^{-3} \sum_{\substack{k<l<r \\ k,l,r\neq i}} G_k^{-1}G_l^{-1}G_r^{-1}$$

$$\vdots$$

$$e_{m-1}^{(i)} = (-\omega)^{m-1} G_i^{-(m-1)} \prod_{k\neq i} G_k^{-1} \tag{12}$$

The relationship between the coefficients in Equations (9) and (12) is given by

$$
\begin{aligned}
c_1 &= G_i e_1^{(i)} - \omega G_i^{-1} \\
c_2 &= G_i^2 e_2^{(i)} - \omega e_1^{(i)} \\
c_3 &= G_i^3 e_3^{(i)} - \omega G_i e_2^{(i)} \\
&\vdots \\
c_{m-1} &= G_i^{m-1} e_{m-1}^{(i)} - \omega G_i^{m-3} e_{m-2}^{(i)} \\
c_m &= -\omega G_i^{m-2} e_{m-1}^{(i)}
\end{aligned}
\tag{13}
$$

Substituting these expressions into Equation (10) leads to

$$
\begin{aligned}
&G_i^m + G_i^{m-1} c_1 + \cdots + G_i c_{m-1} + c_m \\
&\quad = G_i^m \left(1 - \omega G_i^{-2}\right)\left(1 + e_1^{(i)} + \cdots + e_{m-1}^{(i)}\right)
\end{aligned}
$$

which is the same as (11). This shows that $\mathbf{g} = \mathbf{g}^*$ and completes the proof.

CHAPTER 8

Special Topics

Several advanced forecast techniques are discussed in this chapter. In Section 8.1 we consider the transfer function model that describes the dynamic relationship between an input time series X_t and an output time series Y_t; this relationship can be used to improve the forecast of a future Y_t. Section 8.2 illustrates the modeling of interventions and the adjustment of outliers in time series. State space models, Kalman filtering, and Bayesian forecasting are described in Section 8.3. Adaptive filtering is discussed in Section 8.4, and methods for forecast evaluation, comparison, and control are given in Section 8.5.

8.1. TRANSFER FUNCTION ANALYSIS

Discussions in the last few chapters concentrated on forecast procedures that relate a single variable to its own past. In this section we relate a variable not only to its own past, but also to the present and past of other variables. The influence of one variable on another can be spread over several time periods. Instantaneous and lagged effects of an input variable X_t on an output variable Y_t can be represented by a model of the form

$$Y_t = v_0 X_t + v_1 X_{t-1} + v_2 X_{t-2} + \cdots = v(B) X_t \tag{8.1}$$

where $v(B) = v_0 + v_1 B + v_2 B^2 + \cdots$ is called the *transfer function*, and the coefficients $v_0, v_1, \ldots$ are called the *impulse response weights*. In theory, this model involves an infinite number of coefficients, which would make estimation impossible. One alternative is to use a truncated form,

$$Y_t = v_0 X_t + v_1 X_{t-1} + \cdots + v_h X_{t-h}$$

where h is to be chosen such that the lagged effects beyond h are negligible. However, this choice is usually difficult and can be avoided by writing $v(B)$ as a rational function:

$$v(B) = \frac{\omega(B)B^b}{\delta(B)} \tag{8.2}$$

The operators $\omega(B) = \omega_0 - \omega_1 B - \cdots - \omega_s B^s$ and $\delta(B) = 1 - \delta_1 B - \cdots - \delta_r B^r$ are polynomials in B, and b is a parameter representing the delay between the variables. Furthermore, it is assumed that the roots of $\delta(B) = 0$ are on or outside the unit circle. The relation between the coefficients v_k and the parameters $\boldsymbol{\omega} = (\omega_0, \ldots, \omega_s)'$, $\boldsymbol{\delta} = (\delta_1, \ldots, \delta_r)'$ and b can be obtained by equating the coefficients of B^k in

$$\delta(B)v(B) = \omega(B)B^b \tag{8.3}$$

Some simple cases are given here for illustration.

(a) $v(B) = \omega_0 - \omega_1 B$; $v_0 = \omega_0$, $v_1 = -\omega_1$, $v_k = 0 \quad k \geqslant 2$

(b) $v(B) = (\omega_0 - \omega_1 B)B^2$; $b = 2$, $v_0 = v_1 = 0$, $v_2 = \omega_0$, $v_3 = -\omega_1$, $v_k = 0 \quad k \geqslant 4$

(c) $v(B) = (\omega_0 - \omega_1 B)/(1 - \delta B)$
$= (\omega_0 - \omega_1 \mathrm{B})(1 + \delta \mathrm{B} + \delta^2 \mathrm{B}^2 + \ldots)$
$= \omega_0 + (\delta\omega_0 - \omega_1)B + (\delta\omega_0 - \omega_1)\delta B^2 + (\delta\omega_0 - \omega_1)\delta^2 B^3 + \cdots$
Hence, $v_0 = \omega_0$, $v_1 = \omega_0\delta - \omega_1$, $v_k = \delta^{k-1}v_1 \quad k \geqslant 1$

The impulse response weights v_k, which correspond to the transfer functions in cases (a) to (c), are shown in Figure 8.1. These special cases show that B^b determines the lag between X and Y and $\delta(B)$ determines the exponential decay.

So far we have assumed that Y_t is completely determined by X_t and its past. Usually such a relationship is masked by added noise N_t, which itself may be autocorrelated. Hence we should consider models of the form

$$Y_t = \frac{\omega(B)B^b}{\delta(B)} X_t + N_t \tag{8.4}$$

where N_t is assumed to be uncorrelated with X_t. In this model it is supposed that the past X_t's influence future Y_t's, but not vice versa. Such models are referred to as *transfer function-noise models*; they are similar to the distributed lag models briefly introduced in Section 2.12. In several respects

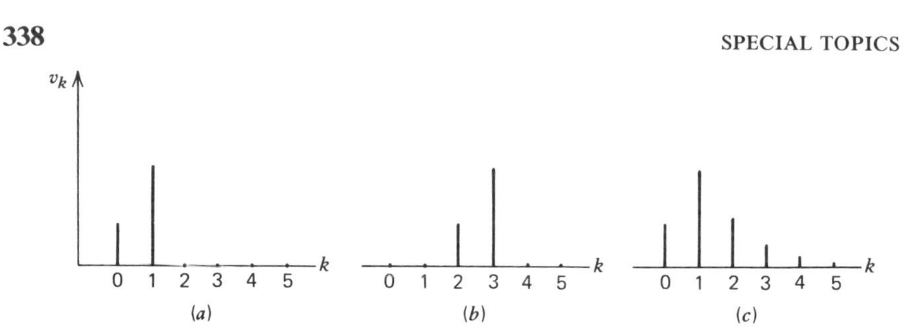

Figure 8.1. Impulse response weights: (a) $v(B) = \omega_0 - \omega_1 B$; ($b$) $v(B) = (\omega_0 - \omega_1 B)B^2$; ($c$) $v(B) = (\omega_0 - \omega_1 B)/(1 - \delta B)$.

transfer function-noise models are generalizations of the usual regression models, since they allow for (1) correlated errors, (2) dynamic relationships between output and input variables, and (3) stochastic inputs.

8.1.1. Construction of Transfer Function-Noise Models

The ACF plays a significant role in the specification of time series models. In the study of the relationship between two time series, a function referred to as the *cross-correlation function* becomes important.

Cross-Correlation Function (CCF)

Suppose that the bivariate process (X_t, Y_t) is such that $x_t = \nabla^d X_t$ and $y_t = \nabla^d Y_t$ form a stationary process. This means that (1) $E(x_t) = \mu_x$, $E(y_t) = \mu_y$, $V(x_t) = \sigma_x^2$ and $V(y_t) = \sigma_y^2$ are constants; (2) the autocovariances $\gamma_x(k) = E(x_t - \mu_x)(x_{t+k} - \mu_x)$ and $\gamma_y(k) = E(y_t - \mu_y)(y_{t+k} - \mu_y)$ depend only on the lag k and not on t; and (3) the cross-covariances between x and y at lag k, given by

$$\gamma_{xy}(k) = E(x_t - \mu_x)(y_{t+k} - \mu_y) \qquad k = 0, 1, 2, \ldots \tag{8.5}$$

and those between y and x at lag k, given by

$$\gamma_{yx}(k) = E(y_t - \mu_y)(x_{t+k} - \mu_x) \qquad k = 0, 1, 2, \ldots \tag{8.6}$$

are not time-dependent.

Since

$$\gamma_{xy}(k) = E(x_t - \mu_x)(y_{t+k} - \mu_y) = E(y_{t+k} - \mu_y)(x_t - \mu_x) = \gamma_{yx}(-k)$$

we consider only the function $\gamma_{xy}(k)$ for $k = 0, \pm 1, \pm 2, \ldots$. From this we define the cross-correlation between x and y at lag k as

$$\rho_{xy}(k) = \frac{\gamma_{xy}(k)}{\sigma_x \sigma_y} \qquad k = 0, \pm 1, \pm 2, \ldots \tag{8.7}$$

Their estimates from a sample of size n are given by

$$r_{xy}(k) = \frac{c_{xy}(k)}{s_x s_y} \qquad k = 0, \pm 1, \pm 2, \ldots \tag{8.8}$$

where

$$c_{xy}(k) = \begin{cases} n^{-1} \sum_{t=1}^{n-k} (x_t - \bar{x})(y_{t+k} - \bar{y}) & k = 0, 1, 2, \ldots \\ n^{-1} \sum_{t=1-k}^{n} (x_t - \bar{x})(y_{t+k} - \bar{y}) & k = 0, -1, -2, \ldots \end{cases} \tag{8.9}$$

$$s_x^2 = n^{-1} \sum_{t=1}^{n} (x_t - \bar{x})^2 \qquad s_y^2 = n^{-1} \sum_{t=1}^{n} (y_t - \bar{y})^2$$

and $\bar{x}$ and $\bar{y}$ are the sample means of the respective series.

In general, the variances and covariances of the sample cross-correlations are not easy to calculate. However, in certain special cases some useful approximations are available [Bartlett (1955)]. Two such cases are considered here.

1. If x_t is a white-noise sequence, y_t is an autocorrelated series with autocorrelations $\rho_y(k)$, and if y_t and x_t are uncorrelated at all lags [$\rho_{xy}(k) = 0$ for all k], then

$$V\left[r_{xy}(k)\right] \cong (n - k)^{-1} \tag{8.10}$$

Furthermore, the correlation between $r_{xy}(k)$ and $r_{xy}(k + l)$ is given by

$$\rho\left[r_{xy}(k), r_{xy}(k + l)\right] \cong \rho_y(l) \tag{8.11}$$

2. If both x_t and y_t are white-noise sequences and x_t and y_t are uncorrelated, then the correlation given in (8.11) is zero.

Specification of the Transfer Function Model

From experience with regression modeling we might be tempted to specify the transfer function $v(B)$ in model (8.4) from the cross-correlations $r_{xy}(k)$. However, as indicated before, the sampling properties of $r_{xy}(k)$ from autocorrelated series are difficult to assess, and the computation of the v_k weights from $r_{xy}(k)$ is not always easy. Box and Jenkins (1976) suggest specification through "*prewhitening*" as an alternative. This can be described as follows.

Step 1. Prewhiten the input series and obtain an ARMA model,

$$\phi_x(B)x_t = \theta_x(B)\alpha_t \tag{8.12}$$

for the suitably differenced x_t series. In other words, we generate the white-noise series α_t with variance σ_α^2 by passing the x_t series through the filter $\theta_x^{-1}(B)\phi_x(B)$.

Step 2. Transform the y_t series using the same filter, and generate the series

$$\beta_t = \theta_x^{-1}(B)\phi_x(B)y_t \tag{8.13}$$

The cross-correlations between the α_t and β_t series play an important role in the specification of the transfer function model. This can be seen from the following discussion. From (8.4) we obtain

$$y_t = v(B)x_t + n_t \tag{8.14}$$

where $y_t = \nabla^d Y_t$, $x_t = \nabla^d X_t$, and $n_t = \nabla^d N_t$ are stationary. If the filter $\theta_x^{-1}(B)\phi_x(B)$ is applied on both sides of (8.14), we get

$$\beta_t = v(B)\alpha_t + n_t^* \tag{8.15}$$

where $n_t^* = \theta_x^{-1}(B)\phi_x(B)n_t$. This implies that the transfer function between X and Y is the same as that between α and β. Multiplying both sides of (8.15) by α_{t-k} and taking expectations, we obtain

$$E(\alpha_{t-k}\beta_t) = \left[v_0 E(\alpha_{t-k}\alpha_t) + \cdots + v_k E\left(\alpha_{t-k}^2\right) + \cdots\right] + E(\alpha_{t-k}n_t^*)$$

which yields

$$\gamma_{\alpha\beta}(k) = v_k \sigma_\alpha^2$$

All other terms disappear from the equation, since $E(\alpha_{t-k}\alpha_{t-j}) = 0$ for

$j \neq k$, and $E(\alpha_{t-k} n_t^*) = 0$ for all k (since x_t and n_t are uncorrelated). Thus,

$$v_k = \frac{\gamma_{\alpha\beta}(k)}{\sigma_\alpha^2} = \frac{\sigma_\beta}{\sigma_\alpha} \frac{\gamma_{\alpha\beta}(k)}{\sigma_\alpha \sigma_\beta} = \frac{\sigma_\beta}{\sigma_\alpha} \rho_{\alpha\beta}(k) \qquad k = 0, 1, 2, \ldots \quad (8.16)$$

where σ_β^2 is the variance of the β_t process. The coefficient v_k is proportional to the cross-correlation $\rho_{\alpha\beta}(k)$ between α_t and β_t at lag k. Hence, v_k can be estimated by

$$\hat{v}_k = \frac{s_\beta}{s_\alpha} r_{\alpha\beta}(k) \qquad (8.17)$$

where s_α and s_β denote the estimated standard deviations of the α_t and β_t series.

Step 3. Calculate the sample cross-correlations $r_{\alpha\beta}(k)$, and estimate the weights v_k from (8.17). Assess the significance of the cross-correlations, or equivalently of $\hat{v}_k$, by comparing $r_{\alpha\beta}(k)$ with its standard error $(n - k)^{-1/2}$.

The relationship between the v weights and the parameters b, ω, and δ can be obtained by comparing coefficients of B^k in (8.3):

$$(1 - \delta_1 B - \delta_2 B^2 - \cdots - \delta_r B^r)(v_0 + v_1 B + \cdots)$$

$$= (\omega_0 - \omega_1 B - \cdots - \omega_s B^s) B^b$$

This leads to

$$v_k = \begin{cases} 0 & k = 0, 1, \ldots, b - 1 \\ \delta_1 v_{k-1} + \delta_2 v_{k-2} + \cdots + \delta_r v_{k-r} + \omega_0 & k = b \\ \delta_1 v_{k-1} + \delta_2 v_{k-2} + \cdots + \delta_r v_{k-r} - \omega_{k-b} & \\ \qquad k = b + 1, b + 2, \ldots, b + s & \\ \delta_1 v_{k-1} + \delta_2 v_{k-2} + \cdots + \delta_r v_{k-r} & \\ \qquad k = b + s + 1, b + s + 2, \ldots & \end{cases} \qquad (8.18)$$

Thus the weights v_k are described by (1) b zero values $v_0, v_1, \ldots, v_{b-1}$; (2) an additional $s - r + 1$ values $v_b, v_{b+1}, \ldots, v_{b+s-r}$ following no fixed pattern; and (3) values v_k, for $k \geq b + s - r + 1$, following an rth-order difference equation with starting values $v_{b+s-r+1}, \ldots, v_{b+s}$.

We can determine the orders (r, s, b) from the patterns in the estimated impulse response weights $\hat{v}_k$ or the patterns in the cross-correlations $r_{\alpha\beta}(k)$.

Once the orders are chosen, preliminary estimates $\hat{\boldsymbol{\omega}}$ and $\hat{\boldsymbol{\delta}}$ can be obtained from (8.18). Thus a preliminary estimate of the transfer function $v(B)$ is given by

$$\hat{v}(B) = \hat{\delta}^{-1}(B)\hat{\omega}(B)B^b$$

Specification of the Noise Model

After the form of the transfer function model has been determined, it remains to specify a model for the noise component in (8.14). This can be done as follows.

Step 1. Generate the noise series

$$\hat{n}_t = y_t - \hat{v}(B)x_t = y_t - \hat{\delta}^{-1}(B)\hat{\omega}(B)x_{t-b} \tag{8.19}$$

using the preliminary estimates $\hat{\boldsymbol{\omega}}$ and $\hat{\boldsymbol{\delta}}$.

Step 2. Calculate the SACF and SPACF of $\hat{n}_t$, and specify an appropriate ARMA model,

$$\phi(B)\hat{n}_t = \theta(B)a_t \tag{8.20}$$

Estimation of the Transfer Function-Noise Model

Combining the transfer function model and the noise model leads to

$$y_t = \frac{\omega(B)}{\delta(B)}x_{t-b} + \frac{\theta(B)}{\phi(B)}a_t \tag{8.21}$$

The parameters $\boldsymbol{\beta}' = (\boldsymbol{\omega}', \boldsymbol{\delta}', \boldsymbol{\phi}', \boldsymbol{\theta}') = (\omega_0, \omega_1, \ldots, \omega_s; \delta_1, \ldots, \delta_r; \phi_1, \ldots, \phi_p; \theta_1, \ldots, \theta_q)$ in (8.21) have to be estimated from past data (x_t, y_t), $t = 1, 2, \ldots, n$. Now model (8.21) can also be written as

$$\delta(B)\phi(B)y_t = \phi(B)\omega(B)x_{t-b} + \delta(B)\theta(B)a_t \tag{8.22}$$

which implies that

$$\begin{aligned} a_t = {} & y_t + d_1 y_{t-1} + \cdots + d_{p+r} y_{t-p-r} + c_0 x_{t-b} + c_1 x_{t-b-1} \\ & + \cdots + c_{p+s} x_{t-b-p-s} + b_1 a_{t-1} + b_2 a_{t-2} + \cdots + b_{r+q} a_{t-r-q} \end{aligned} \tag{8.23}$$

The coefficients in $\mathbf{d}$ are obtained from $\boldsymbol{\delta}$ and $\boldsymbol{\phi}$, those of $\mathbf{c}$ from $\boldsymbol{\omega}$ and $\boldsymbol{\phi}$, and those of $\mathbf{b}$ from $\boldsymbol{\delta}$ and $\boldsymbol{\theta}$. Assuming that the errors a_t ($t = 1, 2, \ldots, n$) are normal random variables, we can write the conditional likelihood function as

$$L\left(\boldsymbol{\beta}, \sigma^2 | \mathbf{x}, \mathbf{y}, \mathbf{x}_0, \mathbf{y}_0, \mathbf{a}_0\right) \propto \sigma^{-n} \exp\left(-\frac{1}{2\sigma^2} \sum_{t=1}^{n} a_t^2(\boldsymbol{\beta})\right) \qquad (8.24)$$

Conditional on the starting values $\mathbf{x}_0 = (x_{1-b-(p+s)}, \ldots, x_0)'$, $\mathbf{y}_0 = (y_{1-p-r}, \ldots, y_0)'$, and $\mathbf{a}_0 = (a_{1-r-q}, \ldots, a_0)'$, we can calculate $a_t(\boldsymbol{\beta})$ from (8.23). Usually the calculation of a_t is begun from $t_* = \max\{p + r + 1, b + p + s + 1\}$ and the initial a_t's are set equal to zero. Then the conditional maximum likelihood or least squares estimates $\hat{\boldsymbol{\beta}}$ are obtained by minimizing $\sum_{t=t_*}^{n} a_t^2(\boldsymbol{\beta})$; nonlinear least squares procedures are used to carry out this minimization. An estimate of σ^2 is given by

$$\hat{\sigma}^2 = m^{-1} \sum_{t=t_*}^{n} \hat{a}_t^2 \qquad (8.25)$$

where $\hat{a}_t$ ($t = t_*, \ldots, n$) are the $m = n - t_* + 1$ residuals from the fitted model.

Diagnostic Checks

After estimating the model we perform diagnostic checks to determine its adequacy. The fitted model can be inadequate if either (1) the noise model, (2) the transfer function, or (3) both noise and transfer function models are incorrect.

Box and Jenkins (1976, pp. 392–393) have shown the following results:

1. If the noise model alone is inadequate, then $\rho_a(k) \neq 0$ for some k and $\rho_{\alpha a}(k) = 0$ for all k, where α_t is the prewhitened input series.
2. If the transfer function model is inadequate, then $\rho_a(k) \neq 0$, and $\rho_{\alpha a}(k) \neq 0$ for some k.

Thus, if the residuals $\hat{a}_t$ are still autocorrelated, but the cross-correlations $r_{\alpha\hat{a}}(k)$ ($k = 0, \pm 1, \pm 2, \ldots$) are negligible compared to their standard errors $(n - k)^{-1/2}$, then the noise model needs to be modified. If, in addition, significant cross-correlations $r_{\alpha\hat{a}}(k)$ ($k = 0, \pm 1, \pm 2, \ldots$) are observed, then the transfer function model and possibly the noise model have to be changed.

In addition, various portmanteau tests can be carried out. To test whether the residuals are still autocorrelated, we can calculate

$$Q_1 = m(m+2)\sum_{k=1}^{K}\frac{r_{\hat{a}}^2(k)}{m-k} \tag{8.26}$$

where $r_{\hat{a}}(k)$ $(k = 1, 2, \ldots, K)$ denote the first K sample autocorrelations from the m available residuals. Under the hypothesis of model adequacy, Q_1 follows approximately a χ^2 distribution with $K - p - q$ degrees of freedom. Thus, by referring this statistic to a table of χ^2 percentage points, we obtain a test of model adequacy. To test whether the input and the residuals are correlated, we calculate

$$Q_2 = m(m+2)\sum_{k=0}^{K}\frac{r_{\alpha\hat{a}}^2(k)}{m-k} \tag{8.27}$$

and compare it to the percentage points from a χ^2 distribution with $K + 1 - (r + s + 1) = K - r - s$ degrees of freedom.

8.1.2. Forecasting

Time series models in Chapters 5 and 6 express the time series Y_t as a function of its own past. Transfer function-noise models relate Y_t also to present and past values of an additional time series X_t. Forecasts of future Y's thus depend on the past history of Y as well as of X.

The model in (8.4) can be written as

$$Y_t = (\pi_1 Y_{t-1} + \pi_2 Y_{t-2} + \cdots) + (v_0^* X_t + v_1^* X_{t-1} + \cdots) + a_t \tag{8.28}$$

where the π coefficients are given by

$$1 - \pi_1 B - \pi_2 B^2 - \cdots = \frac{(1-B)^d \phi(B)}{\theta(B)} \tag{8.29}$$

and the v^* coefficients by

$$v_0^* + v_1^* B + \cdots = \frac{(1-B)^d v(B)\phi(B)}{\theta(B)} \tag{8.30}$$

We are interested in predicting Y_{n+l} $(l \geqslant 1)$ as a linear combination of past observations $Y_n, Y_{n-1}, \ldots$ and $X_n, X_{n-1}, \ldots,$ and consider forecasts of the form

$$Y_n(l) = \left(\eta_0^{(1)} Y_n + \eta_1^{(1)} Y_{n-1} + \cdots\right) + \left(\eta_0^{(2)} X_n + \eta_1^{(2)} X_{n-1} + \cdots\right) \quad (8.31)$$

Equivalently, we can express the forecast as a linear combination of the past shocks a_t and α_t:

$$Y_n(l) = \left(\xi_0^{(1)} a_n + \xi_1^{(1)} a_{n-1} + \cdots\right) + \left(\xi_0^{(2)} \alpha_n + \xi_1^{(2)} \alpha_{n-1} + \cdots\right) \quad (8.32)$$

where the η and ξ coefficients are to be determined. They are usually chosen such that the mean square error $E[Y_{n+l} - Y_n(l)]^2$ is minimized.

Equations (8.12) and (8.21) lead to

$$\begin{aligned}
Y_{n+l} &= u(B)\alpha_{n+l} + \psi(B)a_{n+l} \\
&= u_0\alpha_{n+l} + u_1\alpha_{n+l-1} + \cdots + u_{l-1}\alpha_{n+1} + u_l\alpha_n + u_{l+1}\alpha_{n-1} + \cdots \\
&\quad + a_{n+l} + \psi_1 a_{n+l-1} + \cdots + \psi_{l-1}a_{n+1} + \psi_l a_n + \psi_{l+1}a_{n-1} + \cdots
\end{aligned}$$

where

$$u(B) = v(B)(1-B)^{-d}\phi_x^{-1}(B)\theta_x(B) = \frac{\omega(B)B^b\theta_x(B)}{\delta(B)\phi_x(B)(1-B)^d}$$

and

$$\psi(B) = (1-B)^{-d}\phi^{-1}(B)\theta(B)$$

The mean square error

$$\begin{aligned}
E\left[Y_{n+l} - Y_n(l)\right]^2 = E\{&[a_{n+l} + u_0\alpha_{n+l} + \psi_1 a_{n+l-1} + u_1\alpha_{n+l-1} \\
&+ \cdots + \psi_{l-1}a_{n+1} + u_{l-1}\alpha_{n+1}] \\
&+ \left[\left(u_l - \xi_0^{(2)}\right)\alpha_n + \left(u_{l+1} - \xi_1^{(2)}\right)\alpha_{n-1} + \cdots\right] \\
&+ \left[\left(\psi_l - \xi_0^{(1)}\right)a_n + \left(\psi_{l+1} - \xi_1^{(1)}\right)a_{n-1} + \cdots\right]\}^2
\end{aligned}$$

as a function of $\xi_k^{(1)}$ and $\xi_k^{(2)}$, is minimized if $\xi_k^{(1)} = \psi_{l+k}$ and $\xi_k^{(2)} = u_{l+k}$, $k = 0, 1, 2, \ldots$. This leads to the MMSE forecast,

$$Y_n(l) = (u_l\alpha_n + u_{l+1}\alpha_{n-1} + \cdots) + (\psi_l a_n + \psi_{l+1}a_{n-1} + \cdots) \quad (8.33)$$

and the corresponding forecast error

$$\begin{aligned} e_n(l) &= Y_{n+l} - Y_n(l) \\ &= (u_0\alpha_{n+l} + u_1\alpha_{n+l-1} + \cdots + u_{l-1}\alpha_{n+1}) \\ &\quad + (a_{n+l} + \psi_1 a_{n+l-1} + \cdots + \psi_{l-1}a_{n+1}) \end{aligned} \tag{8.34}$$

with variance

$$V[e_n(l)] = \sum_{j=0}^{l-1} \left(u_j^2\sigma_\alpha^2 + \psi_j^2\sigma^2\right) \tag{8.35}$$

The MMSE forecasts can also be written as

$$Y_n(l) = E[Y_{n+l}| Y_n, Y_{n-1},\ldots; X_n, X_{n-1},\ldots] \tag{8.36}$$

since

$$\begin{aligned} &E\left[a_{n+j}| Y_n, Y_{n-1},\ldots; X_n, X_{n-1},\ldots\right] \\ &\quad = E\left[\alpha_{n+j}| Y_n, Y_{n-1},\ldots; X_n, X_{n-1},\ldots\right] = 0 \qquad j > 0 \end{aligned}$$

This representation is very convenient, since the forecasts can be calculated from the difference equation (8.22).

Forecast computations are illustrated on a simple example. We use the notation $E[Y_{n+l}| \cdot]$ and $E[a_{n+l}| \cdot]$ for the conditional expectations in (8.36); and $E[X_{n+l}| \cdot] = E[X_{n+l}|X_n, X_{n-1}, \ldots]$.

Illustration

Consider the model

$$y_t = \frac{\omega_0}{(1-\delta B)}x_t + (1-\theta B)a_t \qquad x_t = (1-\theta_x B)\alpha_t \tag{8.37}$$

where $y_t = (1-B)Y_t$ and $x_t = (1-B)X_t$. This model can be rewritten as

$$(1-B)(1-\delta B)Y_t = (1-B)\omega_0 X_t + (1-\theta B)(1-\delta B)a_t$$

which implies that

$$\begin{aligned} Y_{n+l} &= (1+\delta)Y_{n+l-1} - \delta Y_{n+l-2} + \omega_0 X_{n+l} - \omega_0 X_{n+l-1} + a_{n+l} \\ &\quad - (\delta+\theta)a_{n+l-1} + \delta\theta a_{n+l-2} \end{aligned}$$

This leads to the one-step-ahead forecast

$$\begin{aligned} Y_n(1) &= E[Y_{n+1}|\cdot] \\ &= (1+\delta)Y_n - \delta Y_{n-1} + \omega_0[X_n(1) - X_n] - (\delta+\theta)a_n + \delta\theta a_{n-1} \end{aligned}$$

since

$$E[Y_{n+j}|\cdot] = Y_{n+j} \qquad E[a_{n+j}|\cdot] = a_{n+j} \qquad E[X_{n+j}|\cdot] = X_{n+j} \qquad j \leqslant 0$$

and

$$E[a_{n+j}|\cdot] = 0 \qquad E[X_{n+j}|\cdot] = X_n(j) \qquad j > 0$$

Note that $X_n(1)$ is the MMSE forecast of X_{n+1} that is determined from its own past.

Forecasts for other lead times can be calculated from

$$Y_n(2) = E[Y_{n+2}|\cdot] = (1+\delta)Y_n(1) - \delta Y_n + \omega_0[X_n(2) - X_n(1)] + \delta\theta a_n$$

and

$$\begin{aligned} Y_n(l) &= E[Y_{n+l}|\cdot] \\ &= (1+\delta)Y_n(l-1) - \delta Y_n(l-2) + \omega_0[X_n(l) - X_n(l-1)] \qquad l > 2 \end{aligned} \tag{8.38}$$

The forecast error variances can be obtained from (8.35); the u weights are given by

$$(1-B)(1-\delta B)(u_0 + u_1 B + u_2 B^2 + \cdots) = \omega_0(1 - \theta_x B) \tag{8.39}$$

and the ψ weights by

$$(1-B)(1 + \psi_1 B + \psi_2 B^2 + \cdots) = 1 - \theta B \tag{8.40}$$

A comparison of coefficients of B^j in (8.39) yields $u_0 = \omega_0$, $u_1 = (1 + \delta - \theta_x)\omega_0$, and $u_j = (1+\delta)u_{j-1} - \delta u_{j-2}$ for $j \geqslant 2$. This leads to

$$u_j = A_0 + A_1\delta^j \qquad j \geqslant 0$$

where $A_0 = \omega_0(1-\theta_x)/(1-\delta)$ and $A_1 = \omega_0(\theta_x - \delta)/(1-\delta)$. From (8.40), we obtain $\psi_0 = 1$, $\psi_j = 1 - \theta$ $(j \geqslant 1)$. Thus,

$$V[e_n(1)] = u_0^2\sigma_\alpha^2 + \sigma^2 = \omega_0^2\sigma_\alpha^2 + \sigma^2$$

$$V[e_n(2)] = \left(u_0^2 + u_1^2\right)\sigma_\alpha^2 + \left(1 + \psi_1^2\right)\sigma^2$$

$$= \left[1 + (1 + \delta - \theta_x)^2\right]\omega_0^2\sigma_\alpha^2 + \left[1 + (1 - \theta)^2\right]\sigma^2$$

$$V[e_n(l)] = \sum_{j=0}^{l-1}\left(A_0 + A_1\delta^j\right)^2\sigma_\alpha^2 + \left[1 + (l-1)(1-\theta)^2\right]\sigma^2$$

$$= \left[lA_0^2 + A_1^2\left(\frac{1 - \delta^{2l}}{1 - \delta^2}\right) + 2A_0A_1\left(\frac{1 - \delta^l}{1 - \delta}\right)\right]\sigma_\alpha^2$$

$$+\left[1 + (l-1)(1-\theta)^2\right]\sigma^2$$

8.1.3. Related Models

In Chapter 2 we gave a brief discussion of distributed lag models and regression models with correlated errors. There we assumed a simple lag structure and a first-order autoregressive model for the errors. The transfer function modeling approach extends these ideas and includes more general lag structures and error models.

Various extensions of transfer function-noise models are possible. As illustrated in the following example, such models can be built for seasonal series. One can also consider more than one input series and, in addition, allow situations in which the output series influences the input series [see Tiao and Box (1981), and Granger and Newbold (1977)].

8.1.4. Example

To illustrate transfer function modeling, we consider U.S. monthly housing starts Y_t and U.S. monthly houses sold X_t for the period 1965 to 1974. These data have been analyzed by Hillmer and Tiao (1979). Computer programs such as the TS-package or the PACK-SYSTEM are available for transfer function analysis; here we use the TS-package to prewhiten the series and to estimate the transfer function-noise model.

It is reasonable to assume that past houses sold (X_t) will influence future housing starts (Y_t), but not vice versa. If past housing starts also influence future sales, then the transfer function analysis would not be appropriate, since N_t and X_t are correlated. In such a case we would have to consider multiple (bivariate) time series models.

Specification of the Transfer Function Model

Step 1. Prewhiten the X_t series. The iterative strategy outlined in Chapters 5 and 6 leads to the model

$$(1 - B)(1 - B^{12})X_t = (1 - .20B)(1 - .83B^{12})\alpha_t$$

with $s_\alpha^2 = 13.27$. The parameter estimates are significant; their standard errors are .09 and .05, respectively.

Step 2. Generate the series

$$\beta_t = (1 - .20B)^{-1}(1 - .83B^{12})^{-1}(1 - B)(1 - B^{12})Y_t$$

which results in $s_\beta^2 = 45.60$.

Step 3. The cross-correlations $r_{\alpha\beta}(k)$ for positive and negative lags are shown in Figure 8.2*a*. The insignificant cross-correlations at negative lags indicate that the past Y_t's do not influence future X_t's. The coefficients $\hat{v}_k$ in Table 8.1 are calculated from the cross-correlations at nonnegative lags. We notice a nonzero $\hat{v}_0$ and an exponential decay from lag 1 onwards. Thus we conjecture that $b = 0$, $\omega(B) = \omega_0 - \omega_1 B$, and $\delta(B) = 1 - \delta B$. Alternatively, we could argue that all $\hat{v}$ weights except $\hat{v}_0$ and $\hat{v}_1$ are zero, and hence a model with $b = 0$, $\omega(B) = \omega_0 - \omega_1 B$, and $\delta(B) = 1$ would also be appropriate. However, for illustrative purposes we proceed with the former model.

Specification of the Noise Model

Step 1. Preliminary estimates of $\omega_0, \omega_1, \delta$ are given in Table 8.1; these are used to generate the noise series in (8.19).

Step 2. The SACF of the noise series in Figure 8.2*b* suggests the noise model

$$\hat{n}_t = (1 - \theta B)(1 - \Theta B^{12})a_t$$

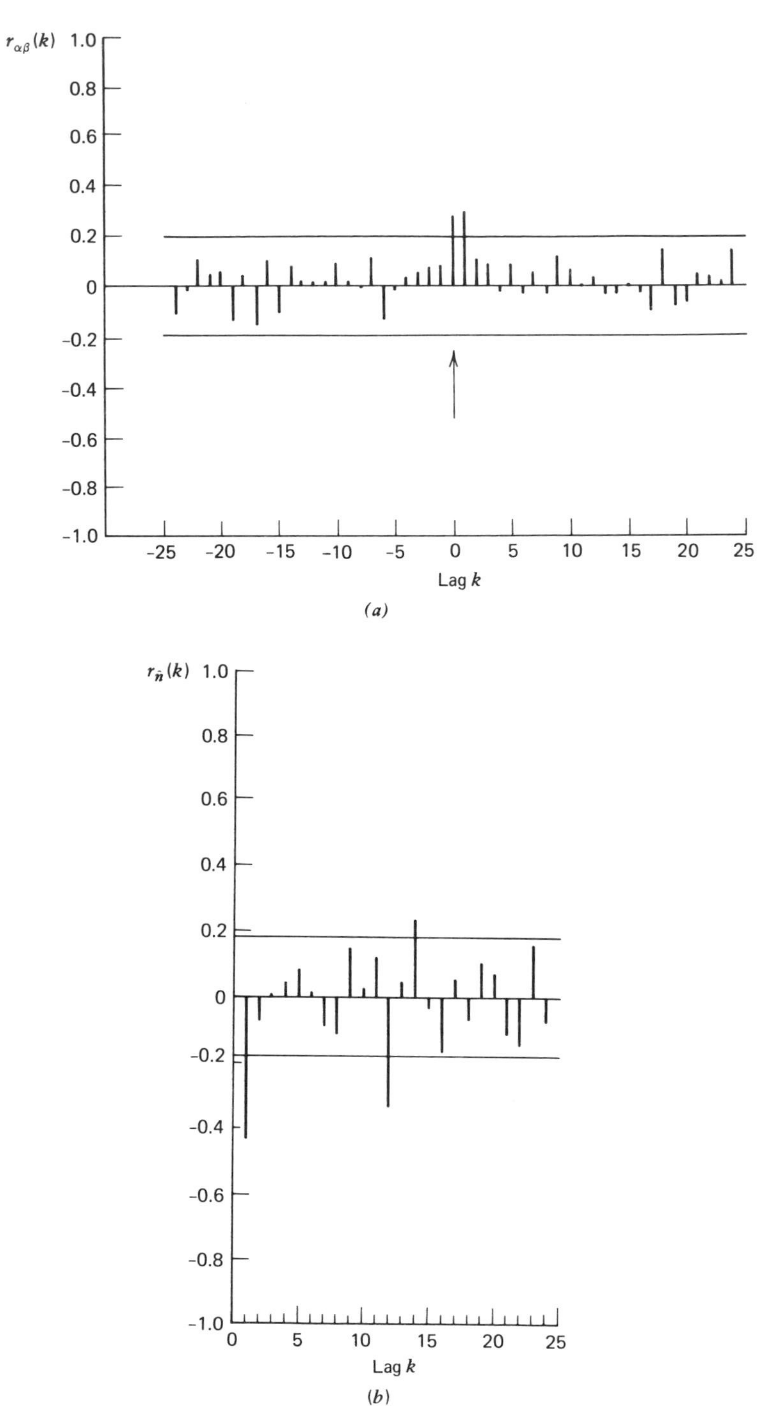

Figure 8.2. Housing starts and houses sold: (*a*) Sample CCF $r_{\alpha\beta}(k)$; (*b*) SACF of the noise series $\hat{n}_t$.

Table 8.1. Calculation of $\hat{v}_k$ Weights and Preliminary Estimates of Transfer Function Parameters

k	0	1	2	3	4	5	6	7	8	9	10
$r_{\alpha\beta}(k)$	.262	.274	.103	.079	−.044	.080	−.043	.054	−.043	.114	.053
$\hat{v}_k$	.486	.508	.191	.146	−.082	.148	−.080	.100	−.080	.211	.098

$$\hat{v}_k = \frac{s_\beta}{s_\alpha} r_{\alpha\beta}(k) = \left(\frac{45.60}{13.27}\right)^{1/2} r_{\alpha\beta}(k)$$

$$(1 - \hat{\delta}B)(\hat{v}_0 + \hat{v}_1 B + \hat{v}_2 B^2 + \cdots) = \hat{\omega}_0 - \hat{\omega}_1 B$$

Comparing coefficients of B^j ($j = 0, 1, 2$) yields

$$\hat{\omega}_0 = \hat{v}_0 \qquad \hat{\omega}_1 = \hat{\delta}\hat{v}_0 - \hat{v}_1 \qquad \hat{v}_2 = \hat{\delta}\hat{v}_1$$

Thus,

$$\hat{\omega}_0 = .486 \qquad \hat{\delta} = \frac{\hat{v}_2}{\hat{v}_1} = \frac{.191}{.508} = .376$$

and

$$\hat{\omega}_1 = .376(.486) - .508 = -.325$$

Estimation and Diagnostic Checking

In the combined transfer function-noise model

$$y_t = \left(\frac{\omega_0 - \omega_1 B}{1 - \delta B}\right) x_t + (1 - \theta B)(1 - \Theta B^{12}) a_t$$

where

$$y_t = (1 - B)(1 - B^{12}) Y_t \qquad \text{and} \qquad x_t = (1 - B)(1 - B^{12}) X_t$$

the parameter estimates and their standard errors are

$$\hat{\omega}_0 = .56(.13) \qquad \hat{\omega}_1 = -.53(.22) \qquad \hat{\delta} = .30(.14)$$

$$\hat{\theta} = .70(.07) \qquad \hat{\Theta} = .81(.06) \qquad \hat{\sigma}^2 = 29.10$$

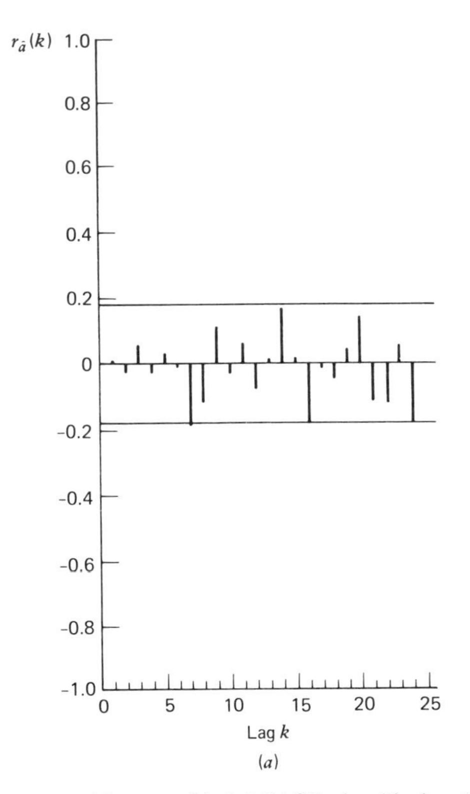

Figure 8.3. Housing starts and houses sold: (*a*) SACF of residuals $r_{\hat{a}}(k)$; (*b*) Sample CCF $r_{\alpha\hat{a}}(k)$.

The residual autocorrelations $r_{\hat{a}}(k)$ in Figure 8.3*a* and the cross-correlations $r_{\alpha\hat{a}}(k)$ in Figure 8.3*b* show no evidence of serious model inadequacy. The portmanteau statistics, $Q_1 = 26.79$ and $Q_2 = 23.75$, are not significant compared with the upper 5 percentage points of the χ^2 distributions with $K - p - q = 24 - 2 = 22$, and $K + 1 - (r + s + 1) = 22$ degrees of freedom, respectively.

Forecasts

The fitted model can be rewritten as

$$(1 - \hat{\delta}B)(1 - B)(1 - B^{12})Y_t = (\hat{\omega}_0 - \hat{\omega}_1 B)(1 - B)(1 - B^{12})X_t$$
$$+ (1 - \hat{\delta}B)(1 - \hat{\theta}B)(1 - \hat{\Theta}B^{12})a_t$$

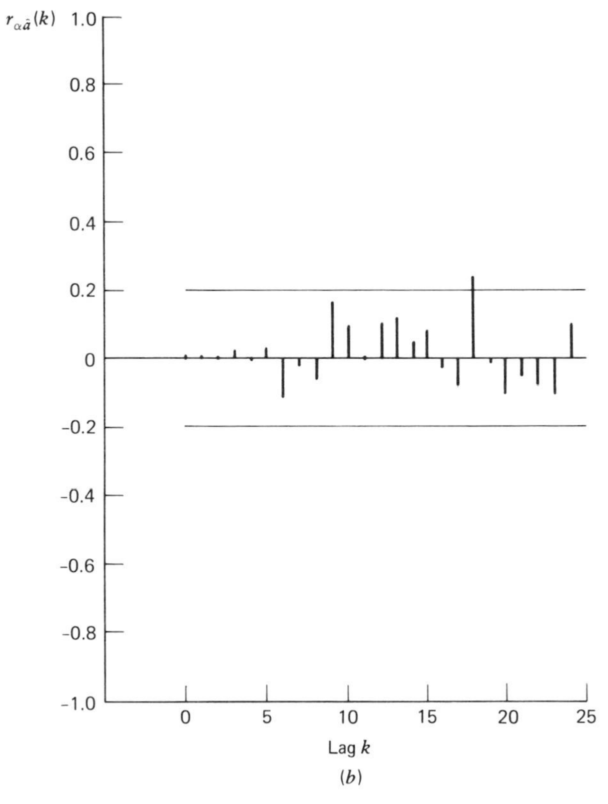

Figure 8.3. (*Continued*).

Thus the forecasts can be generated from

$$\begin{aligned}
\hat{Y}_n(l) = {} & 1.30[Y_{n+l-1}] - .30[Y_{n+l-2}] + [Y_{n+l-12}] \quad 1.30[Y_{n+l-13}] \\
& + .30[Y_{n+l-14}] + .56[X_{n+l}] - .03[X_{n+l-1}] - .53[X_{n+l-2}] \\
& - .56[X_{n+l-12}] + .03[X_{n+l-13}] + .53[X_{n+l-14}] + [a_{n+l}] \\
& - 1.00[a_{n+l-1}] + .21[a_{n+l-2}] - .81[a_{n+l-12}] \\
& + .81[a_{n+l-13}] - .17[a_{n+l-14}]
\end{aligned}$$

where $[\cdot] = E(\cdot \mid Y_n, Y_{n-1}, \ldots; X_n, X_{n-1}, \ldots)$ denotes the conditional expectation in (8.36). The variance of the forecast error is given by

$$V[e_n(l)] = \sigma_\alpha^2(u_0^2 + \cdots + u_{l-1}^2) + \sigma^2(1 + \psi_1^2 + \cdots + \psi_{l-1}^2)$$

where the u weights are obtained from

$$(1 - B)(1 - B^{12})(u_0 + u_1 B + u_2 B^2 + \cdots)(1 - .30B)$$
$$= (.56 + .53B)(1 - .20B)(1 - .83B^{12})$$

and the ψ weights from

$$(1 + \psi_1 B + \psi_2 B^2 + \cdots)(1 - B)(1 - B^{12}) = (1 - .70B)(1 - .81B^{12})$$

These weights and the standard errors

$$\text{SE}[e_n(l)] = \{V[e_n(l)]\}^{1/2}$$

are estimated and shown in Table 8.2 for $l = 1, \ldots, 12$.

Forecasts of future housing starts can also be generated from past data alone; this was done in Chapter 6. In Table 8.2 we compare the standard errors of the forecasts from the univariate time series model with the ones from the transfer function model. We notice that the standard errors from the transfer function analysis are somewhat smaller, especially for short lead

Table 8.2. Standard Errors (SE) of Forecasts, and the u and ψ Weights

l	u_{l-1}	ψ_{l-1}	SE from Transfer Function	SE from Single Series
1	0.56	1	5.77	6.51
2	1.15	.3	7.31	8.04
3	1.22	.3	8.71	9.32
4	1.24	.3	9.94	10.44
5	1.25	.3	11.05	11.46
6	1.25	.3	12.06	12.39
7	1.25	.3	12.99	13.26
8	1.25	.3	13.86	14.07
9	1.25	.3	14.68	14.84
10	1.25	.3	15.46	15.57
11	1.25	.3	16.19	16.27
12	1.25	.3	16.90	16.94

times. This indicates that the inclusion of information from a relevant variable (X_t) can make the forecasts of future Y_t's more precise.

8.2. INTERVENTION ANALYSIS AND OUTLIERS

8.2.1. Intervention Analysis

Time series are frequently affected by policy changes and other events usually referred to as *interventions*. Three examples of such intervention effects are: (1) the effect on inflation of the creation of the Canadian Anti-Inflation Board in November 1975, (2) the impact on the number of traffic fatalities of introducing the 55 miles-per-hour speed limit in the United States in 1974; and (3) the influence on product sales of a change in advertising strategy. Interventions can affect the response in several ways. They can change the level of a series either abruptly or after some delay, change the trend, or lead to other, more complicated, effects. Transfer function models studied in Section 8.1 can be used to determine whether there is evidence that such a change in the series has actually occurred and, if so, its nature and magnitude.

The Student t distribution is traditionally used to estimate and test for a change in the mean level. Such a test may not be adequate for situations where the data occur in the form of time series, because (1) successive observations may be correlated and (2) the effect may not be a step change as postulated by the t test.

Box and Tiao (1975) have provided a strategy for modeling the effect of interventions. They consider transfer function-noise models of the form

$$Y_t = v(B)I_t + z_t \tag{8.41}$$

where I_t is an indicator sequence reflecting the absence and presence of an intervention. Before the intervention has occurred, z_t and Y_t are the same and can be represented by an ARIMA model. In transfer function modeling, the form of $v(B) = \omega(B)B^b/\delta(B)$ is usually determined empirically. However, in intervention analysis it is postulated based on the expected form of the change. In the simplest case of a step change at time T, the transfer function is $v(B) = \omega_0/(1 - B)$, and $I_t(T) = 1$ if $t = T$ and zero otherwise (see Fig. 8.4*a*). If an initial increase ω_0 is followed by a gradual decrease, and if no lasting effect is expected, the appropriate model is $v(B) = \omega_0/(1 - \delta B)$, and $I_t(T) = 1$ if $t = T$ and zero otherwise (see Fig. 8.4*b*). In addition, if there is a lasting effect ω_1, then the transfer function becomes $v(B) = [\omega_0/(1 - \delta B)] + [\omega_1/(1 - B)]$ (see Fig. 8.4*c*.)

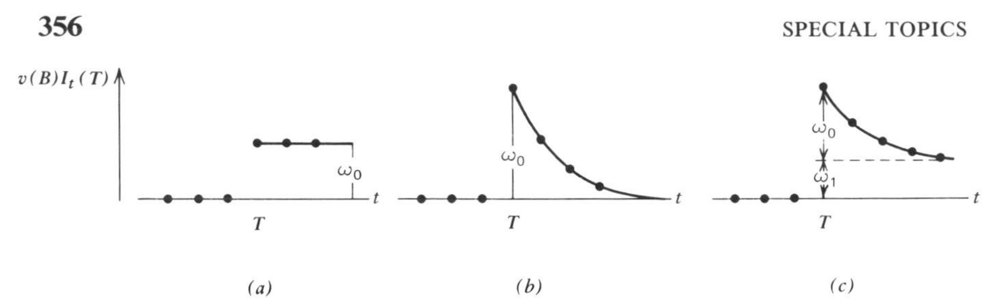

Figure 8.4. Responses to various intervention models $v(B)I_t(T)$, where $I_t(T) = 1$ if $t = T$ and zero otherwise: $(a)\ v(B) = \omega_0/(1 - B)$; $(b)\ v(B) = \omega_0/(1 - \delta B)$; $(c)\ v(B) = [\omega_0/(1 - \delta B)] + [\omega_1/(1 - B)]$.

The parameter estimates and their standard errors can be obtained as in the previous section.

Additional Comments

It should be emphasized that intervention analysis assumes that (1) the time series model parameters are the same before and after the intervention and (2) no other events or interventions coincide with the particular one being considered. If other interventions occur at roughly the same time, the effects of the interventions are confounded.

Model (8.41) can also be extended to allow the inclusion of other exogenous variables and more than one intervention. For further discussion see Abraham (1980, 1983), Box and Tiao (1975) and Bhattacharyya and Layton (1979).

8.2.2. Outliers

Business and economic time series are sometimes influenced by nonrepetitive interventions such as strikes, outbreaks of war, and sudden changes in the market structure of a commodity. If the timing of such interventions is known, intervention models can be used to account for their effects. However, in practice the timing is frequently unknown. Since the effects of interventions can bias the parameter estimates, forecasts, and seasonal adjustments, it is important to develop procedures that can help detect and remove such effects. This is known as the problem of outliers or spurious observations.

Abraham and Box (1979), Fox (1972), and Martin (1980) discuss two characterizations of outliers in the context of time series models:

1. Aberrant observation model (AO: additive outlier):

$$z_t^* = z_t + \omega I_t(T) \qquad \phi(B)z_t = \theta(B)a_t \tag{8.42}$$

2. Aberrant innovation model (AI: innovational outlier):

$$z_t^* = z_t + \phi^{-1}(B)\theta(B)\omega I_t(T) = \phi^{-1}(B)\theta(B)[a_t + \omega I_t(T)] \tag{8.43}$$

Here z_t^* denotes the observed time series, z_t the underlying process without the impact of outliers, and $I_t(T) = 1$ if $t = T$ and zero otherwise. In the first model, only the level of the Tth observation is affected. In the second model, the outlier affects the shock at time T, which in turn influences $z_T, z_{T+1}, \ldots$. The procedures suggested by Abraham and Box (1979) and Martin (1980) involve considerable computation. Here we describe a somewhat simpler method [see Hillmer, Bell, and Tiao (1982)].

To motivate this procedure, let us assume that the time T of the intervention and the time series parameters $\boldsymbol{\phi}$ and $\boldsymbol{\theta}$ are known. Let $e_t = \pi(B)z_t^*$, where $\pi(B) = \phi(B)\theta^{-1}(B) = 1 + \pi_1 B + \pi_2 B^2 + \cdots$. Then we have

(AO) $$e_t = \omega\pi(B)I_t(T) + a_t = \omega x_{1t} + a_t$$

(AI) $$e_t = \omega I_t(T) + a_t = \omega x_{2t} + a_t \tag{8.44}$$

where $x_{1t} = \pi(B)I_t(T)$ and $x_{2t} = I_t(T)$. The least squares estimates of the intervention impact ω, and their variances can be obtained as follows:

(AO) $$\hat{\omega}_{OT} = \frac{\sum e_t x_{1t}}{\sum x_{1t}^2} = \frac{\sum_{j\geq 0} \pi_j e_{T+j}}{\sum_{j\geq 0} \pi_j^2} = \frac{\pi(F)e_T}{\eta^2}$$

$$V(\hat{\omega}_{OT}) = \frac{\sigma^2}{\sum x_{1t}^2} = \frac{\sigma^2}{\eta^2} \tag{8.45}$$

where $\pi_0 = 1$, $\eta^2 = \Sigma_{j\geq 0}\pi_j^2$, and $F = B^{-1}$.

(AI) $$\hat{\omega}_{IT} = \frac{\Sigma e_t x_{2t}}{\Sigma x_{2t}^2} = e_T \qquad V(\hat{\omega}_{IT}) = \sigma^2 \tag{8.46}$$

Our notation reflects the fact that the estimates depend upon T.

In the AO model, $\hat{\omega}_{OT}$ is a linear combination of the shocks future to T, and its variance can be much smaller than σ^2. In the AI model, $\hat{\omega}_{IT}$ is the residual at T with variance σ^2. Significance tests for outliers can be performed based on the standardized estimates $\hat{\lambda}_{1T} = \eta\hat{\omega}_{OT}/\sigma$ and $\hat{\lambda}_{2T} = \hat{\omega}_{IT}/\sigma$.

In practice, T as well as the time series parameters are unknown and have to be replaced by estimates. Since the presence of outliers can seriously bias the estimates, Hillmer, Bell, and Tiao (1982) suggest the following iterative strategy.

Step 1. Suppose that there are no outliers, and model the series z_t^*. Compute the residuals $\hat{e}_t$ from the estimated model, and calculate

$$\hat{\sigma}^2 = \frac{1}{n} \sum_{t=1}^{n} \hat{e}_t^2$$

as an initial estimate of σ^2.

Step 2. Calculate $\hat{\lambda}_{it}$, $i = 1, 2$ and $t = 1, 2, \ldots, n$, and $|\hat{\lambda}_T| = \max_t \max_i |\hat{\lambda}_{i\,t}|$. If $|\hat{\lambda}_T| = |\hat{\lambda}_{1\,T}| > c$, where c is a predetermined positive constant usually taken as 3, then there is the possibility of an additive outlier at T. Its effect can be estimated by $\hat{\omega}_{OT}$ and removed by defining new residuals $\tilde{e}_t - \hat{e}_t - \hat{\omega}_{OT}\hat{\pi}(B)I_t(T)$ for $t \geqslant T$. If $|\hat{\lambda}_T| = |\hat{\lambda}_{2T}| > c$, then there is the possibility of an innovational outlier. Its effect can be eliminated by defining the new residual $\tilde{e}_T = \hat{e}_T - \hat{\omega}_{IT}$. A new estimate $\hat{\sigma}^2$ can be calculated from the modified residuals.

Step 3. Use the modified residuals to compute revised estimates $\hat{\lambda}_{it}$ and $\hat{\sigma}^2$. The initial estimates of $\boldsymbol{\phi}$ and $\boldsymbol{\theta}$ are kept unchanged. Repeat step 2.

Step 4. Suppose that step 3 terminated with k AO or AI outliers at times $T_1, T_2, \ldots, T_k$. Now treat these times as if they were known, and estimate $\omega_1, \ldots, \omega_k$ and the time series model parameters in

$$z_t^* = \sum_{j=1}^{k} \omega_j L_j(B) I_t(T_j) + \phi^{-1}(B)\theta(B)a_t$$

For AO outliers at time T_j, $L_j(B) = 1$, and for AI outliers, $L_j(B) = \phi^{-1}(B)\theta(B)$. Nonlinear estimation procedures have to be used. The new residuals are given by

$$\hat{e}_t^{(1)} = \hat{\theta}^{-1}(B)\hat{\phi}(B)\left[z_t^* - \sum_{j=1}^{k} \hat{\omega}_j L_j(B) I_t(T_j)\right]$$

and a revised estimate of σ^2 can be calculated.

Steps 2 to 4 are repeated until all outliers are identified and their effects simultaneously estimated. This procedure can be implemented rather easily into existing transfer function software.

8.3. THE STATE SPACE FORECASTING APPROACH, KALMAN FILTERING, AND RELATED TOPICS

In this section we give a brief description of a very general approach to forecasting known as the *state space forecasting* approach. It includes the traditional regression and ARIMA time series models, Bayesian forecasting [see Harrison and Stevens (1976)], and models with time-varying coefficients as special cases.

State space models of random processes are based on the so-called Markov property, which implies the independence of the future of a process from its past, given the present state. In such a system the *state* of the process summarizes all the information from the past that is necessary to predict the future. Let us assume that the unknown state of the system at time t is given by a vector $\mathbf{S}_t$ referred to as the *state vector*. Then a state space model is described by two equations: (1) a *measurement equation*, which describes the generation of the observations from a given state vector, and (2) a *system equation*, which describes the evolution of the state vector.

Measurement equation:

$$\mathbf{y}_t = \mathbf{H}_t\mathbf{S}_t + \boldsymbol{\varepsilon}_t \tag{8.47}$$

System equation:

$$\mathbf{S}_{t+1} = \mathbf{A}\mathbf{S}_t + \mathbf{G}\mathbf{u}_t + \mathbf{a}_{t+1} \tag{8.48}$$

In Equation (8.47), $\mathbf{S}_t$ is an unobservable state vector that describes the state of the system at time t; $\mathbf{y}_t$ is an observation vector; $\mathbf{H}_t$ is a known matrix; and $\boldsymbol{\varepsilon}_t$ is the measurement noise, which is a white-noise process with mean vector zero and covariance matrix $E(\boldsymbol{\varepsilon}_t\boldsymbol{\varepsilon}_t') = \mathbf{R}_1$.

Equation (8.48) describes the evolution of the state vector; $\mathbf{u}_t$ is a vector of known inputs; $\mathbf{A}$ and $\mathbf{G}$ are known matrices; and $\mathbf{a}_t$ is the process noise, which is a white-noise process with mean vector zero and covariance matrix $E(\mathbf{a}_t\mathbf{a}_t') = \mathbf{R}_2$.

The system in (8.47) and (8.48) contains two error terms: the measurement and the process noise. The errors in observing the true state of the system are given by $\boldsymbol{\varepsilon}_t$, whereas the shocks in the evolution of the states are given by $\mathbf{a}_t$. It is assumed that these two error terms are uncorrelated at all lags; $E(\boldsymbol{\varepsilon}_t\mathbf{a}_{t-k}') = \mathbf{0}$ for all k.

In our discussion we have assumed that the measurement and system equations are linear. Extensions to nonlinear models and models in which the parameters $\mathbf{A}$, $\mathbf{G}$, $\mathbf{R}_1$, and $\mathbf{R}_2$ are time-dependent are possible but are not pursued here.

Special cases of Equations (8.47) and (8.48) lead to (1) the regression models discussed in Chapter 2, (2) ARIMA time series models described in Chapters 5 and 6, (3) models considered in Bayesian forecasting, and (4) models with time-varying coefficients. Two examples are discussed below.

Example 1: Regression Models

Consider the state space model in which there is no process noise, that is, $E(\mathbf{a}_t\mathbf{a}'_t) = \mathbf{0}$. Furthermore, assume that there are no input variables $\mathbf{u}_t$ and that $\mathbf{A} = \mathbf{I}$. In such a case the system equation reduces to time-constant states, $\mathbf{S}_{t+1} = \mathbf{S}_t$. Assume that the dimension of the observation vector is 1. If now $\mathbf{H}_t = \mathbf{x}'_t$ and the coefficients $\boldsymbol{\beta}$ take the place of the state vector, then the state space model in (8.47) and (8.48) reduces to the regression model $y_t = \mathbf{x}'_t\boldsymbol{\beta} + \varepsilon_t$.

Example 2: ARIMA Time Series Models

Consider the model in which there are no input variables $\mathbf{u}_t$ and no measurement noise, that is, $E(\boldsymbol{\varepsilon}_t\boldsymbol{\varepsilon}'_t) = \mathbf{0}$. Furthermore, assume that $\mathbf{H}_t = \mathbf{I}$, the identity matrix. Then the state space model simplifies to

$$\mathbf{y}_{t+1} = \mathbf{A}\mathbf{y}_t + \mathbf{a}_{t+1} \tag{8.49}$$

which is a multivariate version of the first-order autoregressive model.

Mixed ARMA models can be obtained by introducing additional state variables and augmenting the state vector. For example, a state space representation of the univariate ARMA(p, q) process

$$y_t = \phi_1 y_{t-1} + \cdots + \phi_p y_{t-p} + a_t - \theta_1 a_{t-1} - \cdots - \theta_q a_{t-q}$$

is given by

$$y_t = [1 \quad 0 \quad \cdots \quad 0]\mathbf{S}_t$$

$$\begin{bmatrix} S_{t+1,1} \\ S_{t+1,2} \\ \vdots \\ S_{t+1,k} \end{bmatrix} = \begin{bmatrix} \phi_1 & & & \\ \phi_2 & & \mathbf{I}_{k-1} & \\ \vdots & & & \\ \phi_k & 0 & \cdots & 0 \end{bmatrix} \begin{bmatrix} S_{t,1} \\ S_{t,2} \\ \vdots \\ S_{t,k} \end{bmatrix} + \begin{bmatrix} 1 \\ -\theta_1 \\ \vdots \\ -\theta_{k-1} \end{bmatrix} a_{t+1} \tag{8.50}$$

where $k = \max\{p, q+1\}$, $\phi_j = 0$ for $j > p$, $\theta_j = 0$ for $j > q$, $\mathbf{I}_{k-1}$ is the $(k-1) \times (k-1)$ identity matrix, $\mathbf{S}_t = (S_{t,1}, \ldots, S_{t,k})'$ is the augmented state vector, the new process noise has covariance matrix $\mathbf{R}_2 = \sigma_a^2\boldsymbol{\theta}\boldsymbol{\theta}'$, and $\boldsymbol{\theta} = (1, -\theta_1, \ldots, -\theta_{k-1})'$. The equivalence is easily shown by repeated substitution.

State space models are not unique, and many equivalent state space representations can be found for the same ARIMA model. For example, consider the ARIMA(0, 1, 1) model. It could be represented as a special case of (8.50); see Exercise 8.7. Or for $\theta \geqslant 0$, we could write it as

$$y_t = S_t + \varepsilon_t$$

$$S_{t+1} = S_t + a_{t+1} \tag{8.51}$$

By simple substitution we find that

$$y_{t+1} = y_t + \varepsilon_{t+1} - \varepsilon_t + a_{t+1}$$

Since the ACF of the first differences $y_{t+1} - y_t$ is such that $\rho_1 = -\sigma_\varepsilon^2/(2\sigma_\varepsilon^2 + \sigma_a^2)$ and $\rho_k = 0$ for $k > 1$, the model in (8.51) is an equivalent representation of an ARIMA (0, 1, 1) model with a nonnegative moving average coefficient ($\theta \geqslant 0$).

8.3.1. Recursive Estimation and Kalman Filtering

The state vector at time t summarizes the information from the past that is necessary to predict the future. Thus, before forecasts of future observations can be calculated, it is necessary to make inferences about the state vector $\mathbf{S}_t$.

In order to specify the distribution of the state vectors $\mathbf{S}_t$ in (8.48), it is necessary to start with a distribution for the state vector at time zero. Let us denote the probability distribution of $\mathbf{S}_0$ by $p(\mathbf{S}_0)$. This, together with the system equation in (8.48), determines the distribution of the state vectors $p(\mathbf{S}_t)$, $t = 0, 1, 2, \ldots$. We refer to these distributions as the *prior distributions*, since they represent our "belief" prior to having observed the data $\mathbf{y}_t, \mathbf{y}_{t-1}, \ldots$.

After the history of the process $\mathbf{Y}_t = \{\mathbf{y}_t, \mathbf{y}_{t-1}, \ldots\}$ has been observed, we want to revise our prior distribution of the unknown state vector $\mathbf{S}_t$. The new or revised information about $\mathbf{S}_t$ is expressed by the conditional distribution $p(\mathbf{S}_t|\mathbf{Y}_t)$. This is also called the *posterior distribution*, since it measures the information about $\mathbf{S}_t$ after the observations in $\mathbf{Y}_t$ have become available.

From the model in (8.47), (8.48), and the prior distribution $p(\mathbf{S}_0)$, we can derive recursive equations that propagate the conditional distributions $p(\mathbf{S}_t|\mathbf{Y}_t) \to p(\mathbf{S}_{t+1}|\mathbf{Y}_t) \to p(\mathbf{S}_{t+1}|\mathbf{Y}_{t+1}) \to$ etc. If we assume that the errors $\boldsymbol{\varepsilon}_t$

and $\mathbf{a}_t$ are normally distributed and that the prior distribution $p(\mathbf{S}_0)$ is normal with mean vector $\hat{\mathbf{S}}_{0|0}$ and covariance matrix $\mathbf{P}_{0|0}$, then the conditional distributions $p(\mathbf{S}_t|\mathbf{Y}_t)$ and $p(\mathbf{S}_{t+1}|\mathbf{Y}_t)$ are also normal and completely characterized by their first two moments. Let us denote the mean vector and the covariance matrix of the distribution $p(\mathbf{S}_t|\mathbf{Y}_t)$ by $\hat{\mathbf{S}}_{t|t}$ and $\mathbf{P}_{t|t}$, and those of $p(\mathbf{S}_{t+1}|\mathbf{Y}_t)$ by $\hat{\mathbf{S}}_{t+1|t}$ and $\mathbf{P}_{t+1|t}$. Then the moments of these distributions can be updated recursively. These updating equations, which are commonly referred to as the *Kalman filter*, are given below:

$$\hat{\mathbf{S}}_{t+1|t} = \mathbf{A}\hat{\mathbf{S}}_{t|t} + \mathbf{G}\mathbf{u}_t$$

$$\mathbf{P}_{t+1|t} = \mathbf{A}\mathbf{P}_{t|t}\mathbf{A}' + \mathbf{R}_2$$

$$\hat{\mathbf{S}}_{t+1|t+1} = \hat{\mathbf{S}}_{t+1|t} + \mathbf{k}_{t+1}\left(\mathbf{y}_{t+1} - \mathbf{H}_{t+1}\hat{\mathbf{S}}_{t+1|t}\right)$$

$$\mathbf{P}_{t+1|t+1} = \mathbf{P}_{t+1|t} - \mathbf{k}_{t+1}\mathbf{H}_{t+1}\mathbf{P}_{t+1|t}$$

and

$$\mathbf{k}_{t+1} = \mathbf{P}_{t+1|t}\mathbf{H}'_{t+1}\left(\mathbf{H}_{t+1}\mathbf{P}_{t+1|t}\mathbf{H}'_{t+1} + \mathbf{R}_1\right)^{-1} \tag{8.52}$$

The first two equations in (8.52) are prediction equations that specify the one-step-ahead prediction of the state vector and its covariance matrix. The third and fourth equations update the mean and the covariance matrix of the state vector after the new observation $\mathbf{y}_{t+1}$ has become available. The revised estimate $\hat{\mathbf{S}}_{t+1|t+1}$ is the sum of the projected estimate using observations up to time $t, \hat{\mathbf{S}}_{t+1|t}$, and a linear combination of the most recent one-step-ahead forecast errors. These forecast errors are also called the innovations, since they represent the new information brought by $\mathbf{y}_{t+1}$, in addition to the information contained in the past history. The matrix $\mathbf{k}_{t+1}$ is called the *Kalman gain*; it determines how much weight should be given to the most recent forecast errors.

The equations in (8.52) are solved recursively, starting with the mean $\hat{\mathbf{S}}_{0|0}$ and the covariance matrix $\mathbf{P}_{0|0}$ of the prior distribution $p(\mathbf{S}_0)$. Values for these initial conditions have to be specified. If one is relatively uncertain about this distribution, one can start with a diagonal covariance matrix $\mathbf{P}_{0|0}$ with large elements. Then if n is moderately large, the initial choice for $\hat{\mathbf{S}}_{0|0}$ is dominated by the information from the data. Apart from these initial conditions, one has to specify the matrices $\mathbf{H}_t$, $\mathbf{A}$, $\mathbf{G}$, $\mathbf{R}_1$, and $\mathbf{R}_2$.

Originally, Kalman (1960) and Kalman and Bucy (1961) derived these recursive updating equations by a method of orthogonal projection. Here we have adopted a Bayesian approach, in which the state vector $\mathbf{S}_0$ is assigned a prior distribution with mean vector $\hat{\mathbf{S}}_{0|0}$ and covariance matrix $\mathbf{P}_{0|0}$. Then $\hat{\mathbf{S}}_{t|t}$ is the mean of the posterior distribution. [For further discussion, see Ho and Lee (1964) and the review of Kalman filtering by Mehra (1979).] Duncan and Horn (1972) use yet another approach and derive the same equations from a regression (least squares) point of view. In their approach, $\hat{\mathbf{S}}_{0|0}$ and $\mathbf{P}_{0|0}$ are taken as fixed parameters.

8.3.2. Bayesian Forecasting

In the context of Bayesian forecasting, Harrison and Stevens (1976) adopt a state space representation and treat the coefficients in the general linear model as the unknown states:

$$\mathbf{y}_t = \mathbf{H}_t \boldsymbol{\beta}_t + \boldsymbol{\varepsilon}_t$$

$$\boldsymbol{\beta}_{t+1} = \mathbf{A}\boldsymbol{\beta}_t + \mathbf{a}_{t+1} \tag{8.53}$$

Since the coefficients $\boldsymbol{\beta}_t$ themselves vary over time, they refer to this model as the *dynamic linear model*. It is a special case of the model in Equations (8.47) and (8.48), with $\mathbf{S}_t = \boldsymbol{\beta}_t$ and no input variables $\mathbf{u}_t$. Here $\mathbf{H}_t$ is a matrix of independent variables. The Kalman filter equations in (8.52) can be used to derive the estimates $\hat{\boldsymbol{\beta}}_{t|t}$ and their covariance matrix $\mathbf{P}_{t|t}$.

The objective of Bayesian forecasting is to derive the predictive distribution of a future observation $\mathbf{y}_{t+l} = \mathbf{H}_{t+l}\boldsymbol{\beta}_{t+l} + \boldsymbol{\varepsilon}_{t+l}$. For this we have to make inferences about the future parameters $\boldsymbol{\beta}_{t+l}$ and the independent variables in $\mathbf{H}_{t+l}$. The first two equations in (8.52) can be used recursively to derive the mean $\hat{\boldsymbol{\beta}}_{t+l|t}$ and the covariance matrix $\mathbf{P}_{t+l|t}$ of the posterior distribution of

$$\boldsymbol{\beta}_{t+l} = \mathbf{A}^l \boldsymbol{\beta}_t + \mathbf{a}_{t+l} + \mathbf{A}\mathbf{a}_{t+l-1} + \cdots + \mathbf{A}^{l-1}\mathbf{a}_{t+1}$$

They are given by

$$\hat{\boldsymbol{\beta}}_{t+l|t} = \mathbf{A}\hat{\boldsymbol{\beta}}_{t+l-1|t} = \cdots = \mathbf{A}^l \hat{\boldsymbol{\beta}}_{t|t}$$

and

$$\mathbf{P}_{t+l|t} = \mathbf{A}^l \mathbf{P}_{t|t}(\mathbf{A}')^l + \sum_{j=0}^{l-1} \mathbf{A}^j \mathbf{R}_2 (\mathbf{A}')^j \tag{8.54}$$

Let us assume that the future independent variables $\mathbf{H}_{t+l}$ are known. Then the l-step-ahead forecast of $\mathbf{y}_{t+l}$ is given by

$$\hat{\mathbf{y}}_{t+l|t} = E(\mathbf{y}_{t+l}|\mathbf{Y}_t) = \mathbf{H}_{t+l}\hat{\boldsymbol{\beta}}_{t+l|t} = \mathbf{H}_{t+l}\mathbf{A}^l\hat{\boldsymbol{\beta}}_{t|t} \tag{8.55}$$

and its covariance matrix by

$$\begin{aligned} \mathbf{V}(\mathbf{y}_{t+l}|\mathbf{Y}_t) &= \mathbf{H}_{t+l}\mathbf{V}\left(\hat{\boldsymbol{\beta}}_{t+l|t}\right)\mathbf{H}'_{t+l} + \mathbf{R}_1 \\ &= \mathbf{H}_{t+l}\left[\mathbf{A}^l\mathbf{P}_{t|t}(\mathbf{A}')^l + \sum_{j=0}^{l-1}\mathbf{A}^j\mathbf{R}_2(\mathbf{A}')^j\right]\mathbf{H}'_{t+l} + \mathbf{R}_1 \end{aligned} \tag{8.56}$$

In the case when $\mathbf{H}_{t+l}$ is unknown, one has to incorporate the variability that arises from the uncertainty in $\mathbf{H}_{t+l}$. However, the model then becomes nonlinear and the variance approximations to (8.56) become more complicated [for details, see Harrison and Stevens (1976)].

Regression models with constant and time-varying coefficients (static and dynamic regression) arise as special cases of the dynamic linear model in (8.53) and are discussed in the next section. Also polynomial trend models can be written in this form. The *steady model* $y_t = \mu_t + \varepsilon_t$, where $\mu_{t+1} = \mu_t + a_{t+1}$ [see (8.51)], and the *linear growth model* $y_t = \mu_t + \varepsilon_t$, where $\mu_{t+1} = \mu_t + \beta_{t+1} + a_{1,t+1}$ and $\beta_{t+1} = \beta_t + a_{2,t+1}$, are discussed in Exercises 8.10 and 8.11.

8.3.3. Models with Time-Varying Coefficients

Let us consider now a special one-dimensional case of the model in (8.53) and study the regression model with time-varying coefficients:

$$\begin{aligned} y_t &= \mathbf{x}'_t\boldsymbol{\beta}_t + \varepsilon_t \\ \boldsymbol{\beta}_{t+1} &= \boldsymbol{\beta}_t + \mathbf{a}_{t+1} \end{aligned} \tag{8.57}$$

For simplicity we have set $\mathbf{A} = \mathbf{I}$. Then the coefficients vary according to a multivariate version of the random walk model. Since the dimension of the measurement equation is 1, we can reparameterize the covariance matrix $\mathbf{R}_2$ as $\mathbf{R}_2 = \mathbf{V}(\mathbf{a}_t) = \sigma^2\boldsymbol{\Omega}$, where $\sigma^2 = V(\varepsilon_t) = R_1$. Then $\mathbf{V}(\mathbf{a}_t)/V(\varepsilon_t) = \boldsymbol{\Omega}$ is the variance ratio matrix. The diagonal elements in this matrix measure the variation in the shocks that drive the coefficients $\boldsymbol{\beta}_t$, as compared to the variance of the measurement noise.

The Kalman filter equations in (8.52) can be used to derive the first two moments of the conditional distributions of $\boldsymbol{\beta}_t$. Here we also reparameterize the covariance matrices $\mathbf{P}_{t+1|t}$ and $\mathbf{P}_{t+1|t+1}$ by dividing through by σ^2. Then the conditional distributions are

$$(\boldsymbol{\beta}_{t+1}|\mathbf{Y}_t) \sim N\left(\hat{\boldsymbol{\beta}}_{t+1|t}, \sigma^2\mathbf{P}_{t+1|t}\right)$$

$$(\boldsymbol{\beta}_{t+1}|\mathbf{Y}_{t+1}) \sim N\left(\hat{\boldsymbol{\beta}}_{t+1|t+1}, \sigma^2\mathbf{P}_{t+1|t+1}\right)$$

with updating equations

$$\hat{\boldsymbol{\beta}}_{t+1|t} = \hat{\boldsymbol{\beta}}_{t|t}$$

$$\mathbf{P}_{t+1|t} = \mathbf{P}_{t|t} + \boldsymbol{\Omega}$$

$$\hat{\boldsymbol{\beta}}_{t+1|t+1} = \hat{\boldsymbol{\beta}}_{t+1|t} + \mathbf{k}_{t+1}\left(y_{t+1} - \mathbf{x}'_{t+1}\hat{\boldsymbol{\beta}}_{t+1|t}\right)$$

$$\mathbf{P}_{t+1|t+1} = \mathbf{P}_{t+1|t} - \mathbf{k}_{t+1}\mathbf{x}'_{t+1}\mathbf{P}_{t+1|t}$$

$$\mathbf{k}_{t+1} = \left(\mathbf{x}'_{t+1}\mathbf{P}_{t+1|t}\mathbf{x}_{t+1} + 1\right)^{-1}\mathbf{P}_{t+1|t}\mathbf{x}_{t+1} \tag{8.58}$$

Let us set $\hat{\boldsymbol{\beta}}_t = \hat{\boldsymbol{\beta}}_{t|t}$ and $\mathbf{P}_t = \mathbf{P}_{t|t}$ to simplify the notation. Then Equations (8.58) can be summarized as

$$\hat{\boldsymbol{\beta}}_{t+1} = \hat{\boldsymbol{\beta}}_t + \left[1 + \mathbf{x}'_{t+1}(\mathbf{P}_t + \boldsymbol{\Omega})\mathbf{x}_{t+1}\right]^{-1}(\mathbf{P}_t + \boldsymbol{\Omega})\mathbf{x}_{t+1}\left(y_{t+1} - \mathbf{x}'_{t+1}\hat{\boldsymbol{\beta}}_t\right)$$

$$\mathbf{P}_{t+1} = (\mathbf{P}_t + \boldsymbol{\Omega}) - \left[1 + \mathbf{x}'_{t+1}(\mathbf{P}_t + \boldsymbol{\Omega})\mathbf{x}_{t+1}\right]^{-1}(\mathbf{P}_t + \boldsymbol{\Omega})\mathbf{x}_{t+1}\mathbf{x}'_{t+1}(\mathbf{P}_t + \boldsymbol{\Omega}) \tag{8.59}$$

In the regression model with *constant coefficients* ($\boldsymbol{\Omega} = \mathbf{0}$), they simplify to

$$\hat{\boldsymbol{\beta}}_{t+1} = \hat{\boldsymbol{\beta}}_t + [1 + \mathbf{x}'_{t+1}\mathbf{P}_t\mathbf{x}_{t+1}]^{-1}\mathbf{P}_t\mathbf{x}_{t+1}\left(y_{t+1} - \mathbf{x}'_{t+1}\hat{\boldsymbol{\beta}}_t\right)$$

$$\mathbf{P}_{t+1} = \mathbf{P}_t - [1 + \mathbf{x}'_{t+1}\mathbf{P}_t\mathbf{x}_{t+1}]^{-1}\mathbf{P}_t\mathbf{x}_{t+1}\mathbf{x}'_{t+1}\mathbf{P}_t \tag{8.60}$$

These updating equations are the same as Plackett's (1950) expressions for recursive least squares, provided the initial estimates at time p (where p is the number of coefficients) are chosen as $\hat{\boldsymbol{\beta}}_p = (\mathbf{X}'_p\mathbf{X}_p)^{-1}\mathbf{X}'_p\mathbf{y}_p$ and $\mathbf{P}_p = (\mathbf{X}'_p\mathbf{X}_p)^{-1}$, where $\mathbf{X}_p$ and $\mathbf{y}_p$ are the design matrix and the observation vector from the first p equations of the regression model ($t = 1, \ldots, p$).

A Special Case

Additional insight into the updating equations, (8.59) and (8.60), can be gained by considering the case of just one independent variable, $y_t = \beta_t x_t + \varepsilon_t$.

For *time-varying coefficients* [$\beta_{t+1} = \beta_t + a_{t+1}$, with $V(a_t) = \sigma^2\omega$, where ω is the variance ratio $V(a_t)/V(\varepsilon_t)$], the updating equations are

$$P_{t+1} = (P_t + \omega) - \frac{x_{t+1}^2(P_t + \omega)^2}{1 + x_{t+1}^2(P_t + \omega)} = \frac{P_t + \omega}{1 + x_{t+1}^2(P_t + \omega)}$$

and

$$\hat{\beta}_{t+1} = \hat{\beta}_t + P_{t+1}x_{t+1}(y_{t+1} - x_{t+1}\hat{\beta}_t) \tag{8.61}$$

For *constant coefficients* ($\omega = 0$), they are

$$P_{t+1} = \frac{P_t}{1 + x_{t+1}^2 P_t}$$

$$\hat{\beta}_{t+1} = \hat{\beta}_t + P_{t+1}x_{t+1}(y_{t+1} - x_{t+1}\hat{\beta}_t) \tag{8.62}$$

Now through repeated substitution in the first equation of (8.62), we find that

$$P_{t+1} = \left(1 + P_1 \sum_{j=2}^{t+1} x_j^2\right)^{-1} P_1$$

Furthermore, substituting $P_1 = 1/x_1^2$, which is the standardized variance of the coefficient estimate in the regression model from just one observation, we find that

$$P_{t+1} = \frac{1}{\sum_{j=1}^{t+1} x_j^2}$$

and

$$\hat{\beta}_{t+1} = \hat{\beta}_t + \frac{x_{t+1}}{\sum_{j=1}^{t+1} x_j^2}(y_{t+1} - x_{t+1}\hat{\beta}_t) \tag{8.63}$$

In the constant linear regression model, the adjustment weights for $x_{t+1}(y_{t+1} - x_{t+1}\hat{\beta}_t)$ depend only on the past x's. In the regression model with time-varying coefficients, the weights P_{t+1} also depend on the parameter $\omega > 0$; they are always larger than the ones for constant coefficients. This indicates that, as expected, the time-varying coefficient model gives more weight to the most recent forecast error. Thus, the time-varying coefficient model revises the coefficients faster than the constant coefficient least squares procedure.

Selection of Ω

In order to calculate the parameter estimates in (8.59) and the forecasts of future observations from (8.55), one has to specify the variance ratio matrix $\mathbf{\Omega}$. One can either pick the variances of the process noise a priori (usually quite difficult) or use past data to estimate these unknown parameters. Several estimation approaches are discussed in the literature [see Harrison and Stevens (1976) and Mehra (1979) for a review]. Here we discuss a *maximum likelihood* approach [see Schweppe (1965), Garbade (1977), Ledolter (1979)]; a short description is given below.

It follows from (8.55) and (8.56) that for the regression model with random walk coefficients in (8.57) the conditional distribution $p(y_t|\mathbf{Y}_{t-1})$ is normal with mean $\mathbf{x}'_t\hat{\boldsymbol{\beta}}_{t-1}$ and variance $\sigma^2 f_t$, where $f_t = 1 + \mathbf{x}'_t(\mathbf{P}_{t-1} + \mathbf{\Omega})\mathbf{x}_t$. Therefore, the joint probability density of $y_1, \ldots, y_n$ is

$$p(y_1, y_2, \ldots, y_n|\sigma^2, \mathbf{\Omega}) = p(y_1)\prod_{t=2}^{n} p(y_t|\mathbf{Y}_{t-1})$$

$$= \prod_{t=1}^{n} \frac{1}{\sigma(2\pi f_t)^{1/2}} \exp\left\{-\frac{1}{2\sigma^2 f_t}(y_t - \mathbf{x}'_t\hat{\boldsymbol{\beta}}_{t-1})^2\right\}$$

Here we have assumed that the starting values $\hat{\boldsymbol{\beta}}_{0|0}$ and $\mathbf{P}_{0|0}$, which are necessary to calculate $p(y_1)$ and to start the recursions in (8.59), are known. Then for given data on $\mathbf{x}$ and $\mathbf{y}$, the log-likelihood function of σ^2 and $\mathbf{\Omega}$ is

$$l(\sigma^2, \mathbf{\Omega}|\text{ data}; \hat{\boldsymbol{\beta}}_{0|0}, \mathbf{P}_{0|0}) \propto -n\ln\sigma - \frac{1}{2}\sum_{t=1}^{n}\ln f_t - \frac{1}{2\sigma^2}\sum_{t=1}^{n}\frac{(y_t - \mathbf{x}'_t\hat{\boldsymbol{\beta}}_{t-1})^2}{f_t} \tag{8.64}$$

Maximizing this function with respect to σ^2 leads to the estimate

$$\hat{\sigma}^2 = \frac{1}{n}\sum_{t=1}^{n}\frac{(y_t - \mathbf{x}'_t\hat{\boldsymbol{\beta}}_{t-1})^2}{f_t} \tag{8.65}$$

Substituting this estimate into (8.64), we obtain

$$l_c\left(\boldsymbol{\Omega} \mid \text{data}; \hat{\boldsymbol{\beta}}_{0|0}, \mathbf{P}_{0|0}\right) \propto -\frac{1}{2} \sum_{t=1}^{n} \ln f_t - n \ln \hat{\sigma} \tag{8.66}$$

This is referred to as the *concentrated log-likelihood function*. For given $\boldsymbol{\Omega}$, Equations (8.59) are used to update the parameter estimates and to calculate $f_t = 1 + \mathbf{x}_t'(\mathbf{P}_{t-1} + \boldsymbol{\Omega})\mathbf{x}_t$ and $\hat{\sigma}^2$. Values of $\boldsymbol{\Omega}$ that maximize l_c can be found numerically. However, if there is more than one parameter, the maximization with respect to elements of the symmetric nonnegative definite matrix $\boldsymbol{\Omega}$ is quite tedious. Furthermore, for moderate sample sizes the log-likelihood function may be quite flat. Thus, in practice, simplifications are necessary. For example, as a first step beyond a constant coefficient model, one could assume that the coefficients vary independently and follow random walks with possibly different variances. In this case the maximization has to be performed with respect to the nonnegative diagonal elements in $\boldsymbol{\Omega}$.

Additional Comments

In model (8.57) the coefficients change according to a multivariate random walk. This is easily extended to regression models in which the coefficients follow multivariate ARIMA processes.

Here we have considered only the regression model. However, an extension to time series models with time-varying coefficients is also possible. For strictly autoregressive models the generalization is straightforward; in this case $\mathbf{x}_t = (y_{t-1}, \ldots, y_{t-p})'$. However, models with time-varying moving average coefficients cannot be written in a convenient state space representation, and linear approximations have to be considered [for details see Ledolter (1981)].

8.4. ADAPTIVE FILTERING

Forecasters are frequently concerned that the coefficients in their forecast models are not constant but vary over time. This concern has encouraged statisticians and operations researchers to develop forecast techniques that allow for time-varying coefficients. In Section 8.3.3 we have illustrated a model-based approach and have parameterized the time-varying coefficients through stochastic processes. We have shown how to estimate the coefficients in such models and how to forecast future observations. This approach discounts the past observations somewhat faster than ordinary least squares.

Another approach for modeling time-varying coefficients has been developed by Makridakis and Wheelwright (1977, 1978). Their approach, known as *adaptive filtering*, consists of a heuristic recursive algorithm that revises the coefficient estimates as each new observation becomes available. The weights in these recursions are chosen such that the coefficient estimates adapt more quickly to changes in the underlying parameters. Here we describe adaptive filtering only in the autoregressive model; for extensions to ARIMA models, refer to the original papers.

In the autoregressive forecasting model

$$y_t = \sum_{i=1}^{p} \beta_i y_{t-i} + a_t = \mathbf{y}'_{(t)}\boldsymbol{\beta} + a_t \tag{8.67}$$

where $\mathbf{y}_{(t)} = (y_{t-1}, \ldots, y_{t-p})'$, the estimate of the coefficient vector $\boldsymbol{\beta} = (\beta_1, \ldots, \beta_p)'$ at time $t + 1$ is updated according to

$$\hat{\boldsymbol{\beta}}_{t+1} = \hat{\boldsymbol{\beta}}_t + 2\alpha \mathbf{y}_{(t+1)}\left(y_{t+1} - \mathbf{y}'_{(t+1)}\hat{\boldsymbol{\beta}}_t\right) \tag{8.68}$$

The coefficient α is a learning constant and determines how much weight should be given to the most recent one-step-ahead forecast error $y_{t+1} - \mathbf{y}'_{(t+1)}\hat{\boldsymbol{\beta}}_t$ when revising the estimate. One suggestion is to choose this constant in the range $0 < \alpha \leqslant \alpha_{\max}$, where

$$\alpha_{\max} = \frac{1}{\max\limits_{p+1 \leqslant t \leqslant n} \mathbf{y}'_{(t)}\mathbf{y}_{(t)}}$$

and n is the sample size. Makridakis and Wheelwright (1978) have also made other suggestions, including a standardization of the observations.

To start the revisions, one needs an initial estimate of $\boldsymbol{\beta}$. One could start with the ordinary least squares estimate $\hat{\boldsymbol{\beta}}_p$ that is derived from the first p equations of the model in (8.67) and then apply the updating equations in (8.68) sequentially. One complete pass through the series $y_{p+1}, \ldots, y_n$ is called an *iteration*. Usually several iterations are required; the last estimate $\hat{\boldsymbol{\beta}}_n$ from an iteration is used as the starting value $\hat{\boldsymbol{\beta}}_p$ in the subsequent iteration. If the relative change in the mean square error of one-step-ahead forecasts,

$$\text{MSE} = \frac{\sum\limits_{t=p+1}^{n} \left(y_t - \mathbf{y}'_{(t)}\hat{\boldsymbol{\beta}}_{t-1}\right)^2}{n - p}$$

from one iteration to the next is smaller than a predetermined constant, the iterations are stopped, and the estimate $\hat{\boldsymbol{\beta}}_n$ from the last iteration is used in the prediction of all future observations. After a new observation y_{n+1} is obtained, the parameter estimate $\hat{\boldsymbol{\beta}}_{n+1}$ can be calculated and the forecasts can be revised.

The parameter estimates and forecasts from adaptive filtering depend on the learning constant α. A small α should be used for stable series, and a larger value for series that exhibit coefficient variability. The results of simulation studies [Ledolter and Kahl (1982)] have shown that the correct choice of the learning constant is important and that usually the upper bound $\alpha_{\max}$ will lead to large forecast errors. If sufficient data are available, one should estimate the learning constant. For example, one could choose the learning constant α that minimizes the mean square error of past one-step-ahead forecasts.

Adaptive filtering has been questioned on both theoretical and empirical grounds. A major criticism of this technique has been the fact that it does not make explicit reference to a model for the coefficient changes and that it does not discuss the assumptions that are necessary to yield the revision rules in (8.68).

A comparison of the updating equations in (8.68) with the Kalman filter in Section 8.3 provides additional insight into adaptive filtering. The updating equations for the estimates in the (auto)regressive model with random walk coefficients (8.59) and the recursion in (8.68) are quite similar. The main difference, however, is that while the learning constant α is constant, the adjustment weights

$$\left[1 + \mathbf{y}'_{(t+1)}(\mathbf{P}_t + \Omega)\mathbf{y}_{(t+1)}\right]^{-1}(\mathbf{P}_t + \Omega)$$

in (8.59) are data-dependent and are functions of the covariance matrix of the coefficients [see Nau and Oliver (1979) for additional discussion].

8.5. FORECAST EVALUATION, COMPARISON, AND CONTROL

The forecast system introduced in Chapter 1 consists of two phases: model building and forecasting. In the first phase, models are constructed through a three-stage iterative strategy. In the second phase, forecasts are generated and the stability of the system is checked. These two phases are not quite as separable as it might appear from our earlier discussion. In fact, forecasts may also be used to validate and compare models and to control the system.

The residuals, residual autocorrelations, mean square error, and so on, are used to select models at the diagnostic checking stage. This is part of the modeling phase and is very important. However, the residuals or the historic one-step-ahead forecast errors depend on estimates of unknown parameters. Thus the selected model might fit very well the data from which the estimates are calculated. However, when the forecasts are compared with future data that are not used for estimation, the agreement need not be as good. Hence, comparisons of forecasts with actual observations can be additional useful tools for model evaluation and selection [see Box and Tiao (1976)]. In practical situations it may be unreasonable to expect many future observations. However, in long series one can use the initial part for model construction and the remaining part as a holdout period for forecast evaluation and comparison. Such an approach is pursued in this section.

Suppose we have a time series $z_1, z_2, \ldots, z_n, z_{n+1}, \ldots, z_{n+m}$ and have constructed a model from the first n observations. Then we can generate l-step-ahead forecasts $\hat{z}_n(l)$ $(l = 1, 2, \ldots, m)$ from time origin n and the corresponding forecast errors $e_n(l) = z_{n+l} - \hat{z}_n(l)$. Furthermore, we can calculate successive one-step-ahead forecasts $\hat{z}_{n+t-1}(1)$ of z_{n+t} from origin $n + t - 1$ and the corresponding forecast errors $e_{n+t-1}(1) = z_{n+t} - \hat{z}_{n+t-1}(1)$ $(t = 1, 2, \ldots, m)$. The carets indicate that the forecasts depend on the parameter estimates obtained from the first n data values. However, in long series the estimation error is negligible, and the parameter estimates can be taken as the true values.

In the context of the time series models in Chapter 5, we have written the l-step-ahead forecast error as a linear combination of the future shocks,

$$e_n(l) = a_{n+l} + \psi_1 a_{n+l-1} + \cdots + \psi_{l-1} a_{n+1}$$

where ψ_j $(j = 1, 2, \ldots, l - 1)$ are the ψ weights of the model and the a_t's are independent identically distributed $N(0, \sigma^2)$ variables. Usually this equality is only approximate, since the parameters are estimated. Hence if we set $\mathbf{e} = [e_n(1), \ldots, e_n(m)]'$, and $\mathbf{a} = (a_{n+1}, \ldots, a_{n+m})'$, then

$$\mathbf{e} = \mathbf{\Psi a} \tag{8.69}$$

where

$$\mathbf{\Psi} = \begin{bmatrix} 1 & 0 & 0 & \cdots & 0 \\ \psi_1 & 1 & 0 & \cdots & 0 \\ \psi_2 & \psi_1 & 1 & \cdots & 0 \\ \vdots & \vdots & \vdots & & \vdots \\ \psi_{m-1} & \psi_{m-2} & \psi_{m-3} & \cdots & 1 \end{bmatrix}$$

The forecast errors $e_n(l)$ for fixed time origin n but for different lead times are correlated. However, the consecutive one-step-ahead forecast errors $e_{n+t-1}(1) = z_{n+t} - \hat{z}_{n+t-1}(1) = a_{n+t}$ $(t = 1, 2, \ldots, m)$ are uncorrelated and become very important for forecast evaluation. For notational convenience, we write $e_{n+t} = e_{n+t-1}(1)$ for $t = 1, 2, \ldots, m$.

8.5.1. Forecast Evaluation

Good forecast models should lead to small uncorrelated one-step-ahead forecast errors. Various informal checks based on these errors can be performed.

Bias

The mean error

$$\text{ME} = \bar{e} = \frac{\sum_{t=1}^{m} e_{n+t}}{m}$$

is an estimate of forecast bias. Since the quantity $T = \sqrt{m}\,\bar{e}/\sigma$ has an approximate $N(0, 1)$ distribution, an overall test for forecast bias can be obtained by referring T to the standard normal table. In practice, σ^2 must be estimated from n data values to which g parameters have been fitted. Hence, a better approximation is to refer $\hat{T} = \sqrt{m}\,\bar{e}/\hat{\sigma}$ to a t table with $n - g$ degrees of freedom. When n is large, this refinement will make little difference.

Sum of Squares

Since $\sum_{t=1}^{m} a_{n+t}^2/\sigma^2$ follows a chi-square distribution with m degrees of freedom, we can obtain an overall test for the appropriateness of the model during the holdout period $(n + 1, \ldots, n + m)$ by comparing

$$Q_e = \frac{\sum_{t=1}^{m} e_{n+t}^2}{\sigma^2} \tag{8.70}$$

to the percentage points of a chi-square distribution with m degrees of freedom. To account for the error in the estimation of σ^2, a better approximation may be to refer

$$\hat{Q}_e = \frac{\sum_{t=1}^{m} e_{n+t}^2}{m\hat{\sigma}^2}$$

to an F table with m and $n - g$ degrees of freedom.

Autocorrelations

The one-step-ahead forecast errors are supposed to be uncorrelated. Thus, we expect their sample autocorrelations r_k ($k = 1, 2, \dots$) to be close to zero. To check for significant autocorrelations, we compare r_k with its standard error $m^{-1/2}$.

Actual Values Versus Forecasts

A plot of z_{n+t} versus $\hat{z}_{n+t-1}(1)$ ($t = 1, 2, \dots, m$) can also be quite useful. If the forecasts and the observations are identical, all points will fall on a 45° line through the origin. Departures from this line indicate model inadequacy. For example, the forecasts are biased if most of the points are bclow (above) this line.

Onc could also fit a linear regression of z_{n+t} on $\hat{z}_{n+t-1}(1)$ [i.e., $z_{n+t} = \beta_0 + \beta_1 \hat{z}_{n+t-1}(1) + \varepsilon_{n+t}$]. Departures of the estimate $\hat{\beta}_0$ from zero and $\hat{\beta}_1$ from 1 are signs of poor forecasts. Regression tests may be performed to check for such departures. One can also calculate the sample correlation coefficient between the actual observations and the forecasts. Small values indicate inaccurate forecasts.

Histogram

To check whether the one-step-ahead forecast errors are normally distributed, one can construct a histogram of the standardized errors $e_{n+t}/\hat{\sigma}$ and compare it with a standard normal distribution.

8.5.2. Forecast Comparison

The informal checks introduced above can also be used to compare different forecast methods. The comparison can be carried out in terms of the following summary statistics.

1. Mean error

$$\text{ME} = \frac{\sum_{t=1}^{m} e_{n+t}}{m}$$

2. Mean percent error

$$\text{MPE} = \frac{100}{m} \sum_{t=1}^{m} \frac{e_{n+t}}{z_{n+t}}$$

3. Mean square error

$$\text{MSE} = \frac{\sum_{t=1}^{m} e_{n+t}^2}{m}$$

4. Mean absolute error

$$\text{MAE} = \frac{\sum_{t=1}^{m} |e_{n+t}|}{m}$$

5. Mean absolute percent error

$$\text{MAPE} = \frac{100}{m} \sum_{t=1}^{m} \left| \frac{e_{n+t}}{z_{n+t}} \right| \tag{8.71}$$

The first two statistics measure forecast bias and should be close to zero. The other three measure forecast accuracy; methods that yield small values for these statistics should be chosen.

Example

The car sales data were analyzed in Chapters 4 and 6. We considered

1. General exponential smoothing (Sec. 4.3.2)
2. Winters' additive smoothing method (Sec. 4.4.1)
3. Seasonal ARIMA models (Sec. 6.7.4)

In each case the first 96 observations were used to choose the smoothing constants and to construct the models. One-step-ahead forecasts and the corresponding errors for the next 12 periods are shown in Table 8.3. The summary statistics in (8.71) are also given for each method. Based on ME, MPE, and MSE, the ARIMA model performs best. However, the values of MAE and MAPE in the Winters' method are slightly smaller. It appears that in this example the ARIMA model and Winters' smoothing method are both appropriate.

8.5.3. Forecast Control

Usually it is assumed that the model structure and the parameters stay the same during the forecast period; this implies that the forecast-generating process is in control. If this assumption is correct, then the one-step-ahead forecast errors e_{n+t}, $t = 1, 2, \ldots,$ are from an $N(0, \sigma^2)$ distribution that remains constant over time. Hence, approximately 95 percent of the forecast

Table 8.3. Comparison of the Forecasts from General Exponential Smoothing, Winters' Additive Seasonal Forecast Procedure, and the Seasonal ARIMA Model— Car Sales

		General Exponential Smoothing		Winters' Additive Procedure		Seasonal ARIMA Model	
Time	z_{96+t}	$\hat{z}_{96+t-1}(1)$	$e_{96+t-1}(1)$	$\hat{z}_{96+t-1}(1)$	$e_{96+t-1}(1)$	$\hat{z}_{96+t-1}(1)$	$e_{96+t-1}(1)$
97	13.210	14.263	−1.053	14.190	−.980	12.911	.299
98	14.251	15.435	−1.184	14.654	−.403	13.193	1.058
99	20.139	19.625	.514	20.053	.086	21.960	−1.821
100	21.725	24.160	−2.435	22.321	−.596	22.185	−.460
101	26.099	24.764	1.335	23.465	2.634	24.033	2.066
102	21.084	21.856	−.722	21.610	−.526	23.039	−1.955
103	18.024	17.071	.953	16.698	1.326	16.293	1.731
104	16.722	14.677	2.045	14.723	1.999	15.531	1.191
105	14.385	15.723	−1.338	13.731	.654	15.257	−.872
106	21.342	17.489	3.853	18.000	3.342	18.614	2.728
107	17.180	18.568	−1.417	19.229	−2.049	19.082	−1.902
108	14.577	14.650	−1.991	16.605	−2.028	16.275	−1.698
	ME	−0.12		0.29		0.03	
	MPE	−1.59		0.73		−0.18	
	MSE	3.23		2.86		2.66	
	MAE	1.57		1.39		1.48	
	MAPE	8.80		7.61		8.03	

errors should lie within the interval $(-2\sigma, 2\sigma)$. However, if there is a change in the structure, the distribution will shift; in particular, its mean may change. Consequently, a larger proportion of the errors will lie outside the 2σ limits. In such a case the forecast-generating process must be adjusted.

Although it is useful to compare the individual errors with the 2σ limits, it could take a long time until small changes in the mean are recognized. Averages or smoothed errors are more sensitive to such changes.

To "track" the performance of the forecasting system as new observations become available, one can consider the average or the cumulative sum of the forecast errors. The *cumulative sum of the errors*,

$$\mathrm{CE}_{n+t} = \sum_{j=1}^{t} e_{n+j} = e_{n+t} + \mathrm{CE}_{n+t-1} \tag{8.72}$$

where $\mathrm{CE}_n \equiv 0$, is preferred here, since it is more easily updated as new forecast errors become available. Alternatively, one could use the *smoothed error* [see Gilchrist (1976)]

$$\begin{aligned}\mathrm{SE}_{n+t} &= \gamma e_{n+t} + (1-\gamma)\mathrm{SE}_{n+t-1} \\ &= \gamma \sum_{j=0}^{t-1} (1-\gamma)^j e_{n+t-j} + (1-\gamma)^t \mathrm{SE}_n \end{aligned} \tag{8.73}$$

where $\mathrm{SE}_n \equiv 0$.

If the process is under control, then these statistics should vary around mean zero. Thus one needs a measure of their variability to check whether the mean has changed from zero. The standard errors of the statistics, (8.72) and (8.73), depend on the standard deviation of the one-step-ahead forecast errors. This can be estimated by the sample standard deviation of the N most recent errors:

$$\hat{\sigma}_{n+t} = \left[\frac{\sum_{j=t-N+1}^{t} e_{n+j}^2}{N} \right]^{1/2}$$

However, the computation and data-storage requirement make this estimate less attractive than the one based on the mean absolute deviation [see Sec. 3.8, Eq. (3.71)]:

$$\tilde{\sigma}_{n+t} = 1.25\,\hat{\Delta}_{n+t} \tag{8.74}$$

where $\hat{\Delta}_{n+t}$ may be taken as

$$\hat{\Delta}_{n+t} = \frac{\sum_{j=t-N+1}^{t} |e_{n+j}|}{N}$$

or as

$$\hat{\Delta}_{n+t} = \gamma |e_{n+t}| + (1-\gamma)\,\hat{\Delta}_{n+t-1} \tag{8.75}$$

In practice, the statistic based on the smoothed absolute errors in (8.75) is preferred.

Either ratio $\mathrm{CE}_{n+t}/\hat{\Delta}_{n+t}$ or $\mathrm{SE}_{n+t}/\hat{\Delta}_{n+t}$ may be used to monitor the stability of the forecast-generating process. These ratios are usually referred to as *tracking signals*. The process can be considered out of control if the tracking signals exceed certain limits.

In the context of general exponential smoothing, it has been suggested [see, for example, Montgomery and Johnson (1976, pp. 163–167)] to consider the system out of control if

$$\left| \frac{\mathrm{CE}_{n+t}}{\hat{\Delta}_{n+t}} \right| > c_1 \tag{8.76}$$

where the constant c_1 is somewhere between 4 and 6.

Alternatively, one can base the decision on the smoothed errors and question the adequacy of the forecasting process whenever

$$\left| \frac{\mathrm{SE}_{n+t}}{\hat{\Delta}_{n+t}} \right| > c_2 \tag{8.77}$$

The tracking signal $\mathrm{SE}_{n+t}/\hat{\Delta}_{n+t}$ is between -1 and $+1$, and the constant c_2 is usually somewhere between .2 and .5. Whenever the tracking signal exceeds the control limits for two or three consecutive periods, corrective action is called for, and the forecast system should be checked.

8.5.4. Adaptive Exponential Smoothing

A tracking signal outside the control limits may be an indication of parameter changes in the model. To keep up with these changes, one can revise the smoothing constants. We discuss this approach in the context of simple exponential smoothing. In this case the smoothing equation is

$$S_t = \alpha_t z_t + (1 - \alpha_t) S_{t-1} = S_{t-1} + \alpha_t e_t \tag{8.78}$$

where $e_t = z_t - S_{t-1}$ is the one-step-ahead forecast error and α_t is a smoothing constant that changes with time. Various suggestions for choosing α_t have been made.

Trigg and Leach (1967) recommend

$$\alpha_t = \left| \frac{\mathrm{SE}_t}{\hat{\Delta}_t} \right| \tag{8.79}$$

where SE_t is the smoothed forecast error in (8.73) and $\hat{\Delta}_t$ is the smoothed absolute forecast error in (8.75). The smoothing constant γ in these equations is somewhere between .05 and .20. If the process is under control, then the value of the tracking signal (and also the smoothing constant) will be small. When there is bias, α_t increases and approaches 1 as a limiting case.

Whybark (1973) recommends changing the smoothing constant only if the errors exceed certain specified control limits. This scheme uses three different smoothing constants, $0 \leqslant \alpha_B \leqslant \alpha_M \leqslant \alpha_H$. The recommended values are $\alpha_B = .2$, $\alpha_M = .4$, and $\alpha_H = .8$. If δ_t denotes the indicator variable,

$$\delta_t = \begin{cases} 1 & \text{if } |e_t| > 4\sigma, \text{ or } |e_{t-1}| > 1.2\sigma \text{ and } |e_t| > 1.2\sigma \\ & \text{and both errors with the same sign} \\ 0 & \text{otherwise} \end{cases}$$

then the smoothing constant is adjusted according to

$$\alpha_t = \begin{cases} \alpha_H & \text{if } \delta_t = 1 \\ \alpha_M & \text{if } \delta_t = 0 \text{ and } \delta_{t-1} = 1 \\ \alpha_B & \text{otherwise} \end{cases} \tag{8.80}$$

In other words, if the forecast error in a single period exceeds $\pm 4\sigma$, or if two consecutive errors exceed $\pm 1.2\sigma$, then α is increased from the base value α_B to the high value α_H for one period, is reduced to α_M for the next, and is then reset to α_B.

References

Abraham, B. (1980). Intervention analysis and multiple time series, *Biometrika*, **67**, 73–80.

______(1983). Intervention model analysis, *Encyclopedia of Statistical Sciences*, Wiley, New York, (forthcoming).

Abraham, B., and G. E. P. Box (1978). Deterministic and forecast adaptive time dependent models, *Appl. Statist.*, **27**, 120–130.

______(1979). Bayesian analysis of some outlier problems in time series, *Biometrika*, **66**, 229–236.

Abraham, B., and C. Chatterjee (1982). Seasonal adjustment with X-11 ARIMA and forecast efficiency, *Tech. Rep. STAT-82-05*, Department of Statistics, University of Waterloo, Waterloo, Ontario.

Ansley, C. F. (1979). An algorithm for the exact likelihood of a mixed autoregressive moving average process, *Biometrika*, **66**, 59–65.

Ansley, C. F., and P. Newbold (1979). On the finite sample distribution of residual autocorrelations in autoregressive moving average models, *Biometrika*, **66**, 547–553.

______(1980). Finite sample properties of estimates for autoregressive moving average models, *J. Econometrics*, **13**, 159–183.

Bartlett, M. S. (1946). On the theoretical specification of sampling properties of autocorrelated time series, *J. Roy. Statist. Soc., Ser. B*, **8**, 27–41.

______(1955). *Stochastic Processes*, Cambridge University Press, Cambridge.

Bass, F. M., and D. G. Clarke (1972). Testing distributed lag models of advertising effect, *J. Market. Res.*, **9**, 298–308.

Bhattacharyya, M. N., and A. P. Layton (1979). Effectiveness of seat belt legislation in the Queensland road toll—an Australian case study in intervention analysis, *J. Amer. Statist. Assoc.*, **74**, 596–603.

Blattberg, R. C., and A. P. Jeuland (1981). A micromodeling approach to investigate the advertising-sales relationship, *Manage. Sci.*, **27**, 988–1005.

Bongard, J. (1960). Some remarks on moving averages, *Seasonal Adjustment on Electronic Computers*, OECD, Paris, pp. 361–390.

Bowerman, B. L., and R. T. O'Connell (1979). *Time Series and Forecasting: An Applied Approach*, Duxbury Press, North Scituate, MA.

Box, G. E. P., and D. R. Cox (1964). An analysis of transformations, *J. Roy. Statist. Soc., Ser. B*, **26**, 211–243; discussion, 244-252.

Box, G. E. P., S. C. Hillmer, and G. C. Tiao (1978). Analysis and modeling of seasonal time series, in *Seasonal Analysis of Economic Time Series* (A. Zellner, Ed.), U.S. Department of Commerce, Bureau of the Census, Washington, DC, pp. 309–333.

Box, G. E. P., and G. M. Jenkins (1976). *Time Series Analysis: Forecasting and Control*, 2nd ed., Holden-Day, San Francisco.

Box, G. E. P., and P. Newbold (1971). Some comments on a paper of Coen, Gomme and Kendall, *J. Roy. Statist. Soc., Ser. A*, **134**, 229–240.

Box, G. E. P., and D. A. Pierce (1970). Distribution of residual autocorrelations in autoregressive moving average time series models, *J. Amer. Statist. Assoc.*, **65**, 1509–1526.

Box, G. E. P., and G. C. Tiao (1975). Intervention analysis with applications to economic and environmental applications, *J. Amer. Statist. Assoc.*, **70**, 70–79.

______(1976). Comparisons of forecasts and actuality, *Appl. Statist.*, **25**, 195–200.

Brown, R. G. (1962). *Smoothing, Forecasting and Prediction of Discrete Time Series*, Prentice-Hall, Englewood Cliffs, NJ.

Brown, R. G., and R. F. Meyer (1961). The fundamental theorem of exponential smoothing, *Oper. Res.*, **9**, 673–685.

Burman, J. P. (1965). Moving seasonal adjustment of economic time series, *J. Roy. Statist. Soc., Ser. A*, **128**, 534–538.

______(1967). Assessment of a seasonal adjustment procedure by spectral analysis, *Statistician*, **17**, 247–256.

______(1979). Seasonal adjustment—a survey, in *TIMS Studies in the Management Sciences*; Vol. 12, *Forecasting* (S. Makridakis and S. C. Wheelwright, Eds.), North-Holland, Amsterdam, pp. 45–57.

Clarke, D. G. (1976). Econometric measurement of the duration of advertising effect on sales, *J. Market. Res.*, **13**, 345–357.

Cleveland, W. P. (1972). Analysis and forecasting of seasonal time series, unpublished Ph.D. dissertation, University of Wisconsin, Madison.

Cleveland, W. P., and G. C. Tiao (1976). Decomposition of seasonal time series: a model for the Census X-11 program, *J. Amer. Statist. Assoc.*, **71**, 581–587.

Cleveland, W. S., D. M. Dunn, and I. J. Terpenning (1979). SABL—a resistant seasonal adjustment procedure with graphical methods for interpretation and diagnosis, in *Seasonal Analysis of Economic Time Series* (A. Zellner, Ed.), U.S. Department of Commerce, Bureau of the Census, Washington, DC, pp. 201–231.

Cogger, K. O. (1974). The optimality of general-order exponential smoothing, *Oper. Res.*, **22**, 858–867.

Computer programs for regression and time series analysis:

BMDP: Biomedical Computer Programs, Series P, 1981 edition. Department of Biomathematics, University of California, Los Angeles.

GLIM: Royal Statistical Society Working Party on Statistical Computing, London.

IDA: Interactive Data Analysis, SPSS Inc., Chicago, IL.

IMSL: International Mathematical and Statistical Libraries, Houston, TX.

MINITAB: Interactive (and Batch) Statistical Computing. Department of Statistics, Pennsylvania State University, University Park, PA.

PACK-SYSTEM: Time Series Programs, developed by D. J. Pack; available from Automatic Forecast Systems, Inc., Hartboro, PA.

SAS: Statistical Analysis System. SAS Institute Inc., Raleigh, NC.

SPSS: Statistical Package for the Social Sciences. SPSS Inc., Chicago, IL.

TS-Package: Time Series Programs, developed by A. I. McLeod; Department of Statistics, University of Waterloo, Ontario, Canada.

WMTS-Package: The Wisconsin Multiple Time Series Program, developed by G. C. Tiao; Department of Statistics, University of Wisconsin, Madison, WI.

Conley, D. L., G. S. Krahenbuhl, L. N. Burkett, and A. L. Millar (1981). Physiological correlates of female road racing performance, *Res. Quart. Exercise Sport*, **52**, 441–448.

Dagum, E. B. (1975). Seasonal factor forecasts from ARIMA models, paper presented at the 40th session of the International Statistical Institute, Warsaw, Poland.

Davies, N., C. M. Triggs, and P. Newbold (1977). Significance levels of the Box-Pierce portmanteau statistic in finite samples, *Biometrika*, **64**, 517–522.

Dent, W. (1977). Computation of the exact likelihood function of an ARIMA process, *J. Statist. Comput. Simulation*, **5**, 193–206.

Dobbie, J. M. (1963). Forecasting periodic trends by exponential smoothing, *Oper. Res.*, **11**, 908–918.

Draper, N., and H. Smith (1981). *Applied Regression Analysis*, 2nd ed., Wiley, New York.

Duncan, D. B., and S. D. Horn (1972). Linear dynamic regression estimation from the viewpoint of regression analysis, *J. Amer. Statist. Assoc.*, **67**, 815–821.

Durbin, J. (1960). The fitting of time series models, *Rev. Int. Statist. Inst.*, **28**, 233–244.

Durbin, J., and G. S. Watson (1950). Testing for serial correlation in least squares regression I, *Biometrika*, **37**, 409–428.

______(1951). Testing for serial correlation in least squares regression II, *Biometrika*, **38**, 159–178.

______(1971). Testing for serial correlation in least squares regression III, *Biometrika*, **58**, 1–19.

Erickson, G. M. (1981). Using ridge regression to estimate directly lagged effects in marketing, *J. Amer. Statist. Assoc.*, **76**, 766–773.

Fox, A. J. (1972). Outliers in time series, *J. Roy. Statist. Soc., Ser. B*, **34**, 350–363.

Garbade, K. (1977). Two methods for examining the stability of regression coefficients, *J. Amer. Statist. Assoc.*, **72**, 54–63.

Gilchrist, W. (1976). *Statistical Forecasting*, Wiley, New York.

Godolphin, E. J., and P. J. Harrison (1975). Equivalence theorems for polynomial-projecting predictors, *J. Roy. Statist. Soc., Ser. B*, **37**, 205–215.

Granger, C. W. J., and P. Newbold (1974). Spurious regressions in econometrics, *J. Econometrics*, **2**, 111–120.

______(1976). Forecasting transformed series, *J. Roy. Statist. Soc., Ser. B*, **38**, 189–203.

______(1977). *Forecasting Economic Time Series*, Academic Press, New York.

Harrison, P. J., and C. F. Stevens (1976). Bayesian forecasting, *J. Roy. Statist. Soc., Ser. B*, **38**, 205–228.

Helmer, R. M., and J. K. Johansson (1977). An exposition of the Box-Jenkins transfer function analysis with an application to the advertising-sales relationship, *J. Market. Res.*, **14**, 227–239.

Henderson, H. V., and P. F. Velleman (1981). Building multiple regression models interactively, *Biometrics*, **37**, 391–411.

Hillmer, S. C., W. R. Bell, and G. C. Tiao (1982), Modeling considerations in the seasonal adjustment of economic time series, Tech. Rep. No. 665, Department of Statistics, University of Wisconsin, Madison.

Hillmer, S. C., and G. C. Tiao (1979). Likelihood function of stationary multiple autoregressive moving average models, *J. Amer. Statist. Assoc.*, **74**, 652–660.

______(1982). An ARIMA-model-based approach to seasonal adjustment, *J. Amer. Statist. Assoc.*, **77**, 63–70.

Ho, Y. C., and R. C. K. Lee (1964). A Bayesian approach to problems in stochastic estimation and control, *IEEE Trans. Automatic Control*, **9**, 333–339.

Hoerl, A. E. (1962). Application of ridge regression to regression problems, *Chem. Eng. Prog.*, **58**, 54–59.

Hoerl, A. E., and R. W. Kennard (1970a). Ridge regression: biased estimation for non-orthogonal problems, *Technometrics*, **12**, 55–67.

______(1970b). Ridge regression: applications to non-orthogonal problems, *Technometrics*, **12**, 68–82; correction, **12**, p. 723.

Hogg, R. V., and A. T. Craig (1978). *Introduction to Mathematical Statistics*, 4th ed., Macmillan, New York.

Holt, C. C. (1957). Forecasting trends and seasonals by exponentially weighted moving averages, O. N. R. Memorandum, No. 52, Carnegie Institute of Technology.

Houston, F. S., and D. L. Weiss (1975). Cumulative advertising effects: the role of serial correlation, *Decision Sci.*, **6**, 471–481.

Kalman, R. E. (1960). A new approach to linear filtering and prediction problems, *J. Basic Eng.*, **82**, 35–45.

Kalman, R. E., and R. S. Bucy (1961). New results in linear filtering and prediction theory, *J. Basic Eng.*, **83**, 95–107.

Kendall, M. G. (1976). *Time Series*, 2nd ed., Griffin & Co., London.

Ledolter, J. (1975). Topics in time series analysis, unpublished Ph.D. dissertation, University of Wisconsin, Madison.

______(1979). A recursive approach to parameter estimation in regression and time series models, *Commun. Statist.*, **A8**, 1227–1245.

______(1981). Recursive estimation and adaptive forecasting in ARIMA models with time varying coefficients, in *Applied Time Series II* (D. F. Findley, Ed.), Academic Press, New York, pp. 449–472.

Ledolter, J., and B. Abraham (1981). Parsimony and its importance in time series forecasting, *Technometrics*, **23**, 411–414.

______(1983). Some comments on the initialization of exponential smoothing, *J. of Forecasting*, **2**, (forthcoming).

Ledolter, J., and G. E. P. Box (1978). Conditions for the optimality of exponential smoothing forecast procedures, *Metrika*, **25**, 77–93.

Ledolter, J., and D. R. Kahl (1982). An empirical evaluation of adaptive filtering, *Proceedings of American Institute for Decision Sciences*, San Francisco, Vol. 2, 342–344.

Levinson, N. (1946). The Wiener RMS (root mean square) error criterion in filter design and prediction, *J. Math. Phys.*, **25**, 261–278.

Ljung, G. M., and G. E. P. Box (1978). On a measure of lack of fit in time series models, *Biometrika*, **65**, 297–303.

______(1979). The likelihood function of stationary autoregressive-moving average models, *Biometrika*, **66**, 265–270.

Lovell, M. C. (1963). Seasonal adjustment of economic time series and multiple regression analysis, *J. Amer. Statist. Assoc.*, **58**, 993–1010.

Makridakis, S., and S. C. Wheelwright (1977). Adaptive filtering: an integrated autoregressive/moving average filter for time series forecasting, *Oper. Res. Quart.*, **28**, 425–437.

______(1978). *Forecasting—Methods and Applications*, Wiley, New York.

Mallows, C. L. (1973). Some Comments on C_p, *Technometrics*, **15**, 661–675.

Marquardt, D. W. (1963). An algorithm for least squares estimation of non-linear models, *J. Soc. Ind. Appl. Math.*, **11**, 431–441.

Martin, R. D. (1980). Robust estimation of autoregressive models, in *Directions in Time Series* (D. R. Brillinger and G. C. Tiao, Eds.), Institute of Mathematical Statistics, Hayward, CA, pp. 228–254.

McKenzie, E. (1974). A comparison of standard forecasting systems with the Box-Jenkins approach, *Statistician*, **23**, 107–116.

______(1976). An analysis of general exponential smoothing, *Oper. Res.*, **24**, 131–140.

McLeod, A. I. (1977a). Topics in time series and econometrics, unpublished Ph.D. dissertation, University of Waterloo, Waterloo, Ontario.

______(1977b). Improved Box-Jenkins estimators, *Biometrika*, **64**, 531–534.

______(1978). On the distribution of residual autocorrelations in Box-Jenkins models, *J. Roy. Statist. Soc., Ser. B*, **40**, 296–302.

Mehra, R. K. (1979). Kalman filters and their applications to forecasting, in *TIMS Studies in the Management Sciences*; Vol. 12, *Forecasting* (S. Makridakis and S. C. Wheelwright, Eds.), North-Holland, Amsterdam, pp. 75–94.

Mesnage, M. (1968). Elimination des variations saisonnieres: la nouvelle methode de l'OSCE, *Etudes et Enquetes Statistiques*, **1**, 7–78.

Montgomery, D. C., and L. A. Johnson (1976). *Forecasting and Time Series Analysis*, McGraw-Hill, New York.

Muth, J. F. (1960). Optimal properties of exponentially weighted forecasts, *J. Amer. Statist. Assoc.*, **55**, 299–306.

Narula, S. C., and J. F. Wellington (1977). Prediction, linear regression and the minimum sum of relative errors, *Technometrics*, **19**, 185–190.

Nau, R. F., and R. M. Oliver (1979). Adaptive filtering revisited, *J. Oper. Res. Soc.*, **30**, 825–831.

Neter, J., and W. Wasserman (1974). *Applied Linear Statistical Models*, Irwin, Homewood, IL.

Newbold, P. (1974). The exact likelihood function for a mixed autoregressive moving average process, *Biometrika*, **61**, 423–426.

Palda, K. (1964). *The Measurement of Cumulative Advertising Effects*, Prentice-Hall, Englewood Cliffs, NJ.

Pearson, E. S., and H. O. Hartley (1966). *Biometrika Tables for Statisticans*, Volume 1, 3rd ed., Cambridge University Press, Cambridge.

Pierce, D. A. (1978). Seasonal adjustment when both deterministic and stochastic seasonality are present, in *Seasonal Analysis of Economic Time Series* (A. Zellner, Ed.), U.S. Department of Commerce, Bureau of the Census, Washington, DC, pp. 242–269.

Plackett, R. L. (1950). Some theorems on least squares, *Biometrika*, **37**, 149–157.

Plosser, C. (1979). Short-term forecasting and seasonal adjustment, *J. Amer. Statist. Assoc.*, **74**, 15–24.

Pollay, R. W. (1979). Lydiametrics: applications of econometrics to the history of advertising, *J. Advert. Hist.*, **1**, 3–18.

Rao, C. R. (1965). *Linear Statistical Inference and Its Applications*, Wiley, New York.

Ryan, T. A., B. L. Joiner, and B. F. Ryan (1976). *MINITAB Student Handbook*, Duxbury Press, North Scituate, MA.

Schweppe, F. C. (1965). Evaluation of likelihood functions for Gaussian signals, *IEEE Trans. Inform. Theory*, **11**, 61–70.

Selby, S. M. (Ed.) (1965). *Standard Mathematical Tables*, 14th ed., Chemical Rubber Company, Cleveland, OH.

Shiskin, J., A. H. Young, and J. C. Musgrave (1967). The X-11 variant of the Census Method-II Seasonal Adjustment Program, Tech. Paper No. 15, U.S. Department of Commerce, Bureau of the Census, Washington, DC.

Theil, H. (1971). *Principles of Econometrics*, Wiley, New York.

Tiao, G. C., and G. E. P. Box (1981). Modeling multiple time series with applications, *J. Amer. Statist. Assoc.*, **76**, 802–816.

Tiao, G. C., G. E. P. Box, M. R. Grupe, G. B. Hudak, W. R. Bell, and I. Chang (1979). *The Wisconsin Multiple Time Series (WMTS-1) Program: A Preliminary Guide*, Department of Statistics, University of Wisconsin, Madison, WI.

Tiao, G. C., and S. C. Hillmer (1978). Some considerations of decomposition of a time series, *Biometrika*, **65**, 497–502.

Tintner, G. (1940). *The Variate Difference Method*, Principia Press, Bloomington, IN.

Trigg, D. W., and A. G. Leach (1967). Exponential smoothing with an adaptive response rate, *Oper. Res. Quart.*, **18**, 53–59.

Tukey, J. W. (1961). Discussion emphasizing the connection between analysis of variance and spectrum analysis, *Technometrics*, **3**, 189–219.

Wallis, K. F. (1974). Seasonal adjustment and relations between variables, *J. Amer. Statist. Assoc.*, **69**, 18–31.

Weiss, D. W., F. S. Houston, and P. Windal (1978). The periodic pain of Lydia E. Pinkham, *J. Bus.*, **51**, 91–101.

Whittle, P. (1963). *Prediction and Regulation*, Van Nostrand, New York.

Whybark, D. C. (1973). A comparison of adaptive forecasting techniques, *Logistics Transp. Rev.*, **9**, 13–26.

Wichern, D. W. (1973). The behavior of the sample autocorrelation function for an integrated moving average process, *Biometrika*, **60**, 235–239.

Winters, P. R. (1960). Forecasting sales by exponentially weighted moving averages, *Manage. Sci.*, **6**, 324–342.

Wold, H. (1938). *A Study in the Analysis of Stationary Time Series* (2nd ed. 1954), Almquist and Wicksell, Uppsala.

Working, H. (1960). Note on the correlation of first differences of averages in a random chain, *Econometrica*, **28**, 916–918.

Yaglom, A. M. (1955). The correlation theory of processes whose nth difference constitute a stationary process, *Mat. Sb. N.S.*, **37**(79), 141–196. Translated in *Amer. Math. Soc. Transl., Ser. 2*, **8**(1958), 87–141.

Yamamoto, F. (1981). Predictions of multivariate autoregressive moving average models, *Biometrika*, **68**, 485–492.

Yule, G. U. (1921). On the time-correlation problems with special reference to the variate difference correlation method, *J. Roy. Statist. Soc.*, **84**, 497–526.

______(1927). On a method of investigating periodicities in disturbed series, with special reference to Wölfer's sunspot numbers, *Philos. Trans. Roy. Soc. London, Ser. A*, **226**, 267–298.

Exercises

CHAPTER 2

2.1. The model $y_t = \beta_0 + \beta_1 x_{t1} + \beta_2 x_{t2} + \varepsilon_t$ is fitted to $n = 10$ observations. It is found that

$$\Sigma x_{t1} = \Sigma x_{t2} = \Sigma x_{t1} x_{t2} = 0$$

$$\Sigma x_{t1}^2 = 20 \qquad \Sigma x_{t2}^2 = 40$$

$$\Sigma y_t = 10 \qquad \Sigma y_t^2 = 165$$

$$\Sigma x_{t1} y_t = \Sigma x_{t2} y_t = 40$$

(a) Calculate the least squares estimates $\hat{\beta}_0, \hat{\beta}_1, \hat{\beta}_2$.
(b) Calculate the ANOVA table; calculate and interpret R^2.
(c) Compute the standard errors of the least squares estimates $\hat{\beta}_0$, $\hat{\beta}_1$, and $\hat{\beta}_2$, and calculate and interpret the t statistics.
(d) Test whether $\beta_1 = \beta_2 = 0$ (use $\alpha = .05$).

2.2. Consider the linear trend model $y_t = \beta_0 + \beta_1 t + \varepsilon_t$; $t = 1, \ldots, n$. Write down the normal equations, and derive the least squares estimates. *Hint*: $\Sigma_{i=1}^{n} i^2 = n(n+1)(2n+1)/6$.

2.3. You are given past data on per-capita beer consumption (Y), per-capita real income (X_1), and relative price of beer (X_2). Using $n = 17$ observations, you are fitting the linear regression model $y_t = \beta_0 + \beta_1 x_{t1} + \beta_2 x_{t2} + \varepsilon_t$.

(a) Using a regression program you find the following results:

	$\hat{\beta}$	Std. Error
X_1 (income)	1.14	.16
X_2 (price)	−.83	.20
constant	1.37	.35

Test H_0: $\beta_1 = 0$ vs. H_1: $\beta_1 \neq 0$ at the $\alpha = .05$ level.

(b) A partial ANOVA output is given below

Source	SS	df	MS
Regression			
Error	34		
Total (corr. for mean)	100		

Complete the table, calculate R^2, and test H_0: $\beta_1 = \beta_2 = 0$ at the $\alpha = .05$ level.

2.4. A response variable Y depends on two controllable variables X_1 and X_2. Observations on Y are made at four points:

X_1	$\cos\theta$	$-\sin\theta$	$\sin\theta$	$-\cos\theta$
X_2	$\sin\theta$	$-\cos\theta$	$\cos\theta$	$-\sin\theta$
Y	y_1	y_2	y_3	y_4

(a) Obtain expressions for the least squares estimates of β_1 and β_2 in the model

$$y_t = \beta_0 + \beta_1 x_{t1} + \beta_2 x_{t2} + \varepsilon_t$$

Use the fact that $\cos^2\theta + \sin^2\theta = 1$ and $\sin 2\theta = 2\sin\theta\cos\theta$.

(b) Show that $V(\hat{\beta}_1) = V(\hat{\beta}_2) = \sigma^2/(2\cos^2 2\theta)$

2.5. A linear regression model $y_t = \beta_0 + \beta_1 x_{t1} + \beta_2 x_{t2} + \varepsilon_t$ is fitted to $n = 20$ data points. It is found that SSTO (corrected for mean) = 200 and SSR = 66.

(a) Specify the ANOVA table.

(b) Calculate and interpret R^2.

(c) Test whether $\beta_1 = \beta_2 = 0$ (use $\alpha = .05$).

(d) The simpler model $y_t = \beta_0 + \beta_1 x_{t1} + \varepsilon_t$ is considered, and it is found that SSR = 50. Test whether $\beta_2 = 0$.

(e) How could the residuals from the model in (d) indicate whether the variable X_2 is needed?

2.6. A linear regression model $y_t = \beta_0 + \beta_1 x_{t1} + \beta_2 x_{t2} + \beta_3 x_{t3} + \varepsilon_t$ is fitted to $n = 10$ observations. It is found that SSTO $= 100$, $R^2 = .88$, and $\text{SSR}(X_1) = 82$, $\text{SSR}(X_2) = 40$, $\text{SSR}(X_1|X_2) = 45$, $\text{SSR}(X_3|X_2) = 1$, and $\text{SSR}(X_2|X_1, X_3) = 2$.

(a) Test H_0: $\beta_1 = \beta_2 = \beta_3 = 0$ at the $\alpha = .05$ level.

(b) Calculate partial F tests for X_1, X_2, X_3. Interpret the meaning of these tests.

(c) Test H_0: $\beta_2 = \beta_3 = 0$ at the $\alpha = .05$ level.

(d) Which conclusions about the importance of these three variables would you reach from the above tests?

2.7. A dependent variable Y was observed three times at $x = 0$, three times at $x = 1$, three times at $x = 2$, three times at $x = 3$, and five times at $x = 4$. A simple linear regression model was fitted; the partially completed ANOVA table is given below:

Source	SS	df	MS	F
Regression	73	1		
Error	30			
Lack of fit				
Pure error	9			
Total (corr. for mean)	103			

(a) Complete the ANOVA table.

(b) Perform any appropriate significance tests (use $\alpha = .05$), carefully stating your hypotheses and conclusions.

(c) Give an estimate of σ^2.

2.8. It was found that a first-order (auto)regression model, $z_t - 100 = \beta(z_{t-1} - 100) + \varepsilon_t$, described the behavior of a particular sales series. The last 10 standardized observations $y_t = z_t - 100$ (i.e., deviations from the mean 100) were recorded, and it was found that $y_1 y_2 + y_2 y_3 + \cdots + y_9 y_{10} = 160$, $y_1^2 + y_2^2 + \cdots + y_{10}^2 = 216$, and $y_{10} = 4$.

(a) Calculate the least squares estimate $\hat{\beta}$.

(b) Using your estimate $\hat{\beta}$, predict the future observations z_{11} and z_{12}.

(c) Construct a 95% prediction interval for z_{11}. (Assume that $s^2 = 1$.)

2.9. In a regression study it is quoted that the regression model has led to a 70% reduction in the residual standard deviations. Calculate the adjusted R_a^2.

2.10. Consider the regression model $\mathbf{y} = \mathbf{X}\boldsymbol{\beta} + \boldsymbol{\varepsilon}$, where $\boldsymbol{\beta} = (\beta_0, \beta_1, \ldots, \beta_p)'$.

(a) Show that $\hat{\mathbf{y}}'\hat{\mathbf{y}} = \hat{\boldsymbol{\beta}}'\mathbf{X}'\mathbf{y}$.

(b) Prove that the F statistic $F = \text{MSR}/\text{MSE}$ can be written as

$$F = \frac{R^2}{1 - R^2} \frac{n - p - 1}{p}$$

where R^2 is the coefficient of determination.

2.11. Assume that the true model is given by $y_t = \beta_0 + \beta_1 x_t + \varepsilon_t$. Suppose that the predictor variable x_t is omitted from the model and that $y_t = \beta_0 + \varepsilon_t$ is fitted instead. Show that the least squares estimate $\bar{y}$ is a biased estimate of β_0; derive an expression for the bias.

2.12. Let us assume that the true regression model is given by $\mathbf{y} = \mathbf{X}_1\boldsymbol{\beta}_1 + \mathbf{X}_2\boldsymbol{\beta}_2 + \boldsymbol{\varepsilon}$, where $\mathbf{X}_1$ is an $n \times p_1$ matrix of p_1 independent variables, $\mathbf{X}_2$ is an $n \times p_2$ matrix of p_2 independent variables, and $\boldsymbol{\beta}_1$ and $\boldsymbol{\beta}_2$ are $p_1 \times 1$ and $p_2 \times 1$ vectors of coefficients. When fitting the model, we omitted the independent variables in $\mathbf{X}_2$ and derived the least squares estimate $\hat{\boldsymbol{\beta}}_1$ from the model $\mathbf{y} = \mathbf{X}_1\boldsymbol{\beta}_1 + \boldsymbol{\varepsilon}$.

(a) Calculate $E(\hat{\boldsymbol{\beta}}_1)$.

(b) Under what conditions is $\hat{\boldsymbol{\beta}}_1$ an unbiased estimate of $\boldsymbol{\beta}_1$?

2.13. Simple linear regression models were fitted to two *independent* sets of data. The following statistics were calculated:

	Regression 1	Regression 2
Model	$y = \alpha_1 + \beta_1 x + \varepsilon$	$y = \alpha_2 + \beta_2 x + \varepsilon$
Sample size	15	15
$\hat{\alpha}$	−1.24	0.34
$\hat{\beta}$	2.21	3.59
$\Sigma(x_i - \bar{x})^2$	15,240	15,001
SSE	21,357	29,064

(a) Estimate $V(\hat{\beta}_1 - \hat{\beta}_2)$, assuming $\sigma_1^2 = \sigma_2^2$ where $\sigma_i^2 = V(\varepsilon)$ in regression i, and test whether $\beta_1 = \beta_2$ (use $\alpha = .10$).

(b) Estimate $V(\hat{\beta}_1 - \hat{\beta}_2)$, assuming $\sigma_1^2 \neq \sigma_2^2$. Does your conclusion change?

2.14. The variables Y and Z are related to the same input variable X by the models

$$E(y) = \beta_0 + \beta_1 + \beta_2 x \qquad \text{and} \qquad E(z) = \beta_0 - \beta_1 + \beta_2 x$$

The random variables Y and Z have constant variance σ^2 and are uncorrelated with each other. A set of m independent observations on Z are taken at $x = -c$, and m observations are taken at $x = +c$; $n - 2m$ observations on Y are taken at $x = 0$.

(a) Set up the regression model in matrix form.

(b) Set up the normal equations for the estimate of $\boldsymbol{\beta} = (\beta_0, \beta_1, \beta_2)'$, based on all observations. Derive $\mathbf{V}(\hat{\boldsymbol{\beta}})$, and find the value of m that minimizes the variance of $\hat{y} = \hat{\beta}_0 + \hat{\beta}_1 + \hat{\beta}_2 x$ at $x = c$ and for $n = 7$.

2.15. Construct a regression model(s) for predicting the sale price of a house Y (in thousands of dollars) as a function of property taxes X_1 (in hundreds of dollars), lot size X_2 (in thousands of square feet), living space X_3 (in thousands of square feet), and age of the house X_4 (in years). The data are from Narula and Wellington (1977).

X_1	X_2	X_3	X_4	Y
4.9176	3.4720	0.9980	42	25.9
5.0208	3.5310	1.5000	62	29.5
4.5429	2.2750	1.1750	40	27.9
4.5573	4.0500	1.2320	54	25.9
5.0597	4.4550	1.1210	42	29.9
3.8910	4.4550	0.9880	56	29.9
5.8980	5.8500	1.2400	51	30.9
5.6039	9.5200	1.5010	32	28.9
15.4202	9.8000	3.4200	42	84.9
14.4598	12.8000	3.0000	14	82.9
5.8282	6.4350	1.2250	32	35.9
5.3003	4.9883	1.5520	30	31.5
6.2712	5.5200	0.9750	30	31.0
5.9592	6.6660	1.1210	32	30.9
5.0500	5.0000	1.0200	46	30.0
5.6039	9.5200	1.5010	32	28.9
8.2464	5.1500	1.6640	50	36.9
6.6969	6.9020	1.4880	22	41.9
7.7841	7.1020	1.3760	17	40.5
9.0384	7.8000	1.5000	23	43.9
5.9894	5.5200	1.2560	40	37.5

X_1	X_2	X_3	X_4	Y
7.5422	4.0000	1.6900	22	37.9
8.7951	9.8900	1.8200	50	44.5
6.0931	6.7265	1.6520	44	37.9
8.3607	9.1500	1.7770	48	38.9
8.1400	8.0000	1.5040	3	36.9
9.1416	7.3262	1.8310	31	45.8
12.0000	5.0000	1.2000	30	41.0

Predict the sale price of your house, which is 20 years old, has a lot size of 6000 ft^2, living space of 1500 ft^2, and tax assessment of \$1000. Calculate a 95% prediction interval.

2.16. Discuss whether each of the following statements is true. Also indicate the reasons for your decision.

(a) The residuals from the general linear regression model $\mathbf{y} = \mathbf{X}\boldsymbol{\beta} + \boldsymbol{\varepsilon}$ are always independent.

(b) The regression sum of squares can be written as $\mathbf{y}'\mathbf{X}(\mathbf{X}'\mathbf{X})^{-1}\mathbf{X}'\mathbf{y} - n\bar{y}^2$.

(c) By adding an additional variable to the regression model, we will always decrease the mean square error.

(d) If we find that several t statistics $t_{\hat{\beta}_i} = \hat{\beta}_i/s_{\hat{\beta}_i}$ are insignificant, we should immediately drop all insignificant variables from the model.

(e) If all columns in the design matrix $\mathbf{X}$ are orthogonal, the sequential and the partial F tests will be the same.

(f) $R^2 = \Sigma(\hat{y}_t - y_t)^2/\Sigma(y_t - \bar{y})^2$.

(g) In the simple linear regression model $y_t = \beta_0 + \beta_1 x_t + \varepsilon_t$, the least squares estimate of β_1 is given by $\hat{\beta}_1 = r(s_x/s_y)$, where s_x, s_y are the sample standard deviations and r is the sample correlation coefficient.

(h) When estimating β_1 in model (g), it is best to choose the x levels such that their distances from $\bar{x}$ are large.

(i) If the variance of ε_t in model (g) is proportional to x_t^2, we should consider a logarithmic transformation of the dependent variable.

(j) If we have data from three different factories, we can always take care of possible factory "level" differences by using three levels of a new variable called a "dummy variable."

(k) The model $y = \beta_0 x_1^{\beta_1} x_2^{\beta_2} \varepsilon$ can be fitted by ordinary least squares.

(l) By using appropriate transformations, we can use least squares to estimate the coefficients in the model

$$y_t = \{1 + [\exp(\beta_0 + \beta_1 x_t)]\varepsilon_t\}^{-1}.$$

(m) The lack-of-fit test in Section 2.11 cannot be applied if there is more than one independent variable.

(n) The selection criteria of Section 2.8 for finding the "best" regression equation lead to the same set of independent variables.

(o) The sample correlation between X and Y is .95. This implies that X causes Y.

(p) If the F statistic is very large, then there is no need to look at the residuals.

(q) Suppose that $e_1, e_2, \ldots, e_n$ are the residuals obtained by fitting the regression equation $y_t = \beta_1 x_t + \varepsilon_t$ $(1 \leqslant t \leqslant n)$. A student claims that $\Sigma_{t=1}^{n} e_t = 0$, because the sum of the residuals from a regression is always zero.

2.17. A certain response y is measured at 2-min time intervals. At time 0, a "treatment" is applied, and it is hypothesized that this causes a "jump" in the response curve, as shown in the graph. The data are given next to the figure.

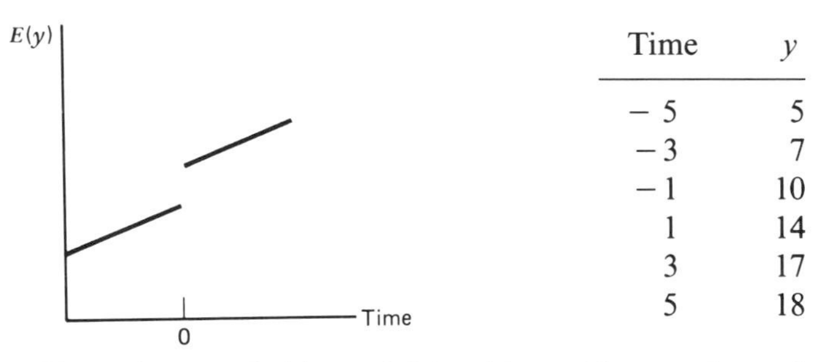

Time	y
− 5	5
− 3	7
− 1	10
1	14
3	17
5	18

(a) Formulate a suitable model for this problem, and test the significance of the jump.

(b) Assume that the slopes before and after the change are not the same. How can you model this situation? Perform the appropriate tests.

2.18. A linear regression model $y_t = \beta_0 + \beta_1 x_t + \varepsilon_t$ is fitted. A residual analysis reveals that the errors follow a random walk (i.e., $\varepsilon_t = \varepsilon_{t-1} + a_t$, where $\{a_t\}$ is an uncorrelated sequence). The investigator claims that the usual regression analysis (estimates and tests concerning β_0, β_1) is not affected. Comment on this. If you feel that the analysis is affected, discuss how a remedy could be found.

2.19. Consider quarterly Iowa nonfarm income, which is listed in series 1 of the Data Appendix. Plot the original data y_t and $\ln y_t$ against time. Consider a linear trend model of the form $\ln y_t = \beta_0 + \beta_1 t + \varepsilon_t$. Determine the least squares estimates, and check the adequacy of the fitted model. Since the observations are collected sequentially in time, you should check whether the residuals are autocorrelated. If you are satisfied with your model, obtain the forecasts and the prediction intervals for the next four quarters.

2.20. Consider the simple linear regression model $y_t = \beta x_t + \varepsilon_t$, where the errors ε_t are serially correlated. In particular, let us assume that $\text{Corr}(\varepsilon_t, \varepsilon_{t-1}) = \rho \neq 0$ but $\text{Corr}(\varepsilon_t, \varepsilon_{t-k}) = 0$ for $k \geqslant 2$.

(a) Consider the usual least squares estimator. Derive its expected value. Is it unbiased? Derive its variance.

(b) Using your result in part (a), discuss why the usual t statistic for testing the hypothesis $\beta = 0$ can lead to incorrect conclusions. Discuss the cases $\rho > 0$ and $\rho < 0$ separately.

2.21. Consider the simple linear regression model through the origin, $y_t = \beta x_t + \varepsilon_t$ $(t = 1, 2, \ldots, n)$. Assume that $V(\varepsilon_t) = \sigma^2/x_t^2$.

(a) Consider the usual least squares estimate $\hat{\beta}$. Is it unbiased? Calculate its variance.

(b) Consider the weighted least squares estimate $\hat{\beta}^* = \Sigma\omega_t x_t y_t / \Sigma\omega_t x_t^2$, where $\omega_t = x_t^2$. Is this estimator unbiased? Derive its variance.

(c) Compare the variances of $\hat{\beta}$ and $\hat{\beta}^*$. Which estimator would you prefer? *Hint*: You may use the fact that for any $a_1, \ldots, a_n$ and $b_1, \ldots, b_n$; $\Sigma a_t^2 \Sigma b_t^2 \geqslant [\Sigma a_t b_t]^2$.

(d) Weighted regression is one approach to achieve homoscedastic (equal-variance) errors. Can you describe another approach? For example, if you knew that $V(y_t) = \eta_t^2$, where $\eta_t = E(y_t)$ is the mean level, how would you proceed?

2.22. Assume that the model is given by $y_{ij} = \beta_0 + \beta_1 x_j + \varepsilon_{ij}$ for $i = 1, 2, \ldots, n_j$ and $j = 1, 2, \ldots, c$. There are n_j observations on y corresponding to each given x_j. A researcher who has not studied regression analysis collected the observations and computed $\bar{y}_1, \bar{y}_2, \ldots, \bar{y}_c$, where

$$\bar{y}_j = \sum_{i=1}^{n_j} y_{ij}/n_j$$

The researcher then discarded the original data; all that remains is

$(x_1, \bar{y}_1), (x_2, \bar{y}_2), \ldots, (x_c, \bar{y}_c)$ and $n_1, n_2, \ldots, n_c$. At this point the researcher has come to you for help. Is it possible to compute the least squares estimates of β_0 and β_1? If so, how can it be done?

CHAPTER 3

3.1. A business forecaster uses simple exponential smoothing to predict future sales and chooses $\alpha = .1$ as the smoothing constant. Using the last 30 observations, the forecaster predicts sales two-steps-ahead and finds that $\hat{z}_{30}(2) = 102.5$. After z_{31} has become available, $z_{31} = 105$, the forecaster wants to revise the forecast for z_{32}. Calculate the revised forecast $\hat{z}_{31}(1)$.

3.2. (a) A business forecaster uses simple exponential smoothing to predict future sales, and uses $\alpha = .2$ as the smoothing constant. Using *monthly* observations of the last 5 years, the forecaster predicts *total sales* for the next year (January–December) to be 960 units. In January, the actual sales are 90 units. What is the revised forecast for the rest of the year (the remaining 11 months)?

(b) In (a), the forecaster is not certain which α to use. Describe a method for determining α.

(c) The business forecaster is also asked to predict the sales of a competitor. A plot of past data shows a linear trend that may be changing. Should the forecaster use simple exponential smoothing? Which procedure would you recommend?

3.3. The one-step-ahead forecast in simple exponential smoothing is given by

$$\hat{z}_n(1) = \alpha\left[z_n + (1-\alpha)z_{n-1} + (1-\alpha)^2 z_{n-2} + \cdots\right]$$

Show that if we substitute the forecast $\hat{z}_n(1)$ for the unknown observation z_{n+1}, then the two-step-ahead forecast is

$$\hat{z}_n(2) = \alpha\left[\hat{z}_n(1) + (1-\alpha)z_n + (1-\alpha)^2 z_{n-1} + \cdots\right] = \hat{z}_n(1)$$

By induction, show that $\hat{z}_n(l) = \hat{z}_n(1)$ for $l > 0$.

3.4. Use simple exponential smoothing to forecast the annual lumber production in Table 3.1. Determine the optimal smoothing constant, calculate the forecasts, and compare your results with the ones from the globally constant mean model.

3.5. The smoothed statistic S_t is used to forecast the demand for a textile product that follows the mean model $z_t = \mu + \varepsilon_t$, with $E(\varepsilon_t) = 0$, $V(\varepsilon_t) = \sigma^2$, and $E(\varepsilon_t \varepsilon_{t-k}) = 0$ for $k \neq 0$. Alternatively, one could take the moving average of the most recent N observations,

$$\bar{z}_t^{(N)} = \frac{1}{N}[z_t + z_{t-1} + \cdots + z_{t-N+1}]$$

as the forecast. Compare these two forecasts. In particular, consider their updating equations and their variances. Show that for the smoothing constant $\alpha = 2/(N+1)$, their variances are the same [$\alpha = 2/(N+1)$ is called the equivalent smoothing constant].

3.6. Suppose simple exponential smoothing is being used to forecast the process $z_t = \mu + \varepsilon_t$. For a *single period* t_0, the mean of the process shifts to the level $\mu_1 = \mu + \delta$. Calculate $E(S_n)$, where S_n is the smoothed statistic and $n > t_0$. Comment on its behavior as n increases. Illustrate this graphically. Can you think of cases where such a situation would occur?

3.7. Repeat Exercise 3.6, but assume that the mean has shifted to μ_1 for all $t \geqslant t_0$.

3.8. The forecasts from double exponential smoothing are discussed in Section 3.6. Calculate and plot the forecast weights in

$$\hat{z}_n(1) = \sum_{j \geqslant 1} \pi_j z_{n+1-j} \qquad \text{and} \qquad \hat{z}_n(2) = \sum_{j \geqslant 1} \pi_j^{(2)} z_{n+1-j}$$

for the smoothing constant $\alpha = .10$.

3.9. It is found that the monthly sales of a particular brand of dog food can be represented by a locally constant linear trend model $z_{n+j} = \beta_0 + \beta_1 j + \varepsilon_{n+j}$. The first 2 years of data ($n = 24$) are used to smooth the coefficients, and it is found that $\hat{\beta}_{0,n} = 30$ and $\hat{\beta}_{1,n} = 2$. Data for the next 3 months became available and are given by $z_{25} = 28$, $z_{26} = 31$, $z_{27} = 36$.

(a) Consider Holt's method with smoothing constants $\alpha_1 = .2$ and $\alpha_2 = .1$. Update the estimates and the forecasts for $l = 1, 2$.

(b) Consider double exponential smoothing with smoothing constant $\alpha = .1$. Update the coefficients and the forecasts, and calculate the smoothed statistics S_t, $S_t^{[2]}$ for $t = 25, 26, 27$.

3.10. Double exponential smoothing: Calculate the variance of the l-step-ahead forecast error ($l = 1, \ldots, 4$) and the 95% prediction intervals for the next four observations. Also calculate the prediction interval for the sum and the average of the next four observations. Assume that the mean absolute deviation is known and $\alpha = .1$.

3.11. Derive expressions for $\sum_{j \geq 0} j^k \omega^j$, for $k = 1, \ldots, 4$. You may use the fact that

$$\sum_{j \geq 0} \omega^j = \frac{1}{1 - \omega} \qquad \text{and} \qquad \sum_{j \geq 0} j^k \omega^j = \omega \frac{\partial}{\partial \omega} \sum_{j \geq 0} j^{k-1} \omega^j$$

3.12. Show that the kth-order smoothed statistic (for $k \geq 2$) can be written as

$$S_n^{[k]} = \frac{(1 - \omega)^k}{(k - 1)!} \sum_{j \geq 0} \left[\prod_{i=1}^{k-1} (j + i) \right] \omega^j z_{n-j}$$

3.13. Discuss whether the following statements are true.

(a) In simple exponential smoothing, the forecasts of z_{n+l} from a fixed forecast origin n for lead times $l = 1, 2, 3$ are different.

(b) In double exponential smoothing, the forecasts of z_{n+l} from a fixed forecast origin n for lead times $l = 1, 2, 3$ lie on a straight line that passes through the first two forecasts.

(c) In simple exponential smoothing, the prediction intervals for future observations are the same.

3.14. Consider the monthly sales of computer software packages for undergraduate college-level curriculum development. The sale of these programs started in 1976. The data for January 1976 to December 1981 ($n = 72$) are given below (read across):

2	2	4	12	0	81	58	59	25	22	87	35
22	23	10	6	7	31	132	21	48	56	87	48
69	97	155	96	62	100	61	101	79	72	49	112
104	216	115	215	178	233	239	217	196	228	164	151
194	152	114	144	151	205	292	180	213	220	193	235
317	305	298	315	231	361	353	256	257	340	438	442

(a) Plot the data.

(b) Use the first 5 years ($n = 60$), and discuss whether simple, double, or triple exponential smoothing is appropriate for this data set.

(c) Determine optimal smoothing constants.

(d) Use the data for 1981, and compare the one-step-ahead forecasts from simple, double, and triple exponential smoothing. Calculate the forecast errors $z_t - \hat{z}_{t-1}(1)$ for $t = 61, \ldots, 72$ for each method. Compare the methods in terms of their mean square forecast error

$$\text{MSE} = \frac{1}{12} \sum_{t=61}^{72} [z_t - \hat{z}_{t-1}(1)]^2$$

Methods with small mean square error are preferable.

Note: Here we compare the forecast methods on "new" data, that is, data that have not been used to determine the smoothing constant. The mean square error is just one possible measure for evaluating forecast methods. Additional criteria are discussed in Section 8.5.

CHAPTER 4

4.1. Consider the seasonal models $z_{n+j} = \mathbf{f}'(j)\boldsymbol{\beta} + \varepsilon_{n+j}$, where

(a) $\mathbf{f}(j) = \left(1, \sin \frac{\pi}{2} j, \cos \frac{\pi}{2} j\right)'$

(b) $\mathbf{f}(j) = \left(1, j, \sin \frac{\pi}{2} j, \cos \frac{\pi}{2} j\right)'$

(c) $\mathbf{f}(j) = \left(1, j, \sin \frac{\pi}{2} j, \cos \frac{\pi}{2} j, j \sin \frac{\pi}{2} j, j \cos \frac{\pi}{2} j\right)'$

(d) $\mathbf{f}(j) = \left(1, j, \sin \frac{2\pi}{12} j, \cos \frac{2\pi}{12} j, \sin \frac{4\pi}{12} j, \cos \frac{4\pi}{12} j\right)'$

Specify the transition matrices in $\mathbf{f}(j) = \mathbf{L}\mathbf{f}(j-1)$.

4.2. Consider general exponential smoothing to forecast from the seasonal model $z_{n+j} = \mathbf{f}'(j)\boldsymbol{\beta} + \varepsilon_{n+j}$, where

$$\mathbf{f}(j) = \left(1, \sin \frac{2\pi}{12} j, \cos \frac{2\pi}{12} j\right)'$$

(a) Specify the transition matrix and the updating equations for the coefficients.

(b) Use the results in Section 3.8 to calculate the variance of the l-step-ahead forecast (in particular, use $l = 1, 2, 3, 4$).

4.3. Consider general exponential smoothing in the linear trend model with one seasonal indicator, $z_{n+j} = \beta_0 + \beta_1 j + \delta \text{IND}_j + \varepsilon_{n+j}$, where $\text{IND}_j = 1$ if j is odd, and 0 otherwise.

(a) Write down the transition matrix $\mathbf{L}$ in $\mathbf{f}(j) = \mathbf{L}\mathbf{f}(j-1)$.

(b) Show without using the result in (4.28) that $\mathbf{f}_* = \lim_{\omega \to 0} \mathbf{F}^{-1}\mathbf{f}(0) = (1, \frac{1}{2}, -\frac{1}{2})'$.

(c) Substituting $\mathbf{f}_*$ into $\hat{\boldsymbol{\beta}}_{n+1} = \mathbf{L}'\hat{\boldsymbol{\beta}}_n + \mathbf{f}_*[z_{n+1} - \hat{z}_n(1)]$, show that $\hat{z}_n(1) = \mathbf{f}'(1)\hat{\boldsymbol{\beta}}_n = z_n + z_{n-1} - z_{n-2}$ and $\hat{z}_n(2) = 2z_n - z_{n-2}$

4.4. Show the results in Table 4.5.

4.5. The monthly U.S. retail sales of passenger cars (in thousands) for the period January 1955 to December 1977 are given below (read across):

440	477	637	652	661	681	647	659	655	576	509	631
432	448	545	564	560	540	535	568	421	424	404	514
437	439	573	549	556	517	543	492	495	464	409	512
382	334	401	418	424	411	400	371	317	321	335	511
421	425	498	575	584	586	567	534	458	535	429	431
430	494	597	647	647	596	547	525	459	548	543	544
414	375	480	496	544	572	501	471	371	550	558	526
506	473	592	635	644	602	614	540	374	678	638	644
554	498	624	759	715	692	709	553	404	715	640	712
612	552	637	812	781	754	724	649	565	659	564	757
667	631	799	896	841	842	834	767	590	746	794	909
733	715	913	820	745	809	688	668	565	855	798	726
612	558	736	774	815	855	696	594	615	735	684	673
702	700	851	815	900	872	829	728	663	980	864	750
730	726	809	855	901	898	761	655	808	923	797	720
624	686	746	799	811	925	762	639	581	755	539	538
694	749	899	883	889	957	819	724	884	1050	958	741
721	814	913	900	1032	1025	905	813	877	1070	1030	848
875	920	1142	1024	1144	1084	960	836	873	977	911	694
679	683	778	816	882	811	811	811	726	757	604	508
578	684	670	660	741	770	794	684	726	889	744	701
679	758	947	914	921	956	865	762	792	868	840	807
725	811	1084	1027	1054	1118	913	931	829	1014	881	795

Use a subset of the data (for example, the period January 1961 to December 1974; $n = 168$), and consider the following forecasting methods.

(a) Globally constant linear trend model with seasonal indicators.

(b) Globally constant linear trend model with added harmonics. Determine the number of harmonics needed.

(c) General exponential smoothing for the locally constant version of model (a). Determine the smoothing constant α.

(d) General exponential smoothing for the locally constant version of model (b). Determine the smoothing constant α.

(e) Winters' multiplicative seasonal forecasting procedure. Determine the optimal smoothing constants.

Determine whether the models (a)–(e) are appropriate. Look at the autocorrelations of the residuals or of the one-step-ahead forecast errors. In addition, compare the forecasting performance of the various methods during the period January 1975 to December 1977. Update the forecasts, and calculate the one-step-ahead forecast errors $z_{n+1} - \hat{z}_n(1), \ldots, z_{n+36} - \hat{z}_{n+35}(1)$. Compare the methods in terms of their mean square errors (see Exercise 3.14).

4.6. Use the first 120 observations of the housing start series given in series 6 of the Data Appendix (January 1965 to December 1974). Repeat Exercise 4.5; use the data for 1975 to update the forecasts and evaluate the methods.

4.7. Let us assume that we use $N = 2q + 1$ observations to fit the model

$$z_t = \beta_0 + \sum_{i=1}^{q} (\beta_{1i} \sin 2\pi f_i t + \beta_{2i} \cos 2\pi f_i t) + \varepsilon_t$$

where $f_i = i/N$ is the ith harmonic of the fundamental frequency $1/N$.

(a) Show that the least squares estimates are given by

$$\hat{\beta}_0 = \bar{z} \qquad \hat{\beta}_{1i} = \frac{2}{N} \sum_{t=1}^{N} z_t \sin 2\pi f_i t \qquad \hat{\beta}_{2i} = \frac{2}{N} \sum_{t=1}^{N} z_t \cos 2\pi f_i t$$

You may use the fact that:

$$\sum_{t=1}^{N} \cos 2\pi f_i t \cos 2\pi f_j t = \begin{cases} 0 & \text{if } i \neq j \\ N/2 & \text{if } i = j \neq 0 \\ N & \text{if } i = j = 0 \end{cases}$$

$$\sum_{t=1}^{N} \cos 2\pi f_i t \sin 2\pi f_j t = 0 \qquad \text{for all } i, j$$

$$\sum_{t=1}^{N} \sin 2\pi f_i t \sin 2\pi f_j t = \begin{cases} 0 & \text{if } i \neq j \\ N/2 & \text{if } i = j \neq 0 \\ 0 & \text{if } i = j = 0 \end{cases}$$

(b) Show that the regression sum of squares contribution of the ith harmonic is given by $I(f_i) = (N/2)(\hat{\beta}_{1i}^2 + \hat{\beta}_{2i}^2)$, $i = 1, 2, \ldots, q$, and show that it is independent of all other harmonics.
Note: The quantity $I(f_i)$ is the intensity at frequency f_i, and the collection of the $q = (N - 1)/2$ intensities is called the *periodogram*. It expresses the decomposition of the total variation

$$\sum_{t=1}^{N} (z_t - \bar{z})^2 = \sum_{i=1}^{q} I(f_i)$$

into the contributions at the various harmonic frequencies. The periodogram is used to detect and estimate the amplitude of a sine component of known frequency and to test for randomness of a series. The definition of the periodogram assumes that the frequencies are harmonics of the fundamental frequency. If we allow the frequency f to vary continuously from 0 to 0.5, then

$$I(f) = (N/2)\left(\hat{\beta}_{1f}^2 + \hat{\beta}_{2f}^2\right) \qquad 0 \leqslant f \leqslant .5$$

is called the *sample spectrum*. It provides the basis for the theory of *spectral analysis* of time series.

4.8. Consider the monthly housing start series listed as series 6 in the Data Appendix. Calculate, plot, and interpret the 12-term centered moving averages.

4.9. Consider the quarterly expenditures for new plant and equipment listed as series 7 in the Data Appendix. Calculate and plot the four-term centered moving averages.

4.10. Consider the demand for repair parts series listed as series 10 in the Data Appendix.

(a) Calculate and plot the (3×3)-term moving averages.

(b) Calculate and plot Henderson's five-term moving averages.

CHAPTER 5

5.1. (a) The autocorrelation function of a time series of $n = 100$ sales data and their first differences were calculated and are given below:

Lag	1	2	3	4	5	6	7
z_t	.97	.94	.90	.86	.81	.76	.71
$z_t - z_{t-1}$	−.53	.41	−.35	.21	−.16	.15	.08

Which time series model(s) would you consider for this data set? State your reasons.

(b) For another data set ($n = 64$), the autocorrelation function is given by

Lag	1	2	3	4	5	6	7
z_t	−.42	.18	−.02	.07	−.06	.14	.05

Which model(s) would you entertain? Why?

5.2. Consider the model $z_t = \beta_0 + \beta_1 t + a_t$, where $\{a_t\}$ is a sequence of uncorrelated random variables. Is this model stationary? Explain.

5.3. Assume that monthly sales of certain products follow the stochastic model $(1 - B)^2 z_t = (1 - \theta B)a_t$, where $|\theta| < 1$.

(a) Is the model stationary? Why?

(b) Is the model invertible? Why?

(c) Derive the autocorrelation function of the second differences.

(d) Derive the π weights. Derive the ψ weights.

(e) Calculate the one-step-ahead forecast $z_t(1)$ and its variance.

5.4. Suppose $y_t = x_t + z_t$, where x_t is an autoregressive process $(1 - \phi B)x_t = a_t$ and z_t is a white-noise process with variance σ_z^2. Furthermore, assume that x_t and z_t are uncorrelated [i.e., $\text{Cov}(x_t, z_{t-k}) = 0$ for all k].

(a) Derive $\rho_1, \rho_2, \rho_{10}$ of y_t.

(b) Which ARMA model would you consider for y_t?

5.5. (a) Define ϕ_{kk}, the kth-order partial autocorrelation of a process z_t.

(b) Obtain ϕ_{11}, ϕ_{22}, and ϕ_{33} for the following models:

(i) $(1 - 1.2B + .8B^2)z_t = a_t$

(ii) $(1 - .7B)z_t = a_t$

(iii) $(1 - .7B)z_t = (1 - .5B)a_t$

(iv) $z_t = (1 - .5B)a_t$

5.6. Consider the noninvertible MA(1) process $z_t = a_t - a_{t-1}$, where $V(a_t) = \sigma^2$. Show that the variance of the sample mean from a sample of size n is given by

$$\text{Var}\left(\frac{1}{n}\sum_{t=1}^{n} z_t\right) = \frac{2\sigma^2}{n^2}$$

Comment on this result.

5.7. [H. Working (1960).] Assume the random walk model $z_t = z_{t-1} + a_t$. Consider the first differences m-steps-apart:

$$_t\Delta_m = z_t - z_{t-m} \tag{1}$$

and the first differences of consecutive averages:

$$_t\Delta_m^* = \frac{1}{m}[z_t + z_{t+1} + \cdots + z_{t+m-1}]$$

$$- \frac{1}{m}[z_{t-m} + z_{t-m+1} + \cdots + z_{t-1}] \tag{2}$$

Derive the variances of (1) and (2) and show that, for large m, their ratio $V(_t\Delta_m^*)/V(_t\Delta_m)$ approaches 2/3.

Hint: $\sum_{i=1}^{n} i^2 = \frac{n(n+1)(2n+1)}{6}$.

5.8. Let z_{ti} represent the annual cost of manufacturing a particular product i in year t. Assume that the company manufactures 100 different products ($i = 1, 2, \ldots, 100$). Suppose that the annual cost can be represented by the time series model $z_{ti} = (1 - \theta_i B)a_{ti}$, where $V(a_{ti}) = \sigma_i^2$, $i = 1, 2, \ldots, 100$. Discuss the appropriate model for the total annual cost $\Sigma_{i=1}^{100} z_{ti}$ under each of the following conditions:

(a) The annual costs of different products are uncorrelated (i.e., for $i \neq j$, z_{ti} and z_{t_*j} are uncorrelated, for all t and t_*).

(b) The annual costs are contemporaneously correlated (i.e., for $i \neq j$, z_{ti} and z_{t_*j} are correlated for $t = t_*$ but not for $t \neq t_*$).

5.9. An ARIMA(0, 1, 1) model was fitted to $n = 100$ observations. The autocorrelations of the residuals $\hat{a}_t$ are given below

Lag k	1	2	3	4	5
$r_{\hat{a}}(k)$	−.32	.20	.05	−.06	−.08

Would you judge this model adequate? If not, which other model(s) would you consider?

5.10. Consider the following models:

$$(1 - \phi B)(z_t - \mu) = a_t \tag{1}$$

$$(1 - \phi_1 B - \phi_2 B^2)(z_t - \mu) = a_t \tag{2}$$

$$(1 - \phi B)(1 - B) z_t = \theta_0 + (1 - \theta B) a_t \tag{3}$$

$$(1 - B)^2 z_t = (1 - \theta_1 B - \theta_2 B^2 - \theta_3 B^3) a_t \tag{4}$$

(a) For each model, write down the forecast equation that generates the forecasts.

(b) Derive the variance of the l-step-ahead forecast error (use $l = 1, 2, 3$).

5.11. Consider the stationary and invertible process $(1 - \phi B) z_t = (1 - \theta B) a_t$.

(a) Obtain the MMSE forecasts $z_t(1)$ and $z_t(2)$ from the autoregressive form of the model.

(b) Find $V[e_t(l)]$, the variance of the l-step-ahead forecast error.

(c) Discuss the behavior of the forecasts and of $V[e_t(l)]$ as $\phi \to 1$.

5.12. It was found that a first-order autoregressive model $z_t = \theta_0 + \phi z_{t-1} + a_t$ describes the behavior of a particular sales series. A standard computer program was used to estimate θ_0 and ϕ, and it was found that $\hat{\theta}_0 = 50$ and $\hat{\phi} = .60$. The last observation at time 100 was 115 ($z_{100} = 115$). Calculate the forecasts for periods 101, 102, and 103.

5.13. An ARIMA(0, 1, 1) model $z_t = z_{t-1} + a_t - \theta a_{t-1}$ was fitted to 100 sales records $z_1, \ldots, z_{100}$. It was found that $\hat{\theta} = .6$. A statistician had forecast the next 10 observations and found that $\hat{z}_{100}(10) = 26$. A new observation becomes available, $z_{101} = 24$. Update the forecasts, and derive $\hat{z}_{101}(1)$, $\hat{z}_{101}(2)$. Calculate 95% prediction intervals (assume $\sigma^2 = 1$).

5.14. An ARIMA(1, 1, 0) model is fitted to the past 50 observations, and it is found that $\hat{\phi} = .40$ and $\hat{\sigma} = .18$. The last two observations are $z_{49} = 33.4$, $z_{50} = 33.9$.

(a) Calculate the MMSE forecasts and 95% prediction intervals for the next five periods.

(b) A new observation $z_{51} = 34.2$ is observed. Update the forecasts.

5.15. Consider the model $(1 - B)^2 z_t = (1 - .81B + .38B^2)a_t$. Suppose that you are at time $t = 100$. Obtain the MMSE forecast for lead times $l = 1, 2, \ldots, 10$. The past 10 observations are given below.

t	z_t	t	z_t
91	15.1	96	16.5
92	15.8	97	17.1
93	15.9	98	16.9
94	15.2	99	17.3
95	15.9	100	18.2

Note: The forecasts depend on a_{100} and a_{99}. You may use the following procedure to obtain these values. Set a_{91} and $a_{92} = 0$, and derive the forecast $z_{92}(1) = 2z_{92} - z_{91}$; calculate $a_{93} = z_{93} - z_{92}(1)$. Then derive the forecast $z_{93}(1)$, calculate $a_{94} = z_{94} - z_{93}(1)$, and continue until a_{99} and a_{100} are reached.

5.16. Consider the ARIMA(0, 1, 1) model $(1 - B)z_t = (1 - .8B)a_t$.

(a) Show that this model can be written as $z_t = \bar{z}_{t-1} + a_t$, where $\bar{z}_{t-1} = \sum_{j=1}^{\infty} \pi_j z_{t-j}$. Derive the coefficients π_j and show that $\sum_{j=1}^{\infty} \pi_j = 1$.

(b) Write the one- and two-step-ahead forecasts in the form

$$z_t(1) = \sum_{j=1}^{\infty} \pi_j z_{t-j+1} \quad \text{and} \quad z_t(2) = \sum_{j=1}^{\infty} \pi_j^{(2)} z_{t-j+1}$$

Express $\pi_j^{(2)}$ in terms of the π_j weights.

(c) Obtain the covariance matrix of $[e_t(1), e_t(2)]$, where $e_t(1), e_t(2)$ are the one- and two-step-ahead forecast errors.

5.17. Show that the covariance between forecast errors from different origins is given by

$$\operatorname{Cov}\left[e_t(l), e_{t-j}(l)\right] = \sigma^2 \sum_{i=j}^{l-1} \psi_i \psi_{i-j} \qquad \text{for } l > j$$

For $l \leqslant j$, this covariance is zero.

5.18. Show that the covariance between forecast errors from the same origin with different lead times is given by

$$\operatorname{Cov}\left[e_t(l), e_t(l+j)\right] = \sigma^2 \sum_{i=0}^{l-1} \psi_i \psi_{i+j}$$

5.19. Let us assume that you want to predict the sum of the next s observations $\Sigma_{l=1}^{s} z_{t+l}$. Then the minimum mean square error forecast is given by $\Sigma_{l=1}^{s} z_t(l)$. Using the result in Exercise 5.18, calculate the variance of the forecast error

$$\sum_{l=1}^{s} z_{t+l} - \sum_{l=1}^{s} z_t(l) = \sum_{l=1}^{s} e_t(l)$$

In particular, assume that $s = 4$ and that z_t follows an ARIMA(0, 1, 1) model.

5.20. Specify the form and the starting values of the eventual forecast function for the following models:

(a) $(1 - B)z_t = (1 - \theta_1 B - \theta_2 B^2)a_t$

(b) $(1 - \phi B)(1 - B)^3 z_t = (1 - \theta B)a_t$

For model (a), with $\theta_2 = 0$, specify the updating equation for the coefficient in the eventual forecast function. Discuss the case when $\theta_1 \to 1$.

5.21. Consider the trend model $z_t = \mu_t + \beta t + a_t$, where the intercept μ_t follows a random walk model $\mu_t = \mu_{t-1} + \varepsilon_t$. $\{a_t\}$ and $\{\varepsilon_t\}$ are uncorrelated white-noise sequences with $E(a_t) = E(\varepsilon_t) = 0$; $V(a_t) = \sigma^2$; $V(\varepsilon_t) = \sigma_\varepsilon^2$; $E(a_t \varepsilon_{t-k}) = 0$ for all k.

(a) Is the stochastic process for z_t stationary? Explain.

(b) Derive the lag 1, 2, and 3 autocorrelations for the first differences $w_t = z_t - z_{t-1}$.

(c) What ARIMA model is implied by the above process?

(d) Consider the case $\beta = 0$. Does exponential smoothing lead to MMSE forecasts?

5.22. The observations are assumed to follow the model $z_t = \mu_t + a_t$, where the mean level μ_t is stochastic and changes according to $\mu_t = \mu_{t-1} + \varepsilon_t$. $\{a_t\}, \{\varepsilon_t\}$ are two unobservable uncorrelated white-noise processes, with variances 1 and .05, respectively.

A statistician who derived the MMSE forecasts $z_t(l)$ for this process lost most of the past data. The only information left consists of the last observation $z_n = 100.5$ and the most recent forecast error $z_n - z_{n-1}(1) = 1$.

(a) Derive the MMSE forecasts $z_n(1)$, $z_n(2)$, $z_n(3)$.

(b) Compute the variance of the forecast errors.

5.23. Consider the first-order autoregressive process $z_t = \phi z_{t-1} + a_t$. Assume that a_t are independent $N(0, \sigma^2)$. For known σ^2,

(a) Derive the conditional log-likelihood function of ϕ (conditional on the nonstochastic starting value z_1), $l(\phi | z_2, z_3, \ldots, z_n; z_1$ fixed).

(b) Derive the conditional maximum likelihood estimator of ϕ.

(c) Calculate the exact log-likelihood function $l^*(\phi | z_1, z_2, \ldots, z_n)$, unconditional on the starting value z_1.

5.24. The data given below are monthly approved Master Card applications for the period January 1976 to November 1980 (read across). Construct appropriate time series model(s), and derive predictions for the next 4 months.

17,400	15,000	17,047	19,159	24,185	30,420
29,866	21,575	18,865	18,487	27,407	27,410
25,126	18,458	20,210	21,870	14,587	17,499
23,292	18,691	17,007	14,870	26,572	27,138
13,357	17,809	14,477	15,831	19,393	17,070
17,874	14,883	19,013	13,886	16,468	11,019
10,889	16,658	14,616	16,159	17,202	12,464
10,193	11,322	9,971	11,611	11,619	12,401
9,196	12,443	11,677	11,185	11,015	9,738
9,161	11,155	8,820	10,867	10,601	

5.25. Consider the annual farm parity ratios for 1910 to 1978 (read across; data source: *Historical Statistics of the United States* and *Statistical Abstract of the United States* (100th ed.), published by U.S. Department of Commerce). The farm parity is a ratio of two price indices; the first consists of items commonly bought by farmers (food, clothing, shelter, materials needed in farming), the second includes the items sold by farmers.

107	96	98	101	98	94	103	120	119	110
99	80	87	89	89	95	91	88	91	92
83	67	58	64	75	88	92	93	78	77
81	93	105	113	108	109	113	115	110	100
101	107	100	92	89	84	83	82	85	81
80	79	80	78	76	77	80	74	73	74
72	70	74	88	81	76	71	66	70	

Construct appropriate time series models. Calculate forecasts and prediction intervals for the next 5 years.

5.26. Consider the data in Exercise 3.14.

(a) Determine the appropriate ARIMA model.

(b) Calculate the forecasts and 95% prediction intervals for the next 3 months.

(c) Calculate the forecast for the combined sales in the next quarter, and calculate a 95% prediction interval.

5.27. To gain experience in time series model building, consider one or more of the data series listed in the *Survey of Current Business* published by U.S. Department of Commerce. Also, model stock-price closing series, and check whether the random walk hypothesis is appropriate.

5.28. Consider (1) the number of quarterly business mergers (Y), (2) quarterly averages of the monthly Standard & Poor's stock price index (X), and (3) quarterly averages of the monthly Standard & Poor's industrial bond index (Z).

The data for the period 1947–1977 are listed below (read downwards, left to right). Construct appropriate ARIMA models, and calculate forecasts for the next four observations.

Y	X	Z	Y	X	Z	Y	X	Z
98	14.91	2.47	64	20.01	2.54	109	31.37	2.71
97	14.28	2.45	46	21.81	2.58	130	34.62	2.73
84	15.13	2.47	43	22.47	2.73	173	37.50	2.85
125	15.09	2.68	66	23.21	2.75	157	39.79	2.93
60	14.18	2.71	72	23.57	2.83	167	45.46	3.03
55	15.92	2.66	69	24.23	2.83	186	46.85	3.07
52	15.83	2.73	69	24.19	2.85	157	48.20	3.06
52	15.43	2.70	74	25.36	2.87	175	50.13	3.22
31	14.84	2.58	76	25.34	2.94	182	51.44	3.39
26	14.29	2.57	55	26.23	2.97	159	49.41	3.68
31	14.93	2.49	87	24.66	3.20	159	47.13	3.66
33	15.92	2.46	55	24.04	3.12	119	49.82	3.71
46	16.83	2.43	91	24.44	3.02	153	49.86	3.96
58	18.18	2.46	74	26.13	2.79	154	43.71	3.87
51	18.32	2.50	72	28.87	2.72	149	44.32	3.48

Y	X	Z	Y	X	Z	Y	X	Z
159	46.41	3.45	243	91.48	4.25	215	108.77	7.34
129	50.79	3.74	260	92.65	4.31	207	106.40	7.05
152	55.92	3.95	238	92.18	4.40	367	116.90	6.96
195	59.14	4.01	267	97.63	4.58	498	120.77	7.14
200	61.59	4.23	268	97.90	4.80	505	122.23	7.17
219	63.06	4.35	292	94.39	4.95	442	127.27	7.08
221	62.02	4.42	209	87.15	5.28	501	128.87	7.13
221	60.19	4.41	226	84.87	5.32	431	120.27	7.19
233	59.79	4.29	276	93.03	5.03	388	118.00	7.44
199	59.06	4.15	372	98.58	5.27	312	114.73	7.64
191	58.67	4.19	377	102.10	5.60	395	106.77	7.63
217	65.58	4.12	468	103.00	5.97	346	102.17	7.89
232	69.82	4.17	478	99.40	6.02	286	85.20	8.23
280	70.58	4.27	544	107.03	6.16	206	77.51	8.41
225	73.98	4.26	716	108.83	6.03	187	87.90	8.38
236	73.81	4.27	669	114.70	6.27	204	100.19	8.38
198	65.43	4.15	652	109.80	6.57	226	98.28	8.48
208	60.52	4.20	524	111.27	6.77	259	98.83	8.46
211	62.38	4.09	506	103.70	7.04	230	111.73	8.30
220	68.54	4.03	625	103.83	7.31	252	114.33	8.28
205	73.13	4.07	389	97.36	7.66	302	116.93	8.22
247	74.50	4.18	422	86.71	7.85	280	114.43	8.05
189	77.39	4.21	260	86.35	7.99	293	113.07	8.12
251	82.15	4.23	280	94.72	7.55	248	109.50	8.10
213	85.28	4.29	290	106.13	6.95	365	107.83	8.02
176	87.72	4.26	299	112.10	7.32	266	103.33	8.16
214	89.61	4.27						

CHAPTER 6

6.1. Compute the ACF for the following two models

(a) $z_t = (1 - \theta_1 B - \theta_{12} B^{12} - \theta_{13} B^{13}) a_t$

(b) $z_t = (1 - \theta B)(1 - \Theta B^{12}) a_t$

Discuss how one could distinguish between these two models.

6.2. Derive the ACF for the following two models

(a) $z_t = (1 - \theta_1 B - \theta_2 B^2)(1 - \Theta B^{12})a_t$

(b) $z_t = (1 - \theta_1 B - \theta_2 B^2 - \theta_{12} B^{12})a_t$

Discuss how one could distinguish between these two models.

6.3. Assume that we use seasonal differences $y_t = z_t - z_{t-12}$ to transform a seasonal time series z_t.

(a) Assume that the sample autocorrelation function of the transformed series dies down fairly quickly and that the sample partial autocorrelation has a spike at lag 1. Suggest a tentative model for the time series z_t.

(b) Assume that after fitting the tentative model, the autocorrelation function of the residuals has a spike at lag 12. Suggest a new, improved model.

6.4. Show the updating equations in (6.13). *Hint*: Express z_{t+1+l} in (6.12) in relation to time origin t, and also in relation to time origin $t + 1$. This leads to the equation

$$\beta_1^{(t+1)} \sin\frac{2\pi}{12}l + \beta_2^{(t+1)} \cos\frac{2\pi}{12}l$$

$$= \beta_1^{(t)} \sin\frac{2\pi}{12}(l+1) + \beta_2^{(t)} \cos\frac{2\pi}{12}(l+1) + \psi_l a_{t+1}$$

The result follows by solving the equation system for $l = 1, 2$.

6.5. Show the updating equations in (6.16) and (6.21), using the same approach as in Exercise 6.4.

6.6. Consider the $(0, 1, 1)(0, 1, 1)_{12}$ model. Show that the one-step-ahead forecast can be written as

$$z_t(1) = \text{EWMA}(z_t) + \text{SEWMA}[z_{t-11} - \text{EWMA}(z_{t-12})]$$

where

$$\text{EWMA}(z_t) = (1 - \theta) \sum_{j \geq 0} \theta^j z_{t-j}$$

$$\text{SEWMA}(z_t) = (1 - \Theta) \sum_{j \geq 0} \Theta^j z_{t-12j}$$

Interpret this result.

6.7. An economist uses the following seasonal model to describe quarterly time series data:

$$(1 - B)(1 - B^4)z_t = (1 - \Theta B^4)a_t$$

(a) Discuss the autocorrelation structure of the stationary difference.

(b) What is the eventual forecast function?

(c) Derive the one-step-ahead forecast in terms of past observations.

6.8. Consider the following models: $(0, 1, 1)(1, 0, 0)_{12}$; $(0, 1, 0)(1, 0, 1)_{12}$; and $(1, 0, 0)(0, 1, 1)_{12}$.

(a) Write down the forecast equations and discuss the forecasts from these models.

(b) Obtain the eventual forecast functions, and specify the time periods at which these functions begin to hold (you may want to read Chap. 7 before attempting this problem).

6.9. Consider the quarterly earnings per share for General Motors stock (adjusted for stock splits and dividends) given below:

1951	0.53	0.52	0.34	0.50
1952	0.47	0.53	0.44	0.64
1953	0.57	0.60	0.52	0.54
1954	0.71	0.89	0.60	0.83
1955	1.14	1.27	0.90	0.99
1956	1.01	0.79	0.48	0.74
1957	0.93	0.78	0.43	0.85
1958	0.65	0.52	0.22	0.83
1959	1.03	1.05	0.47	0.51
1960	1.14	1.01	0.30	0.90
1961	0.65	0.88	0.30	1.20
1962	1.23	1.41	0.64	1.55
1963	1.44	1.62	0.72	1.77
1964	1.87	2.10	0.77	1.30
1965	2.22	2.23	0.91	2.05
1966	2.08	1.91	0.35	1.90
1967	1.35	1.82	0.51	1.97
1968	1.46	1.88	0.62	2.05
1969	1.82	1.56	0.79	1.77
1970	1.21	1.64	−0.28	−0.49

1971	2.12	1.97	0.75	1.88
1972	2.26	2.52	0.41	2.32
1973	2.84	2.78	0.92	1.80
1974	0.41	1.05	0.05	1.76
1975	0.20	1.14	0.84	2.14
1976	2.78	3.16	1.37	2.77
1977	3.14	3.82	1.40	3.26
1978	3.03	3.83	1.84	3.54
1979	4.39	4.13	0.06	1.46

Determine the appropriate seasonal ARIMA model, and forecast the next four observations ($\hat{z}_n(l)$, $l = 1, 2, 3, 4$). Calculate 90% prediction intervals for the future observations.

6.10. Consider the monthly arrivals of U.S. citizens from foreign travel (in thousands) for the period 1971–1978 (read across).

550	444	517	563	573	595
897	1065	768	647	544	427
655	579	618	765	704	749
1055	1130	844	771	664	543
663	589	713	780	775	790
993	1172	761	751	630	594
620	601	720	767	706	724
906	1054	753	599	571	518
627	531	553	624	625	701
872	1003	653	658	606	514
571	493	585	590	617	711
825	936	683	687	535	468
588	511	618	645	643	710
919	1002	719	760	575	511
633	570	711	706	718	785
1024	1077	742	740	612	584

Determine the appropriate seasonal ARIMA model, and forecast the next 12 observations.

6.11. Consider the unemployed U.S. civilian labor force (in thousands) for the period 1970–1978 (read across).

3406	3794	3733	3552	3384	4669
4510	4220	4292	4259	4607	4636
5414	5442	5175	4694	4394	5490
5330	5061	4840	4570	4815	4695
5447	5412	5215	4697	4344	5426
5173	4857	4658	4470	4266	4116
4675	4845	4512	4174	3799	4847
4550	4208	4165	3763	4056	4058
5008	5140	4755	4301	4144	5380
5260	4885	5202	5044	5685	6106
8180	8309	8359	7820	7623	8569
8209	7696	7522	7244	7231	7195
8174	8033	7525	6890	6304	7655
7577	7322	7026	6833	7095	7022
7848	8109	7556	6568	6151	7453
6941	6757	6437	6221	6346	5880
6897	6735	6479	5685	5457	6326
6438	5931	5797	5460	5629	5725

(a) Plot the data, and determine an appropriate ARIMA model for this series. Derive forecasts for the next 12 observations.

(b) Assume you wanted to use Winters' forecasting method described in Chapter 4. Discuss the magnitude of the smoothing constants you could expect.

6.12. Consider the monthly percentage changes in Canadian wages and salaries (series 13; Data Appendix). Use the data from February 1967 to December 1974 to determine the appropriate model(s). Calculate the forecasts and prediction intervals for the next 12 periods.

6.13. Consider the beer shipments in series 8 of the Data Appendix. Determine the appropriate seasonal ARIMA models, and calculate forecasts for the next 13 periods.

6.14. Consider the series of monthly traffic fatalities given in series 5 of the Data Appendix. Use the first 108 observations (January 1960 to December 1968) to determine the appropriate seasonal ARIMA models.

(a) Calculate the forecasts for the next 12 periods, and give 95% prediction intervals.

(b) Update your forecasts with the additional observations from 1969; that is, calculate $\hat{z}_{109}(l)$, $\hat{z}_{110}(l), \ldots, \hat{z}_{120}(l)$, $l = 1, 2, 3$.

6.15. Consider the gasoline demand data in series 12 of the Data Appendix. Use the first 180 observations (January 1960 to December 1974) to determine the appropriate seasonal ARIMA model(s).

(a) Calculate forecasts and prediction intervals for the next 12 months.

(b) Use the data for 1975 to update your forecasts.

CHAPTER 7

7.1. Specify the ARIMA models that lead to the same forecasts as general exponential smoothing with discount coefficient ω and the following forecast functions.

(a) $\mathbf{f}(l) = (1, l)'$

(b) $\mathbf{f}(l) = (1, l, \frac{1}{2}l^2)'$

(c) $\mathbf{f}(l) = (1, \phi^l)'$

(d) $\mathbf{f}(l) = \left(1, \sin\frac{\pi}{2}l, \cos\frac{\pi}{2}l\right)'$

(e) $\mathbf{f}(l) = \left(1, l, \sin\frac{\pi}{2}l, \cos\frac{\pi}{2}l\right)'$

(f) $\mathbf{f}(l) = \left(1, l, \sin\frac{\pi}{2}l, \cos\frac{\pi}{2}l, l\sin\frac{\pi}{2}l, l\cos\frac{\pi}{2}l\right)'$

(g) $\mathbf{f}(l) = \left(1, l, \sin\frac{2\pi}{12}l, \cos\frac{2\pi}{12}l, \sin\frac{4\pi}{12}l, \cos\frac{4\pi}{12}l\right)'$

(h) $\mathbf{f}(l) = \left(1, l, \sin\frac{4\pi}{12}l, \cos\frac{4\pi}{12}l\right)'$

CHAPTER 8

8.1. Calculate and plot the impulse response weights for the following transfer function models:

(a) $v(B) = \omega_0/(1 - \delta_1 B)$

(b) $v(B) = (\omega_0 - \omega_1 B)/(1 - \delta_1 B)$

(c) $v(B) = (\omega_0 - \omega_1 B - \omega_2 B^2)/(1 - \delta_1 B)$

(d) $v(B) = (\omega_0 - \omega_1 B)/(1 - \delta_1 B - \delta_2 B^2)$

Use $\omega_0 = 1$; $\omega_1 = -2$; $\omega_2 = -1$; $\delta_1 = .6$; $\delta_2 = -.5$.

8.2. *Gas furnace example*. Box and Jenkins (1976, p. 407) model the relationship between the CO_2 gas content (Y) and the methane feed rate (X). They obtain the following transfer function model:

$$Y_t - 53.51 = \frac{-(.53 + .37B + .51B^2)}{1 - .57B}(X_{t-3} + 0.057)$$
$$+ \frac{a_t}{1 - 1.53B + .63B^2}; \quad \hat{\sigma}^2 = .0561$$

and

$$(1 - 1.97B + 1.37B^2 - .34B^3)(X_t + 0.057) = \alpha_t; \hat{\sigma}_\alpha^2 = .0353$$

The last eight observations on X and Y are as follows:

t	X_t	Y_t	t	X_t	Y_t
206	−.569	60.5	202	−2.330	58.0
205	−1.261	60.4	201	−2.473	55.6
204	−1.739	60.0	200	−2.499	53.5
203	−2.053	59.5	199	−2.378	52.4

Calculate the forecasts and 95% prediction intervals for the next six periods.

8.3. *Forecasts with leading indicators*. Box and Jenkins (1976, p. 409) obtain the following transfer function model between sales (Y) and a leading indicator (X):

$$(1 - B)Y_t = .035 + \frac{4.82B^3}{1 - .72B}(1 - B)X_t + (1 - .54B)a_t;$$
$$\hat{\sigma}^2 = .0484$$

and

$$(1 - B)X_t = (1 - .32B)\alpha_t; \quad \hat{\sigma}_\alpha^2 = .0676$$

(a) Indicate how one could calculate the forecasts.
(b) Derive 95% prediction intervals for the next six observations.

8.4. Using the data in Exercise 5.28, construct a transfer function model relating business mergers (Y) and the Standard and Poor's stock price index (X).

8.5. Consider the following intervention models:

(a) $\omega_0 I_t(T)$ (b) $\dfrac{\omega_0}{1-B} I_t(T)$

(c) $\dfrac{\omega_0}{1-\delta B} I_t(T)$ (d) $\left(\dfrac{\omega_0}{1-B} + \dfrac{\omega_1}{1-\delta B}\right) I_t(T)$

(e) $\left(\omega_0 + \dfrac{\omega_1 B}{1-\delta B}\right) I_t(T)$ (f) $\dfrac{\omega_0}{(1-\delta B)(1-B)} I_t(T)$

where $I_t(T) = 1$ if $t = T$ and 0 otherwise. Graph the effect of these interventions, and discuss possible applications of the various intervention models.

8.6. Comment on the difference between the additive outlier (AO) and the innovational outlier (AI) models. Discuss situations in which one would expect each type of outlier.

8.7. Represent the ARIMA(0, 1, 1) model as a special case of the state space model in (8.50). Here $k = \max\{p, q+1\} = 2$ and $\phi_1 = 1$.

8.8. Represent the ARIMA(1, 1, 1) model as a special case of the state space model in (8.50).

8.9. Consider the simple linear regression model $y_t = \beta_0 + \beta_1 x_t + \varepsilon_t$. Assume that the parameters are constant. Specify the updating equations for the least squares estimators.

8.10. Consider the model $y_t = \mu_t + \varepsilon_t$, where $\mu_{t+1} = \mu_t + a_{t+1}$. Without loss of generality assume that $\sigma_\varepsilon^2 = 1$ and $\sigma_a^2 = \omega$.

(a) Show that this model is a special case of the state space model in Equation (8.53).

(b) Show that in the steady state (i.e., after P_t has reached its steady-state limit) the application of the Kalman filter equations in (8.61) and (8.54) leads to the same updating equations as simple exponential smoothing. Determine the smoothing constant α as a function of ω.

(c) Expressing this model as an ARIMA(0, 1, 1) model, use the results in Chapter 7 to show that the forecasts must be equivalent to the ones from simple exponential smoothing.

8.11. Consider the linear growth model [see Harrison and Stevens (1976)]

$$y_t = \mu_t + \varepsilon_t$$

where

$$\mu_{t+1} = \mu_t + \beta_{t+1} + a_{1,t+1}$$
$$\beta_{t+1} = \beta_t + a_{2,t+1}$$

(a) Show that this model is a special case of the state space model in (8.53).

(b) Show that in the steady state [i.e., after the Kalman gain vector $\mathbf{k}_{t+1}$ in (8.52) has reached a steady-state limit], the updating equations and forecasts for this model are the same as the ones for Holt's linear trend model in (3.41) and (3.42). *Hint*: The steady-state gain vector $\mathbf{k}$ is a function of the parameters in $E(\mathbf{a}_t\mathbf{a}_t')=\mathbf{R}_2$, where $\mathbf{a}_t=(a_{1,t}+a_{2,t},a_{2,t})'$. Assume that $\sigma_\varepsilon^2=1$. To show the relationship, set $\mathbf{P}_{t+1|t+1}=\mathbf{P}_{t|t}$ in (8.52) and solve for the elements in $\mathbf{R}_2$.

(c) Show that this model is equivalent to the ARIMA(0, 2, 2) model.

8.12. Consider the growth rates in series 2 of the Data Appendix. In Chapter 3 we used simple exponential smoothing and found that $\alpha = .11$. Use the last 20 observations to calculate the tracking signals in (8.76) and (8.77).

8.13. Consider a sequence of daily stock prices. For example, you might want to take the last 120 observations of any major stock.

(a) Consider the following forecast methods:

(i) Simple and double exponential smoothing with smoothing constant $\alpha = .1$.

(ii) Simple and double exponential smoothing with smoothing constants determined from the first 100 observations.

(iii) Naive or random walk forecast, $\hat{z}_t(l) = z_t$.

(b) Calculate the one- and two-step-ahead forecast errors,

$$z_{100+t} - \hat{z}_{100+t-1}(1) \qquad t = 1,\ldots, 20$$

and

$$z_{100+t} - \hat{z}_{100+t-2}(2) \qquad t = 2,\ldots, 20$$

Use the summary statistics in (8.71) to compare these methods. Interpret the results.

8.14. Consider the electricity usage data in series 3 of the Data Appendix.

(a) Use the first 96 observations and consider the following forecast methods:

(i) Winters' additive forecast method.

(ii) General exponential smoothing with seasonal indicators or trigonometric functions.

(iii) ARIMA time series models [see Equation (6.30)].

(b) For each forecast method, calculate the one-step-ahead forecast errors $e_{96+t-1}(1) = z_{96+t} - \hat{z}_{96+t-1}(1)$ for $t = 1, 2, \ldots, 10$.

(c) Evaluate these forecast methods in terms of the summary statistics in (8.71).

8.15. Repeat Exercise 8.14 for the gas usage data in series 11 of the Data Appendix. The ARIMA model is given in (6.31).

8.16. Repeat Exercise 8.14 for the housing start series (series 6 of the Data Appendix). Use the first 120 observations to construct the models. Calculate $e_{120+t-1}(1)$ for $t = 1, \ldots, 12$, and evaluate the forecast methods. The ARIMA model is given in (6.32).

8.17. Consider monthly percentage changes in Canadian wages and salaries (series 13 of the Data Appendix). Repeat Exercise 8.14 using the first 95 observations from February 1967 to December 1974 to construct the models. Use the last nine observations to calculate $e_{95+t-1}(1)$ for $t = 1, \ldots, 9$, and evaluate the methods.

Data Appendix

Here we list the series that are used in this book. Read across in all tables.

Series 1. Quarterly Iowa nonfarm income (in millions of dollars); 1948–1979.

Series 2. Quarterly growth rates of Iowa nonfarm income; second quarter 1948 through fourth quarter 1979.

Series 3. Monthly average residential electricity usage in Iowa City (in kilowatt-hours); January 1971 to October 1979.

Series 4. Monthly car sales in Quebec, Canada (in thousands of units); 1960–1968.

Series 5. Monthly traffic fatalities in Ontario; 1960–1974.

Series 6. Monthly U.S. housing starts of privately owned single-family structures (in thousands of units); 1965–1975.

Series 7. Quarterly new plant and equipment expenditures in U.S. industries (in billions of dollars); 1964–1976.

Series 8. Four-week totals for beer shipments (in thousands of units).

Series 9. Yield data; monthly differences between the yield on mortgages and the yield on government loans in the Netherlands; January 1961 to March 1974.

Series 10. Demand for repair parts for a large heavy equipment manufacturer in Iowa (in thousands of dollars); January 1972 to October 1979.

Series 11. Monthly average residential gas usage in Iowa City (in hundreds of cubic feet); January 1971 to October 1979.

Series 12. Monthly gasoline demand in Ontario (in millions of gallons); 1960–1975.

Series 13. Monthly percentage changes in Canadian wages and salaries; February 1967 to September 1975.

Series 14. Monthly sales of U.S. houses (in thousands of units); 1965–1975.

Series 1. Quarterly Iowa Nonfarm Income (in millions of dollars), First Quarter 1948 to Fourth Quarter 1979

601	604	620	626	641	642	645	655
682	678	692	707	736	753	763	775
775	783	794	813	823	826	829	831
830	838	854	872	882	903	919	937
927	962	975	995	1001	1013	1021	1028
1027	1048	1070	1095	1113	1143	1154	1173
1178	1183	1205	1208	1209	1223	1238	1245
1258	1278	1294	1314	1323	1336	1355	1377
1416	1430	1455	1480	1514	1545	1589	1634
1669	1715	1760	1812	1809	1828	1871	1892
1946	1983	2013	2045	2048	2097	2140	2171
2208	2272	2311	2349	2362	2442	2479	2528
2571	2634	2684	2790	2890	2964	3085	3159
3237	3358	3489	3588	3624	3719	3821	3934
4028	4129	4205	4349	4463	4598	4725	4827
4939	5067	5231	5408	5492	5653	5828	5965

Series 2. Quarterly Growth Rates of Iowa Nonfarm Income, Second Quarter 1948 to Fourth Quarter 1979

	0.50	2.65	0.97	2.40	0.16	0.47	1.55
4.12	−0.59	2.06	2.17	4.10	2.31	1.33	1.57
0.00	1.03	1.40	2.39	1.23	0.36	0.36	0.24
−0.12	0.96	1.91	2.11	1.15	2.38	1.77	1.96
−1.07	3.78	1.35	2.05	0.60	1.20	0.79	0.69
−0.10	2.04	2.10	2.34	1.64	2.70	0.96	1.65
0.43	0.42	1.86	0.25	0.08	1.16	1.23	0.57
1.04	1.59	1.25	1.55	0.68	0.98	1.42	1.62
2.83	0.99	1.75	1.72	2.30	2.05	2.85	2.83
2.14	2.76	2.62	2.95	−0.17	1.05	2.35	1.12
2.85	1.90	1.51	1.59	0.15	2.39	2.05	1.45
1.70	2.90	1.72	1.64	0.55	3.39	1.52	1.98
1.70	2.45	1.90	3.95	3.58	2.56	4.08	2.40
2.47	3.74	3.90	2.84	1.00	2.62	2.74	2.96
2.39	2.51	1.84	3.42	2.62	3.02	2.76	2.16
2.32	2.59	3.24	3.38	1.55	2.93	3.10	2.35

Series 3. Monthly Average Residential Electricity Usage in Iowa City (in kilowatt-hours), January 1971 to October 1979

454	421	389	368	460	386	501	606	539	381	396	452
470	485	456	414	388	502	543	642	617	454	429	473
504	454	424	406	384	466	668	662	672	426	415	442
483	431	402	399	374	430	666	663	501	392	400	501
543	473	451	444	384	489	763	711	598	406	415	440
523	502	439	420	387	453	630	637	576	411	455	512
530	507	436	407	392	531	710	658	500	414	418	520
535	503	464	414	383	472	676	622	652	474	422	501
536	506	477	422	385	438	546	642	566	399		

Series 4. Monthly Car Sales in Quebec, Canada (in thousands of units), January 1960 to December 1968

6.550	8.728	12.026	14.395	14.587	13.791	9.498	8.251	7.049	9.545	9.364	8.456
7.237	9.374	11.837	13.784	15.926	13.821	11.143	7.975	7.610	10.015	12.759	8.816
10.677	10.947	15.200	17.010	20.900	16.205	12.143	8.997	5.568	11.474	12.256	10.583
10.862	10.965	14.405	20.379	20.128	17.816	12.268	8.642	7.962	13.932	15.936	12.628
12.267	12.470	18.944	21.259	22.015	18.581	15.175	10.306	10.792	14.752	13.754	11.738
12.181	12.965	19.990	23.125	23.541	21.247	15.189	14.767	10.895	17.130	17.697	16.611
12.674	12.760	20.249	22.135	20.677	19.933	15.388	15.113	13.401	16.135	17.562	14.720
12.225	11.608	20.985	19.692	24.081	22.114	14.220	13.434	13.598	17.187	16.119	13.713
13.210	14.251	20.139	21.725	26.099	21.084	18.024	16.722	14.385	21.342	17.180	14.577

Series 5. Monthly Traffic Fatalities in Ontario, January 1960 to December 1974

61	65	55	56	91	80	135	129	129	130	109	126
73	68	74	95	105	108	127	108	126	154	127	103
95	59	68	82	92	124	139	167	138	146	128	145
91	66	89	98	113	130	127	157	157	136	145	112
71	95	95	105	116	104	128	181	130	124	123	152
85	86	101	105	135	142	138	190	129	175	146	179
103	104	80	108	125	134	150	149	148	167	160	168
99	102	98	104	114	165	184	161	205	189	152	146
72	83	97	105	134	145	172	175	149	166	142	146
111	84	116	102	153	176	145	175	147	174	169	131
81	85	107	111	135	114	150	156	146	166	154	130
92	80	87	101	130	151	195	224	191	188	169	161
121	89	105	134	155	203	192	195	225	186	172	157
130	84	122	116	147	183	211	256	207	196	153	154
94	89	118	101	150	150	191	214	173	170	175	123

Series 6. Monthly U.S. Housing Starts of Privately Owned Single-Family Structures (in thousands of units), January 1965 to December 1975

52.149	47.205	82.150	100.931	98.408	97.351	96.489	88.830	80.876	85.750	72.351	61.198
46.561	50.361	83.236	94.343	84.748	79.828	69.068	69.362	59.404	53.530	50.212	37.972
40.157	40.274	66.592	79.839	87.341	87.594	82.344	83.712	78.194	81.704	69.088	47.026
45.234	55.431	79.325	97.983	86.806	81.424	86.398	82.522	80.078	85.560	64.819	53.847
51.300	47.909	71.941	84.982	91.301	82.741	73.523	69.465	71.504	68.039	55.069	42.827
33.363	41.367	61.879	73.835	74.848	83.007	75.461	77.291	75.961	79.393	67.443	69.041
54.856	58.287	91.584	116.013	115.627	116.946	107.747	111.663	102.149	102.882	92.904	80.362
76.185	76.306	111.358	119.840	135.167	131.870	119.078	131.324	120.491	116.990	97.428	73.195
77.105	73.560	105.136	120.453	131.643	114.822	114.746	106.806	84.504	86.004	70.488	46.767
43.292	57.593	76.946	102.237	96.340	99.318	90.715	79.782	73.443	69.460	57.898	41.041
39.791	39.959	62.498	77.777	92.782	90.284	92.782	90.655	84.517	93.826	71.646	55.650

Series 7. Quarterly New Plant and Equipment Expenditures in U.S. Industries (in billions of dollars), First Quarter 1964 to Fourth Quarter 1976

10.00	11.85	11.70	13.42
11.20	13.63	13.65	15.93
13.33	16.05	15.92	18.22
14.46	16.69	16.20	18.12
15.10	16.85	16.79	19.03
16.04	18.81	19.25	21.46
17.47	20.33	20.26	21.66
17.68	20.60	20.14	22.79
19.38	22.01	21.86	25.20
21.50	24.73	25.04	28.48
24.10	28.16	28.23	31.92
25.82	28.43	27.79	30.74
25.87	29.70	30.41	34.52

Series 8. Beer Shipments—Four-Week Totals (in thousands of units) Read downwards, left to right.

Year 1	Year 2	Year 3	Year 4
18.705	19.598	20.816	21.260
20.232	21.463	23.743	24.109
20.467	23.287	25.152	26.320
22.123	24.065	28.804	27.701
25.036	27.447	31.158	34.502
26.839	30.413	31.540	33.297
29.640	32.307	32.849	31.252
30.935	32.974	33.748	35.173
28.278	29.973	31.910	36.207
24.235	23.986	27.609	31.511
22.370	26.953	25.170	28.560
21.224	24.250	24.040	26.828
21.061	23.518	25.368	26.660

Series 9. **Yield data—Monthly Differences Between the Yield on Mortgages and the Yield on Government Loans in the Netherlands, January 1961 to March 1974**

0.66	0.70	0.74	0.63	0.70	0.66	0.61	0.52	0.60	0.61	0.70	1.10
1.17	1.23	0.85	0.78	0.71	0.55	0.56	0.74	0.80	0.75	0.74	0.79
0.78	1.00	1.05	1.09	1.05	0.75	0.73	0.77	0.77	0.84	0.66	0.68
0.67	0.56	0.62	0.73	0.70	0.74	0.93	1.00	1.50	1.30	1.18	1.15
1.34	1.37	1.13	1.04	0.92	1.15	0.99	1.32	1.46	1.24	1.01	1.04
1.08	0.94	0.81	1.00	0.98	1.02	1.16	0.96	1.23	1.10	1.02	1.08
1.30	0.97	0.96	0.80	0.62	0.51	0.56	0.84	0.87	0.87	0.76	0.86
0.81	0.77	0.74	0.80	0.78	0.72	0.66	0.92	0.99	0.98	0.70	0.65
0.78	0.57	0.41	0.61	0.85	0.85	1.11	1.05	0.96	1.31	1.49	1.35
1.32	1.24	1.47	1.32	1.23	1.33	1.48	1.49	1.48	1.49	1.55	1.73
1.70	1.43	1.44	1.37	1.20	1.19	1.39	1.41	1.40	1.39	1.62	1.59
1.36	1.31	0.99	0.89	0.87	0.94	1.03	1.27	1.20	1.10	0.93	1.00
1.04	1.10	1.10	1.09	1.05	0.70	0.88	0.81	1.08	1.39	1.16	0.49
0.74	0.90	0.91									

Series 10. **Monthly Demand for Repair Parts for a Large Heavy-Equipment Manufacturer in Iowa (in thousands of dollars), January 1972 to October 1979**

954	765	867	940	913	1014	801	990	712	959	828	965
915	891	991	971	1129	1091	1195	1295	1046	1121	1033	1222
1199	1012	1404	1137	1421	1162	1639	1545	1420	1916	1491	1295
1764	1727	1654	1811	1520	1635	1984	1898	1853	2015	1709	1667
1625	1562	2121	1783	1474	1657	1746	1763	1517	1457	1388	1501
1227	1342	1666	2091	1629	1451	1727	1736	1952	1420	1345	842
1576	1485	1928	2072	1887	2175	2199	1961	2091	1993	1595	1372
1607	1871	2594	1743	2267	2602	2565	2567	2344	2805		

Series 11. **Monthly Average Residential Gas Usage in Iowa City (in hundreds of cubic feet), January 1971 to October 1979**

302	262	218	175	100	77	43	47	49	69	152	205
246	294	242	181	107	56	49	47	47	71	151	244
280	230	185	148	98	61	46	45	55	48	115	185
276	220	181	151	83	55	49	42	46	74	103	200
237	247	215	182	80	46	65	40	44	63	85	185
247	231	167	117	79	45	40	38	41	69	152	232
282	255	161	107	53	40	39	34	35	56	97	210
260	257	210	125	80	42	35	31	32	50	92	189
256	250	198	136	73	39	32	30	31	45		

Series 12. Monthly Gasoline Demand in Ontario (in millions of gallons), January 1960 to December 1975

87.695	86.890	96.442	98.133	113.615	123.924	128.924	134.775	117.357	114.626	107.677	108.087
92.188	88.591	98.683	99.207	125.485	124.677	132.543	140.735	124.008	121.194	111.634	111.565
101.007	94.228	104.255	106.922	130.621	125.251	140.318	146.174	122.318	128.770	117.518	115.492
108.497	100.482	106.140	118.581	132.371	132.042	151.938	150.997	130.931	137.018	121.271	123.548
109.894	106.061	112.539	125.745	136.251	140.892	158.390	148.314	144.148	140.138	124.075	136.485
109.895	109.044	122.499	124.264	142.296	150.693	163.331	165.837	151.731	142.491	140.229	140.463
116.963	118.049	137.869	127.392	154.166	160.227	165.869	173.522	155.828	153.771	143.963	143.898
124.046	121.260	138.870	129.782	162.312	167.211	172.897	189.689	166.496	160.754	155.582	145.936
139.625	137.361	138.963	155.301	172.026	165.004	185.861	190.270	163.903	174.270	160.272	165.614
146.182	137.728	148.932	156.751	177.998	174.559	198.079	189.073	175.702	180.097	155.202	174.508
154.277	144.998	159.644	168.646	166.273	190.176	205.541	193.657	182.617	189.614	174.176	184.416
158.167	156.261	176.353	175.720	193.939	201.269	218.960	209.861	198.688	190.474	194.502	190.755
166.286	170.699	181.468	174.241	210.802	212.262	218.099	229.001	203.200	212.557	197.095	193.693
188.992	175.347	196.265	203.526	227.443	233.038	234.119	255.133	216.478	232.868	221.616	209.893
194.784	189.756	193.522	212.870	248.565	221.532	252.642	255.007	206.826	233.231	212.678	217.173
199.024	191.813	195.997	208.684	244.113	243.108	255.918	244.642	237.579	237.579	217.775	227.621

Series 13. Monthly Percentage Changes in Canadian Wages and Salaries, February 1967 to September 1975

	0.71	1.47	1.61	2.97	3.58	0.94	0.68	0.32	−0.67	0.02	−2.56
−0.84	0.38	1.18	1.95	4.08	2.41	0.55	1.16	1.70	0.13	0.63	−2.19
−0.32	1.48	0.84	1.45	3.00	3.24	−0.62	0.35	3.26	−0.53	0.90	−1.31
−0.91	0.46	0.50	0.96	2.23	3.27	−1.95	0.65	3.05	0.23	−0.45	−1.78
−1.44	1.49	1.22	2.22	4.76	3.04	−2.39	0.89	3.33	−0.14	−0.47	−1.79
−0.49	0.58	2.11	0.69	3.81	3.32	−2.26	1.46	3.81	0.96	−0.03	0.50
−1.70	2.51	0.75	2.21	3.15	3.05	−2.36	0.26	4.94	1.44	−0.12	−1.08
0.19	1.03	2.12	1.58	4.51	2.96	0.30	1.66	4.65	−0.43	−0.43	1.19
−2.93	0.21	2.09	1.60	4.64	2.98	0.86	−1.88	5.01			

Series 14. Monthly Sales of U.S. Houses (in thousands of units), January 1965 to December 1975

38	44	53	49	54	57	51	58	48	44	42	37
42	43	53	49	49	40	40	36	29	31	26	23
29	32	41	44	49	47	46	47	43	45	34	31
35	43	46	46	43	41	44	47	41	40	32	32
34	40	43	42	43	44	39	40	33	32	31	28
34	29	36	42	43	44	44	48	45	44	40	37
45	49	62	62	58	59	64	62	50	52	50	44
51	56	60	65	64	63	63	72	61	65	51	47
54	58	66	63	64	60	53	52	44	40	36	28
36	42	53	53	55	48	47	43	39	33	30	23
29	33	44	54	56	51	51	53	45	45	44	38

Table Appendix

Table A. Tail Areas of the Student t Distribution
Table of $t_\alpha(\nu)$ such that $\Pr[t(\nu) > t_\alpha(\nu)] = \alpha$, where ν is the number of degrees of freedom

	α					
ν	.25	.10	.05	.025	.01	.005
1	1.00	3.08	6.31	12.71	31.82	63.66
2	0.82	1.89	2.92	4.30	6.96	9.92
3	0.76	1.64	2.35	3.18	4.54	5.84
4	0.74	1.53	2.13	2.78	3.75	4.60
5	0.73	1.48	2.02	2.57	3.36	4.03
6	0.72	1.44	1.94	2.45	3.14	3.71
7	0.71	1.42	1.90	2.36	3.00	3.50
8	0.71	1.40	1.86	2.31	2.90	3.36
9	0.70	1.38	1.83	2.26	2.82	3.25
10	0.70	1.37	1.81	2.23	2.76	3.17
11	0.70	1.36	1.80	2.20	2.72	3.11
12	0.70	1.36	1.78	2.18	2.68	3.06
13	0.69	1.35	1.77	2.16	2.65	3.01
14	0.69	1.34	1.76	2.14	2.62	3.00
15	0.69	1.34	1.75	2.13	2.60	2.95
16	0.69	1.34	1.75	2.12	2.58	2.92
17	0.69	1.33	1.74	2.11	2.57	2.90
18	0.69	1.33	1.73	2.10	2.55	2.88
19	0.69	1.33	1.73	2.09	2.54	2.86
20	0.69	1.33	1.72	2.09	2.53	2.84
21	0.69	1.32	1.72	2.08	2.52	2.83
22	0.69	1.32	1.72	2.07	2.51	2.82
23	0.69	1.32	1.71	2.07	2.50	2.81
24	0.69	1.32	1.71	2.06	2.49	2.80
25	0.68	1.32	1.71	2.06	2.48	2.79
26	0.68	1.31	1.71	2.06	2.48	2.78
27	0.68	1.31	1.70	2.05	2.47	2.77
28	0.68	1.31	1.70	2.05	2.47	2.76
29	0.68	1.31	1.70	2.04	2.46	2.76
30	0.68	1.31	1.70	2.04	2.46	2.75
40	0.68	1.30	1.68	2.02	2.42	2.70
60	0.68	1.30	1.67	2.00	2.39	2.66
120	0.68	1.29	1.66	1.98	2.36	2.62
∞[a]	0.67	1.28	1.64	1.96	2.33	2.58

[a] Normal.

Table B. Upper Percentage Points of the F Distribution
Table of $F_\alpha(\nu_1, \nu_2)$ such that $\Pr[F(\nu_1, \nu_2) > F_\alpha(\nu_1, \nu_2)] = \alpha$

5%

ν_2 \ ν_1	1	2	3	4	5	6	7	8	9	10	12	15	20	24	30	40	60	120	∞
1	161.4	199.5	215.7	224.6	230.2	234.0	236.8	238.9	240.5	241.9	243.9	245.9	248.0	249.1	250.1	251.1	252.2	253.3	254.3
2	18.51	19.00	19.16	19.25	19.30	19.33	19.35	19.37	19.38	19.40	19.41	19.43	19.45	19.45	19.46	19.47	19.48	19.49	19.50
3	10.13	9.55	9.28	9.12	9.01	8.94	8.89	8.85	8.81	8.79	8.74	8.70	8.66	8.64	8.62	8.59	8.57	8.55	8.53
4	7.71	6.94	6.59	6.39	6.26	6.16	6.09	6.04	6.00	5.96	5.91	5.86	5.80	5.77	5.75	5.72	5.69	5.66	5.63
5	6.61	5.79	5.41	5.19	5.05	4.95	4.88	4.82	4.77	4.74	4.68	4.62	4.56	4.53	4.50	4.46	4.43	4.40	4.36
6	5.99	5.14	4.76	4.53	4.39	4.28	4.21	4.15	4.10	4.06	4.00	3.94	3.87	3.84	3.81	3.77	3.74	3.70	3.67
7	5.59	4.74	4.35	4.12	3.97	3.87	3.79	3.73	3.68	3.64	3.57	3.51	3.44	3.41	3.38	3.34	3.30	3.27	3.23
8	5.32	4.46	4.07	3.84	3.69	3.58	3.50	3.44	3.39	3.35	3.28	3.22	3.15	3.12	3.08	3.04	3.01	2.97	2.93
9	5.12	4.26	3.86	3.63	3.48	3.37	3.29	3.23	3.18	3.14	3.07	3.01	2.94	2.90	2.86	2.83	2.79	2.75	2.71
10	4.96	4.10	3.71	3.48	3.33	3.22	3.14	3.07	3.02	2.98	2.91	2.85	2.77	2.74	2.70	2.66	2.62	2.58	2.54
11	4.84	3.98	3.59	3.36	3.20	3.09	3.01	2.95	2.90	2.85	2.79	2.72	2.65	2.61	2.57	2.53	2.49	2.45	2.40
12	4.75	3.89	3.49	3.26	3.11	3.00	2.91	2.85	2.80	2.75	2.69	2.62	2.54	2.51	2.47	2.43	2.38	2.34	2.30
13	4.67	3.81	3.41	3.18	3.03	2.92	2.83	2.77	2.71	2.67	2.60	2.53	2.46	2.42	2.38	2.34	2.30	2.25	2.21
14	4.60	3.74	3.34	3.11	2.96	2.85	2.76	2.70	2.65	2.60	2.53	2.46	2.39	2.35	2.31	2.27	2.22	2.18	2.13
15	4.54	3.68	3.29	3.06	2.90	2.79	2.71	2.64	2.59	2.54	2.48	2.40	2.33	2.29	2.25	2.20	2.16	2.11	2.07
16	4.49	3.63	3.24	3.01	2.85	2.74	2.66	2.59	2.54	2.49	2.42	2.35	2.28	2.24	2.19	2.15	2.11	2.06	2.01
17	4.45	3.59	3.20	2.96	2.81	2.70	2.61	2.55	2.49	2.45	2.38	2.31	2.23	2.19	2.15	2.10	2.06	2.01	1.96
18	4.41	3.55	3.16	2.93	2.77	2.66	2.58	2.51	2.46	2.41	2.34	2.27	2.19	2.15	2.11	2.06	2.02	1.97	1.92
19	4.38	3.52	3.13	2.90	2.74	2.63	2.54	2.48	2.42	2.38	2.31	2.23	2.16	2.11	2.07	2.03	1.98	1.93	1.88
20	4.35	3.49	3.10	2.87	2.71	2.60	2.51	2.45	2.39	2.35	2.28	2.20	2.12	2.08	2.04	1.99	1.95	1.90	1.84
21	4.32	3.47	3.07	2.84	2.68	2.57	2.49	2.42	2.37	2.32	2.25	2.18	2.10	2.05	2.01	1.96	1.92	1.87	1.81
22	4.30	3.44	3.05	2.82	2.66	2.55	2.46	2.40	2.34	2.30	2.23	2.15	2.07	2.03	1.98	1.94	1.89	1.84	1.78
23	4.28	3.42	3.03	2.80	2.64	2.53	2.44	2.37	2.32	2.27	2.20	2.13	2.05	2.01	1.96	1.91	1.86	1.81	1.76
24	4.26	3.40	3.01	2.78	2.62	2.51	2.42	2.36	2.30	2.25	2.18	2.11	2.03	1.98	1.94	1.89	1.84	1.79	1.73
25	4.24	3.39	2.99	2.76	2.60	2.49	2.40	2.34	2.28	2.24	2.16	2.09	2.01	1.96	1.92	1.87	1.82	1.77	1.71
26	4.23	3.37	2.98	2.74	2.59	2.47	2.39	2.32	2.27	2.22	2.15	2.07	1.99	1.95	1.90	1.85	1.80	1.75	1.69
27	4.21	3.35	2.96	2.73	2.57	2.46	2.37	2.31	2.25	2.20	2.13	2.06	1.97	1.93	1.88	1.84	1.79	1.73	1.67
28	4.20	3.34	2.95	2.71	2.56	2.45	2.36	2.29	2.24	2.19	2.12	2.04	1.96	1.91	1.87	1.82	1.77	1.71	1.65
29	4.18	3.33	2.93	2.70	2.55	2.43	2.35	2.28	2.22	2.18	2.10	2.03	1.94	1.90	1.85	1.81	1.75	1.70	1.64
30	4.17	3.32	2.92	2.69	2.53	2.42	2.33	2.27	2.21	2.16	2.09	2.01	1.93	1.89	1.84	1.79	1.74	1.68	1.62
40	4.08	3.23	2.84	2.61	2.45	2.34	2.25	2.18	2.12	2.08	2.00	1.92	1.84	1.79	1.74	1.69	1.64	1.58	1.51
60	4.00	3.15	2.76	2.53	2.37	2.25	2.17	2.10	2.04	1.99	1.92	1.84	1.75	1.70	1.65	1.59	1.53	1.47	1.39
120	3.92	3.07	2.68	2.45	2.29	2.17	2.09	2.02	1.96	1.91	1.83	1.75	1.66	1.61	1.55	1.50	1.43	1.35	1.25
∞	3.84	3.00	2.60	2.37	2.21	2.10	2.01	1.94	1.88	1.83	1.75	1.67	1.57	1.52	1.46	1.39	1.32	1.22	1.00

Table B. Upper Percentage Points of the F Distribution

Table of $F_\alpha(\nu_1, \nu_2)$ such that $\Pr[F(\nu_1, \nu_2) > F_\alpha(\nu_1, \nu_2)] = \alpha$

1%

ν_2 \ ν_1	1	2	3	4	5	6	7	8	9	10	12	15	20	24	30	40	60	120	∞
1	4052	4999.5	5403	5625	5764	5859	5928	5982	6022	6056	6106	6157	6209	6235	6261	6287	6313	6339	6366
2	98.50	99.00	99.17	99.25	99.30	99.33	99.36	99.37	99.39	99.40	99.42	99.43	99.45	99.46	99.47	99.47	99.48	99.49	99.50
3	34.12	30.82	29.46	28.71	28.24	27.91	27.67	27.49	27.35	27.23	27.05	26.87	26.69	26.60	26.50	26.41	26.32	26.22	26.13
4	21.20	18.00	16.69	15.98	15.52	15.21	14.98	14.80	14.66	14.55	14.37	14.20	14.02	13.93	13.84	13.75	13.65	13.56	13.46
5	16.26	13.27	12.06	11.39	10.97	10.67	10.46	10.29	10.16	10.05	9.89	9.72	9.55	9.47	9.38	9.29	9.20	9.11	9.02
6	13.75	10.92	9.78	9.15	8.75	8.47	8.26	8.10	7.98	7.87	7.72	7.56	7.40	7.31	7.23	7.14	7.06	6.97	6.88
7	12.25	9.55	8.45	7.85	7.46	7.19	6.99	6.84	6.72	6.62	6.47	6.31	6.16	6.07	5.99	5.91	5.82	5.74	5.65
8	11.26	8.65	7.59	7.01	6.63	6.37	6.18	6.03	5.91	5.81	5.67	5.52	5.36	5.28	5.20	5.12	5.03	4.95	4.86
9	10.56	8.02	6.99	6.42	5.06	5.80	5.61	5.47	5.35	5.26	5.11	4.96	4.81	4.73	4.65	4.57	4.48	4.40	4.31
10	10.04	7.56	6.55	5.99	5.64	5.39	5.20	5.06	4.94	4.85	4.71	4.56	4.41	4.33	4.25	4.17	4.08	4.00	3.91
11	9.65	7.21	6.22	5.67	5.32	5.07	4.89	4.74	4.63	4.54	4.40	4.25	4.10	4.02	3.94	3.86	3.78	3.69	3.60
12	9.33	6.93	5.95	5.41	5.06	4.82	4.64	4.50	4.39	4.30	4.16	4.01	3.86	3.78	3.70	3.62	3.54	3.45	3.36
13	9.07	6.70	5.74	5.21	4.86	4.62	4.44	4.30	4.19	4.10	3.96	3.82	3.66	3.59	3.51	3.43	3.34	3.25	3.17
14	8.86	6.51	5.56	5.04	4.69	4.46	4.28	4.14	4.03	3.94	3.80	3.66	3.51	3.43	3.35	3.27	3.18	3.09	3.00
15	8.68	6.36	5.42	4.89	4.56	4.32	4.14	4.00	3.89	3.80	3.67	3.52	3.37	3.29	3.21	3.13	3.05	2.96	2.87
16	8.53	6.23	5.29	4.77	4.44	4.20	4.03	3.89	3.78	3.69	3.55	3.41	3.26	3.18	3.10	3.02	2.93	2.84	2.75
17	8.40	6.11	5.18	4.67	4.34	4.10	3.93	3.79	3.68	3.59	3.46	3.31	3.16	3.08	3.00	2.92	2.83	2.75	2.65
18	8.29	6.01	5.09	4.58	4.25	4.01	3.84	3.71	3.60	3.51	3.37	3.23	3.08	3.00	2.92	2.84	2.75	2.66	2.57
19	8.18	5.93	5.01	4.50	4.17	3.94	3.77	3.63	3.52	3.43	3.30	3.15	3.00	2.92	2.84	2.76	2.67	2.58	2.49
20	8.10	5.85	4.94	4.43	4.10	3.87	3.70	3.56	3.46	3.37	3.23	3.09	2.94	2.86	2.78	2.69	2.61	2.52	2.42
21	8.02	5.78	4.87	4.37	4.04	3.81	3.64	3.51	3.40	3.31	3.17	3.03	2.88	2.80	2.72	2.64	2.55	2.46	2.36
22	7.95	5.72	4.82	4.31	3.99	3.76	3.59	3.45	3.35	3.26	3.12	2.98	2.83	2.75	2.67	2.58	2.50	2.40	2.31
23	7.88	5.66	4.76	4.26	3.94	3.71	3.54	3.41	3.30	3.21	3.07	2.93	2.78	2.70	2.62	2.54	2.45	2.35	2.26
24	7.82	5.61	4.72	4.22	3.90	3.67	3.50	3.36	3.26	3.17	3.03	2.89	2.74	2.66	2.58	2.49	2.40	2.31	2.21
25	7.77	5.57	4.68	4.18	3.85	3.63	3.46	3.32	3.22	3.13	2.99	2.85	2.70	2.62	2.54	2.45	2.36	2.27	2.17
26	7.72	5.53	4.64	4.14	3.82	3.59	3.42	3.29	3.18	3.09	2.96	2.81	2.66	2.58	2.50	2.42	2.33	2.23	2.13
27	7.68	5.49	4.60	4.11	3.78	3.56	3.39	3.26	3.15	3.06	2.93	2.78	2.63	2.55	2.47	2.38	2.29	2.20	2.10
28	7.64	5.45	4.57	4.07	3.75	3.53	3.36	3.23	3.12	3.03	2.90	2.75	2.60	2.52	2.44	2.35	2.26	2.17	2.06
29	7.60	5.42	4.54	4.04	3.73	3.50	3.33	3.20	3.09	3.00	2.87	2.73	2.57	2.49	2.41	2.33	2.23	2.14	2.03
30	7.56	5.39	4.51	4.02	3.70	3.47	3.30	3.17	3.07	2.98	2.84	2.70	2.55	2.47	2.39	2.30	2.21	2.11	2.01
40	7.31	5.18	4.31	3.83	3.51	3.29	3.12	2.99	2.89	2.80	2.66	2.52	2.37	2.29	2.20	2.11	2.02	1.92	1.80
60	7.08	4.98	4.13	3.65	3.34	3.12	2.95	2.82	2.72	2.63	2.50	2.35	2.20	2.12	2.03	1.94	1.84	1.73	1.60
120	6.85	4.79	3.95	3.48	3.17	2.96	2.79	2.66	2.56	2.47	2.34	2.19	2.03	1.95	1.86	1.76	1.66	1.53	1.38
∞	6.63	4.61	3.78	3.32	3.02	2.80	2.64	2.51	2.41	2.32	2.18	2.04	1.88	1.79	1.70	1.59	1.47	1.32	1.00

Reproduced with permission from E. S. Pearson and H. O. Hartley, *Biometrika Tables for Statisticians*, Vol. 1, Cambridge University Press, New York, 1954.

Table C. Tail Areas of the Chi-Square Distribution
Table of $\chi^2_\alpha(k)$ such that $\Pr[\chi^2(k) > \chi^2_\alpha(k)] = \alpha$

	α														
k	0.995	0.99	0.975	0.95	0.9	0.75	0.5	0.25	0.1	0.05	0.025	0.01	0.005	0.001	k
1	—	—	—	—	0.016	0.102	0.455	1.32	2.71	3.84	5.02	6.63	7.88	10.8	1
2	0.010	0.020	0.051	0.103	0.211	0.575	1.39	2.77	4.61	5.99	7.38	9.21	10.6	13.8	2
3	0.072	0.115	0.216	0.352	0.584	1.21	2.37	4.11	6.25	7.81	9.35	11.3	12.8	16.3	3
4	0.207	0.297	0.484	0.711	1.06	1.92	3.36	5.39	7.78	9.49	11.1	13.3	14.9	18.5	4
5	0.412	0.554	0.831	1.15	1.61	2.67	4.35	6.63	9.24	11.1	12.8	15.1	16.7	20.5	5
6	0.676	0.872	1.24	1.64	2.20	3.45	5.35	7.84	10.6	12.6	14.4	16.8	18.5	22.5	6
7	0.989	1.24	1.69	2.17	2.83	4.25	6.35	9.04	12.0	14.1	16.0	18.5	20.3	24.3	7
8	1.34	1.65	2.18	2.73	3.49	5.07	7.34	10.2	13.4	15.5	17.5	20.1	22.0	26.1	8
9	1.73	2.09	2.70	3.33	4.17	5.90	8.34	11.4	14.7	16.9	19.0	21.7	23.6	27.9	9
10	2.16	2.56	3.25	3.94	4.87	6.74	9.34	12.5	16.0	18.3	20.5	23.2	25.2	29.6	10
11	2.60	3.05	3.82	4.57	5.58	7.58	10.3	13.7	17.3	19.7	21.9	24.7	26.8	31.3	11
12	3.07	3.57	4.40	5.23	6.30	8.44	11.3	14.8	18.5	21.0	23.3	26.2	28.3	32.9	12
13	3.57	4.11	5.01	5.89	7.04	9.30	12.3	16.0	19.8	22.4	24.7	27.7	29.8	34.5	13
14	4.07	4.66	5.63	6.57	7.79	10.2	13.3	17.1	21.1	23.7	26.1	29.1	31.3	36.1	14
15	4.60	5.23	6.26	7.26	8.55	11.0	14.3	18.2	22.3	25.0	27.5	30.6	32.8	37.7	15
16	5.14	5.81	6.91	7.96	9.31	11.9	15.3	19.4	23.5	26.3	28.8	32.0	34.3	39.3	16

	α														
k	0.995	0.99	0.975	0.95	0.9	0.75	0.5	0.25	0.1	0.05	0.025	0.01	0.005	0.001	k
17	5.70	6.41	7.56	8.67	10.1	12.8	16.3	20.5	24.8	27.6	30.2	33.4	35.7	40.8	17
18	6.26	7.01	8.23	9.39	10.9	13.7	17.3	21.6	26.0	28.9	31.5	34.8	37.2	42.3	18
19	6.84	7.63	8.91	10.1	11.7	14.6	18.3	22.7	27.2	30.1	32.9	36.2	38.6	43.8	19
20	7.43	8.26	9.59	10.9	12.4	15.5	19.3	23.8	28.4	31.4	34.2	37.6	40.0	45.3	20
21	8.03	8.90	10.3	11.6	13.2	16.3	20.3	24.9	29.6	32.7	35.5	38.9	41.4	46.8	21
22	8.64	9.54	11.0	12.3	14.0	17.2	21.3	26.0	30.8	33.9	36.8	40.3	42.8	48.3	22
23	9.26	10.2	11.7	13.1	14.8	18.1	22.3	27.1	32.0	35.2	38.1	41.6	44.2	49.7	23
24	9.89	10.9	12.4	13.8	15.7	19.0	23.3	28.2	33.2	36.4	39.4	43.0	45.6	51.2	24
25	10.5	11.5	13.1	14.6	16.5	19.9	24.3	29.3	34.4	37.7	40.6	44.3	46.9	52.6	25
26	11.2	12.2	13.8	15.4	17.3	20.8	25.3	30.4	35.6	38.9	41.9	45.6	48.3	54.1	26
27	11.8	12.9	14.6	16.2	18.1	21.7	26.3	31.5	36.7	40.1	43.2	47.0	49.6	55.5	27
28	12.5	13.6	15.3	16.9	18.9	22.7	27.3	32.6	37.9	41.3	44.5	48.3	51.0	56.9	28
29	13.1	14.3	16.0	17.7	19.8	23.6	28.3	33.7	39.1	42.6	45.7	49.6	52.3	58.3	29
30	13.8	15.0	16.8	18.5	20.6	24.5	29.3	34.8	40.3	43.8	47.0	50.9	53.7	59.7	30

Table D. Significance Points of d_L and d_U in the Durbin-Watson Test: 5%

n	$p = 1$		$p = 2$		$p = 3$		$p = 4$		$p = 5$	
	d_L	d_U	d_L	d_U	d_L	d_U	d_L	d_U	d_L	d_U
15	1.08	1.36	0.95	1.54	0.82	1.75	0.69	1.97	0.56	2.21
16	1.10	1.37	0.98	1.54	0.86	1.73	0.74	1.93	0.62	2.15
17	1.13	1.38	1.02	1.54	0.90	1.71	0.78	1.90	0.67	2.10
18	1.16	1.39	1.05	1.53	0.93	1.69	0.82	1.87	0.71	2.06
19	1.18	1.40	1.08	1.53	0.97	1.68	0.86	1.85	0.75	2.02
20	1.20	1.41	1.10	1.54	1.00	1.68	0.90	1.83	0.79	1.99
21	1.22	1.42	1.13	1.54	1.03	1.67	0.93	1.81	0.83	1.96
22	1.24	1.43	1.15	1.54	1.05	1.66	0.96	1.80	0.86	1.94
23	1.26	1.44	1.17	1.54	1.08	1.66	0.99	1.79	0.90	1.92
24	1.27	1.45	1.19	1.55	1.10	1.66	1.01	1.78	0.93	1.90
25	1.29	1.45	1.21	1.55	1.12	1.66	1.04	1.77	0.95	1.89
26	1.30	1.46	1.22	1.55	1.14	1.65	1.06	1.76	0.98	1.88
27	1.32	1.47	1.24	1.56	1.16	1.65	1.08	1.76	1.01	1.86
28	1.33	1.48	1.26	1.56	1.18	1.65	1.10	1.75	1.03	1.85
29	1.34	1.48	1.27	1.56	1.20	1.65	1.12	1.74	1.05	1.84
30	1.35	1.49	1.28	1.57	1.21	1.65	1.14	1.74	1.07	1.83
31	1.36	1.50	1.30	1.57	1.23	1.65	1.16	1.74	1.09	1.83
32	1.37	1.50	1.31	1.57	1.24	1.65	1.18	1.73	1.11	1.82
33	1.38	1.51	1.32	1.58	1.26	1.65	1.19	1.73	1.13	1.81
34	1.39	1.51	1.33	1.58	1.27	1.65	1.21	1.73	1.15	1.81
35	1.40	1.52	1.34	1.58	1.28	1.65	1.22	1.73	1.16	1.80
36	1.41	1.52	1.35	1.59	1.29	1.65	1.24	1.73	1.18	1.80
37	1.42	1.53	1.36	1.59	1.31	1.66	1.25	1.72	1.19	1.80
38	1.43	1.54	1.37	1.59	1.32	1.66	1.26	1.72	1.21	1.79
39	1.43	1.54	1.38	1.60	1.33	1.66	1.27	1.72	1.22	1.79
40	1.44	1.54	1.39	1.60	1.34	1.66	1.29	1.72	1.23	1.79
45	1.48	1.57	1.43	1.62	1.38	1.67	1.34	1.72	1.29	1.78
50	1.50	1.59	1.46	1.63	1.42	1.67	1.38	1.72	1.34	1.77
55	1.53	1.60	1.49	1.64	1.45	1.68	1.41	1.72	1.38	1.77
60	1.55	1.62	1.51	1.65	1.48	1.69	1.44	1.73	1.41	1.77
65	1.57	1.63	1.54	1.66	1.50	1.70	1.47	1.73	1.44	1.77
70	1.58	1.64	1.55	1.67	1.52	1.70	1.49	1.74	1.46	1.77
75	1.60	1.65	1.57	1.68	1.54	1.71	1.51	1.74	1.49	1.77
80	1.61	1.66	1.59	1.69	1.56	1.72	1.53	1.74	1.51	1.77
85	1.62	1.67	1.60	1.70	1.57	1.72	1.55	1.75	1.52	1.77
90	1.63	1.68	1.61	1.70	1.59	1.73	1.57	1.75	1.54	1.78
95	1.64	1.69	1.62	1.71	1.60	1.73	1.58	1.75	1.56	1.78
100	1.65	1.69	1.63	1.72	1.61	1.74	1.59	1.76	1.57	1.78

Source: J. Durbin and G. S. Watson (1951), Testing for serial correlation in least squares regression II, *Biometrika*, **38**, 159–178.

Table D. Significance Points of d_L and d_U in the Durbin-Watson Test: 1%

n	$p = 1$		$p = 2$		$p = 3$		$p = 4$		$p = 5$	
	d_L	d_U	d_L	d_U	d_L	d_U	d_L	d_U	d_L	d_U
15	0.81	1.07	0.70	1.25	0.59	1.46	0.49	1.70	0.39	1.96
16	0.84	1.09	0.74	1.25	0.63	1.44	0.53	1.66	0.44	1.90
17	0.87	1.10	0.77	1.25	0.67	1.43	0.57	1.63	0.48	1.85
18	0.90	1.12	0.80	1.26	0.71	1.42	0.61	1.60	0.52	1.80
19	0.93	1.13	0.83	1.26	0.74	1.41	0.65	1.58	0.56	1.77
20	0.95	1.15	0.86	1.27	0.77	1.41	0.68	1.57	0.60	1.74
21	0.97	1.16	0.89	1.27	0.80	1.41	0.72	1.55	0.63	1.71
22	1.00	1.17	0.91	1.28	0.83	1.40	0.75	1.54	0.66	1.69
23	1.02	1.19	0.94	1.29	0.86	1.40	0.77	1.53	0.70	1.67
24	1.04	1.20	0.96	1.30	0.88	1.41	0.80	1.53	0.72	1.66
25	1.05	1.21	0.98	1.30	0.90	1.41	0.83	1.52	0.75	1.65
26	1.07	1.22	1.00	1.31	0.93	1.41	0.85	1.52	0.78	1.64
27	1.09	1.23	1.02	1.32	0.95	1.41	0.88	1.51	0.81	1.63
28	1.10	1.24	1.04	1.32	0.97	1.41	0.90	1.51	0.83	1.62
29	1.12	1.25	1.05	1.33	0.99	1.42	0.92	1.51	0.85	1.61
30	1.13	1.26	1.07	1.34	1.01	1.42	0.94	1.51	0.88	1.61
31	1.15	1.27	1.08	1.34	1.02	1.42	0.96	1.51	0.90	1.60
32	1.16	1.28	1.10	1.35	1.04	1.43	0.98	1.51	0.92	1.60
33	1.17	1.29	1.11	1.36	1.05	1.43	1.00	1.51	0.94	1.59
34	1.18	1.30	1.13	1.36	1.07	1.43	1.01	1.51	0.95	1.59
35	1.19	1.31	1.14	1.37	1.08	1.44	1.03	1.51	0.97	1.59
36	1.21	1.32	1.15	1.38	1.10	1.44	1.04	1.51	0.99	1.59
37	1.22	1.32	1.16	1.38	1.11	1.45	1.06	1.51	1.00	1.59
38	1.23	1.33	1.18	1.39	1.12	1.45	1.07	1.52	1.02	1.58
39	1.24	1.34	1.19	1.39	1.14	1.45	1.09	1.52	1.03	1.58
40	1.25	1.34	1.20	1.40	1.15	1.46	1.10	1.52	1.05	1.58
45	1.29	1.38	1.24	1.42	1.20	1.48	1.16	1.53	1.11	1.58
50	1.32	1.40	1.28	1.45	1.24	1.49	1.20	1.54	1.16	1.59
55	1.36	1.43	1.32	1.47	1.28	1.51	1.25	1.55	1.21	1.59
60	1.38	1.45	1.35	1.48	1.32	1.52	1.28	1.56	1.25	1.60
65	1.41	1.47	1.38	1.50	1.35	1.53	1.31	1.57	1.28	1.61
70	1.43	1.49	1.40	1.52	1.37	1.55	1.34	1.58	1.31	1.61
75	1.45	1.50	1.42	1.53	1.39	1.56	1.37	1.59	1.34	1.62
80	1.47	1.52	1.44	1.54	1.42	1.57	1.39	1.60	1.36	1.62
85	1.48	1.53	1.46	1.55	1.43	1.58	1.41	1.60	1.39	1.63
90	1.50	1.54	1.47	1.56	1.45	1.59	1.43	1.61	1.41	1.64
95	1.51	1.55	1.49	1.57	1.47	1.60	1.45	1.62	1.42	1.64
100	1.52	1.56	1.50	1.58	1.48	1.60	1.46	1.63	1.44	1.65

Source: J. Durbin and G. S. Watson (1951), Testing for serial correlation in least squares regression II, *Biometrika*, **38**, 159–178.

Author Index

Subject Index

Applied Probability and Statistics (Continued)

FLEISS • Statistical Methods for Rates and Proportions, *Second Edition*
FRANKEN, KÖNIG, ARNDT, and SCHMIDT • Queues and Point Processes
GALAMBOS • The Asymptotic Theory of Extreme Order Statistics
GIBBONS, OLKIN, and SOBEL • Selecting and Ordering Populations: A New Statistical Methodology
GNANADESIKAN • Methods for Statistical Data Analysis of Multivariate Observations
GOLDBERGER • Econometric Theory
GOLDSTEIN and DILLON • Discrete Discriminant Analysis
GREENBERG and WEBSTER • Advanced Econometrics: A Bridge to the Literature
GROSS and CLARK • Survival Distributions: Reliability Applications in the Biomedical Sciences
GROSS and HARRIS • Fundamentals of Queueing Theory
GUPTA and PANCHAPAKESAN • Multiple Decision Procedures: Theory and Methodology of Selecting and Ranking Populations
GUTTMAN, WILKS, and HUNTER • Introductory Engineering Statistics, *Third Edition*
HAHN and SHAPIRO • Statistical Models in Engineering
HALD • Statistical Tables and Formulas
HALD • Statistical Theory with Engineering Applications
HAND • Discrimination and Classification
HILDEBRAND, LAING, and ROSENTHAL • Prediction Analysis of Cross Classifications
HOAGLIN, MOSTELLER, and TUKEY • Understanding Robust and Exploratory Data Analysis
HOEL • Elementary Statistics, *Fourth Edition*
HOEL and JESSEN • Basic Statistics for Business and Economics, *Third Edition*
HOLLANDER and WOLFE • Nonparametric Statistical Methods
IMAN and CONOVER • Modern Business Statistics
JAGERS • Branching Processes with Biological Applications
JESSEN • Statistical Survey Techniques
JOHNSON and KOTZ • Distributions in Statistics
Discrete Distributions
Continuous Univariate Distributions—1
Continuous Univariate Distributions—2
Continuous Multivariate Distributions
JOHNSON and KOTZ • Urn Models and Their Application: An Approach to Modern Discrete Probability Theory
JOHNSON and LEONE • Statistics and Experimental Design in Engineering and the Physical Sciences, Volumes I and II, *Second Edition*
JUDGE, HILL, GRIFFITHS, LÜTKEPOHL and LEE • Introduction to the Theory and Practice of Econometrics
JUDGE, GRIFFITHS, HILL and LEE • The Theory and Practice of Econometrics
KALBFLEISCH and PRENTICE • The Statistical Analysis of Failure Time Data
KEENEY and RAIFFA • Decisions with Multiple Objectives
LAWLESS • Statistical Models and Methods for Lifetime Data
LEAMER • Specification Searches: Ad Hoc Inference with Nonexperimental Data
McNEIL • Interactive Data Analysis
MAINDONALD • Statistical Computation
MANN, SCHAFER and SINGPURWALLA • Methods for Statistical Analysis of Reliability and Life Data
MARTZ and WALLER • Bayesian Reliability Analysis
MIKÉ and STANLEY • Statistics in Medical Research: Methods and Issues with Applications in Cancer Research
MILLER • Survival Analysis
MILLER, EFRON, BROWN, and MOSES • Biostatistics Casebook
MONTGOMERY and PECK • Introduction to Linear Regression Analysis
NELSON • Applied Life Data Analysis